FOOD, FERTILIZER AND AGRICULTURAL RESIDUES

Proceedings of the 1977 Cornell Agricultural Waste Management Conference

RAYMOND C. LOEHR
Editor

Director, Environmental Studies Program
College of Agriculture and Life Sciences
Cornell University
Ithaca, New York

ANN ARBOR SCIENCE
PUBLISHERS INC
P.O. BOX 1425 • ANN ARBOR, MICH. 48106

230 Collingwood, P. O. Box 1425, Ann Arbor, Michigan 48106

Library of Congress Catalog Card No. 77-85092
ISBN 0-250-40190-8

Manufactured in the United States of America

PREFACE

The quantity and quality of the world's food supply is of national and international concern. When producing the required food, fertilizers and other chemicals are needed and residues, such as manure, crop residues and food processing solids, are produced. Such materials, if used incorrectly or disposed of improperly, result in environmental problems.

The causes of such problems are excesses of production that have not been effectively utilized. Recycling, reprocessing and proper utilization are positive approaches that provide for beneficial use as opposed to the more traditional approach of waste treatment and disposal. The challenge is to consider these and other residues of man as resources and to evaluate how they can be used for food, animal and energy production.

This volume explores ways to utilize the nutrient and energy resources in municipal and agricultural residues, decrease the losses of fertilizer nutrients and avoid environmental problems caused by residues of food production.

Major topics include municipal sludge management and utilization, land application of wastewaters, methane production from agricultural wastes, animal waste management alternatives, fertilizer application rates and timing, economic and regulatory aspects and use of agricultural residues as energy sources. Laboratory, pilot plant and full-scale studies are documented.

The success of the Conference from which these papers originated was due in equal measure to the quality of the authors and their studies, the interest and discussion of other participants and the skill of the moderators. Important assistance was provided by Ms. Colleen Raymond who had a major role in helping plan the Conference and prepare the papers for publication.

The 1977 Waste Management Conference, "Food, Fertilizer and Agricultural Residues," was sponsored by the New York State College of Agriculture and Life Sciences, a Statutory College of the State University at Cornell University, Ithaca, New York. The Conference was cosponsored by the

American Society of Agronomy and the American Society of Agricultural Engineering. This Conference was the ninth in the annual series of conferences sponsored by the College of Agriculture and Life Sciences on various waste management topics.

The assistance of individuals who served on the program committee such as Dr. David Bouldin, Dr. Douglas Haith, Dr. David Ludington and Dr. Robert Zall and others who assisted with the Conference is gratefully acknowledged.

Any opinions, findings, conclusions or recommendations expressed in this publication are those of the authors and do not necessarily reflect the views of the College of Agriculture and Life Sciences or the other sponsoring organizations.

Raymond C. Loehr, Director
Environmental Studies Program
New York State College of
Agriculture and Life Sciences
Conference Chairman

CONTENTS

SECTION I

OPENING REMARKS
AND
KEYNOTE ADDRESS

1

OPENING REMARKS

D. W. Barton
New York State Agricultural Experiment Station
Geneva, New York

It is a pleasure to welcome you on behalf of the New York State College of Agriculture and Life Sciences and its cosponsors, the American Society of Agricultural Engineers and the American Society of Agronomy, to the Ninth Annual Waste Management Conference sponsored by the College.

The program theme, "Food, Fertilizer and Agricultural Residues," reflects society's growing concern for disposal of wastes, as we gradually come to a realization that we do not live in an infinite environment. Our lands, our water and our air can become overburdened with the waste products of modern civilization unless such residues are managed properly.

It was in the first year of these conferences (1969) that the Secretary of Agriculture and the Office of Science and Technology published a Report to the President entitled "Control of Agriculture-Related Pollution." While the President's charge to the Secretary had been to study only agriculturally related pollution, it is significant to note that the Departments of Agriculture, Commerce, Defense, Health, Education and Welfare, Housing and Urban Development, Interior and Transportation were involved in the study. That report covered eight pollutants considered to be of major concern: sediment, animal wastes, wastes from industrial processing of raw agricultural products, plant nutrients, forest and crop residues, inorganic salts and minerals, pesticides in the environment, and air pollution in relation to agriculture. While the theme of this program is more restrictive in title, I am sure that more than half of the agricultural pollutants identified by the report as major concerns will be topics of discussion here. As you know, other conferences in this series have dwelt on specific other pollutant problems.

It is quite appropriate, too, that the theme of this conference goes beyond the limited perspective of the panel preparing the 1969 report. Animal wastes were identified as major problems, with livestock and poultry wastes particularly identified. Eutrophication, fish kills, nitrate contamination, off-flavors, odors, recreational depreciation and insect reproduction are associated problems. This ninth annual conference has rightly recognized human and municipal waste as a part of the overall waste burden that has implications for agriculture, and for society, and which causes the same side effects and responds to similar treatment processes.

The 1969 report also discussed wastes from industrial processing of raw agricultural products, excluding garbage disposal, which involved the processing and conversion of these wastes into organic forms of fertility. This conference does recognize that agricultural, industrial, municipal and other wastes should be considered jointly in an integrated plan for improving and maintaining the quality of our environment.

Research is the key factor in providing the principles and methods for waste management of the present and future. You are here to consider the state of present waste management knowledge and to get insights as to where research may lead us in the future. This program will have many examples of the kind of team effort involving multiple disciplines of science which has characterized research on waste management. It also is attuned to the times in that it recognizes that no research involving wastes and by-products can meet its national commitment without due consideration to the implications of the process on energy use or energy generation.

There is another tradition of this conference that makes it stand out as an example. It provides the mix of scientific disciplines, combining people involved in advising on field action programs with the key persons from governmental regulation and decision-making and with representatives from agencies that provide program funding.

The purpose of this conference is to explore approaches to (a) utilizing the nutrient and energy resources in municipal and agricultural residues, (b) decreasing the losses of fertilizer nutrients, and (c) avoiding environmental problems caused by residues of food production. In a world increasingly challenged with problems of population expansion, use of land, water and air resources, and the need for ever-increasing production of food, the importance of the purposes of this conference is growing day by day.

In closing, it is appropriate to recall a message to Congress by President Theodore Roosevelt in 1907: "To waste, to destroy, our natural resources, to skin and exhaust the land instead of using it so as to increase its usefulness, will result in undermining in the days of our children the very prosperity which we ought by right to hand down to them amplified and developed."

2

PERSPECTIVES ON FERTILIZER USE, RESIDUE UTILIZATION AND FOOD PRODUCTION

R. White-Stevens
Bureau of Conservation and Environmental Science
Cook College
Rutgers–The State University
New Brunswick, New Jersey

INTRODUCTION

In concert with the population explosion and the energy crisis the man-food equation constitutes the third major political-socioeconomic problem confronting mankind. Actually the three are integrated into a single entity, and the resolution of both food and energy requires that man reduce his population to a steady state.

Assuming that some strategy will be evolved and applied to curb the population within the next century, and assuming that a combination of conservation and developments (*e.g.*, nuclear fusion) in energy supply will meet the demand, then it can be virtually guaranteed that technical agriculture will be able to provide the calories, protein, vitamins and minerals to meet the total needs of perhaps 10 to 12 billion humans at least at subsistence level by the year 2025.

At present about 1.4 billion humans are continually on the verge of starvation or marasmus–total caloric insufficiency. About 2.8 billion humans suffer from varying degrees of protein deficiency, and at least 250 million children are victims of kwashiorkor, or protein deficiency, in Asia, in Africa and in Latin America. Many adults also suffer protein starvation, but it is particularly vicious among children, for although many of those who develop frank symptoms of kwashiorkor die by age five, those who survive the anemia and frequent virulent infections which follow the edema and the pigmented skin do not recover from the mental impairment induced by the lack of essential amino acids at the crucial age when brain cells are forming.

The quantitative and qualitative protein requirements of man have now been established by the WHO and FAO of the UN as 60 grams per capita per day for a 70-kg human of which 7 g per capita per day must be animal protein and 17 g pulse protein and the balance (36 g), can be from vegetable, cereal or fruit protein. The qualitative protein requirements in terms of specific amino acids have also been defined and appear to be universal for all the human race.[1]

Based on these daily requirements it is possible to estimate the present and foreseeable future protein needs of mankind. The picture is not encouraging and is, in fact, quite Malthusian in prospect. Although the present total world protein product is sufficient to meet the minimal requirement of all the people on earth for animal, pulse and total protein, the actual distribution of protein is seriously maladjusted so that only in North America, western Europe and Oceania is the daily protein intake adequate or above. The reasons for this maldistribution are multiple and complex and include economic, political and sociological customs and mores, but the basic reason for the prevalence of protein deficiency is nitrogen deficiency in the arable crop-producing soils of these overpopulated hungry lands.

Tanner[2] considered the protein needs of humans in terms of what he termed the Fertilizer Nitrogen Equivalent Diet (FNED) and its relation to annual income per capita in some 19 different countries. He found that FNED rises with income to a steady state at about 80 lb N per capita per annum.

Relating the WHO-FAO human protein daily requirement to nitrogen, the annual minimal human N intake is 8.8 lb N total. Assuming a conversion efficiency of 15% for animal protein and 50% for plant protein, this computes to 38 lb N fertilizer per capita per annum for human subsistence. For the present population of 4.2 billion, this minimal demand requires about 80 million m.ton of fertilizer nitrogen. Assuming that the 3.3 billion acres of arable land receive 10 lb N/ac in atmospheric precipitation, which amounts to 16 million m.ton, 64 million m.ton must be supplied by man either from natural organic wastes or by fixation of atmospheric nitrogen from legumes or from chemical manufacturing process.

Worldwide the fixation of nitrogen on arable lands is presently estimated to be (in megatons per year):

Legumes	35
Nonlegumes	9
Grasslands	45
Subtotal	79
Industrial	47
Total	126

which is, in fact, more than enough (197%) to meet the minimal protein requirements of all humans on earth. However, at least 500 million humans consume large daily intakes of animal protein, which in turn results in at least a sevenfold increase in demand for vegetable protein. Thus in the United States daily protein intake is close to 120 grams per capita per day of which 70g is animal protein or 10 times the minimal requirement. This calculates to a FNED of 73 lb per capita per annum which for a population of 212 million requires about 8 million m.ton of fertilizer nitrogen/yr. In 1973-74 the nitrogen fertilizer applied to all crops was 9-10 million m.ton to which should be added the atmospheric-precipitated N of about 2 million tons and symbiotic and nonsymbiotic N fixation of about 2 million tons plus organic manures that contribute about 2 million tons of N for a total applied N in the United States of 15-16 million m.ton. From this, however, must be subtracted loss to insect and disease of about 10% (even with control), leaching and erosion of another 10% and denitrification of about 10%, an aggregate of at least 4.5 million m.ton, leaving 10.6 million m.ton for food and fiber use.

Overall this appears to be a favorable nitrogen balance for U.S. soils, and in some areas it no doubt is. In many areas, however, applications of nitrogen are minimal or below; leaching, erosion and denitrification are of major proportions; animal and crop detritus are not returned to the soil regularly; and the fertility steadily declines. Around the world the soil nitrogen economy is seriously unbalanced and the crop- and livestock-carrying capacity of the land has been depleted to the point where major nutrient restoration is mandatory to meet the bare subsistence needs of food, fiber, fuel and shelter of the burgeoning population.

There is therefore a direct relation, as Tanner[2] proposed, between the nitrogen level of the soil and the living status of the people who dwell and feed upon it. In the U.S. and the developed countries the FNED is adequate at present levels and would appear to be so through the year 2000, but beyond that date considerable increases in artificial nitrogen fixation will be essential to provide a bare subsistence level of life status for the human population without consideration of raising mankind to the level now enjoyed in America, actually a completely futile and hopeless endeavor.

MANUFACTURED FERTILIZERS

The increase of manufactured fertilizers in the western world, particularly in North America, constitutes an essential factor in the rising food, fiber and shelter supply and the progressive standard of living in these areas. In the U.S. at least 40% of the agricultural production rests upon the regular use of manufactured and processed fertilizers.[3] The USDA-SRS[4] reports consumption of mixed fertilizer (N-P_2O_5-K_2O) increased from 8.6 million tons in

1940 to 42.5 million tons in 1973, a fivefold increase in 33 years. Shapley[5] estimates world industrial fixation of nitrogen, most of which is employed in agricultural fertilizers (about 68%) to be (in m.tons):

1950	2 x 10^6
1960	9.7
1970	40
2000	120-300 (estimated)

This constitutes an annually compounded U.S. increase of 5% for complete fertilizer which would attain a level of 1.59 billion tons by the year 2000, while worldwide industrial N-fixation by 2000 would reach the astronomical level of 3.433 x 10^9 m.tons N if growth continues from 1970 to 2000 at the 16% annual compounded rate increase experienced from 1950 to 1970. This is a quite impossible level and is, in fact, unnecessary. It would exceed the natural world N-fixation rate by 22 times. There would not be sufficient hydrogen gas nor available energy to meet this demand, *i.e.*, 1.309 x 10^{14} ft^3 of natural gas and 618 x 10^6 ton fuel oil equivalent in energy. Also if we take Tanner's[2] 80 lb (36.3kg) N-fertilizer equivalent diet (NFED) per capita as the maximum annual level required, such a fixed N production would supply 94.6 billion humans at a protein-N dietary level of the present U.S. population.

The maximum estimated figure of Shapley[5], 300 x 10^6 m.ton of industrially fixed N plus the natural agricultural annual fixed N of 89 x 10^6 m.ton, would feed 10.72 billion humans at the U.S. dietary protein level, again an overly optimistic figure.

The continually increasing level of fertilizer consumption in both the U.S. and the world evolves from:

1. the burgeoning population now over 4 billion with a current N demand of 145 million m.ton/yr;
2. improving the diet of humans, particularly among developing nations; and
3. increasing export demands for food, fiber and forestry products from the agricultural nations.

Although the prospects for maintaining the soil nitrogen supply by enhanced biological fixation are quite promising, reliance on industrial atmospheric fixation will unquestionably increase. However, atmospheric nitrogen occurs in the diatomic ($N \equiv N$) form and therefore requires hydrogen and enormous amounts of energy to free and break the triple bond and convert it into ammonia.[6]

Presently natural gas is the favored form of energy employed and is used for 70-80% of the N-fixation in the U.S. Even so, this comprises less than 2.5% of the natural gas consumed annually in the U.S. Nevertheless, the rising cost of oil and natural gas has incurred corresponding price increases in industrial-fixed N, with the result that available fertilizer supply is about 5% below demand. To a lesser extent these energy deficiencies have also increased the cost of phosphate and potash. Ultimately these cost pressures will become reflected in the price and availability of food at the consumer level.

Should the nation turn to nuclear fusion energy, as it inevitably will, then electrical power will substantially decrease in unit cost–perhaps to the point where it would not pay to meter it, hydrogen could be produced by electrolysis of water and N-fixation would become both abundant and cheap.

The use of coal as both an energy and hydrogen source by gasification is exceedingly dubious on economic, environmental and efficiency grounds. It would be better to save fossil fuels for use as basic chemical intermediates than to burn them under conditions where more than 75% of the potential energy escapes to pollute the environment physically, chemically and biologically.

Essential as is the continued and indeed increased use of manufactured and mined fertilizers, the improved efficiency of use and conservation of these natural resources together with their recycling as much as is economically feasible and practicable must be paramount. There are several approaches to the enhanced consideration of soil fertility:

1. Soil and plant tissue tests can be used routinely to assess the acidity and specific soil nutrient requirements by being applied area by area, season by season, crop species by species. These have been worked out for the arable soil types in many states, and such routine assessments enable the grower to meter the needs of his particular crops in a precise and optimum manner.
2. Lime, fertilizer and other soil amendments can be timed to provide optimum fertility with minimal loss to leaching, surface erosion or evaporation. Generally, fertilizer application is best at time of planting or immediately before. This is particularly true with respect to nitrogen and phosphorus. For certain row crops postemergence side dressings of nitrogen are exceedingly efficient but must be made with special equipment and great care.
3. Placement of fertilizers in two bands on either side of a seeded or transplanted row crop, one about 4-6 in. deep, depending on the soil type and its drainage, will generally increase the efficiency of fertilizer use. This requires special planting equipment usually for specific types of row crops, *e.g.*, potatoes, large seeded crops, such as beans, small seeded crops, such as beets, lettuce and carrots, and seedling transplants, such as tomatoes, coles, celery, etc. For nonrow or closely drilled crops, such as forages, grasses and cereals, broadcast fertilizer applications immediately followed by discing or cultivation into the surface 2 in. prior to sowing is still the most efficient procedure.
4. Maintaining the optimum soil reaction (pH) for the particular crop species grown is essential in maintaining fertilizer efficiency. Availability of soil nutrients, particularly such essential minor elements as

zinc, manganese, magnesium, copper, iron and boron, is rigidly controlled by soil reaction. Failure to recognize and respond to these nutrient reactions can readily obviate the abundant use of the major fertilizer nutrients–N, P_2O_5 and K_2O. Furthermore, in the case of legume crops, an excessively acid soil reaction will prevent nodulation by the symbiotic nitrogen fixers and as much as 50 to 100 lb of N/ac can be lost yearly.

5. Inadequate soil drainage will prevent adequate penetration of oxygen, induce anaerobiosis, prevent crop growth, reduce yields and promote denitrification and consequent loss of soil nitrogen to the atmosphere. Conversely, on highly porous, well-drained soils of low organic matter, such as the sandy soils of the coastal plains, fertilizer, particularly nitrogen, is rapidly leached down to the aquifers and not only is lost to crop nutrition but can readily pollute the ground waters with nitrate. For example, on the sassafras silt loam soils of eastern Long Island ground water from as deep as 90 ft has been found to carry 20 ppm of nitrate-nitrogen, a level adequate to grow a good crop of potatoes.
6. Prevention of soil erosion both by wind and water by use of wind breaks and contour cultivation will conserve not only large quantities of fertilizer nutrients but also important organic matter and humus. Being of lower specific gravity than the mineral components of the soil, organic matter tends to blow, wash and erode more rapidly.
7. Virtually any or all soil and crop-management practices which increase yields of crops also improve the efficiency of fertilizer use and conservation. This would include selection of rapid-growth high-yield cultivars, optimum row and plant spacing, irrigation as needed, correct planting and seeding times, and control of insect, disease and weed pests.

Efficient fertilizer use and conservation are coincident with efficient and economic crop production in terms of optimum use of resources, energy and biotic potential. They are integrative functions of all factors optimized simultaneously, and failure to fulfill one factor usually obviates the efficiency of *all* other factors. It is for this reason that reliance on the fortuity of bioligical control of crop pests almost invariably results in yield reduction and lost efficiency if not total crop loss. Yet there are those who deplore the use of so-called artificial fertilizers on the grounds that they pollute the environment by contaminating drinking waters, promoting eutrophication, reducing edible quality, even poisoning foodstuffs and unnecessarily consuming excessive energy.[7]

To deny American farmers the use of artificial fertilizers–which was, at one time, seriously proposed–would require the removal of 100 million Americans, some of whom, in a democracy it would seem, are likely to vote against the idea. It would also become necessary to open up some 250 million acres of new lands, pressing wildlife into further extinction; to return some 20 million nonrural people back to the toil of the soil and stoop labor; to reduce the present standard of living by at least 50% and to retreat to the way of life of a century ago.

The contention that agricultural chemicals, fertilizers, pesticides and livestock supplements and medicaments pollute our environment and poison our food supply is not only crass exaggeration but in most part unmitigated nonsense. It is self-serving propaganda foisted upon an unsuspecting public, generally quite ignorant and largely indifferent to the source and nature of its food supply, by self-appointed custodians of the public welfare, with neither training, education nor practical experience in the business of food and fiber production.

The claim that crop-applied manufactured nitrogen fertilizer penetrates to the deep aquifers and induces blue baby disease (methemoglobimemia) among infants is confronted with the evidence that fatalities from this disease have been less than those from smallpox vaccinations, and those which have occurred over the past 40 years were traced to well water contaminated with livestock or human wastes, not fertilizers.[8-11] Nitrate contamination of aquifers was found by Enfield[12] in a number of watersheds to be no greater today when large amounts of industrial nitrogen are applied to the soil than at the turn of the century when most soils were fertilized with animal manures.

Numerous field studies reveal that descent of nitrate into soil depths is effectively interdicted by crop growth. Where crops have been well fertilized with an adequate level of balanced plant nutrients, their rapidly proliferating roots hungrily absorb soluble nutrients, particularly nitrogen, and prevent their leaching to the aquifers.[13-17]

The claim that artificial nitrogen fertilizer is often applied in excess of crop needs as determined by the yield of grain and stover leaving as much as 40% of the applied nitrogen in the soil "unaccompanied by carbon" ignores the fact that the root system of most plants equals in dry weight of carbonaceous and protein matter that of the above-ground portion of the plant, and that whatever weight of carbon-bound nitrogen is present in the grain and stover is also present in the roots. In general, the amount of fertilizer, particularly nitrogen, applied by farmers is less than the total extracted from the soil, and each successive crop continues to deplete the reserve fertility of the soil. Only when yields clearly commence to decline do most farmers increase their fertilizer levels to offset depletion.

In recent years there has been a gathering interest in support (among urban and nonrural people) for the so-called organic foods grown, it is claimed, entirely in the absence of artificial, *i.e.*, man-made, "chemicals." Such foods are claimed to be more nutritious, taste better and be totally safe, even though the cost is considerably higher. This is a shabby fraud upon the public, although entirely legal. All foods are "organic," with the possible exception of table salt, and all growing crop plants absorb only nutrients in the inorganic form, except small amounts of urea—an organic nitrogen source often, incidentally, manufactured by man. There is no evidence whatsoever that such "organic" foods are nutritionally superior to those produced with artificial fertilizers. There is also no evidence that foods protected by pesticides are more toxic or dangerous than those produced without.[18-21] There is, however, evidence that foods unprotected from pests can carry natural pest-produced toxins and infections which can and do induce disease in man, such as alfatoxins, botulin and salmonellosis.

A vigorous ecological argument against the use of artificial fertilizers and particularly monoculture is that they both tend to depress the complexity and therefore the stability of successive trophic levels of the biosphere. However, Hurd *et al.*,[22] studying the effect of man-applied perturbations to the ecosystems (in this case fertilizers) of old abandoned pastures, one unused for 7, the other for 16 years, found upon comparing the impact upon the diversity response at three trophic levels (grasses, invertebrated herbivores and carnivores) that the applied fertilizer expectably enhanced productivity at the first trophic level (grasses) but more in the younger than in the older field, while conversely diversity increased more in the older field. Hurd *et al.*[22] concluded, "in contrast to current ecological theory, that greater diversity at a trophic level was accompanied by lower stability at the next higher level."

As pointed out by Claus and Bolander[23] greater stability is *not* dependent on species diversity during successional aging, for even if complexity increased, stability decreases; and finally successive trophic levels are not determined as to productivity or diversity by the growth and complexity of the primary trophic level. There are those who vigorously oppose the "Green Revolution" on the grounds that by introducing concentrations of a few new varieties of cereals and by using fertilizers to enhance yields and pesticides to protect them, we are suppressing diversity and encouraging first population growth and then famine, following collapse of the crop system.[24]

Finally, the suggestion that manufactured fertilizers could largely if not totally be replaced by so-called natural organic fertilizers is also based upon an economic unreality if not an outright myth.

LIVESTOCK AND MUNICIPAL ORGANIC WASTES

With the steady increase in the animal protein level of the U.S. diet, which has increased over 50% since 1910, there has been a substantial expansion in the meat-producing livestock population. In spite of the rise in total milk and by-product production there has been a sharp decline in total milk cows over the same period.

Americans are enormous meat consumers. In fact they feed sufficient protein (largely animal) to their pets to provide a subsistence level of protein to over 100 million deprived humans annually.[25]

The resulting animal population generates some two billion tons of raw animal wastes of which about half is distributed diffusely and recycled naturally in pastures, ranges, open fields and woodlands. The remainder, or about 170 million tons of dry manure equivalent, is produced in cattle, sheep and hog feedlots, livestock barns, broiler and laying houses and turkey runs and duck farms and is usable as a fertilizer resource for crops, woodlots, golf courses, parks, and home lawns and gardens. In the dry form (10-20% moisture) the total major nutrient analysis averages 2% N, 0.5% P_2O_5 and 1.5% K_2O, of which about 50% of the N becomes available in the year of application to give an actual nutrient level of 1% N, 0.5% P_2O_5 and 1.5% K_2O for a total of 3%. This amounts to 1.7 x 10^6 tons N, 0.85 x 10^6 tons P_2O_5 and 2.6 x 10^6 tons K_2O equivalent fertilizer per annum of 19% of the fertilizer N, 17% of the fertilizer P_2O_5 and 51% of the fertilizer K_2O applied in an average year. It is assumed that all the N-P-K in the animal waste would be retained through the drying and distribution process and would become available to the fertilized plants.

In actuality, where manures are accumulated in farm storages over the year to be applied to the fields at planting time, either leaching or fermentation occurs and from 25 to 50% of the nitrogen is leached or evaporated before it becomes incorporated in the soil. If it is spread as produced during the winter or noncrop months, nutrients are leached or surface-eroded. Only where animal wastes are dried as produced, under controlled conditions, can the nutrients be largely preserved, and even then some nitrogen is volatilized and lost despite the high expense of handling and drying.

Thus on the basis of plant nutrient costs, a 2:1:2 total analysis farm manure would be worth $16/ton, while on the basis of 50% availability of the N component in the year of application it would be worth about $12/ton.[26]

From the economic point of view it is obvious that animal manures cannot sustain much investment in either processing, packaging, distribution or hauling costs. It is true that manures add organic matter and humus to soil and improve its porosity, its water and nutrient holding capacity and its general structure, but this can be achieved equally effectively by discing in cover-crop and harvested-crops refuse which protect against wind and water erosion as well. However, the livestock husbandman has to dispose of the animal waste produced, and if he has available crop soils, pasture, range or woodlands these areas are the most appropriate and useful.

Apart from the economic impracticality of relying upon livestock and municipal organic wastes as the sole or major source of plant nutrients as advocated by Commoner[7] and Taiganides,[27] it is logistically contraindicated as well. If the 7 million tons of dry sewage sludge produced annually is added to the 170 million tons of dried animal manures, the total of approximately 180 million tons would provide available plant nutrients of about 1.84 x 10^6

tons nitrogen (N), 1.13 x 10^6 tons phosphorus (P_2O_5) and 2.64 x 10^6 tons potash (K_2O). This constitutes 17% of the nitrogen, 18% of the phosphorus and 34% of the potash now applied to the arable soils of the U.S.

As the amounts of N-P-K applied as fertilizer annually are only about 1/3 on average of the amounts recommended for replacement of nutrients removed by the harvested crop, if farmers used manures exclusively, at best they would be applying around 5 to 10% of the major nutrients they need to apply to preclude serious depletion of soil fertility.

The removal rate of N-P-K nutrients from soils is quite substantial. Table I lists examples based on the reports of Nelson *et al.*[28] and Dewey and Almy.[29]

Table I. Removal rate of N-P-K.

			Nutrient Removal (lb/ac)		
Crop	Yield Basis/ac	(Total)	N	P_2O_5	K_2O
Corn (grain)	91.4 bu	(169)	82	39	48
Wheat (grain)	31.8 bu	(76)	40	22	14
Cotton (lint and seed)	3 bale	(225)	120	60	45
Soybeans (seed)	27.8 bu	(160)	105	23	32
Potato	230 cwt	(250)	77	29	144
Sorghum	100 bu	(150)	81	44	25
Hay tons (3 cuts)	4 ton	(390)	175	40	175

From these seven crops and the 1975 production figures for the U.S. (USDA-SRS) we find a total annual removal of nutrients for just these crops alone of: nitrogen (N), 10.237 x 10^6 tons; phosphorus (P_2O_5), 3.532 x 10^6 tons; and potash (K_2O), 6.019 x 10^6 tons. This is 5.56 times the nitrogen, 3.13 times the phosphorus and 2.28 times the potash that would be contributed in that year from all the animal, municipal and industrial wastes combined.

In summary, organic wastes are economically, logistically, nutritionally and practically insufficient and inadequate to meet the present needs of U.S. crop production.

Looking at it in terms of Tanner's[2] Nitrogen Fertilizer Equivalent Diet (NFED), the total available organic waste nitrogen would be sufficient to meet the present food and fiber consumption needs of 46 million modern Americans or 117 million humans at subsistence level. We would, in short, have to dispose of over 100 million citizens or secure food and fiber for them from elsewhere.

In addition to the excess cost, labor and bulky handling of organic residues as fertilizer supplements there are several other disadvantages. These include the presence of soluble salts which become an added burden in irrigated arid western lands and a leachate and washoff problem in the wetter eastern soils unless the applied manure is immediately incorporated into the soil by discing, plow cover or injection. Sewage sludges in particular may carry toxic elements such as heavy metals (Pb, Hg, Zn, Cu, Ni, Cr and especially Cd) and certain nonmetals such as B and Se—both essential trace elements at one end of the scale but distinctly toxic at the other.[26] Thus B, which is essential to plants, though inessential to animals, can readily become phytotoxic. While Se is essential for animals, and probably man in trace levels, it is a known carcinogen in excess dose.

Among the heavy metals—Hg, Cu, Pb, Zn, Ni and Cr—Hg and Pb are not generally accumulated by most plants, and although Hg can become highly mobile and volatile when converted into the alkyl form (methyl mercury) under conditions of anaerobiosis in poorly drained soil, it is not known whether it would be absorbed and accumulated by crop plants and reach humans in the edible component. It is, however, highly toxic. Cu and Zn, both essential trace elements for plants, are accumulated and become phytotoxic although it is most unlikely they would reach humans in toxic doses.

Cadmium is a different matter. It is used in many areas including tires, fungicides and worming medicaments, and sewage sludge often carries a perceptible level ranging from 5 to 2000 ppm in sludges from industrial municipalities and from 5 to 10 ppm in nonindustrial areas. Researchers at Pennsylvania State University have found that both corn and sorghum forage grown on soils treated with sewage sludge from an industrial municipality had over a tenfold increase in Cd. When such plants were fed to small vertebrates (voles), they accumulated the Cd from 5 to 30-fold in their liver and kidneys but not in their muscle tissue.[30]

The toxicology of Cd is not as yet clarified, but it is recognized as a serious cumulative disease in man, first described in Japan as Itai-Itai disease and characterized by bone deformities, muscular atrophy and excruciating pain.

Animal manures are generally low in heavy metals B and Se. However, where Cu and Zn have been fed as supplements and where crops have been treated with Cu, Zn, Ni, Hg and Pb pesticides, appreciable levels can appear in the wastes. Most heavy metals, particularly B, occur in industrial and municipal sludges. In areas where such sludges are employed as landfill, reconstitution of surface mine spoil banks or low-fertility wastelands, these elements can accumulate to substantial levels which render the soil phytotoxic or the vegetation grown thereon toxic to livestock and wildlife. B, for example, which is an essential plant nutrient at soil levels of from 0.1 to 5.0 ppm depending upon the species or cultivar grown, has been found to occur at levels up to 800 ppm in dried industrial and municipal sludge. Where applications of up to 50 tons of such sludge per acre are made on poor low organic soils, the annual addition of 88 lb of B/ac that would result would render such soils phytotoxic for virtually all crop plants within 3 to 5 yr.

Heavy metal and B and Se toxicity tend to become increasingly severe in acid soils, so some relief can be obtained by liming up to a pH of 7 or above.

Emmelin[31] reports in Table II the following ranges of heavy metals in municipal and industrial sewage sludges.

Table II. Ranges of heavy metals in municipal and industrial sewage sludges.[31]

	Concentration of Heavy Metals (mg/kg)	
Element	**Municipal Sludge**	**Industrial Sludge**
Zn[a]	1000-3000	10,000
Cu[a]	500-1500	3,000
Mn[a]	200-500	2,000
Pb	100-300	1,000
Cr	50-200	1,000
Ni	25-100	500
Co	8-20	50
Cd	5-15	25
Hg	4-8	25

[a]Essential plant nutrients.

From these data and those derived from U.S. municipal and industrial sewage sludges it is clear that excessive applications of such sludge to lands producing crops for human or animal food should be restricted to not more than 1 to 5 ton/ac/yr. For woodlands, parklands, ornamental plantings, lawns, fairways and greens somewhat higher applications would be safe. However, even in these cases health risks of runoff and leaching into waterways and aquifers remain.

The contribution of the basic N-P-K nutrients from sewage sludge is, however, so meagre that the total amount of such sludge applied to crop lands would add so little heavy metals that the probability of toxic edible products reaching the consumer is essentially insignificant.

Furthermore, as the waste treatment provisions of PL 92-500 become implemented nationwide the heavy metal content of sewage sludges will presumably decline, although slowly. As part of all sewage is from nonpoint sources that may not be regulated, some heavy metal accumulation will continue to occur. Such modest levels can be alleviated by reduced soil applications, maintaining the pH at 7.0 or above and by treating only soils planted to nonedible crops such as ornamentals, parklands, lawns and woodlands.

Alternative Uses of Animal and Sewage Wastes

Apart from the conventional procedure of returning organic wastes to the soil, there are alternative options which under certain circumstances can be considered for useful disposal of such detritus.

Protein Supplements in Livestock Feeds

Certain animal wastes such as poultry wastes are rich in protein equivalents, *i.e.*, 25 to 30%, and modest amounts have been effectively and successfully fed to ruminants which can utilize some nonprotein nitrogen in their ration.

There are certain problems and restrictions involved. These include the aesthetic distaste of consumers who are expected to eat such meat; the remote possibility of recurrent and transmissable diseases; the necessity and investment for substantial processing to meet government regulations; and the geographic need to have the ruminant feeding operations reasonably close to the poultry manure source. Of the 30×10^6 ton of poultry wastes now produced, as much as $5\text{-}10 \times 10^6$ ton could be employed in ruminant feeds with reasonable safety and economic validity.

Bioconversion

Animal and sewage wastes can be converted into single-cell bacterial and fungal protein, either in the semiwet condition through ensiling, or more probably in a series of fermentation lagoons where the wastes are comminuted, suspended in water and successively digested either aerobically or anaerobically and cultivated into specific species and strains of bacteria, algae or aquatic plants.

The use of algae and aquatic plants such as water narcissus and duckweed is perhaps the most promising option for the economic and effective recycling of organic wastes. There are five distinct returns from such an aquaculture:

1. All organic wastes could be effectively recycled.
2. The algae or aquatic plants could be harvested and either used directly as livestock feed—fresh, ensiled or dried—thereby releasing an equivalent amount of energy and protein from grains and pulses for direct use by humans; or the product could be extracted for protein, vitamins and carbohydrates to supplement human needs.

3. Through a chain of sequential lagoons the organic waste could be converted to green plant tissue, herbivorous microfauna browsed on the plants, carnivorous microfauna fed on the herbivores, krill on the carnivores, dace and other small fish on the krill, and ultimately bass and/or trout on the dace. The top of the aquatic food chain would then be fed directly to humans or processed and fed back to monogastric livestock.
4. Polluted waters would become depolluted chemically, biologically and thermally (where thermal pollution was present) and recycled into the human and livestock water supply.
5. The genetic development of effective and improved aquatic species, particularly algae, would lead to the discovery of new and useful chemicals as pharmaceuticals, pesticides and chemical industrial intermediates as by-products. Such a procedure would become self-supporting if not clearly profitable and would remove one of the most nagging and aggravating environmental problems—the effective recycling of organic wastes. The major problem would be one of the logistics.

Conversion of Organic Wastes into Methane

The process of converting organic wastes by anaerobic fermentation into methane to provide fuel for farm or feedlot stationary mechanical operations has been practiced in India and other developing countries for some years now on a simple and elementary scale with modest success. It has the advantage over burning of sundried animal wastes directly as fuel in that the spent digesta still retains a large proportion of the original plant nutrients which can be spread on the crop lands for supplemental fertility.

However, the logistics and economics are not encouraging. For example, the daily manure production from 100 swine of 150 lb each would upon efficient anaerobic digestion in a 570-ft^3 digester produce about 430 ft^3 of methane gas per day equivalent in heating value to approximately 2 gal of gasoline. But about half of this energy would be consumed in operation of the digester at 98°F and preheating the organic material. The net return of 1 gal of gasoline equivalent per day would take many years to recover the investment. The residue must still be disposed of even though it would be reduced as much as two-thirds in weight and volume.[26]

Pyrolysis

Direct pyrolysis of organic waters, particularly municipal and industrial sewage sludges in combination with other organic wastes (garbage), appears to be logistically and economically sound in large cities to provide energy for heating and/or cooling public buildings. It has proved to be successful in Nashville, Tennessee, but failed in Baltimore, Maryland, for reasons as yet unclear. Continuous availability of supply, storage accumulation facilities and air pollution control constitute the major difficulties in such operations. These can be solved technically, however, and as conventional energy sources become increasingly expensive and organic wastes mount pyrolysis may well prove to be an efficient resolution to the problem, at least in densely populated urban areas. Such a method of disposal is not, however, a conservation of energy but merely a convenient method of disposal.[26]

ENERGY CONSIDERATIONS

The conservation of energy in all the works and walks of American life has become paramount and will continue to be so. Organic wastes are a potential positive source of energy, which can readily be inverted to the negative if not managed logically. Collecting, hauling, storing and particularly drying animal and human wastes generally leads to a negative energy return, particularly where more useful fossil fuels are employed such as gasoline, fuel oil, natural gas or coal in the collection, processing and distribution process. Thus drying cow manure from 88% moisture reveals an energy break-even point at 41% moisture; below that it becomes energy-negative. The biological or chemical oxidation of carbon releases 60 to 70% of the energy which can be converted and conserved biologically as single cell protein and energy more effectively than by pyrolysis.

Using the organic wastes as a substitute for chemical fertilizers will depend upon the distance of distribution and the moisture content. Reducing the moisture content artificially quickly becomes energy-negative, and drying it in the sun is seasonally and geographically limited, evokes air pollution problems and loses substantial nutrient value. Comparative figures indicate that one ton of 12-12-12 fertilizer/ac required 9×10^6 Btu to manufacture and about 4.1×10^6 Btu to transport to farming areas and spread over the land for a total of 13.1×10^6 Btu. Dried barnyard manure (20% moisture) of 2-0.5-1.5 N-P-K analysis would require the equivalent of 50 tons of wet manure/ac with a collection, storing, load and hauling energy expenditure of about 6×10^6 Btu, a saving of approximately 7×10^6 Btu/ac. However, to this must be added the additional labor and time of covering the area with 25 trips at 2 ton/trip to apply the same amount of nutrient as 1 ton/ac of 12-12-12. Obviously today when farm labor and machinery investment and maintenance are expensive, the additional human and mechanical work involved in spreading manure is uneconomic. If the manure is sun-dried to about 30% moisture, the two energy costs would be about equivalent, but it would still take about 10 trips across each acre to apply the manure as opposed to distributing the fertilizer, which in fact would probably cover 10 acres in a single trip in a bulk truck.

NITROGEN FIXATION

Fixation of atmospheric nitrogen is the major factor in the nitrogen cycle as it has been since life commenced on the planet. The natural process, though incessant and continuous, is generally too slow in localized areas of intensive crop production. Including physical fixation by lightning and rain, chemical fixation from volcanoes, natural or man-made smog (peroxyacetyl nitrate), industrial and auto effluents and biological fixation, the total amount of fixation varies from 3 kg/ha at the poles to as much as 60 kg/ha in densely populated industrial areas.

Malo and Purvis[32] reported an annual atmospheric deposition in New Jersey of from 50 to 60 kg N/ha. Such an application of nitrogen would provide more than adequate fertility renewal for wooded, range or wild lands but would fall considerably short of replacement on intensively cultivated soils.

The world nitrogen fixation in 1974 was recently estimated to total 237 million m.ton distributed by source as given in Table III.

Table III. Estimates of world-wide nitrogen fixation–1974.[33]

Source	Metric Tons/yr x 10^6	(%)
Natural	160	67
(Arable Lands)	(89)	(38)
(Forest and Range)	(60)	(25)
(Lightning)	(10)	(4)
(Waters)	(1)	(<1)
Industrial	77	33
(Fertilizer, etc.)	(57)	(24)
(Combustion)	(20)	(9)
Total	237	100

From these data it appears that natural fixation comprises about twice as much fixed-N tonnage (160 x 10^6 m.ton/yr) as does man-fixed-N tonnage (77 x 10^6 N ton/yr). Artificially fixed N is, however, in addition to that N naturally fixed and does constitute a very substantial modification of the nitrogen cycle in natural ecosystems. A closer estimate of specific sources is given in Table IV.

From this table it is obvious that the major source of fixed nitrogen among cropped and livestock soils is in grasslands, some of which is, of course, legume-nitrogen fixation from nodular symbiotic association such as with *Rhizobium* species of bacteria, but nonlegumes contribute a very substantial quota to the natural soil fixation of nitrogen.

Table IV. Estimates of world-wide nitrogen fixation sources–1974.[33]

Source	Metric Tons/yr x 10^6	(%)
Agriculture	89	36.6
(Legumes)	(35)	(14.8)
(Nonlegumes)	(9)	(3.8)
(Grass lands)	(45)	(19.0)
Forests	50	21.1
Unused Lands	10	4.2
Oceans	1	0.4
Lightning	10	4.2
Combustion	20	8.4
Industry	57	24.1
(Fertilizers)	(39)	(16.5)
(Chem. Prod. Mfg.)	(8)	(3.4)
(Losses)	(10)	(4.2)
Total	237	100.0

Brill[34] notes that *Spirillum lipoferum* "infects" various tropical grasses and captures atmospheric nitrogen in substantial quantities in a symbiotic relationship. This important observation has also been recently reported by others.[35-38] Smith *et al.*[38] estimated that *Spirillum lipoferum* fixes atmospheric nitrogen on two species of grasses studied at a rate up to 0.6 kg N/ha/day or close to 100 kg N/ha per season–an astonishing contribution to the nitrogen economy of the soil.

Brill[34] reports *Spirillum lipoferum* has been found associated with such economic grasses as corn and provides significant nitrogen nutrition. If such nitrogen fixers can become genetically modified or "trained" to associate symbiotically with other commercially important grasses such as the cereals (wheat, rice, barley, oats, rye and millet), sorghum and sugar cane, the biological and economic impact would be immense. Nitrogen is the most important single limiting factor in the resolution of world hunger, and the cereal grasses are the basis of the world food supply. To make these basic foodstuffs self-nitrogen-fertilizing would constitute a technical breakthrough comparable to hybrid vigor and would provide immense relief to the world hunger problem. Furthermore, if progress continues in the current research in plant breeding to enhance the inherent protein level in these cereals, as has already been accomplished by the plant breeders in corn and wheat, both protein and caloric deficiencies can be fulfilled.

Brill[34] lists the major nitrogen-fixing microflora and their host habitats. Some important free-living N-fixers are listed in Table V.

Table V. Important free-living N-fixers.

Those requiring organic detritus as energy source	
Achromobacter, sp.	soil aerobe
Azotabacter vinelandi	soil aerobe
Klebsiella pneumoniae	soil aerobe or anaerobe
Beijerinckia, sp.	soil aerobe
Clostridium pasteurianum	soil anaerobe
Those with self-supporting energy source	
Rhodospirillum rubrum	Aquatic both N-fixer and photosynthesizer
Blue green aquatic algae and *Rhodopseudomonas,* sp.	
Symbionts–those accepting energy sources from host	
Spirillum lipoferum	tropical grasses *Digitaria,* sp., *Zea Mays*
Citrobacter freundii	termite intestinal bacterium
Frankia alni	alder
Nostoc muscorum	on tropical herb, *Gunnera macrophylla*
Anabaena azollae	aquatic fern cyanobacterium
Rhizobium japonicum	soybean nodules, aerobe
R. Trifolii	clover nodules
R. Meliloti	alfalfa nodules
R. radicola	bean and pea nodules

There is therefore a wealth of DNA and germ plasm capable of fixing atmospheric nitrogen into forms which can be utilized by plants and animals, both terrestrial and aquatic, if genetic engineering is applied intensively to build in nitrogen-fixing capability.

Massive atmospheric nitrogen fixation, essential as it is to feed, clothe and shelter 7 to 10 billion humans at a reasonable, above-subsistence level by the year 2000, also implies a rise in the denitrification of nitrogen from soils and waters and release of increased levels of nitrous oxide (N_2O) into the atmosphere.[39,40]

NITROGEN OXIDES AND OZONE LAYER

The ozone layer in the stratosphere is essential for the attenuation of ultraviolet light ($<$ 2500 a) which if it should penetrate at full strength to the surface of the earth would destroy much of the terrestrial biosphere. It is generally accepted that as present UV-insolation is the major cause of skin cancer, even small reductions in the ozone layer of the stratosphere would increase dermal carcinoma among humans to a significant and perhaps catastrophic increment.

Nitrogen oxides which reach the stratosphere are believed to react with ozone and reduce it to oxygen.[41-47] Sources of nitrous oxide (N_2O) that may reach the stratosphere are multiple in the modern world. Artificial or man-made sources include auto exhausts, industrial gaseous pollutants and biological denitrification of N fertilizers applied to the soil. There are, of course, natural sources including volcanic activity and denitrification of wetlands, wildlands and the oceans.[40,48-59]

The relative contributions to stratospheric nitrogen fertilizers as compared to uncontrollable natural sources, though uncertain, clearly weigh preponderantly upon natural sources. Assuming that the nitrogen content of the oceans is at a steady state at 1 billion m.tons[60,61] and that the nitrous oxide level is close to saturation, Hahn[54] estimates the annual evolution of N_2O from the oceans as about 100 x 10^6 m.ton, although others consider it to be not more than 40 x 10^6 m.ton/yr.[33] Schuetz *et al.*[59] estimate the total world nitrous oxide evolution to be 130-260 x 10^6 m.ton/yr. The estimate of nitrous oxide production is from 5 to 10% of the annual fixation rate or about 11-24 x 10^6 m.ton of which fixed nitrogen fertilizers would, on this basis, contribute 2-4 x 10^6 m.ton or from about 1.0 to 3.0%.

Low as this level may be in relation to natural evolution of nitrogen oxides, it should not be ignored as of no consequence. The residence time of N_2O in the atmosphere can be up to 10 yr and during that period there would be ample opportunity for it to reduce ozone.

One interesting point is that green plants appear to absorb nitrogen oxide from the atmosphere and metabolize it directly into plant tissue as protein. If this turns out to apply widely among green plant species, then the problem may well be of considerably less concern than it now appears to be. Perhaps the best and most practicable solution to the smog, PAN and other nitrogen oxide atmospheric pollution is to increase green space to the fullest extent possible.

In any case the evolution of nitrous oxides into the air as a direct result of soil-applied nitrogen fertilizers whether organic or inorganic must not be regarded as trivial. A continuous monitoring of NO_X levels in the atmosphere should be maintained until we secure sufficient data to prepare a total world inventory of nitrogen balance between the atmosphere, the lithosphere and the hydrosphere.

CONCLUSION

Fertilizers and the maintenance of soil fertility particularly with respect to nitrogen are essential if we hope to feed, clothe and shelter the human population at even a bare subsistence level. There is no possibility that natural processes can provide such support for the present population, let alone the additional billions anticipated by the year 2000.

Nitrogen is required at a level of about 30 lb per capita per annum for subsistence. This requires about 65 million m.ton of fertilizer nitrogen per year, which will increase to over 130 million m.ton by the year 2000. Although the total natural annual N fixation rate is estimated to be 160 million m.ton, over 100 million tons is in nonarable, nonfood- or fiber-producing lands. Of the remainder over half, or 30 million tons, is not present in the hungry lands or is consumed by luxury living in the developed countries. Therefore, the current industrial fixation of about 40 million tons of nitrogen constitutes the margin of safety from human starvation on an unparalleled scale.

The use of organic wastes, livestock, human and crop processing can contribute from 5 to 10% of the need but is in many cases uneconomic and impracticable. There is more promise in converting these sources of energy and protein by recycling, particularly through aquatic plants, into more acceptable and useful sources of food or feed. Recovering their energy by pyrolysis is wasteful; fermenting the material into methane is inefficient; and spreading it on crop land can lead to problems of toxicity in the edible products.

There is no evidence that foods grown solely on organic wastes are superior in productivity, quality, taste or nutritional value to those raised on industrial fertilizers.

The problem of the evolution of nitrogen oxides from soils fertilized with industrial-fixed nitrogen is no greater than that from those dressed with organic wastes. The evolution of nitrogen oxides is a function of biological activity, climatic edaphic and aeration conditions and the quantity of nitrogen in the soil at any one time regardless of its source. There is some evidence that nitrous oxide, in particular, can migrate to the stratosphere and contribute to the reduction of ozone. A substantial reduction in the ozone layer would lead to greater penetration of ultraviolet light into the troposphere and lithosphere and create serious damage to the biosphere.

There is no unequivocal evidence to date to show that man-made fixed nitrogen applied to the soil has in fact significantly reduced the ozone layer. It should, however, be watched, without panic.

There is evidence that green plants reduce nitrogen oxides in the air by foliar absorption and metabolism.

REFERENCES

1. White-Stevens, R. *Proc. 7th Internat. Cong. Nutrition*, Vol. 4 (New York: Pergamon Press, 1967), pp. 925-935.
2. Tanner, C. C. *Outlook on Agric.* 5(6):235-40 (1968).
3. Council for Agricultural Science and Technology. "The U.S. Fertilizer Situation and Outlook," 1974 Task Force Report, Iowa State University, Ames, Iowa (1974).

4. U.S. Department of Agriculture. Special Circular No. 7, Statistical Reporting Service, Washington, D.C. (1973).
5. Shapley, D. *Science* 195(4279):658 (1977).
6. Pesek, J., G. Stanford and N. C. Case. "Nitrogen Production and Use," in *Fertilizer Technology and Use*, R. A. Olsen *et al.*, Eds. (Madison, Wisconsin: Soil Scientific Society of America, 1971), pp. 217-269.
7. Commoner, B. *The Closing Circle–Nature, Man and Technology* (New York: A. A. Knopf, 1971), pp. 81-93.
8. Smith, G. E. Proc. 22nd Ann. Meeting Soil Conservation Society of America, 1967, pp. 108-114.
9. Smith, G. E. Proc. 16th Ann. California Fertilizer Conf., Fresno, California, March 25, 1968, pp. 34-47.
10. Smith, G. E. Univ. Missouri Coll. Agr. and Missouri Water Poll. Board–Joint Seminar, Columbia, Missouri, April 9, 1969.
11. Smith, G. E. *J. Missouri Ag. Exp. Sta.*, Series 1651.
12. Enfield, G. H. Proc. Ann. Meeting, The Fertilizer Institute, Chicago, Illinois, February 16-17, 1970.
13. Adams, L. J. Univ. Nebraska Agric. Expt. Station Progress Report (1968).
14. Allison, F. E. *Adv. Agron.* 18:219-258 (1966).
15. Pratt, P. F. *et al. Hilgard.* 38(8):265-283 (1967).
16. Robertson, L. S. *Agric. Nitrogen News* (March 4, 1969), p. 38.
17. Welch, L. F. *et al.* Illinois Fertilizer Conference Proceedings (1969), pp. 37-39.
18. Duggan, R. E. *et al. Science* 151(3706):101-104 (1966).
19. Duggan, R. E. *Pesticides Monitoring J.* 1(3):2-8 (1967).
20. Duggan, R. E. *Science* 157:1006 (1967).
21. Duggan, R. E. *et al. Pesticides Monitoring J.* 2(4):153-162 (1969).
22. Hurd, L. E., M. V. Mellinger, L. L. Wolf and S. J. McNaughton. *Science* 173:1134-1136 (1971).
23. Claus, G. and K. Bolander. *Ecological Sanity* (New York: David McKay, 1977).
24. Erhlich, P. R. "Project Survival," *Playboy* (1971), p. 65.
25. Reid, J. T. Proc. Am. Assoc. Advancement Science, 141st Annual Meeting, New York, 1975.
26. Council for Agricultural Science and Technology. "Utilization of Animal Manures and Sewage Sludges in Food and Fiber Production," Task Force Report 41 (1975).
27. Taiganides, E. P. and R. L. St. Roshine. Proc. Symp. Livestock Waste–Center for Tomorrow, American Society of Agricultural Engineers, St. Joseph, Michigan, 1971, pp. 95-98.
28. Nelson, L. B. *et al.*, Eds. "Changing Patterns of Fertilizer Use," Soil Sci. Soc. Amer., Madison, Wisconsin (1968).
29. Dewey and Almy Chem. Div. "Plant Nutrient Removal Chart," W. R. Grace & Co., Nashua, New Hampshire (1972).
30. Williams, P. H., J. H. Shenk and D. E. Baker. *Sci. Agric.* 24(2):5 (1977).
31. Emmelin, L. *Environment* (Sweden) 38 (1973).
32. Malo, B. A. and E. R. Purvis. *Soil. Sci.* 94:242-247 (1964).
33. Council for Agricultural Science and Technology. "Effect of Increased Nitrogen Fixation on Stratospheric Ozone," Task Force Report 53 (1976).

34. Brill, W. J. *Sci. Amer.* 236(3):68-81 (1977).
35. Garcia, M., D. Hubbell and M. Gaskins. "Role of N-Fixing, Blue-Green Algae, & Asymbiotic Bacteria," International Symposium on Environment, Upsala, Sweden (1976).
36. Gaskins, M. H., C. Napoli, D. H. Hubbell. *Agron. Abst.* (1976), p. 71.
37. Rogerson, A. C. *Science* 195:1362 (1977).
38. Smith, R. L. *et al.* *Science* 193:1003 (1976).
39. Blackmer, A. M. and J. M. Bremner. *Geophys. Res. Lett.* 3:739 (1976).
40. Robinson, E. and R. E. Robbins. "Source, Abundance and Fate of Gaseous Atmospheric Pollutants," Standard Research Institute, Stanford, California (1968).
41. Crutzen, P. J. *Quart. J. Res. Meteorol. Soc.* 96:320 (1970).
42. Crutzen, P. J. *Ambio* 3:201 (1974).
43. Crutzen, P. J. *G.A.R.P. Publ. Ser. 16* (1975).
44. Crutzen, P. J. *Geophys. Res. Lett.* 3:169 (1976).
45. Johnston, H. S. *Science* 173:517 (1971).
46. McElroy, M. B. and J. C. McConnell. *J. Atmos. Sci.* 28:1095 (1971).
47. Nicolet, M. and E. Vergison. *Aeronom. Acta.* 90:1 (1971).
48. Broadbent, F. E. and F. E. Clark. *Denitrification–Soil Nitrogen* (Madison, Wisconson: American Society of Agronomy, 1965).
49. Cady, F. B. and W. V. Bartholomew. *Soil Sci. Soc. Amer. Proc.* 24: 477-482 (1960).
50. Cavender, J. H., D. S. Kircher and A. J. Hoffman. U.S. Environmental Protection Agency Publication AP 115 (1973).
51. Goering, J. J. *Deep Sea Res.* 15:157-164 (1968).
52. Goering, J. J., F. A. Richards, L. A. Godispoti and R. C. Dugdale. "Proc. International Symp. Hydrogeochim," E. Ingerson, Ed. *Biogeochem.* 2:12-27 (1973).
53. Grobecker, A. J., S. C. Coronniti and R. H. Cannon, Jr. Dept. Transportation Report of Findings of Climatic Assessment Prog. DOT-TST-75-50, Washington, D.C. (1975).
54. Hahn, J. *Tellus* 26:160-168 (1974).
55. Junge, C. and J. Hahn. *J. Geophys. Res.* 76:8143-8146 (1971).
56. McElroy, M. B. "International Rev. Sci.," Physics Chemistry Series 2 (1976), p. 127.
57. McElroy, M. B., J. W. Elkins, S. C. Wofsy and Y. L. Yung. *Rev. Geophys. Space Phys.* 14:143 (1976).
58. Reiter, R. *Tellus* 22:122-136 (1970).
59. Schuetz, K., C. Junge, R. Beck and B. Allbrecht. *J. Geophys. Res.* 75:2230-2246 (1970).
60. Emery, K. O., W. L. Orr and S. C. Rittenberg. "Nutrient Budgets in the Ocean," in *Essays in the Natural Sciences in Honor of Captain Hancock* (Los Angeles, California: University of California Press, 1955), pp. 299-310.

SECTION II

APPLICATION OF WASTEWATERS TO LAND

3

STOPPING THE ONE-WAY FLOW OF NUTRIENTS FROM FARMS TO CITIES; BENEFITS FOR FARMERS AND CLEANER WATER

M. Gravitz
Clean Water Action Project
Washington, D.C.

INTRODUCTION

This presentation is about one of the largest agricultural pollution problems in the country: human sewage disposal. Municipal sewage is not generally thought of as an agricultural problem, but it is an agricultural problem. The fertilizer and nutrients that farmers put into the production of food eventually reach municipal sewers that can carry them into the nearest river, lake or ocean. In America today the flow of nutrients and organic material is all one way: from the farm to the city. Only a tiny portion of these resources ever find their way back to the farm.[1] As a direct consequence of this one-way path, America suffers from an oversupply of nutrient-rich human sludge and wastewater. One measure of this oversupply can be seen in the increasing eutrophication and premature aging of our rivers, lakes and estuaries. At the same time, certain parts of the country suffer from a growing scarcity of water and a shortage of phosphate and nitrogen fertilizers. In 1974, American farmers spent five billion dollars on artificial fertilizers.[2] About one-third of the energy budget for an average farm is attributable to fertilizer production.[3] Yet, American cities and sewer districts spend hundreds of millions of dollars a year to incinerate, bury, dump or otherwise dispose of phosphates and nitrogen-rich sludge and sewage.

We are also facing a water shortage which has become most visible in California. According to Mayor Bradley's Advisory Committee on Water Resources, Los Angeles will have to institute water rationing in a matter of months. The question is not whether they will ration, but when and how. Areas that have traditionally had water surpluses now face impending shortages. These areas include suburban Chicago, southeast Virginia, metropolitan Washington, Long Island and Cape Cod, among others.[4,5] As the quantity of water wasted by our sewage treatment systems grows, so too do the water shortages.

In the past several years, the federal government has spent close to 18 billion dollars on sewers and sewage treatment plants. Over the next ten years, we may spend an additional four to five billion dollars per year. This spending program, the largest public works program in our nation's history, has been characterized by a philosophy of waste. We waste water, we waste nutrients and we waste money. At the same time, we have in some instances made our environmental problems worse and not better.

RESULTANT PROBLEMS

Some examples of this wasted money, wasted water, wasted nutrients and pollution caused by conventional sewage treatment works include the following:

1. The waste of nutrients may be illustrated by the City of Merrimack, New Hampshire. The City of Merrimack boasts a beer brewery of some size. Brewery waste is enormously rich in nutrients and organic material. Some time ago, citizens in Merrimack, recognizing the value of this resource, sought help from the Environmental Protection Agency of the United States in capturing and reusing the brewery by-products that were sent down to their sewage treatment plant. The Environmental Protection Agency, instead of aiding and encouraging them, took a different approach to the problem, an approach that typified its administration of the $18 billion sewage grants program. EPA insisted that the city incinerate the nutrient-rich sludge in a six-million-dollar incinerator. The incinerator will use 600,000 to 900,000 gal of No. 2 fuel oil per year—enough to heat 500 homes—to burn up some of the best sludge in the U.S. The common rule-of-thumb is that 50% of the costs of running a typical sewage treatment plant are now spent on burning, burying or ocean dumping the sludge and, only in some few cases, using it agriculturally.

2. The Patuxent River flows past a growing metropolitan area into the Chesapeake Bay. Chlorine from several sewage treatment plants along the river has seriously reduced fish population and species diversity. Shellfish reproduction is hindered by chlorine and its reaction in concentrations of only parts per billion.[6] The problem will grow worse. During the summer months in the year 2000, the Patuxent River will be 75% chlorine-polluted sewage.

3. Chlorinated organics, formed when sewage is disinfected so that streams can be used for shellfishing or recreation without a health hazard, jeopardize downstream drinking water systems because some of the materials are carcinogenic. In Montgomery County, Maryland, work on a large sewage treatment plant above Washington D.C.'s water intakes was stopped partly for this reason. Five thousand tons of stable chlorinated organic compounds are released into the nation's waters every year from this source.[7]

4. Nutrients in sewage and urban runoff eutrophy hundreds of lakes and slow-moving rivers and estuaries. The stench of rotting algae hung over some of Washington's famous monuments in the northeastern drought period of the mid-1960s. The river turned pea green and is still that color many miles below the city. Eutrophication endangers water supplies because the more organic material there is in water, the more chloroform and other chlorinated organics are formed when drinking water is chlorinated. The Occaquon Reservoir, which supplies 600,000 residents in the nation's capital, is eutrophying, and chloroform levels in drinking water form a significant long-term health hazard.[8] Many rivers and lakes have an area of water with low or zero dissolved oxygen due to nutrients and organic material. Fish avoid them and often cannot get up rivers to spawning areas or die in massive numbers.

5. Billions of gallons of freshwater in the form of sewage are dumped into the ocean in California, Long Island and Cape Cod, yet some of these areas are having to ration water because of a drought or have falling ground water tables–their only source of fresh water.[9]

6. Centralized conventional sewage treatment systems are driving small towns broke, even though the EPA normally pays for 75% of the costs of a system. The average *local cost* (for the 25% of capital cost EPA does not pay) of 23 systems in small towns in Illinois was $1900/family for capital costs only. Yearly users' fees were $190 to $230 *per capita*, not per household.[10] Dunkirk, Ohio, and Wilton, New York, are both well-publicized examples of where a town is going broke paying for a conventional sewage system that may never have been needed.[11,12]

BASIS OF THESE PROBLEMS

Why is this topic being discussed with agricultural scientists interested in utilizing farm wastes on the farm? It is because the EPA construction grant program is not oriented toward creating a cycle of nutrients and organic material between town and farms. It is instead oriented in almost every conceivable way, even to the incentives it creates for land use patterns, toward reinforcing the destructive one-way flow of nutrients and organic material from farms to towns.*

It is ironical that EPA does not emphasize that sewage is a farm residue and a resource that needs to be recycled, because the legislation creating the sewage construction grant program specifically directs EPA to treat sewage in this fashion. Section 201 of the 1972 Federal Water Pollution Control Act directs EPA to *encourage* the use of sewage and sludge in agriculture, forests and aquaculture (see Appendix A). Before a sewage treatment plant is built, sewage recycling technologies have to be compared with conventional alternatives such as activated sludge or trickling filters. It is clear that the framers of this law wanted EPA working with towns to build sewage recycling systems wherever possible. The mandate of the law on sewage recycling was very broad and very strong. Yet it has been virtually ignored by EPA, by local and state governments, by consulting engineers who choose and design treatment plants for towns, by citizens, and by many scientists who could help guide technical decision making in their towns.

What are the facts of the case? EPA is more than three-fourths of its way through its original $18 billion, and as of June 30, 1976, after three years of spending, EPA has funded only 43 land treatment sewage recycling systems. Spending on these totalled $34 million, 0.009% of EPA's expenditures on new and upgraded sewage treatment systems.** At the same time, EPA spent billions of dollars on thousands of new or improved conventional

---

*The trend toward large regional wastewater treatment plants often means that interceptor sewers cross large areas of undeveloped land. This encourages the urbanization of productive farmland. The 75% federal grant for the capital cost of the facilities encourages oversizing. Then the need to pay for running the overly large facility makes the municipality encourage conversion of farmland to higher property tax uses.

**These figures are near approximations, as accurate as EPA data allow one to get. In order to find the amount of money spent on treatment plants, I looked at all PL 92-500 grants inclusive of June 1976. I counted step 2 (plans and specifications) and step 3 (construction) grants for treatment plant projects involving new plants and upgrading the treatment levels of old ones. Excluded were grants for expansion in capacity only, modification of the facility, or grants only for collection systems. Unfortunately, there was no way of separating the costs on a grant that went to treatment plants and collection systems.

systems. This is an abysmal record and one unlikely to change in the future if one looks at the kinds of sewage treatment plants in the pipeline.*

How does this compare with the number of land treatment systems already operating? Again the record is abysmal. There are already 325 land treatment sewage recycling systems in this country, most of them in the southwest. These land treatment systems are mostly small, treating less than one million gallons of sewage per day and only about 1% of the nation's flow of sewage.[13]

The small number of land treatment systems suggests that this is a new technique for treating sewage. This is not true for there have been land treatment recycling systems in this country and elsewhere since the late 1800s. Human wastes have been recycled back to crops and fish ponds for thousands of years. There are several recycling systems in this country—at Woodland, California; Bakersfield, California; and Abilene, Texas, for example—that have been running since the late 1800s or early 1900s. Most land treatment systems in the U.S. are in the southwest where water is scarce and they are used to growing food fodder and fiber crops. However, many European land treatment systems were in areas not thought of as being water scarce. Berlin started a system in 1896 that encompassed 27,000 acres and treated the sewage from 1.5 million people. Much of that system is still running. Paris had a system that treated 120 mgd. Other sizeable facilities are located in Australia, Great Britain, Hungary, India and Israel.[14]

WHAT CAN BE DONE?

Neither Congress, EPA, USDA nor agribusiness have done much to stem the one-way flow of nutrients and organic material from farms to cities that pollutes urban rivers, costs farmers huge amounts of money for fertilizers and pollutes rural water with runoff and nitrate leaching. The steps that can be taken to change this picture include the following:

1. Some very practical action on the institutional constraints to the use of recycling systems needs to be taken. Most of the constraints to use of sewage recycling are institutional, and not technical problems, although there are still some health effects questions to clarify. While more research certainly is needed, it will never be enough to get sewage recycling systems built. Even physical or natural scientists should apply their talents to institutional problems. Many institutional problems stem from a need for increased public education and exposure to recycling systems.

---

*Two-thirds of the sewage treatment plants projected to be built in the 1974 EPA Needs Survey specified the use of activated sludge technology.

2. To attack these institutional problems there need to be practical, comprehensive bulletins for farmers on how to recycle sewage–what crops they can use, what soils they can do it on, descriptions of application technologies and health safeguards.

3. Different constituencies–farmers, farm communities, food processors and urban dwellers–need to get interested in returning to the land the nutrients and humus they send to municipal sewers. It obviously must be shown that it is in their economic and environmental interest to return municipal wastewater and sludge to land.

4. Interested individuals must get involved with local governments when they are planning future sewage treatment systems or choosing between treatment technologies. They need expertise in agricultural affairs or recycling systems and demonstrated interest in such alternatives. It is possible that their engineering consultants may not have much expertise in this area nor have ever designed and built a sewage recycling system. Although the legislation (Appendix A) requires that land application systems be considered as an alternative, local governments may hire engineering companies with no experience in sewage recycling to examine it as an alternative and choose between it and conventional technologies.

SEWAGE RECYCLING FOR FARMERS

First of all, in order to use the nutrients in sewage it is not enough merely to apply sludge. Sludge contains only 10 to 25% of the nutrients in sewage. Most of the nutrients are contained in sewage effluents. If most of the nutrients are to be recovered, effluents must be utilized by farmers. Treatment plants in the United States handle approximately:

980,000 (short) tons of nitrogen (as NH_3)
825,000 tons of phosphorus (as P_2O_5) and
470,000 tons of potassium (as K_2O) every year.

where 24,000 mgd = U.S. sewage flow
Secondary treated effluents = 17 mg/l total nitrogen, 8 mg/l total phosphorus and 10 mg/l total potassium
tons = short tons = 2000 lb
6.2 million dry tons/yr = U.S. production of sludge
Nitrogen = 3% of dry solids
Phosphorus as P_2O_5 = 2.5% of dry solids
Potassium as K_2O = 0.5% of dry solids

The total value of all these nutrients in effluents and sludge was $630 million in 1975 prices. If the fertilizer value of effluents and sludge had been used by farmers in 1975, 11% of the nation's agricultural demand for nitrogen would have been satisfied, 18% of its demand for phosphorus and 11% of its demand for potassium. Take away the sludge and you still would have satisfied 9% of the nitrogen demand, 15% of the phosphorus demand and 10% of the potassium demand. While these percentages may not seem particularly large, they would represent a significant monetary and energy savings to farmers.

Besides energy and cost advantages to be gained by use of human farm residues on the farm, there are other advantages. Continuous application of inorganic nitrogen to cropland is a vicious cycle insofar as loss of nitrogen is concerned. Nitrate or ammonia fertilizers effectively sustain plant growth, but they do not rebuild the humus lost from the soil. The increased supply of nitrogen available to microbial populations after fertilizer speeds decomposition of humus. Humus maintains the porosity of soil, which in turn allows the soil to remain aerated. As soil porosity and therefore soil oxygen content falls when humus content is depleted, plants may not take up as much of the applied nutrients and sustain optimum crop growth. Farmers then may have to apply more nutrients than before to get the same crop response.[15] Applications of organic material, green manuring or nitrogen in wastewater effluents and sludges can reverse this process.

The rate at which atmospheric nitrogen (N_2) is fixed in inorganic fertilizers (NH_3, NO_3) for agricultural use all over the world has gone from 3.5×10^6 ton/yr in 1950 to 40×10^6 ton/yr in 1974, an eleven-fold increase in less than 25 years. This rapid increase has made some scientists suspect that man's activities are becoming an increasingly important part of the global nitrogen cycle. Production of nitrogen fertilizer projected for the year 2000 will by itself be larger than what the worldwide biological fixation of nitrogen is thought to be. And this excludes all the nitrogen oxides created by combustion in automobiles and stationary sources.[16]

Since there is plenty of nitrogen in the atmosphere that man can use for fertilizers, what is the problem? The potential hazard occurs because as nitrates from inorganic fertilizers are denitrified in the soil, N_2O (gas) as well as elemental nitrogen gas (N_2) is produced. N_2O is changed to NO in the atmosphere and can act as a catalyst to break down the earth's protective shield of ozone. Ozone protects us from the sun's harmful ultraviolet rays. McElroy, a Harvard University atmospheric scientist, has predicted that a 20% decrease in the earth's ozone layer could occur by 2025 due to current and future use of inorganic fertilizers which are releasing larger and larger amounts of N_2O (gas).[16] Not only could the incidence of skin cancer increase, but the productivity of the earth may suffer because many plants and insects are thought to be sensitive to increased ultraviolet radiation.

The increasing use of "virgin" nutrients in agriculture not only can cause environmental pollution when they are manufactured and used, but may have a lasting impact on future generations. Phosphorus is a necessary element for life to exist; it has no substitutes. Commonly it is *the* life-limiting element in natural ecosystems because of the ease with which it forms insoluble salts and is therefore unavailable for uptake. The rate at which the global cycle of phosphorus runs depends on the uplifting of ocean sediments to form land. Since this process is very slow, and since artificial reconcentration of phosphorus is difficult, phosphorus should be treated by man as a crucial nonrenewable resource.

The U.S. Bureau of Mines predicts that the major U.S. reserves of phosphate rock in Florida and Tennessee from which most fertilizer in the U.S. is produced will be exhausted "before or shortly after the end of this century." This will make the U.S., now a major exporter of phosphates, dependent upon imported phosphate fertilizer to maintain its food production. The price of phosphate rock on the world market already has no relationship with the cost of producing it.[17] The importance of an assured supply of phosphate fertilizers to farmers is obvious.

Land treatment or other sewage recycling systems could be a major tool in preserving farmland in urbanizing areas. Urbanization is currently eating up 1.2 million acres of rural land annually.[18] An amazing amount of farmland and forest still lies within what the federal government considers urban area (Standard Metropolitan Statistical Areas). They produce large percentages of our dairy, fruit and vegetable products. These farms need not be lost to the diffuse, land-gobbling process of suburbanization. By using farmland as the areas to return our cities' nutrients and organic materials, farms could be preserved as open space and as productive areas to feed the cities. Farmers do not need to face the choice of selling out and moving further into the country when the city closes in. They could begin to accept the city's wastes and productively use them. They could preserve their livelihood and open space for everyone to enjoy right where they are.

Even two or three years ago conferences like this one on waste or resource recycling were relatively rare. Now they are becoming more commonplace. This indicates that our country is moving into an era of resource recycling. However, there may not be an increase in sewage and sludge recycling unless scientists and other interested individuals get involved in the nitty-gritty problems that serve as constraints to these systems. With greater interest and involvement, sewage, sludge and farm residue recycling could be much more widely used.

REFERENCES

1. Borgstrom, George. "The Breach in the Flow of Mineral Nutrients," *Ambio* 2:129-135 (1973).

2. U.S. Congress, Senate Committee on Agriculture and Forestry. "U.S. and World Fertilizer Situation," (December 1974), p. 5.
3. Pimental, David, *et al.* "Food Production and the Energy Crisis," *Science* 182:445 (1973).
4. U.S. Army Corps of Engineers. "Washington Metropolitan Area Water Supply Study Report," Northeastern U.S. Water Supply (NEWS) Study (November 1975).
5. Frimpter, Michael. "Groundwater Management: Cape Cod, Martha's Vineyard, and Nantucket, Massachusetts," U.S. Department of the Interior, USGS Open File Report (1973).
6. U.S. Environmental Protection Agency, "Task Force Report: Disinfection of Wastewater," Office of Research and Development (July 1975), pp. 13-20.
7. Jolley, Robert. "Chlorine-Containing Organic Constituents in Chlorinated Effluents," *J. Water Poll. Control Fed.* 47:617 (1975).
8. "Chloroform in N. Virginia Water Exceeds Safety Level," *Washington Post* (April 21, 1977).
9. California State Water Resources Control Board. "Policy and Action Plan for Water Reclamation in California" (January 1977), pp. 4-9.
10. Leinecke, Jim, Supervisor, Facilities Planning, Division of Water Pollution Control, Illinois EPA, personal communication.
11. "Dunkirk, Ohio," *60 Minutes,* CBS TV News (May 14, 1976).
12. "Walton, New York," *Wall Street Journal* (July 26, 1976).
13. U.S. Environmental Protection Agency. "Demonstrated Technology and Research Needs for Reuse of Municipal Wastewater," Office of Research and Development, EPA-670/2-75-038, Environmental Protection Technology Series (May 1975).
14. Hartman, Willis, Jr. "An Evaluation of Land Treatment of Municipal Wastewater and Physical Siting of Facility Installations," U.S. Army Corps of Engineers (May 1975).
15. Commoner, Barry. "Threats to the Integrity of the Nitrogen Cycle: Nitrogen Compounds in Soil, Water, Atmosphere and Precipitation," in *The Changing Global Environment,* S. Fred Singer, Ed. (Dordrecht, Holland: D. Reidel Publishing Co., 1975). pp. 341-366.
16. McElroy, Michael, *et al.* "Sources and Sinks for Atmospheric N_2O," *Reviews of Geophysics and Space Physics* (December 1975).
17. U.S. Department of the Interior, Bureau of Mines. "Phosphate Rock," in *Mineral Facts and Problems,* 1975 ed.
18. U.S. Department of Agriculture. "Our Land and Water Resources: Current Uses and Prospective Supplies and Uses," Economic Research Service, Miscellaneous Publication No. 1290 (May 1974).

APPENDIX A. SELECTED SECTIONS OF THE FEDERAL WATER POLLUTION CONTROL ACT AMENDMENTS OF 1972

TITLE II–GRANTS FOR CONSTRUCTION OF TREATMENT WORKS

Purpose

SEC. 201. (a) It is the purpose of this title to require and to assist the development and implementation of waste treatment management plans and practices which will achieve the goals of this Act.

(b) Waste treatment management plans and practices shall provide for the application of the best practicable waste treatment technology before any discharge into receiving waters, including reclaiming and recycling of water, and confined disposal of pollutants so they will not migrate to cause water or other environmental pollution and shall provide for consideration of advanced waste treatment techniques.

(c) To the extent practicable, waste treatment management shall be on an areawide basis and provide control or treatment of all point and nonpoint sources of pollution, including in place or accumulated pollution sources.

(d) The Administrator shall encourage waste treatment management which results in the construction of revenue producing facilities providing for–

(1) the recycling of potential sewage pollutants through the production of agriculture, silviculture, or aquaculture products, or any combination thereof;

(2) the confined and contained disposal of pollutants not recycled;

(3) the reclamation of wastewater; and

(4) the ultimate disposal of sludge in a manner that will not result in environmental hazards.

(e) The Administrator shall encourage waste treatment management which results in integrating facilities for sewage treatment and recycling with facilities to treat, dispose of, or utilize other industrial and municipal wastes, including but not limited to solid waste and waste heat and thermal discharges. Such integrated facilities shall be designed and operated to produce revenues in excess of capital and operation and maintenance costs and such revenues shall be used by the designated regional management agency to aid in financing other environmental improvement programs.

(f) The Administrator shall encourage waste treatment management which combines "open space" and recreational considerations with such management.

(g) (1) The Administrator is authorized to make grants to any State, municipality, or intermunicipal or interstate agency for the construction of publicly owned treatment works.

(2) The Administrator shall not make grants from funds authorized for any fiscal year beginning after June 30, 1974, to any State, municipality, or intermunicipal or interstate agency for the erection, building, acquisition, alteration, remodeling, improvement or extension of treatment works unless the grant applicant has satisfactorily demonstrated to the Administrator that–

(A) alternative waste management techniques have been studied and evaluated and the works proposed for grant assistance will provide for the application of the best practicable waste treatment technology over the life of the works consistent with the purposes of this title; and

(B) as appropriate, the works proposed for grant assistance will take into account and allow to the extent practicable the application of technology at a later date which will provide for the reclaiming or recycling of water or otherwise eliminate the discharge of pollutants.

4

IONIC DISTRIBUTION IN A SPRAY IRRIGATION SYSTEM

P. L. Deese, R. F. Vaccaro, B. H. Ketchum, P. C. Bowker and M. R. Dennett
Woods Hole Oceanographic Institution
Woods Hole, Massachusetts

INTRODUCTION

Early emphasis by the Cape Cod Wastewater Renovation and Retrieval System centered on variations in terrestrial wastewater disposal as an acceptable alternative to marine outfalls. More recently, interest has focused on the additional renovation provided when perennial forage crops are irrigated with domestic wastewater prior to soil infiltration and ground-water recharge. As a conservation measure, agricultural irrigation plus recharge is particularly attractive on Cape Cod which depends entirely on precipitation and recycling for the replenishment of freshwater reserves. Over the past four years, our efforts have been generously supported by the U.S. Environmental Protection Agency and the Commonwealth of Massachusetts.

In a wastewater-spray irrigation system, the agricultural crop is but one of several components which contribute toward wastewater improvement. The overall extent of renovation can be measured by comparing the water quality of the ground-water recharge with that of the incoming irrigation water. However, in an experimental program and for planning purposes, it is also necessary to assess the contribution of each renovative stage. This report aims to evaluate the relative contributions of agricultural uptake, soil attenuation and ground-water dilution for a specific area of Cape Cod. Conclusions are based on a mass balance for the elementary spectrum which characterizes Otis secondary effluent which is stabilized and chlorinated following two weeks of retention in a relatively shallow lagoon.

A lagoon system is essential to a spray irrigation system for a variety of reasons. In winter, ice formation can be detrimental and can interfere with routine irrigation. At such times, additional complications are introduced by plant inactivity and poor soil percolation. During summer, liming and crop harvesting along with normal maintenance requirements interrupt irrigation and can impose a temporary wastewater storage requirement. Even during normal operation, a reserve of relatively stable and homogeneous irrigation water is advantageous. Within temperate latitudes, the winter storage requirement will usually determine the necessary volume of the lagoon component.

The spray irrigation facility shown on Figure 1 includes two experimental agricultural sites located within Otis Air Force Base, Cape Cod. A fraction of the effluent from the trickling filter plant is diverted to a lagoon and ultimately chlorinated before being pumped to the irrigation fields. The chemical composition of Otis secondary effluent is not unlike the typical secondary effluents that are free of industrial waste (Table I).

Differences attributable to rates of irrigation are assessed at Site A which provides three rectangular plots (0.2 ha each) planted in reed canarygrass. The irrigation rates provided approximate 2.5, 5.0 and 7.6 cm per week, respectively, for the east, south and west fields. The effective rates of irrigation on an annual basis are considerably less due to unavoidable schedule interruptions. Site A includes a control plot also planted in reed canarygrass which receives normal rainfall but no sewage effluent. The three irrigated plots are equipped with fixed deflection head sprinklers which permit year-round irrigation.

Crop variations are studied on Site B which is a circular field 100 m in diameter which has been divided in four equal quadrants. Here irrigation is provided by a center-pivot rotary irrigation rig which delivers irrigation water at a normal rate of 5.0 cm per week. The four segments of the circle are planted in smooth brome (southeast), timothy-alfalfa (southwest), timothy (northeast) and reed canary (northwest) and an adjacent plot planted in timothy serves as a nonirrigated control. The rotary rig is not conducive to winter use; therefore a comparison of reed canary response at Site A vs Site B provides a useful measure of year-round as opposed to seasonal irrigation.

Crop, turf, soil, soil-water and ground-water samples are routinely collected from both sites and wells for ground-water sampling are located immediately down gradient for each of the Site A plots. Sampling and chemical analysis for plant nutrients, major ions and trace metals are conducted according to the recommendations of the Environmental Protection Agency.[3]

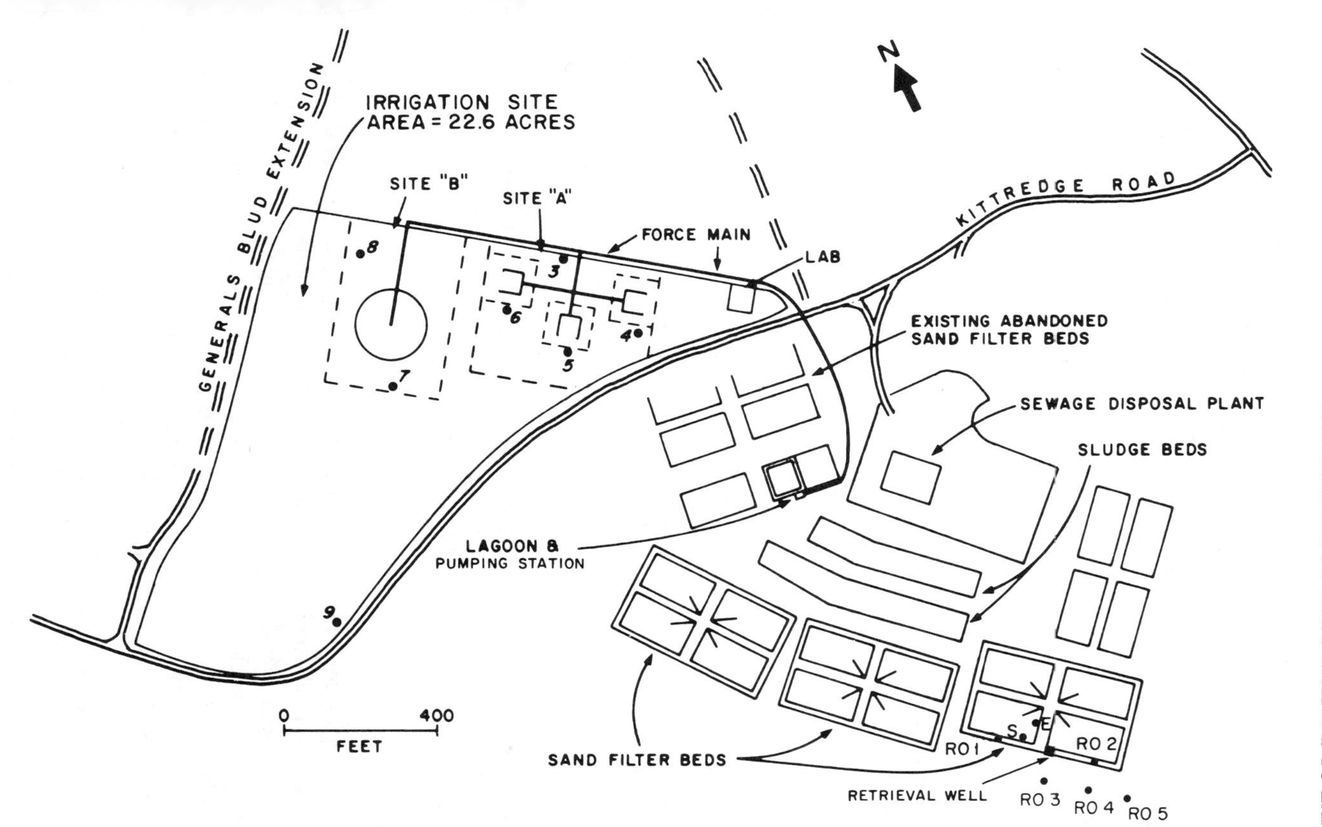

Figure 1. Plan of sewage treatment plant, sand filter beds, lagoon and irrigated sites.

Table I. Wastewater characteristics of Otis treatment plant secondary effluent (concentrations as mg/l; ppm).

Constituent	Typical Secondary Treatment Effluent	Otis Secondary Effluent Mean	Otis Secondary Effluent Standard Deviation
Nitrogen (as N)	20		
Total inorganic	–	17.90	± 3.25
Nitrate	–	9.05	± 4.88
Nitrite	–	0.23	± 0.14
Ammonium	–	7.50	± 6.27
Phosphorus (as P)			
Total dissolved	10	8.49	± 0.83
Orthophosphate	–	7.74	± 2.17
Other Elements[a]			
Cadmium	0.01-0.03	0.00024	± 0.0001
Calcium	24	7.6	± 0.70
Chloride	45	26.78	± 2.39
Chromium (Cr^{+6} + Cr^{+3})	0.02-0.14	<0.01	
Copper	0.07-0.14	0.050	± 0.020
Iron	0.10-4.3	0.502	± 0.07
Lead	0.01-0.03	0.00054	± 0.0001
Magnesium	17	3.79	± 0.17
Manganese	0.02	0.021	± 0.016
Mercury (Hg^{+2})	0.01	<0.00027	± 0.0001
Potassium	14	8.94	± 1.09
Sodium	50	37.91	± 6.34
Zinc	0.20-0.44	0.047	± 0.025

[a]Typical values from Pound and Crites[1] or from Driver *et al.*[2]

Crops

Crop samples collected on Site A show that, within the range observed, both yield and elementary composition, while independent of irrigation rate, are dependent on the total amount of effluent applied (TEA). The former relation is clearly shown in Figure 2 which compares accumulated crop yield and the total hydraulic load (THL) which includes both TEA and precipitation as shown in Figure 3. Yield differences can be expressed in terms of the Yield Ratio as defined below:

$$\text{Yield Ratio} = \frac{\text{Crops harvested, kg/ha}}{\text{Total Hydraulic Load, kg/ha}}$$

Thus, in Figure 3 the Yield Ratio taken as the slope of the regression line is equal to 24×10^{-5} for Site A reed canarygrass.

The results of the experiments conducted on Site B with different species of grasses show that both yield and elementary composition vary with crop

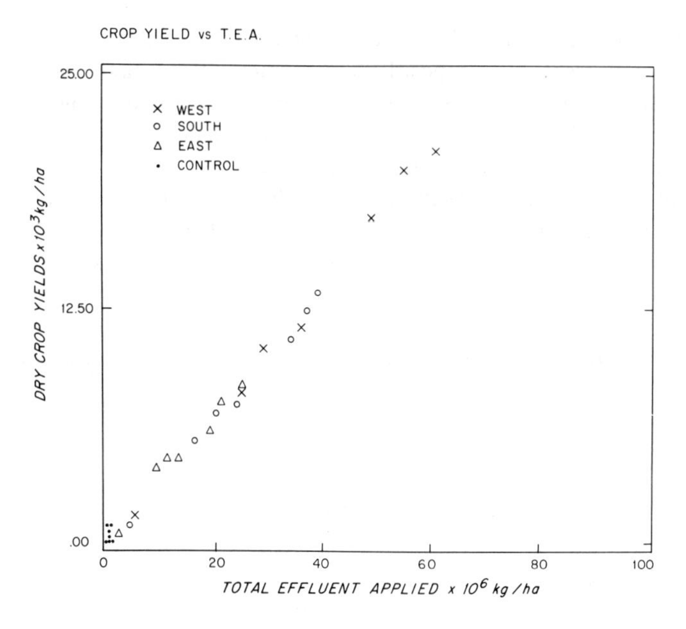

Figure 2. Relations between total effluent applied and total dry crop yields, Site A.

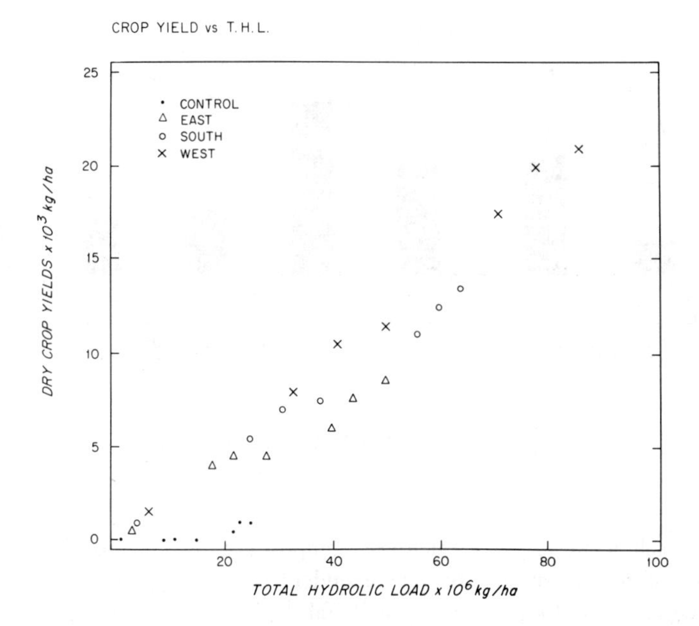

Figure 3. Relations between total hydraulic load and total dry crop yields, Site A.

type. However, further analysis shows that for individual species, the Yield Ratio and elementary compositions are once again relatively constant. Because Site B cannot be irrigated during the three to four coldest winter months, crop yields correlate less well with TEA than for Site A. However, the correlations with THL are quite comparable.

Figure 4 compared the Yield Ratios for the various crops on both Site A and Site B and demonstrates the following. First, one notes the superior yield of the timothy-alfalfa wherein the latter's nitrogen-fixing capability supports our contention that nitrogen is the limiting nutrient in our system. Second, the similarity between Yield Ratios for reed canarygrass on Sites A and B demonstrates that crop yields are species specific in a spray irrigation system.

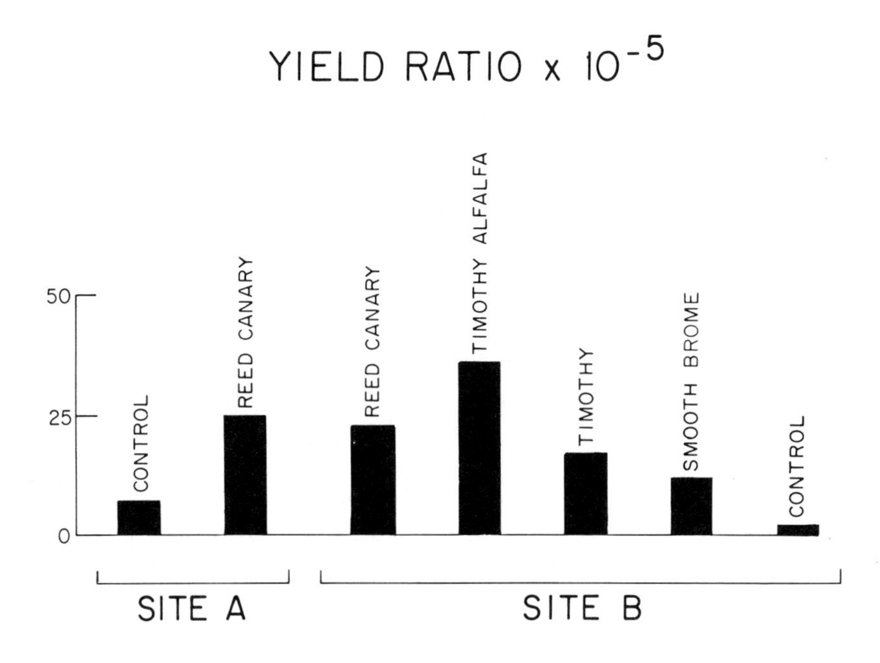

Figure 4. Yield ratios for Site A and Site B.

For each harvest, the elementary composition of the crops was independent of the irrigation *rate* but dependent on crop type (Table II). The observed concentrations of trace metals fall within the accepted range while nutrient and protein levels compare favorably with standard agricultural crops.[4-6]

Table II. Elementary composition of animal forage crops (% dry weight).

	Site A	Site B			
	Reed Canary	Timothy	Smooth Brome	Timothy Alfalfa	Reed Canary
N	1.63	1.43	1.57	2.25	1.59
P	0.407	0.316	0.302	0.433	0.340
Cu	0.0007	0.0006	0.0006	0.0007	0.0006
Zn	0.0029	0.0043	0.0019	0.0024	0.0022
Pb	0.0002	0.0001	0.0002	0.0002	0.0002
Cd	0.00002	0.00002	0.00002	0.00003	0.00003
Mn	0.0047	0.0016	0.0034	0.0024	0.0047
Na	0.041	0.020	0.023	0.332	0.041
K	1.80	1.88	1.82	1.62	1.88
Mg	0.230	0.106	0.088	0.163	0.154
Ca	0.160	0.150	0.151	0.316	0.146
Fe	0.0110	0.0084	0.0094	0.0159	0.0079

A fixed Yield Ratio and a constant elementary composition for each grass species assures a linear relationship between THL and the constituent uptake and removal by the harvestable crops if an average constituent concentration is assigned for the applied effluent.

The Efficiency of Crop Uptake for a given element can be defined as follows:

$$\text{Efficiency of Crop Uptake (\%)} = \frac{\text{mass in harvested crop}}{\text{mass applied in effluent}} \times 100.$$

Figure 5 shows the linear relation between the mass of nitrogen applied and the mass of nitrogen removed with the harvested crop. A similar relation can be seen in the case of lead in Figure 6. The Efficiencies of Crop Uptake for all the ions measured are summarized in Table III.

Apparently the range of irrigation rates so far employed by us have not caused any marked differences in renovation efficiency. However, extensive increases in the rates of application would very likely alter this condition once some critical irrigation rate is exceeded. During the next growing season, we hope to provide a better definition of this upper limit by increasing our maximum rate of irrigation to 10 cm per week while maintaining comparable observations at 5.0 and 7.5 cm per week.

Figure 7 compares the Percent Crop Uptake of several constituents for the three most efficient crops, the reed canarygrass, Site A and Site B, and timothy-alfalfa, Site B. Reed canarygrass on Site B consistently demonstrates a higher Efficiency of Crop Uptake than that of Site A, even though the Yield Ratios are about the same and the gross yield of Site A is larger. The most

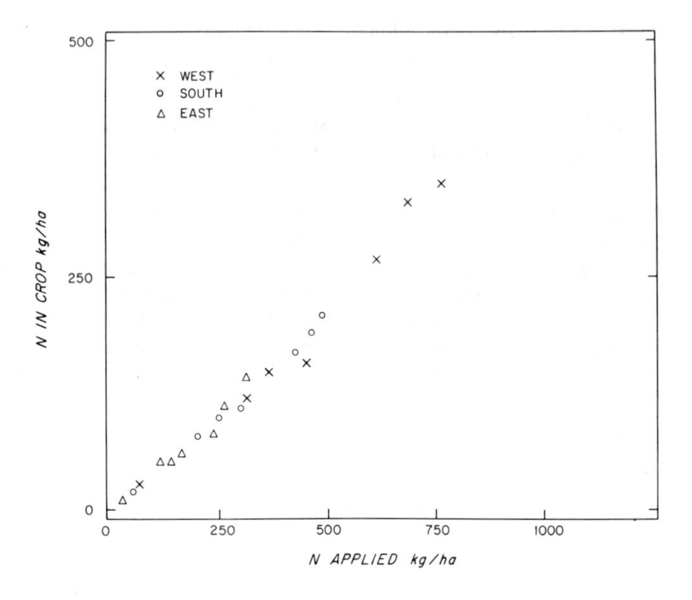

Figure 5. Relations between total nitrogen applied and total nitrogen in harvested crop, Site A.

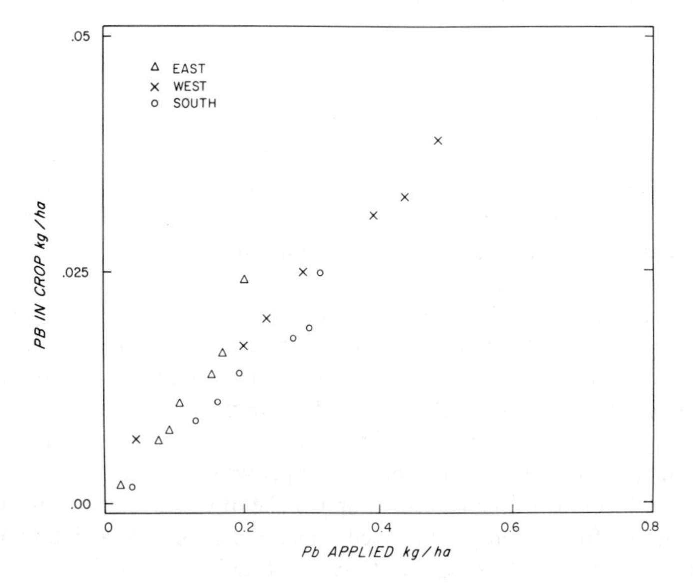

Figure 6. Relations between total lead applied and total lead in harvested crop, Site A.

Table III. Uptake by crops of various constituents from secondary effluent (% amount applied).

	Site A	Site B			
	Reed Canary	Timothy	Smooth Broom	Timothy Alfalfa	Reed Canary
N	46	58	51	156	76
P	16	18	14	42	23
Cu	4	4	4	9	5
Zn	25	32	19	52	33
Pb	7	8	8	22	11
Cd	12	15	11	37	21
Mn	89	39	72	109	140
Na	0.4	0.3	0.2	7.6	0.7
K	68	103	82	155	121
Mg	24	15	11	42	27
Ca	8	10	8	36	11
Fe	8	8	8	28	10

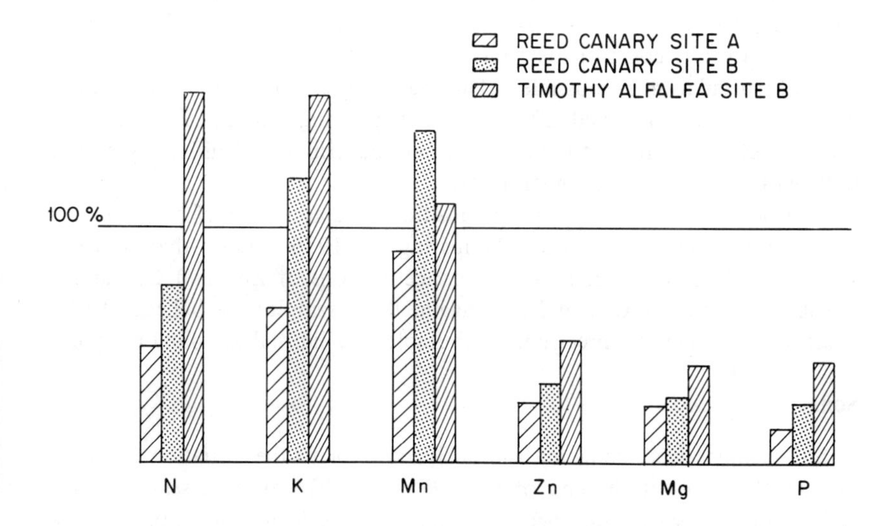

Figure 7. Efficiency of crop uptake of ions as a percent of total applied. Uptake of Na, Ca, Fe, Pb, Cu and Cr were considerably less than uptake of P.

probable explanation for this pattern is that limiting irrigation of Site B to the growing season assures a more efficient uptake of applied constituents on an annual basis. Thus, a more valid assessment of the superiority of Site B timothy-alfalfa is obtained by its comparison with the Site B reed canarygrass rather than the reed canarygrass of Site A. The uptake efficiency of timothy-alfalfa would most likely be reduced given a year-round irrigation regime.

The results of these experiments emphasize the reality of nitrogen limitation in a spray irrigation system utilizing domestic effluents without supplementary nitrogen. Where such irrigation occurs over a drinking water aquifer, as in our case, and where the use of fertilizers should be minimized to prevent possible impaction on the ground water, the contributary role of nitrogen-fixing crops merits careful study. In particular, year-round irrigation experiments with detailed ground-water monitoring, similar to our Site A capability, should be conducted with a nitrogen-fixing crop.

Turf

Gross yields of root, stubble or turf samples collected and analyzed at the end of the 1975 and 1976 growing seasons have shown considerable variation, probably due to large sampling errors. These observations permit only a limited short-term assessment on the role of root growth in the removal of various constituents from the applied wastewater although some relevant speculation is possible. Our data suggest that no large buildup in root structure occurred after the 1975 growing season. Prior to that time, a developing root structure must have contributed significantly to the uptake of irrigation water constituents.

A review of the elementary composition data for the turf samples fails to reveal any clear-cut trends. Comparing the 1976 data to that of 1975 suggests a slight enrichment in the concentrations of Zn^{++} and Cu^{++} and a possible decrease in Cr and Ni^{++}. Again, these samples are believed to be obscured by a lack of precision attributable to a limited sampling program.

Soils

Soil samples were collected and analyzed prior to the initiation of irrigation in 1973 and at the end of the 1975 and 1976 growing seasons. Our early crop-free sand filter bed experiments on soil accumulations over a 30-yr period have indicated that most constituents which are retained in the soil remain within the top several inches.[7] A review of the top 6 in. (15 cm) of the soil on each agricultural plot over their 3-year life span reveals that both the Ca^{++} and Mg^{++} content of the soil have increased sharply as a direct result of lime applications to the fields. Regarding phosphorus, a measurable but less dramatic increase was observed. Soil Na^{++} increased precipitously until 1975 and then leveled off at the highest irrigation rate, while at the lower irrigation rates the rise was more gradual. There have been no significant increases in potassium in the soil, which is to be expected since a large proportion of the potassium is taken up by the plants. Changes in the remaining constituents monitored remain small or negligible as compared to the neighboring background soil.

Figure 8 illustrates the importance of the relation between effluent content and soil background composition. The amount of each constituent found in the top 6 in. of the soil, based on 1973 samples, is expressed as a percentage of the amount applied during one year of irrigation at the highest rate. Evidently, the ambient concentration of many trace metals in the soil is large in comparison to that applied by irrigation, thereby obscuring the origin of the trace metal content of the crops.[8]

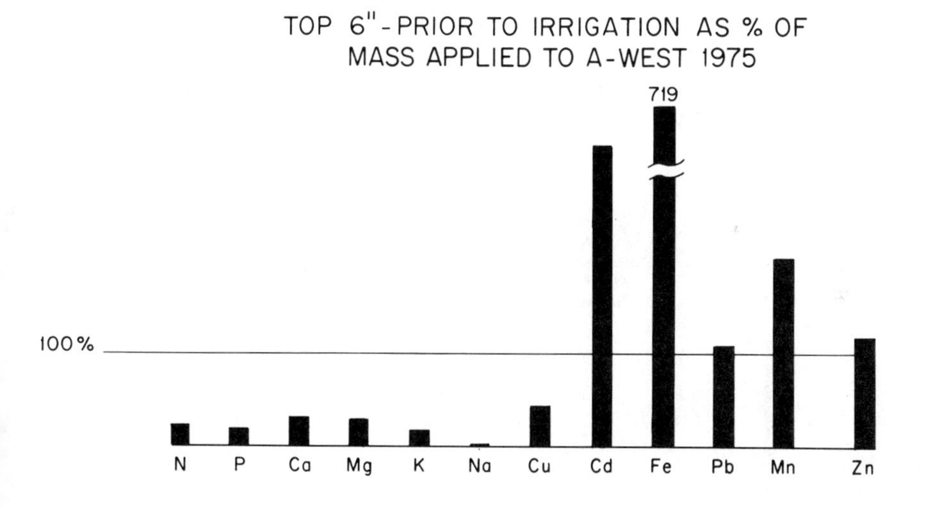

Figure 8. Ionic content of the soil as a percent of the amount applied in one year at 7.6 cm per week irrigation rate.

Ground Water

Ground water samples are collected routinely from wells located immediately down gradient from each of the Site A plots. The wells have 6-ft well points and intersect the top several feet of the aquifer, where the greatest ground-water attenuation might be expected.

While increases with time have been observed for several ground water ions, their concentrations have never exceeded recommended standards for drinking water.[9,10] Further, there has been no measurable increase in the heavy metal (Cd^{++}, Zn^{++}, Cu^{++}, Fe^{+++}, Mn^{++}, Pb^{++}) burden. After irrigating for three growing seasons, there is no evidence the ground water poses any threat to public health.

Sequential changes in ground-water concentrations are shown for all four Site A wells in Figures 9 to 15. Three point running averages are used to smooth out background noise so that overall trends can be more easily identified.

At the highest irrigation rate (west), the Cl^- concentration in the ground water has attained steady state at a significantly increased level of concentration (Figure 9). The detailed stepwise increase of this anion supports our earlier assumption that chlorides are highly conservative and percolate through the soil with little or no attenuation. Demonstration of a less well-defined chloride plateau at the intermediate irrigation rate (south) can be attributed to the increased relative importance of uncontrollable variables such as precipitation, ground-water flow and weather. At the lowest irrigation rate (east), the effect of effluent irrigation is negligible. This general pattern of larger changes at the highest irrigation rate and very

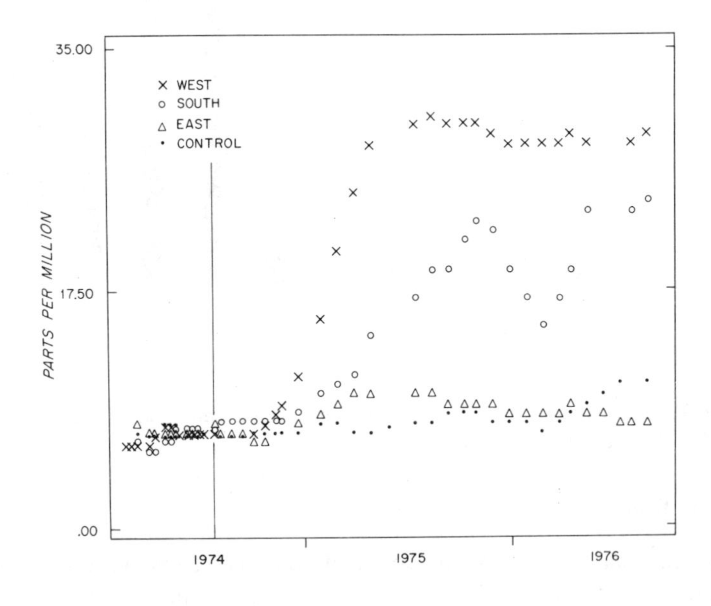

Figure 9. Chloride concentrations in Site A wells over time.

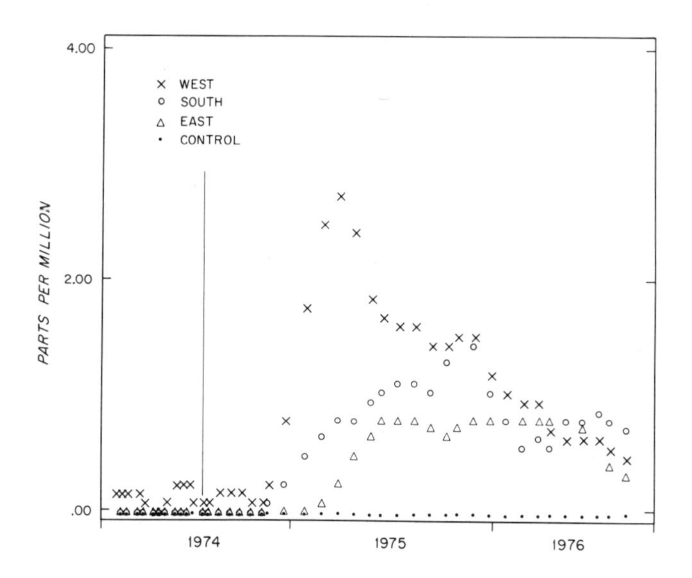

Figure 10. Nitrate concentrations in Site A wells over time.

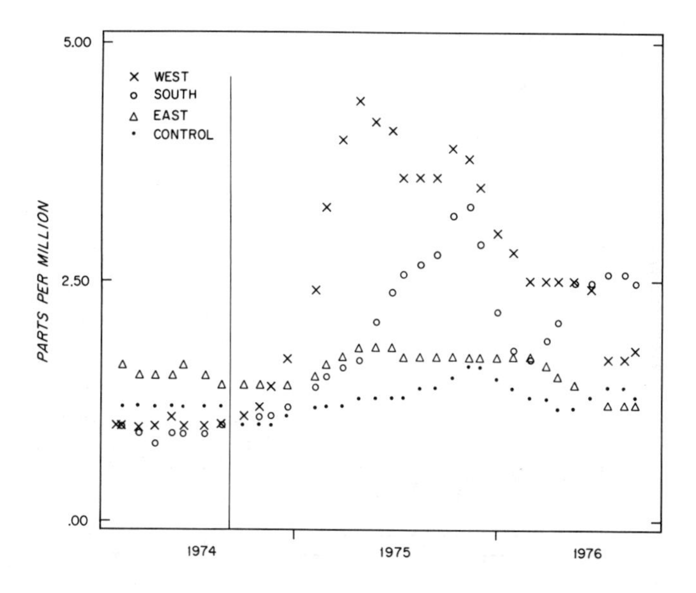

Figure 11. Magnesium concentrations in Site A wells over time.

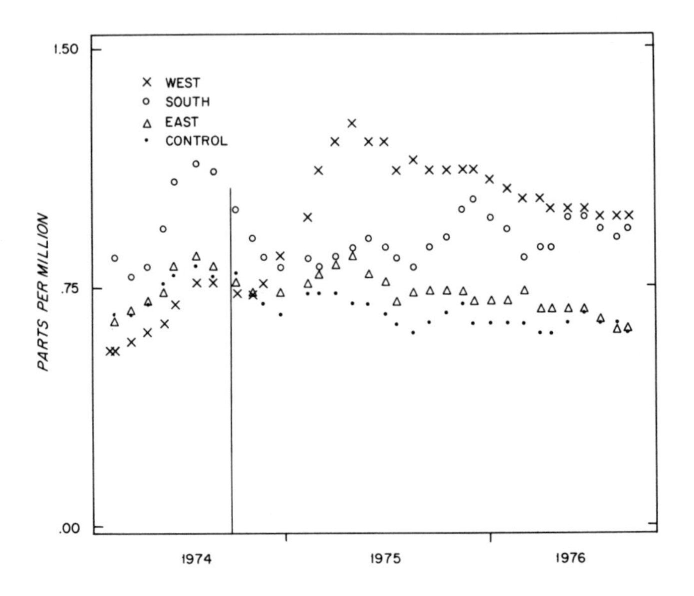

Figure 12. Potassium concentrations in Site A wells over time.

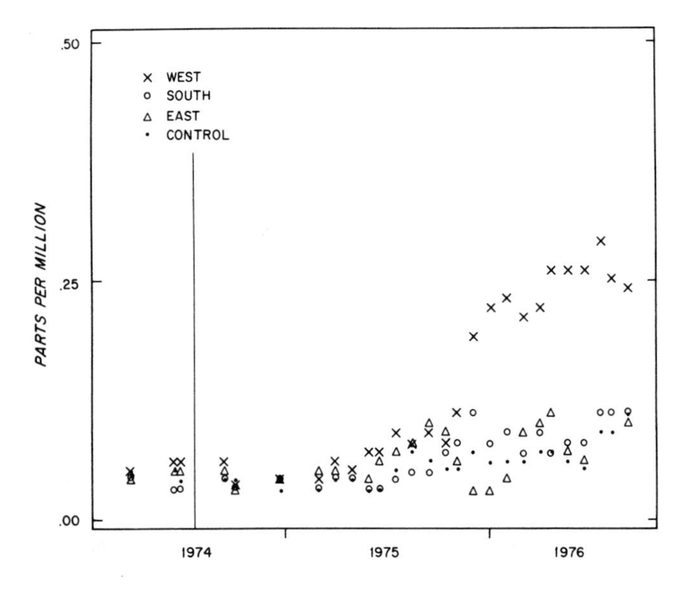

Figure 13. Boron concentrations in Site A wells over time.

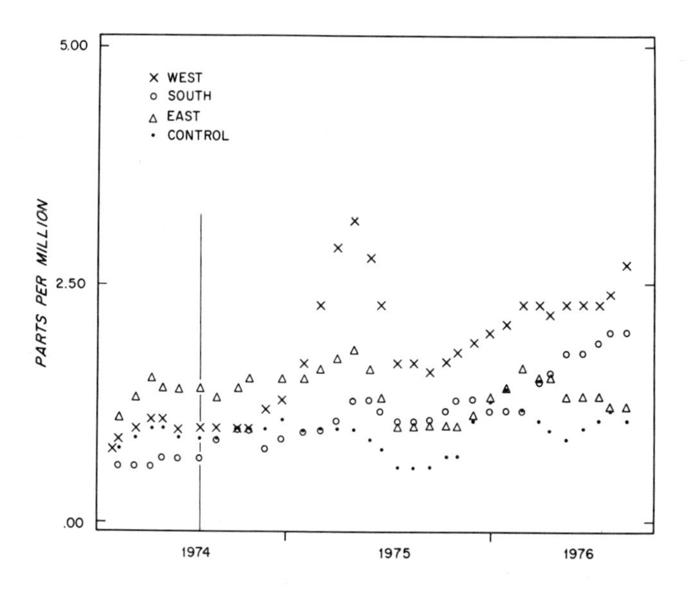

Figure 14. Calcium concentrations in Site A wells over time.

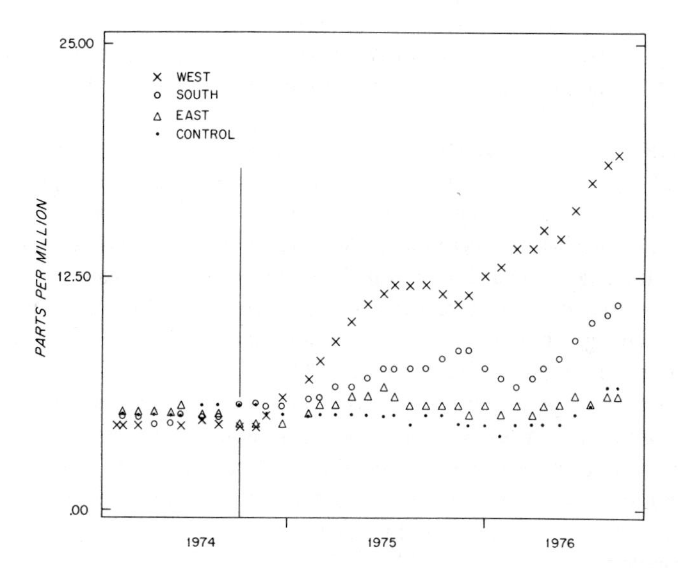

Figure 15. Sodium concentrations in Site A wells over time.

small or undetectable changes at the lowest rate is repeated for most of the ground-water constituents monitored.

Due to a potential health hazard traditionally associated with excessive nitrate in ground waters,[11] the changes in NO_3^-, NO_2^- and NH_4^+ have also been monitored. As expected, well over 95% of the total inorganic nitrogen in the ground water occurs as NO_3^- rather than as NO_2^- or NH_4^+.[8,12,13] Failure to detect increases in either NO_2^- or NH_4^+ suggests that oxidative rather than reducing conditions predominate in the overlying soil and that nitrification is the dominant soil process affecting the transformation of combined nitrogen.

A comparable plot showing NO_3^- changes in ground water (Figure 10) differs markedly from that of chloride. Similar to chlorides, NO_3^- concentrations increased during the early months; however, they later show a precipitous dropoff. The initial NO_3^- peak appears to reflect a variety of contributing factors, particularly an excess of nitrogen applied to unestablished crops during summer 1974, the initial growing season. Since the crops were not well established, large amounts of nitrogen could have passed through the root zone and ultimately entered the ground water. Also, the fields were fertilized in the spring of 1974 in order to help insure the establishment of a healthy crop. Much of this fertilizer may have been released with the first wave of effluent. Finally, the abrupt change in land use from scrub forest to perennial grasses and the subsequent decay of unremoved subsurface debris could have released a shock load of nitrogen which penetrated the ground water. A lack of seasonal variations in ground water NO_3^- suggests that no excess nitrogen is being leached during winter irrigation.

Figures 11 and 12 illustrate that the changes in ground-water concentrations of Mg^{++} and K^+ resemble those of NO_3^-. Since the lime applied prior to irrigation contained Mg^{++}, it is possible that the initial peak was caused by an early wash-through of lime. However, subsequent peaks do not correspond with our seasonal liming schedule and remain unexplained. The contributory role of several recognized ion exchange mechanisms influenced by the change in land use, soil, pH and soil moisture content is being considered in the cases of both Mg^{++} and K^+.[8,14]

An excess of phosphorus in terms of crop requirements is to be expected when domestic wastewaters are used for agricultural purposes. Excess phosphorus at Otis becomes adsorbed on soil particles.[7,15] There has been no significant increase in the $PO_4^{\equiv}$ content of the underlying ground water even though crop uptake accounts for only 30% of the phosphorus delivered by irrigation.

The observed variations in sulfate for Otis ground water are highly irregular and do not appear to correspond to our irrigation schedule.

Alternatively, these changes could reflect sudden background changes in the $SO_4^{=}$ content of the receiving ground water.

Ground-water concentrations of boron and Ca^{++} are shown in Figures 13 and 14. Each of these elements shows an accumulation at the most rapid irrigation rate while the intermediate and lowest rates of irrigation have had no noticeable ground water effect. These patterns suggest that the mean recharge time for these elements may be significantly longer than for Cl^- possibly due to additional interactions within the soil column and that attainment of a steady-state condition under the west plot will develop at some future time.[8] At the lower irrigation rates, a more effective balance between the boron and Ca^{++} supply, plant uptake and soil removal may indefinitely retard the arrival of any detectable amount of these elements. Ground-water increases in Na^+ appear in Figure 15 and are very similar to those of Ca^{++} and boron, except in this case similar response patterns are obtained for both the highest and intermediate rates of irrigation.

Recharge Time

Since Cl^- is one of the least likely ions to be intercepted by crops or soil, its increase in ground water provides a measure of the recharge time for irrigation water. At Otis the arrival of Cl^- in the wells followed an S-shaped curve and the mean arrival time has been defined as the interval between the start of irrigation and the arrival of one-half the maximum concentration of Cl^-. In this manner, our fastest rate of irrigation corresponded to a recharge time of about 200 days at 7.6 cm/day (Figure 16). Greater uncertainty is associated with the intermediate and slowest rates of irrigation; however, a best estimate for arrival under the south plot is about 285 days at 5.0 cm/day (Figure 17).

The lower curves on Figures 16 and 17 represent comparable NO_3^- data from the same wells. These data, unlike those for Cl^-, suggest that continuing dynamic exchanges within the soil and crop complex have delayed the arrival of steady-state conditions. The early buildup of nitrate in the well samples compares favorably with the buildup of Cl^-. Estimates of arrival time based on the maximum concentration attained show good agreement with those obtained from the Cl^- changes.

Mass Balance

Consolidation of the information presented heretofore can provide an estimate of the mass transfer of chemical elements between water, crops and soil. Accordingly, we have developed a steady-state, input-output model comprising two soil phases: saturated and unsaturated. The model relies on the mass transfer of water along with its ionic content to evaluate the

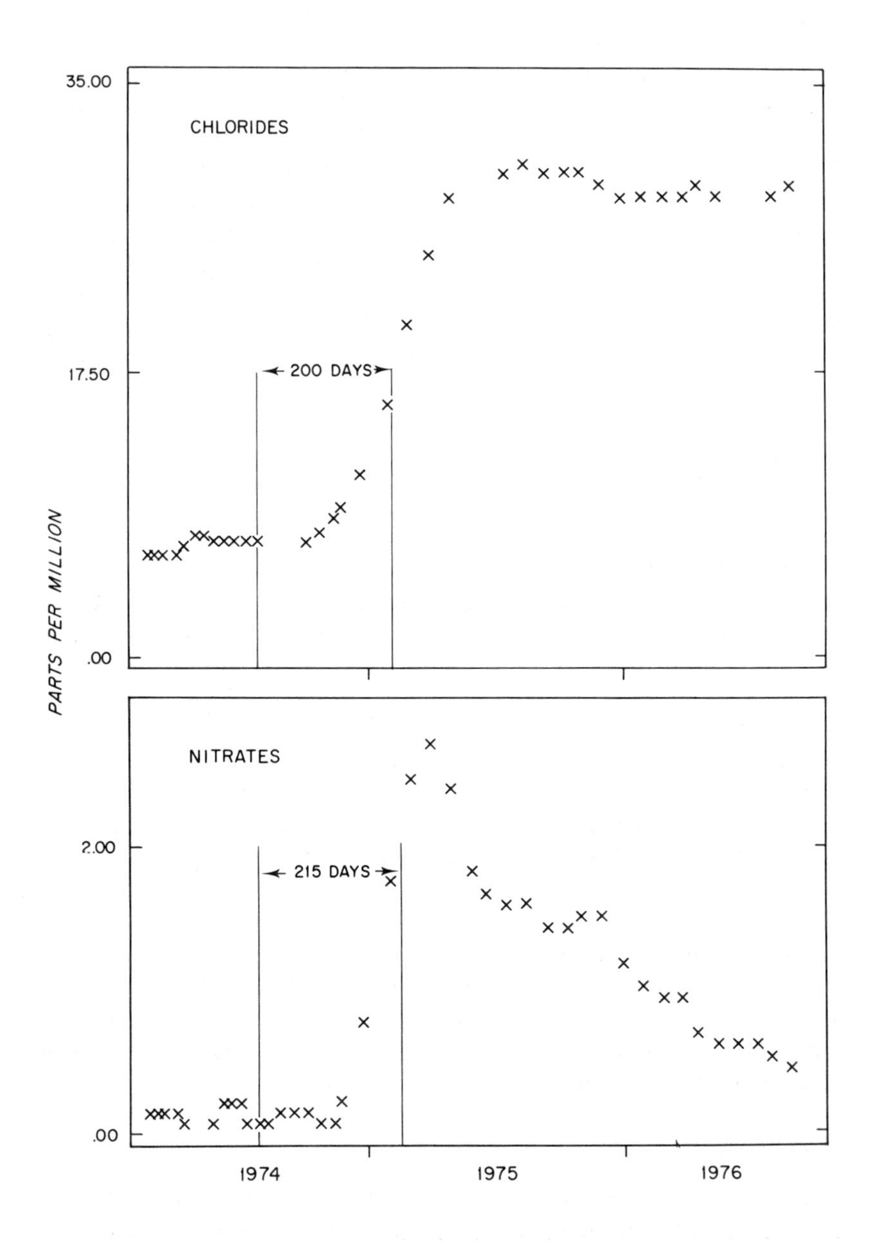

Figure 16. Mean recharge time for the west plot (7.6 cm/wk), Site A.

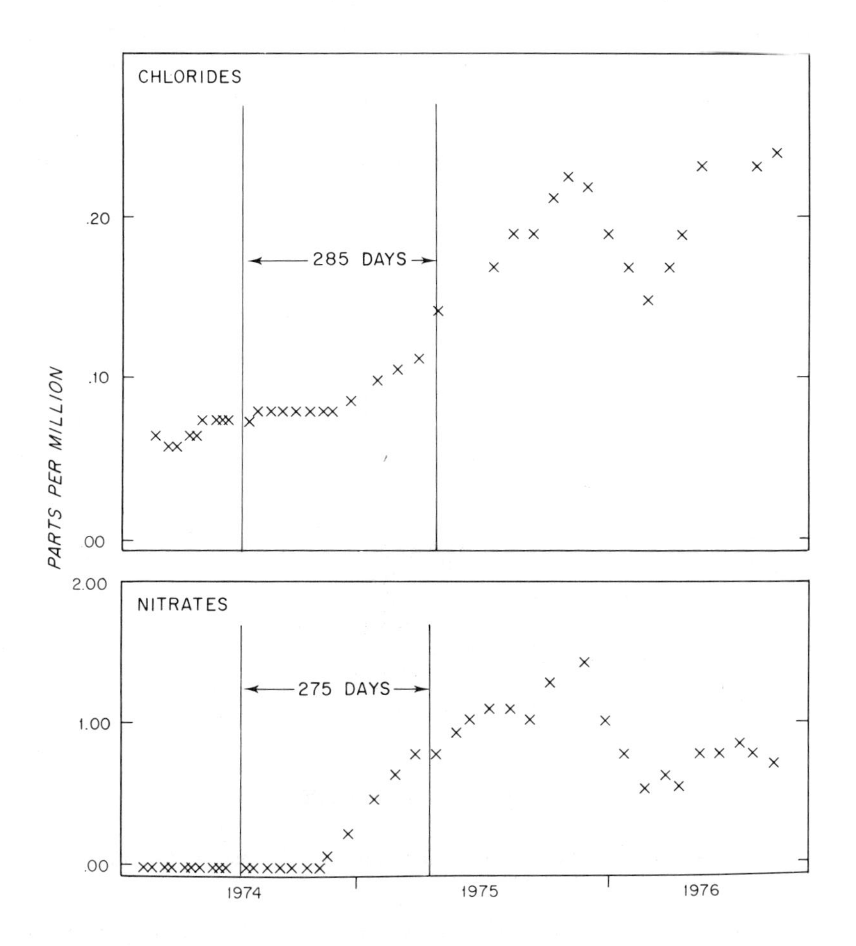

Figure 17. Mean recharge time for the south plot (5.0 cm/wk), Site A.

flux of elements. Elementary masses not accounted for in the ground water are relegated to crops for which the uptake is known, and soil whose relative influence is ultimately estimated by difference.

The first consideration is an approximation of the transfer of water within the system. Here, we rely on the conservative property of Cl^- as a useful indicator of changes in water volume. Once a water balance is established, appropriate ionic concentrations can be assigned to estimate the mass transfers of any particular ion. Figure 18 diagrams the two-phase soil column wherein the V terms refer to water volume and the C terms to the concentration of a given entrained ion. Water input to the unsaturated zone includes

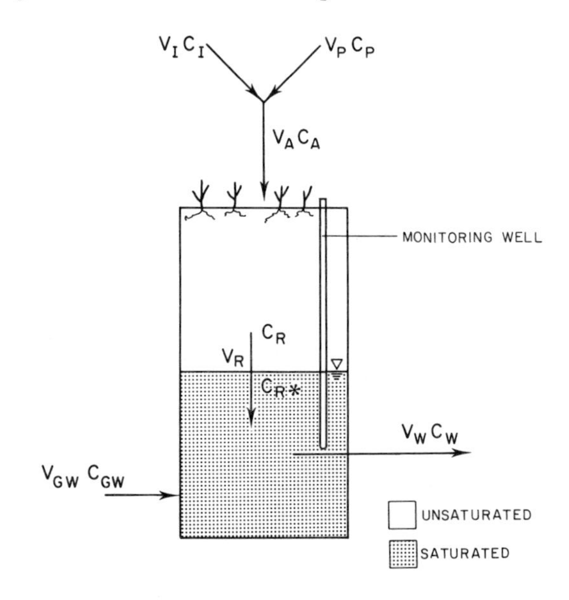

Figure 18. Input-output model for estimating ionic mass recharged.

irrigation (V_I) and precipitation (V_P) whose sum (V_T) is the total hydraulic load. Exit from the unsaturated zone is by recharge (V_R) and evapotranspiration (ET). The latter term does not appear directly in the model but has been estimated by Palmer (1977) from meteorological data and from Cl^- concentrations in the four-foot lysimeters. Losses due to ET are used to estimate the fraction (R) of the hydraulic load which is recharged to the saturated phase. Multiplying the total volume applied (V_T) by the fraction recharged (R) gives the volume of recharge (V_R). Input to the saturated zone is also provided by horizontal movement of ground water (V_{GW}).

Output from the saturated zone is taken as the volume of ground water that has been modified by recharge (V_W) whose mean ionic concentration is that observed in the monitoring wells (C_W). The water balance is summarized by the following equations:

$$V_I + V_P = V_T \tag{1}$$

$$R \cdot V_T = V_R \tag{2}$$

$$V_R + V_{GW} = V_W \tag{3}$$

The mass balance of elements can be derived by multiplying water volumes by the respective concentrations in the applied effluent and rain:

$$(V_I \cdot C_I) + (V_P \cdot C_P) = V_T C_T \tag{4}$$

Our monitoring data include measurements for the parameters V_I, C_I, V_P, C_P; thus C_T can be calculated by combining Equations 1 and 4.

$$\frac{(V_I \cdot C_I) + (V_P \cdot C_P)}{V_I + V_P} = C_T \tag{5}$$

If ionic losses to the soil and crops are negligible, the only change in volume and concentration would result from evapotranspiration and the theoretical transfer of elements from the unsaturated to the saturated layer would be:

$$V_T \cdot C_T = R \cdot V_T \cdot C_T = V_R \cdot C_R \tag{6}$$

C_R is computed by correcting C_T for the fractional loss of water from the unsaturated layer due to evapotranspiration (Equations 2 and 6):

$$\frac{V_T \cdot C_T}{V_R} = \frac{C_T}{R} = C_R \tag{7}$$

The assimilation/adsorption factor (A) is the fraction of each applied constituent removed by the soil and crop system. The fraction in the recharge is thus (1-A). Application of this factor to the total mass applied provides an estimate of the actual concentration in the recharge water, C_R^*, which reflects both soil and crop removal:

$$\frac{(1\text{-}A)\, V_T \cdot C_T}{V_R} = C_R\,(1\text{-}A) = C_R^* \tag{8}$$

The recharged water and the elements it contains joins with and is diluted by the ground water containing the same elements at different concentrations. The concentration of each element monitored in the well is a measure of the mixture of the recharge water and ground water, since:

$$(V_R \cdot C_R^*) + (V_{GW} \cdot C_{GW}) = V_W \cdot C_W \tag{9}$$

In Equation 9 the concentration of the element in the ground water (C_{GW}) and in the well (C_W) are measured directly and the volume of recharge (V_R) is known from Equation 2 when the evapotranspiration loss is known. This leaves two unknowns, the concentration of the recharge water, C_R^*, and the volume of the ground water, V_{GW}. In the case of chlorides, C_R^* can be calculated using Equations 7 and 8 since the fraction A has been estimated at 0.17.[16] Thus, for chlorides, Equation 9 can be solved for V_{GW}.

$$V_R \frac{(C_R^* - C_W)}{(C_W - C_{GW})} = V_{GW} \tag{10}$$

For each plot, the ground-water volume (V_{GW}) is the same for all constituents, so the values obtained with chloride data can be used in Equation 11 to estimate C_R^* for elements whose assimilation/adsorption factor (A) is not known:

$$C_W + \frac{V_{GW}}{V_R}(C_W - C_{GW}) = C_R^* \tag{11}$$

Given C_R^*, the mass of a constituent recharged into the ground water (M_R) for any period of time can be calculated from:

$$V_R \cdot C_R^* = M_R \tag{12}$$

With V_R measured in millions of kg/ha and C_R^* as ppm, the element recharged is given in kg/ha. The remainder of the constituent being examined is presumed to be removed with the harvested crop or adsorbed in the soil column.

In order to calculate values for M_R certain assumptions regarding recharge or arrival times are required. We have used the total volumes measured for the calendar year 1975 for V_I and V_P, and C_I and C_P are the mean concentrations observed over the same period of time (Tables IV and V).

Perhaps the most important assumption necessary for the use of this approach is that the system has attained a steady state whereby C_W would be relatively constant over the period for which calculations were made. The use of a mean concentration for C_W where steady-state conditions have not been attained will inevitably result in an estimation error.

Table IV. Derived partition of soluble ions (ppm) within aqueous components of a spray irrigation system. Concentrations: C_I, in irrigation water; C_P, in precipitation; C_W, in ground water after recharge; C_{GW}, in ground water before recharge; C_R^*, in recharge water.

			South Plot[a]			West Plot[b]		
Element	C_I	C_P	C_W	C_{GW}	C_R	C_W	C_{GW}	C_R^*
Cl^-	38.1	2.8	19.5	6.9	16.2	28.9	7.3	30.1
N	12.11	0.36	0.97	0.02	1.57	1.22	0.18	1.27
$P^{\equiv}$	9.3	0.1	0.005	0.006	0.006	0.006	0.007	0.006
B	0.54	0	0.08	0.04	0.11	0.17	0.05	0.18
Ca^{++}	9.4	~0	1.4	0.7	1.8	2.0	1.0	2.0
Mg^{++}	3.4	~0	2.4	1.0	3.3	3.1	1.0	3.2
K^+	9.0	~0	0.9	0.9	0.9	1.1	0.6	1.1
Na^+	34.5	~0	8.0	5.1	9.8	13.0	4.9	13.4
$SO_4^=$	24.0	~0	8.4	8.3	8.5	11.6	8.0	11.8

[a]South plot irrigation rate *ca.* 5.0 cm per week.
[b]West plot irrigation rate *ca.* 7.5 cm per week.

Table V. Mass balance of water volumes and factors used to estimate evapotranspiration and chloride uptake by crops and soil.

	South Plot	West Plot
Volumes 10^6 kg/ha		
V_I, Irrigation	15.9	25.0
V_P, Precipitation	15.6	14.6
V_R, Recharge	19.5	27.3
V_{GW}, Ground Water	12.2	1.5
V_W, Ground Water Affected	31.7	28.9
Recharge Rate		
R, Fractional Loss of $V_I + V_P$ via Evapotranspiration	0.64	0.69
Assimilation/Adsorption Factor for Chlorides		
A, Fractional Removal by Crops and Soil	0.17	0.17

For our calculations, C_W is based on the mean of concentrations in the test well for each plot for a one-year period advanced for the estimated mean arrival time. As discussed above, mean arrival times have been estimated at 200 days and 285 days for the west and south plots respectively. A sensitivity analysis indicates that if present trends continue to steady state, the estimates of mass recharge (M_R) based on the above C_W are probably good for phosphates and potassium, somewhat high for nitrates and somewhat low for most cations. Tables IV and V give a summary of the values used for our mass balance calculations.

A bar graph (Figure 19) is used to compare the fate of various ions in the system. Here the amount of each constituent recharged (M_R) and/or the amount removed in the harvested crop (M_C) are shown as a percent of the total amount applied by irrigation.

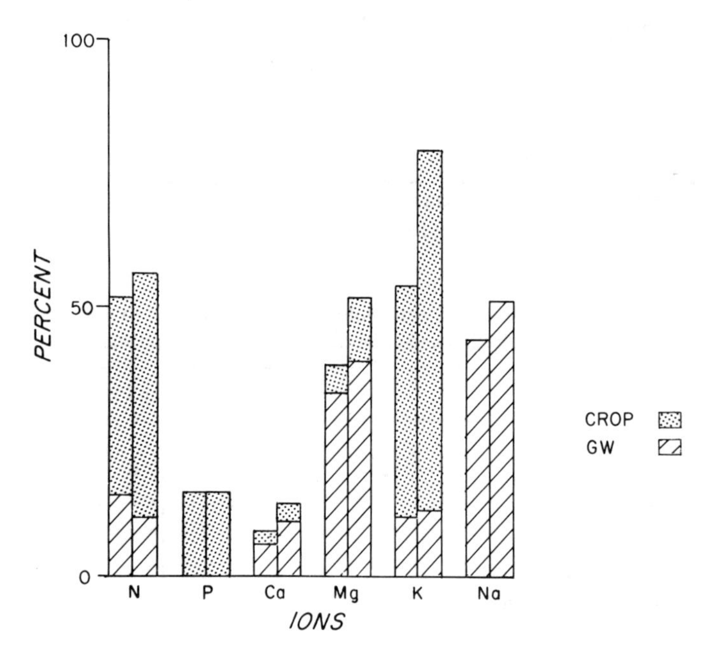

Figure 19. Summary of mass balance approximation. Masses are shown as percent of total mass applied.

For nitrogen, these results confirm that plant uptake and crop harvesting are dominant processes regulating removal from the system. By this method approximately 40 to 50% of the applied nitrogen cannot be accounted for in the ground water or harvested crop. While some of the unaccounted for N may be in the soil-root complex, it is probable that a significant amount has been lost to the atmosphere by denitrification.[13] Phosphate is removed by the plant material to a more limited extent (approximately 15%). Ultimately, a very small amount of the nitrogen and virtually no phosphorus reach the ground water.

Of the major cations studied only potassium is removed by the harvested crop to a significant degree. The percentages of calcium and potassium which reach the ground water are small while well over a third of the applied sodium and magnesium have been accounted for in the ground water. Plant uptake of sodium has been less than that observed for any of the other major cations.

Two other cations, boron and sulfate, were measured in the various water samples but not in the plant materials and therefore do not appear on Figure 2. The applied boron to reach the ground water is estimated at 20% on the south plot and 37% on the west. Similarly the breakthrough of sulfates is 41 and 52%, respectively, under the west and south plots.

The mass balance analysis has not been extended to the heavy metals because there has been no measurable changes for these cations within the ground water. Our crop analyses indicate that an agricultural crop can be expected to remove only about 5 to 15% of the applied heavy metal burden. However, even after prolonged periods of soil application (more than 30 years) a major fraction of the trace metals applied can be accounted for within the uppermost few centimeters even in the absence of an agricultural crop.[7]

CONCLUSIONS

The results of these spray irrigation experiments show that crop yields are species-related and linearly proportional to the total amount of wastewater applied. Neither the rate nor the application pattern of irrigation appear to alter these relationships significantly. Since elementary composition remains relatively constant for most species, removal of a given nutrient is also linearly proportional to the amount of that nutrient applied.

The primary objective of this research has been to assess wastewater renovation rather than to maximize crop yields. However, within the limits of our observations, the degree of renovation contributed by the crops remains remarkably constant. For the three irrigation rates studied, the Yield Ratio (dry crop yield to total hydraulic load) showed no significant variation. However, higher crop yields at higher irrigation rates can only obtain up to some critical combination whereafter a breakdown in the renovation process commences. The optimal irrigation rate will correspond to the highest rate of application that does not diminish the Yield Ratio or unduly increase the flux of elements into the ground water. Recently, we have extended the maximum rate of application from 7.6 to 10.0 cm/wk for a better definition of the upper limit for renovation at this location.

With the exception of phosphorus, turf and soil samples have provided limited information on the impact of three years of irrigation on the soil column. While root growth contributes significantly to nutrient uptake during the initial two years of growth, subsequent nutrient removal is almost entirely associated with the harvested material. Failure to detect a trace metal buildup in the soil to date is attributed to low levels of trace metal application and relatively high background concentrations in the ambient soil.

Ground-water concentrations of several ions increase in proportion to the irrigation rate. Since the percent renovation by the crops is constant, these variations must be attributed to dilution and soil processes. For certain elements, high loading rates can rapidly deplete the adsorptive capacity of the soil. However, it is unlikely that soil saturation will be a critical factor within the near future given the range of application rates being studied.

The importance of the dilution phenomenon is frequently overlooked. As irrigation rates increase and land area decreases, the percent of the total hydraulic load made up of precipitation also decreases. The decreased dilution at higher irrigation rates is partly but not completely balanced by a commensurate decrease in evapotranspiration. The end result is that at higher irrigation rates, ionic concentrations in the recharge water are higher. It may be possible to compensate for decreased dilution at higher rates by maximizing dilution with ground water. This could be accomplished by a facility design which orients long rectangular fields perpendicular to the ground-water gradient. The result would be a larger but less concentrated ground-water plum. Another approach would be to divide the irrigation operation into two or more sites.

In conclusion, spray irrigation of perennial grasses and subsequent ground-water recharge has many advantages over rapid sand filtration, presently the primary land application alternative for the Cape Cod area. The harvesting of hay removes about 50% of the nitrogen and potassium and 15% of the phosphorus applied. Spray irrigation imposes a larger land requirement which leads to a commensurate increase in dilution with precipitation and ground water as well as increased attenuation by soil. At the same time, the actual amount of water recharged is less because of increased evapotranspiration. These factors must clearly be weighed against the higher costs of land acquisition, safe operation and maintenance and the water quantity and quality constraints of the area.

ACKNOWLEDGMENTS

This paper is Woods Hole Oceanographic Institution Contribution No. 3977.

REFERENCES

1. Pound, C. E. and R. W. Crites. "Wastewater Treatment and Reuse by Land Application," Environmental Protection Technology Series EPA-660/2-73-006a, Office of Research and Development, U.S. EPA, Washington, D.C. (1973).
2. Driver, C. H., B. F. Hrutfiord, D. E. Spyridakis, E. B. Welch and D. D. Woolridge. "Assessment of Effectiveness and Effects of Land Methodologies of Wastewater Management," Technical Report, Contract No. DACW 73-73-C-0041, University of Washington, Seattle, Washington (1972).
3. Environmental Protection Agency. "Methods for Chemical Analysis of Water and Wastes," NERC Analytical Control Laboratory, Cincinnati, Ohio (1974).
4. Palazzo, A. J. "Land Application of Wastewater: Forage Growth and Utilization of Applied Nitrogen, Phosphorus and Potassium," in *Land as a Waste Management Alternative,* R. C. Loehr, Ed. (Ann Arbor, Michigan: Ann Arbor Science Publishes, Inc., 1977), pp. 171-180.
5. Sidle, R. C. and L. T. Kardos. "Heavy Metal Relationships in the Penn State 'Living Filter' System," in *Land as a Waste Management Alternative*, R. C. Loehr, Ed. (Ann Arbor, Michigan: Ann Arbor Science Publishers, Inc., 1977), pp. 249-260.
6. Sopper, W. E. and L. T. Kardos. "Vegetation Responses to Irrigation with Treated Municipal Wastewater," in *Recycling Treated Municipal Wastewater and Sludge through Forest and Cropland,* W. E. Sopper and L. T. Kardos, Eds. (University Park, Pennsylvania: The Pennsylvania State University Press, 1973), pp. 271-294.
7. Ketchum, B. H. and R. F. Vaccaro. "Removal of Nutrients and Trace Metals by Spray Irrigation and in a Sand Filter Bed," in *Land as a Waste Management Alternative*, R. C. Loehr, Ed. (Ann Arbor, Michigan: Ann Arbor Science Publishers Inc., 1977), pp. 413-434.
8. Ellis, B. G. "The Soil as a Chemical Filter," in *Recycling Treated Municipal Wastewater and Sludge through Forest and Cropland,* W. E. Sopper and L. T. Kardos, Eds. (University Park, Pennsylvania: The Pennsylvania State University Press, 1973), pp. 46-70.
9. Environmental Protection Agency. "Interim Primary Drinking Water Regulations," 41 FR 28402, Washington, D.C. (1976).
10. U.S. Public Health Service. "Standards for Drinking and Culinary Water Supplied by Carriers Subject to Federal Quarantine Regulations," U.S. PHS, Washington, D.C. (1956).
11. Steel, E. W. *Water Supply and Sewerage.* McGraw-Hill Series in Sanitary Science and Water Resources Engineering, 4th ed. (New York/London/Toronto: McGraw-Hill Book Co., 1960).
12. Clapp, C. E., D. R. Linden, W. E. Larson and J. R. Nylund. "Nitrogen Removal from Municipal Wastewater Effluent by a Crop Irrigation System," in *Land as a Waste Management Alternative*, R. C. Loehr, Ed. (Ann Arbor, Michigan: Ann Arbor Science Publishers, Inc., 1977), pp. 139-150.
13. Wallingford, G. W. "Fate of Nitrogen from Fertilizer Practices," Dept. of Soil Science, Northwest Experiment Station, University of Minnesota, Crookston, Minnesota (1976).

14. Bear, F. E. *Chemistry of the Soil.* (New York: Reinhold Publishing, 1964).
15. Kardos, L. T. and J. E. Hook. "Phosphorus Balance in Sewage Effluent Treated Soils," *J. Environ. Qual.* 5(1):87-90 (1976).
16. Palmer, C. "Hydrogeological Implications of Various Wastewater Management Processes for the Falmouth Area of Cape Cod, Massachusetts," in *Cape Cod Wastewater Renovation and Retrieval System, A Study of Water Treatment and Conservation.* Technical Report, Woods Hole Oceanographic Institution, Woods Hole, Massachusetts (1977).

5

LAND APPLICATION OF FOOD PROCESSING WASTEWATER

D. O. Bridgham, W. A. Britton and B. A. Patrie
Edward C. Jordan Co., Inc.
Portland, Maine

J. A. Lawson
A. E. Staley Manufacturing Co., Inc.
Houlton, Maine

INTRODUCTION

In 1974, the A. E. Staley Company's Houlton, Maine, plant contacted the E. C. Jordan Company requesting a study to determine the feasibility of applying process wastewater onto an adjacent company-owned field. The Houlton plant, which is located in Northern Maine's Aroostook County, chemically modifies raw tapioca and potato starch for use in the food processing and industrial manufacturing industries. In 1974 and now during the nonirrigation season, wastewater is treated in an extended aeration treatment system with subsequent discharge to Meduxnekeag River. This paper will discuss: (1) the findings of the first feasibility study; (2) land application results and problems encountered since 1974; and (3) anticipated future improvements to the overall system.

Prior to 1974, Staley's wastewater treatment system and its waste discharge licenses prevented the plant from operating at peak capacity and did not permit the manufacture of a limited high-phosphorus production product which has subsequently been scheduled during the irrigation season. To overcome these limitations, Staley Company decided to investigate land treatment as a means of pollution control.

To determine the feasibility of land application of wastewater, a two-phase program was carried out. The first phase consisted of a subsurface soil investigation and compilation of basic irrigation data. Utilizing this information during the second phase, an 11-week experimental spray irrigation project was conducted on a portion of the farm. The results of this experimental irrigation operation indicated that the soils were adequate in treating the wastewater. Land treatment of the plant's summertime wastewater has therefore continued since 1974.

SITE EVALUATION

The site consists of an approximately 60-ac field that has slopes of 3 to 6%. About half of the field has been planted to a mixture of reed canarygrass and tall fescue. The other half of the area, which is scheduled to be planted to the canarygrass/fescue mixture during 1978, is an old timothy hayfield (about 8 years old) that at this time also includes an abundance of witchgrass.

During the soils investigation, it was determined that the site's soils were till-derived, well-drained Caribou gravelly silt loam and the moderately well-drained Conant silt loam with small knolls of shallow-to-bedrock Mapleton shaley silt loam soils. In general, the vertically fractured shale bedrock was at 1 to 3 m in the Caribou and Conant soils and at 1/2 to 1 m in the Mapleton soils. Seasonal high water table was at about 1/2 m in the Conant soil and greater than 2 m in the Caribou soil.

Based on the onsite soils investigation and a Soil Conservation Service computer program, it was determined that a grass crop grown on the Caribou and Conant soils in this area could utilize an additional 35 cm of water over and above that supplied by precipitation in a normal year. This figure, which considers only crop need, is based on an irrigation period of 120 days. Due to the shallowness of the bedrock as well as the vertical fractures in the shale, it was determined that the Mapleton soils should not be used in the above application.

IRRIGATION SYSTEM DESIGN

The process wastewater goes through both a primary clarifier and an equalization basin and then into a 130,000-gal aerated stabilization basin. At this point Staley has the option of diverting part or all of the wastewater either to the irrigation system or to the biologic wastewater treatment system and then to the Meduxnekeag River. The 130,000-gal storage vessel provides Staley with approximately a 2-day storage capability. Therefore, it is possible to cease irrigation on rainy days or on extremely cold autumn days.

The wastewater is pumped from the stabilization basin approximately 1/4 mi to the irrigation field. Sufficient laterals cover the irrigation area so that once they are placed down in the spring the company only has to move the spray nozzles. Two sizes of nozzles are used: those having a 72-m diam spray pattern and smaller 36-m spray pattern (Figure 1).

Using a rotational basis, approximately 2.5 cm of water/week is applied to each area. This water is applied at about 0.75 cm/hr.

MONITORING TECHNIQUES

In order to monitor the surface and ground water in the area of the irrigation field, three dikes with culverts and gates and eight monitoring wells were installed, as shown in Figure 1. Five of the test wells were installed in 1974 and the other three in 1975.

The dikes were installed to contain surface runoff from the farm in both the irrigated and nonirrigated areas. To date, there has been no surface runoff as a result of the irrigation.

The monitoring wells are approximately 3-m deep and at or near the bedrock surface. These wells were installed in 18-cm augered holes, back-filled with sand and capped with a concrete plug and compacted soil to prevent surface water from entering the wells without first passing through the soil profile (Figure 2). Each well has its own hand-operated vacuum pump for obtaining the water samples.

Weekly monitoring and analysis of the wastewater for BOD, COD and total coliform have been compared with weekly samples from the monitoring wells. The monitoring wells are also being tested for both nitrogen and phosphorus.

LAND APPLICATION EFFICIENCY

The wastewater being applied (Figure 3) was varied in its composition or strength as exhibited by the following ranges:

BOD: 500 to 5,800 mg/l
COD: 2,000 to 15,000 mg/l
Total coliform: 10,000 to over 3,000,000/100 ml

Staley's wastewater is nutrient-deficient, which is common for food processing wastes. This point had not been addressed during the initial phase of irrigation, which was in September, October, and the first week of November of 1974. This deficiency created a carbon/nitrogen (C/N) imbalance in the soil; therefore, well samples for that period exhibited high BOD, COD and total coliform. Since 1975, nutrient adjustments have been made on the wastewater prior to its land application. This has been done through the addition of urea and phosphoric acid to the water while it is in the aerated stabilization basin. The nutrients have been applied on the ratio

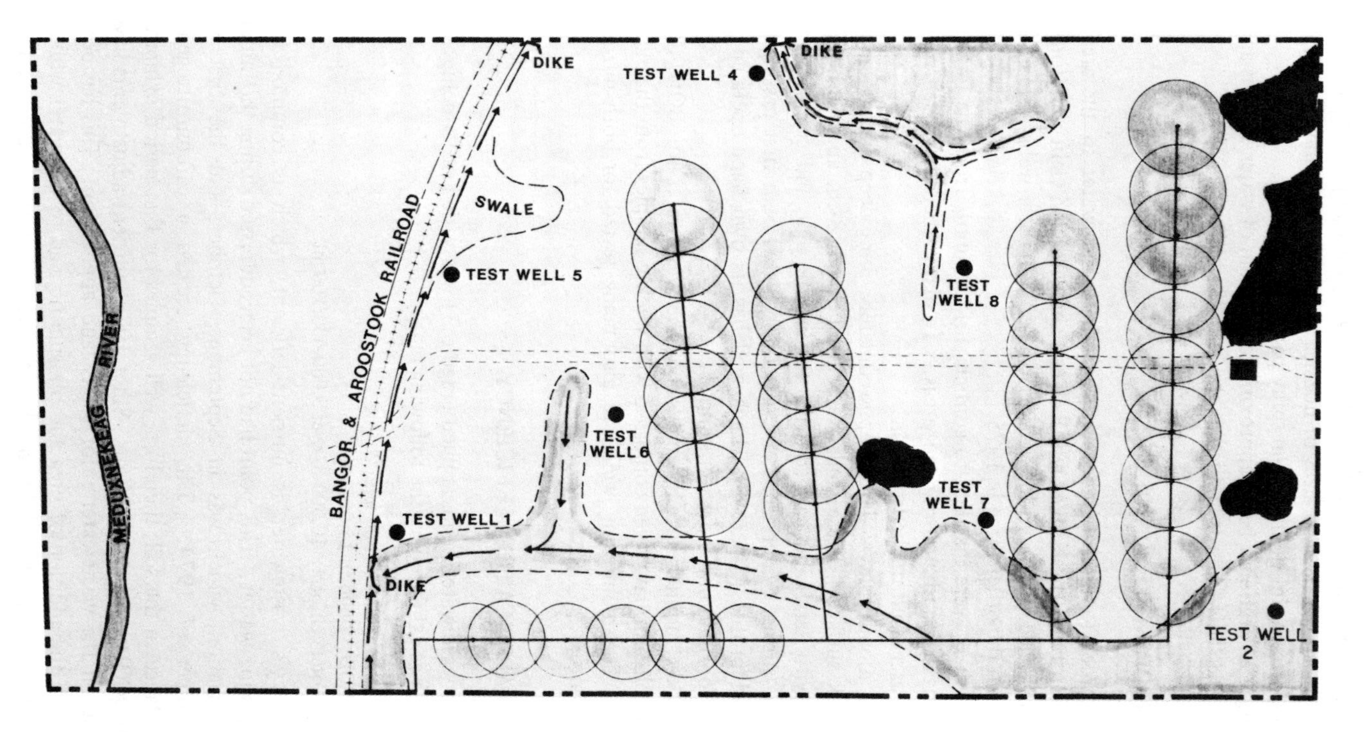

Figure 1. A. E. Staley irrigation system layout.

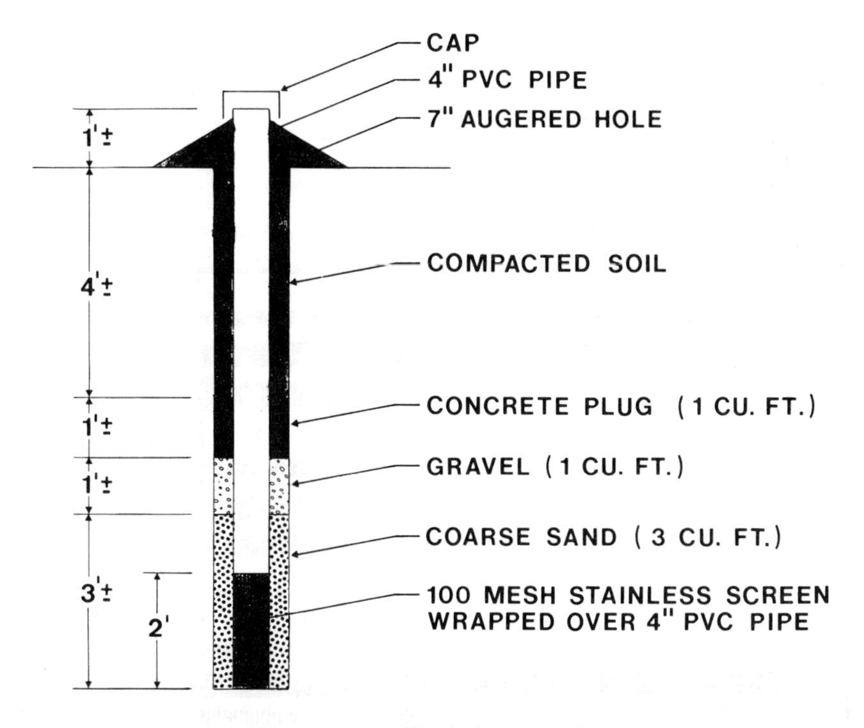

Figure 2. Detail–irrigation test wells.

of (BOD:N:P) 100:5:1. Test well data following the nutrient adjustment have shown marked decreases in BOD, COD and total coliform (Figures 4, 5 and 6). The monitoring well samples have the following characteristics:

BOD: 0.1 to 8.5 mg/l
COD: 5.0 to 62.5 mg/l
Total coliform: 0 to 100/100 ml
NO_3-N: 0.0 to 3.0 mg/l
PO_4-P: 0.03 to 0.2 mg/l

Due to the low BOD readings, as well as the inaccuracy of the test results at those low levels, no BOD analysis was made on the well samples in 1976. It is our opinion, based on results of the monitored well samples, that the soil bacteria and plant growth are adequately assimilating the wastes. This system has a removal efficiency of 97 to 99+% for BOD, COD and total coliform.

Analysis of the soil samples by the Maine Soil Testing Service at the University of Maine has shown that the addition of the wastewater lowers the soil pH. Lime has therefore been applied several times during the 2-1/2-year irrigation project. Phosphorus and potassium levels have remained moderately high to very high during this same period.

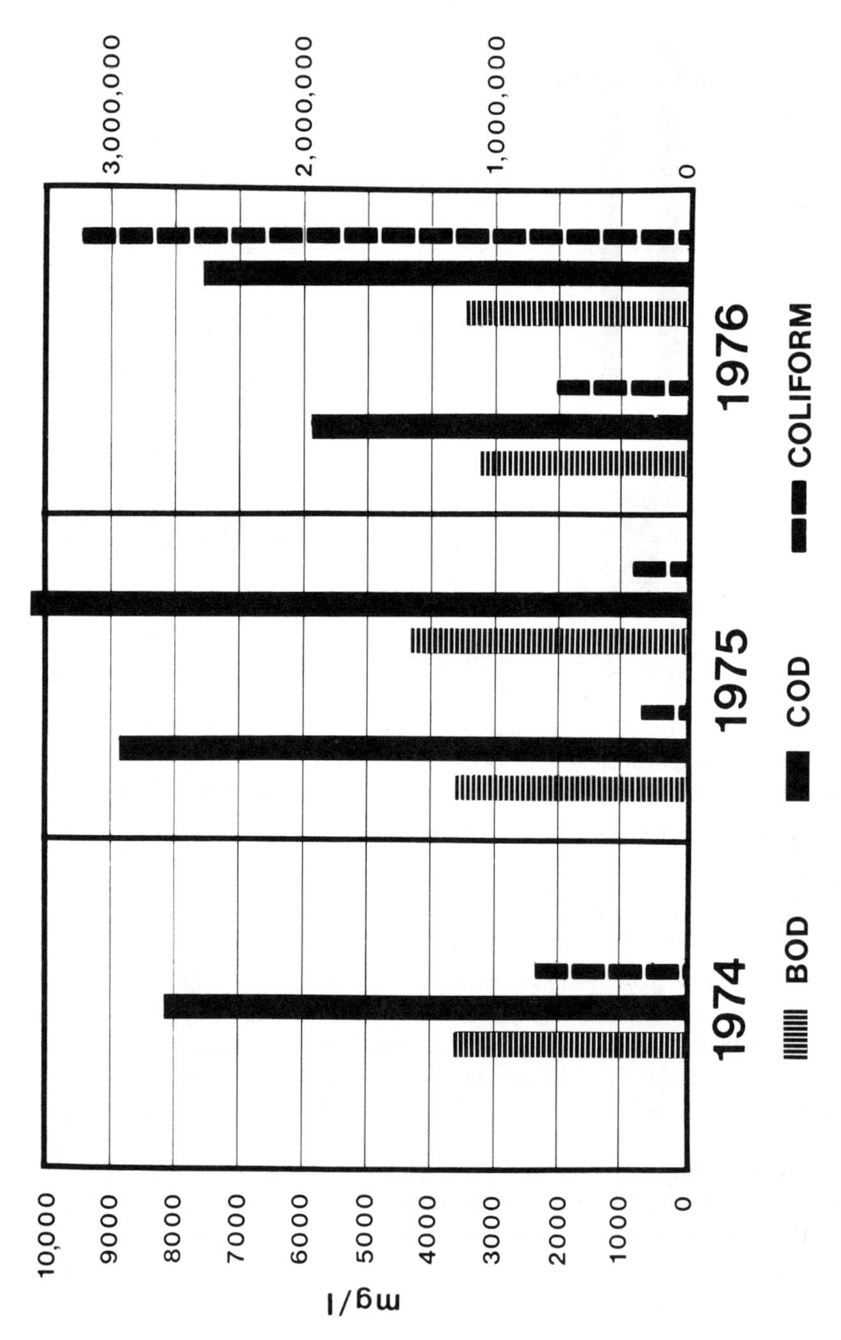

Figure 3. Wastewater characteristics.

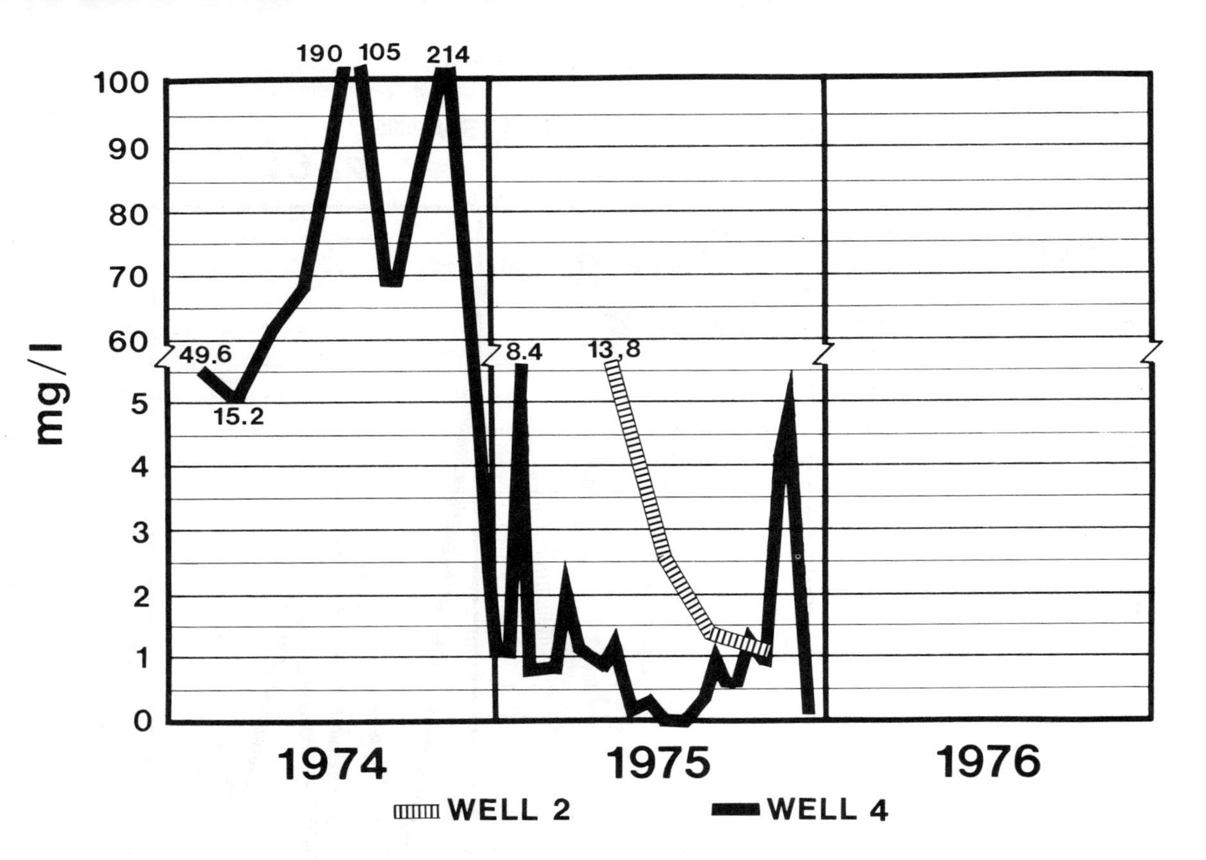

Figure 4. BOD_5—wells 2 and 4.

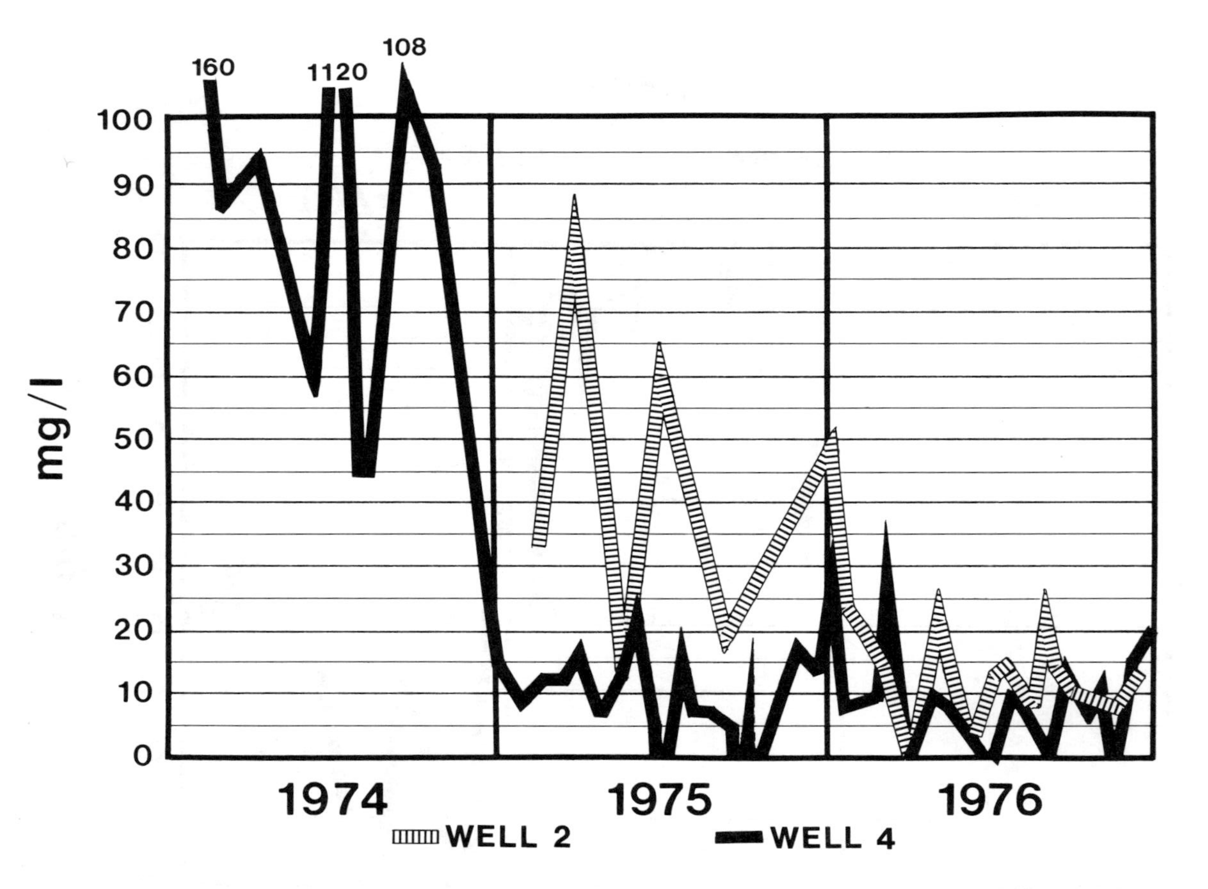

Figure 5. COD–wells 2 and 4.

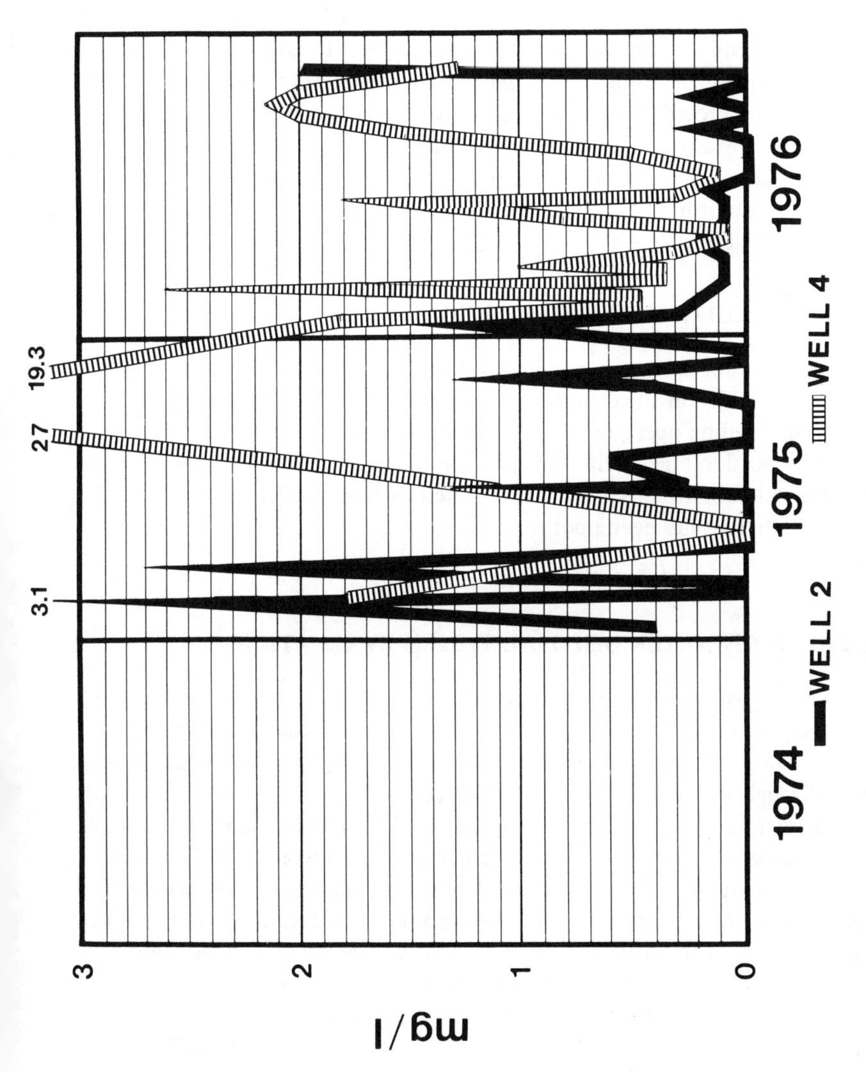

Figure 6. NO_3–wells 2 and 4.

Recently more extensive soil chemistry analysis has confirmed that irrigation of these wastewaters has resulted in an intensive leaching of divalent cations (calcium and magnesium) and a buildup of exchangeable sodium to undesirable levels in some portions of the fields. This has resulted from the presence of approximately a 2% concentration of sodium sulfate in the wastewater. Immediate corrective treatment is required to prevent soil physical degeneration which would result in loss of permeability and its ability to support plant growth. The corrective treatment will be the application of 10 tons/ac of waste cement kiln dust, which we have analyzed and found to have the same efficiency as agricultural lime. This will be followed by soil testing to monitor the results of this corrective action and subsequent irrigation of wastewater.

LAND MANAGEMENT CONSIDERATIONS

The Staley Company is not in the farming business. The company has thus had to rely on local farmers to remove the hay crops and provide for tillage and seeding operations. The farmers in the area are generally potato farmers who don't need the hay and are busy with their own crops when tillage or other contract services are required by Staley. Therefore, the timing of these activities has been poor.

TOTAL WASTEWATER DISPOSAL SYSTEM EVALUATION

Since August of 1974, the majority of the wastewater has been diverted to the irrigation system from June to November of each year. The remainder of the year the wastewater has been treated and discharged to the Meduxnekeag River through Staley's extended aeration system.

The existing extended aeration treatment plant does not have the facilities to produce waste biological sludge. Because of the suspended solids content of the effluent and the limitation of Staley's waste discharge license, the waste treatment system limits the wintertime production capacity of the Staley plant. Therefore, Staley presently is evaluating the feasibility of installing facilities to waste, store and treat excess biological sludge. Waste sludge would be stored in an aerated basin. During the summer season this stored sludge would be diluted with partially treated effluent and/or water and sprayed onto Staley's existing irrigation site.

This system approach should accomplish the following objectives:

1. It would improve waste treatment plant efficiency during the winter season and possibly increase plant production capacity.

2. It would allow constant year-round operation of the extended aeration treatment system, thus avoiding problems with seasonal operating fluctuations.
3. It would provide a mechanism whereby the stored sludge and only a portion of the summer wastewater flow would be applied on the land. This would reduce by a factor of 3 or 4 the hydraulic and salt loadings on the soil.
4. The new system would provide seasonal disposal of particularly difficult to treat wastewaters.

In conclusion, through the development of a land application system, the Staley plant has a mechanism through which it can markedly increase its summer starch production and potentially improve its winter production capability.

6

DEER CREEK LAKE– ON-LAND WASTEWATER TREATMENT SYSTEM

D. J. Lambert
U.S. Army Corps of Engineers
Huntington, West Virginia

H. L. McKim
Wastewater Management Program
U.S. Army Corps of Engineers
Cold Regions Research and Engineering Laboratory
Hanover, New Hampshire

INTRODUCTION

The Deer Creek Lake land treatment system is located 30 miles southwest of Columbus, Ohio, and treats the wastewater from one of the camping sites at the Deer Creek Lake Reservoir (Figure 1). The camping site was opened in the spring of 1974 and accommodates trailers and tents. Sanitary facilities provided for the visitors include a central sewage dump station along with multiple slop drains for trailers, six comfort stations and four wash houses. The wash houses are equipped with two clothes washers and dryers, six shower stalls and six to eight wash basins. The wastewater from these sources feeds into two lift stations located at the camping site, passes through underground drains and is pumped to the stabilization lagoon.

The stabilization lagoon is part of a simple two-stage waste treatment system design which was used prior to the passage of the 1972 amendments to the Federal Water Pollution Control Act. The percolate from this system did not meet future (1977, 1983, 1985) water quality discharge standards; therefore, a choice was made to upgrade the system by use of land treatment to ensure that proper treatment of the wastewater could be accomplished.

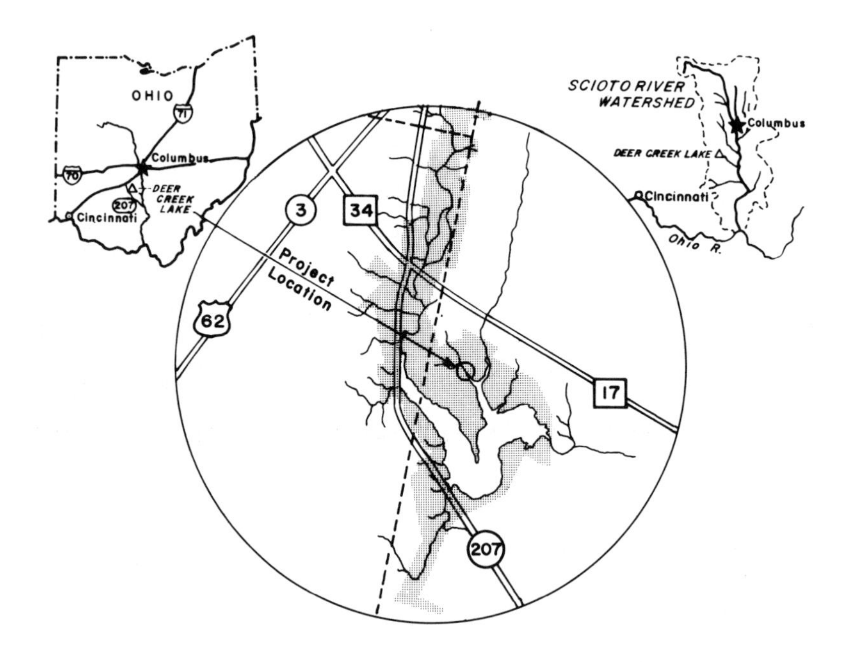

Figure 1. Location map of Deer Creek Lake, Ohio, land treatment system.

DESIGN

The complete treatment system is composed of a stabilization lagoon, chlorine contact chamber, holding basin, pump intake, distribution piping and four 3-ac spray fields (Figure 2). Any runoff outside the spray area is diverted by a berm located around the entire spray area. Runoff from the spray area flows through a series of ditches and is collected at the lower end of the 12-ac spray field. An underdrainage system which collects the water that percolates through the soil was installed at a depth of 3 ft.[1] The soils comprising the majority of the land in the treatment area have neutral to alkaline pH and have moderately slow to slow permeability rates as classified by the Soil Conservation Service (Table I). The clay content below 30 in. always exceeded 36%.

The stabilization lagoon and holding basin were part of the original two-lagoon sewage treatment system installed at Deer Creek Lake in 1973 to handle a flow of 46,000 gpd, which was the anticipated maximum weekend flow to be pumped from the 232 camp sites at the Deer Creek Lake camping area. The primary lagoon was designed to hold 73 days of maximum flow at normal depth and could be raised 2 ft to store an additional 60 days. A

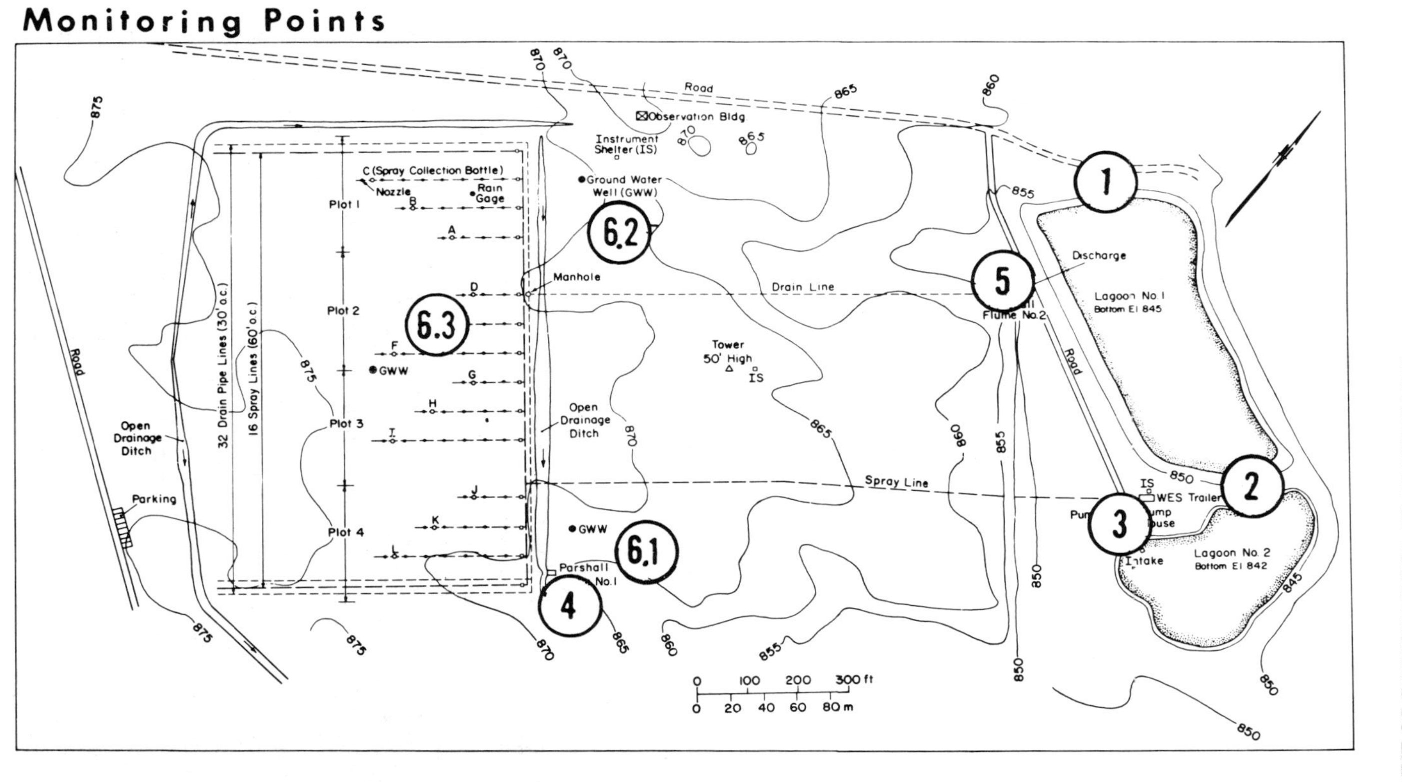

Figure 2. Schematic diagram of the land treatment system at Deer Creek Lake, Ohio.

Table I. Physical and chemical properties of the soil from the Deer Creek Lake land treatment site.

Description (site)	Depth (cm)	pH	Particle Size Distribution (%) Sand	Silt	Clay	Textural Class	$CaCO_3$ Equivalent (%)
Nursery trees	0.0-7.5	7.6	23.0	54.2	22.8	silty loam	4.2
	7.5-15	7.4	23.5	55.2	21.3	silty loam	2.3
	15-30	7.3	18.0	46.4	35.6	silty clay loam	2.1
	30-60	7.2	15.5	43.9	40.6	silty clay	4.7
Alfalfa	0.0-7.5	7.5	21.4	53.4	25.2	silty loam	3.0
	7.5-15	7.5	22.2	50.5	27.3	clay loam	3.1
	15-30	7.6	18.6	45.1	36.3	silty clay loam	4.4
	30-60	7.5	17.2	43.5	39.3	silty clay loam	5.7
Soybeans	0.0-7.5	7.5	24.1	50.2	25.7	silty loam	7.4
	7.5-15	7.6	24.1	50.6	25.3	silty loam	6.2
	15-30	7.2	17.5	43.8	38.7	silty clay loam	2.0
	30-60	7.8	19.1	42.2	38.7	silty clay loam	10.1
Reed canarygrass	0.0-7.5	7.6	28.2	48.2	23.6	loam	8.5
	7.5-15	7.5	24.4	48.3	27.3	clay loam	3.6
	15-30	7.7	19.5	42.6	37.9	silty clay loam	5.0
	30-60	7.8	21.8	42.2	36.0	clay loam	13.5

secondary lagoon, originally intended to polish the effluent from the stabilization lagoon, was converted to a holding basin capable of storing 62 days at maximum flow. The chlorination point is located between the stabilization lagoon and the holding basin.

The partially treated wastewater is pumped from the holding basin at 560 gpm to the 12-ac spray field for an average application rate of 0.1 in./hr. The pumping is limited to 10-hr periods to ensure that more than 1 in. of effluent is not applied per week. The natural slope of the spray field, averaging 1%, was left undisturbed in order to maintain the existing soil profile except for the trenching required to install the underdrain system.

There is a great deal of flexibility designed into the spray system. Any two of the three units in the pump house can accommodate the design flow of 560 gpm, leaving the third as a standby pump (Figure 2, Point 3). Also, the entire flow of 280 gpm from one pump can be sprayed into any of the four equally sized 3-ac parallel sections into which the spray field is divided. Furthermore, each pump is equipped with individual intake screens so that each intake can be back-flushed by the flow from the other pumps should the screens become clogged.

Wastewater application rates are controlled by timers connected to remotely controlled electric solenoid valves. This arrangement limits each section to 15-min applications separated by 15-min rest intervals and provides for alternating flow. This enables an accurate assessment to be made of the cropping and tillage practices in the spray field. Each test section is served by four 4-in. header pipes with 14 risers and sprinklers spaced at 40-ft centers on the headers. All headers and supply lines are placed underground, not only to make the area more aesthetically acceptable, but also to decrease the time involved in the cutting, trimming and harvesting of vegetative growth.

The underdrain collection system will allow early spring application of wastewater. These drains were placed parallel to the spray header lines on 30-ft centers. Each drain consists of a 6-in corrugated flexible plastic pipe with a filter sleeve. All the percolating water can be analyzed and discharged to the reservoir or primary holding lagoon based on the water quality analyses.

MONITORING SYSTEM

The monitoring system serves a dual purpose. It ensures that water entering Deer Creek Lake meets water quality goals and provides a data base that will enable a more accurate assessment of management and cropping practices required for design and construction of cost-effective land treatment systems.

There are eight monitoring points at critical locations in the land treatment system (Figure 2). The flow of raw sewage is measured at point 1, where a device activated by the start-up of the raw sewage pump collects a composite sample for water quality analyses. At point 2, an automatically activated timing device provides for a representative sample of the holding lagoon effluent. Composite samples are collected at point 3 when the pump that transports wastewater to the spray irrigation field is activated. Accurate flow rates and water quality analyses are needed at point 3 to ensure that water and nutrient mass balances for the spray area can be calculated. Points 4 and 5 have automatic flow recording devices to measure overland runoff and subsurface drainage, respectively (Figure 2). Since the system was designed and is operated to ensure that the wastewater applied will infiltrate into the soil, any runoff from the site is attributed to rainfall. When runoff occurs, grab samples are collected at points 4 and 5 and analyzed for water quality. The flow at point 4 determines if the quality of the rainwater runoff from the spray irrigation site meets discharge requirements. The quality of the renovated wastewater at point 5 indicates whether the water meets advanced treatment requirements allowing discharge into Deer Creek Lake (or must be recycled to the stabilization lagoon for further treatment). The Ohio

Department of Natural Resources has established the procedure for determining if water quality criteria have been achieved.

Monitoring points 6.1, 6.2 and 6.3 consist of shallow wells where the level of the phreatic water surface is visible. These wells are sampled to determine the effect of spray irrigation on ground water quality. Analyses of samples from monitoring points will provide information which can be used in the design of more efficient land treatment systems.

CROPPING AND TILLAGE PRACTICES

The type of crops chosen for the four 3-ac test plots included alfalfa, reed canarygrass, soybeans and tree seedlings (Figure 2). Alternative cropping practices and crop sequences are presently being evaluated. The plot used for soybeans was plowed after the fall harvest and the seedbed prepared in mid-April to early May depending on the crop to be used (soybean, alfalfa, etc.). The row crops are planted in middle to late May and cultivated as needed. The alfalfa is harvested three to four times a year at mid- to late-bud stage, whereas the reed canarygrass is cut three to four times a year at late-boot stage or at a height of about 14-16 in. The tree seedlings planted included red oak, white pine, sycamore and sweet gum.

SAMPLE ACQUISITION AND PREPARATION FOR ANALYSIS

Composite soil samples from the 0- to 7.5-, 7.5 to 15-, 15 to 30- and 30 to 60-cm depths were collected from each experimental plot (nursery, alfalfa, soybean and reed canarygrass plots) with a stainless steel probe. The samples were air dried, crushed with a wooden roller and stored in polyethylene bags for analysis.

Four representative 1-m^2 areas were harvested from the alfalfa and reed canarygrass plots in September using hand shears. Dry unit weights were recorded for each site. The average weight of vegetation from the four 1-m^2 areas was used to compute yields (kg/ha). Composite plant samples were taken to determine the moisure percentage. Soybeans were harvested with a conventional combine. Composite seeds and soybean residue samples were obtained for moisture and chemical analysis. Moisture percentages were determined for the plant and seed samples by computing the dried weight at 70°C after drying for three to five days. The dried samples were ground with a stainless steel Wiley Mill and stored in polyethylene containers for chemical analysis.

In 1975 samples of raw sewage (point 1, Figure 2) water samples from the stabilization lagoon (point 2), holding lagoon (point 3), surface runoff (point 4), drainage (point 5), shallow wells (points 6.1, 6.2 and 6.3) and drainage water from each vegetation site (reed canarygrass, soybeans, alfalfa and nursery) were collected by the Ohio Department of Natural Resources and the Corps of Engineers and analyzed by the Soil Microbiology and Biochemistry Laboratory of the Agronomy Department, Ohio State University, the same day as collected. Wastewater samples from points 1, 2 and 3 were collected automatically, whereas grab samples were taken from points 4, 5, 6.1, 6.2 and 6.3, and from the drainage water of each vegetation site. The samples were maintained at 4°C during transportation to the laboratory. BOD and fecal coliform samples were prepared and incubated and pH determined as soon as the samples were received. The remainder of each sample was divided in two parts. One part was preserved with mercuric chloride (40 ppm) and stored at 4°C for the analyses of different forms of nitrogen and total phosphorus. The remaining portion was stored at 4°C for analyses of chlorides, total solids and volatile solids without addition of preservatives. These analyses were usually completed within one week after sampling. Dissolved oxygen, pH and temperature were recorded on site.

RESULTS AND DISCUSSION

Soil Analysis

The results of the *in situ* soil chemical analyses from four different depths (0 to 7.5, 7.5 to 15, 15 to 30 and 30 to 60 cm) are shown in Table II. The majority of chemical constituents were uniformly distributed to a depth of 60 cm. As expected the organic matter and total nitrogen values decreased with depth. The soil pH ranged from neutral to slightly alkaline throughout the soil profile.

Available and total heavy metals concentrations are shown in Tables III and IV. The concentrations of both available and total heavy metals fall within the range of typical soils for this area.

The concentration of heavy metals in domestic secondary wastewater is generally low because most of the heavy metals in sewage are removed by the sludge. However, any heavy metals that may be in the secondary wastewater are retained in the soil by ion exchange, adsorption and chelation with organic matter and with time revert to chemical forms less available to plants. In well-drained soils having neutral to alkaline reaction,the mobility of heavy metals is low.

Crop Analysis

Yields for the different crops and the quantities of different elements (N, P, K) removed by the crop are shown in Table V. Crop yields were low due to a late planting date and, consequently, the total uptake of elements was also low. A control area has been included in future studies so that valid

Table II. Results of soil chemical analyses from soil samples taken at four depths from each of the four test plots in the land treatment area.

Description (site)	Depth (cm)	pH	Organic Matter (%)	Nitrogen NH^{+}_{4} (ppm)	Nitrogen NO^{-}_{3} (ppm)	Nitrogen Total (%)	CEC me/100/g	Exch. Bases K	Exch. Bases Ca (ppm)	Exch. Bases Mg	Available Nutrients/P (ppm)	Soluble Salts (mhos x 10^{-5})
Nursery trees	0.0-7.5	7.9	1.4	3.56	4.07	0.1018	17	419.0	2510	450.5	3.0	16
	7.5-15	7.8	1.1	4.92	2.54	0.1059	13	73.5	2065	395.5	5.5	13
	15-30	7.7	1.3	1.69	2.21	0.0651	20	73.5	2760	767.0	1.5	11
	30-60	7.7	1.0	2.71	2.21	0.0651	25	101.5	2980	1278.0	2.0	14
Alfalfa	0.0-7.5	7.8	1.4	2.03	4.59	0.1208	17	133.0	2550	500.5	5.0	13
	7.5-15	7.8	1.6	2.03	5.26	0.1086	20	95.5	3065	578.5	5.5	14
	15-30	7.9	0.8	1.86	4.42	0.0502	25	94.0	3765	743.0	1.5	14
	30-60	7.9	0.6	3.73	2.38	0.0638	29	98.5	3765	1263.0	1.5	12
Soybean	0.0-7.5	8.0	0.8	3.73	2.89	0.0801	23	144.0	3700	523.5	2.5	13
	7.5-15	8.0	1.0	3.39	2.89	0.0719	23	81.0	3765	557.0	2.5	15
	15-30	6.7	1.0	2.88	2.55	0.0733	24	73.5	3255	1014.0	2.0	10
	30-60	7.6	0.8	3.05	2.04	0.0488	34	141.0	5315	961.5	1.0	16
Reed canarygrass	0.0-7.5	7.7	1.5	2.71	2.89	0.0733	29	126.5	4825	601.0	2.0	8
	7.5-15	7.6	1.2	3.90	1.36	0.0829	20	108.0	3215	533.5	1.5	8
	15-30	7.7	0.8	3.39	0.85	0.0651	28	102.0	4045	958.5	1.5	8
	30-60	7.9	0.5	2.37	1.53	0.0516	33	155.5	5185	796.5	1.0	7

Table III. Available heavy metal concentrations in soil extracted with DTPA.

Description (site)	Depth (cm)	Heavy Metals (ppm)			
		Cu	Ni	Pb	Zn
Nursery trees	0.0-7.5	1.510	<0.10	1.548	1.848
	7.5-15	1.650	<0.10	1.936	2.288
	15-30	1.695	<0.10	1.680	2.572
	30-60	2.21	<0.10	1.680	3.432
Alfalfa	0.0-7.5	2.115	<0.10	2.064	2.244
	7.5-15	1.885	<0.10	1.808	2.728
	15-30	2.025	<0.10	2.196	2.088
	30-60	2.395	<0.10	2.184	1.540
Soybean	0.0-7.5	1.555	<0.10	1.936	1.760
	7.5-15	1.605	<0.10	1.808	1.672
	15-30	2.255	<0.10	2.064	1.388
	30-60	1.835	<0.10	1.548	1.432
Reed canarygrass	0.0-7.5	1.695	<0.10	2.064	1.188
	7.5-15	1.745	<0.10	2.064	1.232
	15-30	2.165	<0.10	1.680	1.276
	30-60	1.745	<0.10	1.548	0.968

Table IV. Total heavy metals concentrations in soil extracted with nitric acid and perchloric acid (1:2).

Description (site)	Depth (cm)	Heavy Metals (ppm)			
		Cu	Ni	Pb	Zn
Nursery trees	0.0-7.5	15.87	20.75	34.67	110.3
	7.5-15	14.25	18.42	31.45	107.5
	15-30	24.75	42.47	37.10	172.0
	30-60	31.07	49.25	40.32	179.0
Alfalfa	0.0-7.5	19.40	21.32	33.05	129.5
	7.5-15	20.10	28.90	33.05	118.5
	15-30	16.35	24.25	33.87	126.8
	30-60	30.82	59.12	40.32	183.0
Soybean	0.0-7.5	17.97	30.05	42.90	118.5
	7.5-15	18.67	31.80	35.47	130.8
	15-30	27.80	49.25	37.97	180.3
	30-60	25.92	54.47	37.90	169.3
Reed canarygrass	0.0-7.5	20.32	33.55	34.67	137.8
	7.5-15	19.85	34.12	32.25	139.0
	15-30	30.12	55.05	40.32	184.5
	30-60	27.80	53.32	39.50	177.5

Table V. Crop yield and nitrogen, phosphorus and potassium removal of the four crops harvested from the test plots.

Crop	Yield[a] (kg/ha)	Nutrients: Nitrogen (kg/ha)	Nutrients: Phosphorus (kg/ha)	Nutrients: Potassium (kg/ha)
Reed canarygrass	1.60	31.5	5.0	38.3
Alfalfa	1.92	50.4	6.6	43.8
Soybean seed	1.79	112.0	8.8	37.2
Soybean residue	1.25	9.5	0.4	17.0

[a]Oven-dry basis.

comparisons with conventional agricultural fertilizer application can be made. Water and nutrient mass balances will be determined in the 1977 studies so that the degree of renovation with respect to the different wastewater constituents can be estimated.

Water Quality Analysis

Average values for the different analytical parameters of the raw sewage and the effluent from the stabilization lagoon are reported in Table VI. The concentrations of the chemical constituents were considerably lower in the stabilization lagoon (point 2) as compared to the raw sewage (point 1). The different forms of nitrogen, particularly the inorganic forms of nitrogen

Table VI. Yearly average concentrations of raw sewage and effluent characteristics in the stabilization lagoon.

Constituents	Raw Sewage Point 1 (ppm)	Stabilization Lagoon Point 2 (ppm)
BOD	134	16
Chloride	269	81
Phosphate	6	0.4
Total nitrogen	50.6	5.4
NH^+_4-N	33.8	0.4
NO_2-N	0.0	0.0
NO_3-N	1.5	0.5
organic-N	15.3	4.5
Total solids	1250	500
Volatile solids	330	200
pH	8.3	9.2
Fecal coliform (MPN/100 ml)	1.3×10^7	7×10^3

(NH_4-N and NO_3-N), and total phosphorus were present in unusually low concentration in the stabilization lagoon (Table VI). This is the result of two factors: (1) low loading rates during the first year of operation and (2) dilution of raw sewage with water already present in the stabilization lagoon before initiation of the spray irrigation program. The concentration of inorganic N and P were further lowered in the holding basin (point 3) by dilution (Table VII). The wastewater from point 3 used in spray irrigation would not currently add significant amounts of nutrients to the soils.

Table VII. Characteristics of wastewater collected from points 3, 4 and 5.

	Range		
Constituents	**Point 3[a] (ppm)**	**Point 4[b] (ppm)**	**Point 5[c] (ppm)**
Chlorides	49-76	6-25	4-73
Phosphate	0.11-0.25	0.00-2.34	0.0-0.10
Total nitrogen	1.9-1.9	0.2-2.5	-
NH_4^+-N	0.0-0.3	0.0-2.7	0.0-0.6
NO_3-N	0.0-0.4	0.0-0.4	0.0-2.0
Total solids	320-430	193-505	93-1000
Volatile solids	86-229	22-249	23-180
pH	8.5-8.7	7.6-8.2	7.1-8.2
BOD	2.6-3.4	2.4-10.0	0.3-7.9
Fecal coliforms (MPN/100 ml)	0-240	30-5.5 x 10^4	0-600

[a]Point 3: Wastewater used as spray irrigation (pumping site).
[b]Point 4: Surface runoff.
[c]Point 5: Drainage water.

BOD values of raw sewage varied from 83 to 253 ppm, while in the stabilization lagoon these values ranged from 10 to 31 ppm. The yearly average values for BOD in the raw sewage and stabilization lagoon was 134 and 16 ppm, respectively. BOD values for the wastewater of the holding basin ranged from 2.6 to 3.4 ppm (Table VII). Storm water (point 4) gave higher BOD values (2.4 to 10 ppm) than might be expected from natural surface waters. Soil organic matter carried with the storm water probably contributed to these high BOD values. BOD values of the drainage water (point 5), shallow test wells and drainage water from different vegetation sites were low except when contaminated with storm water (Tables VII, VIII and IX).

Table VIII. Characteristics of water collected from the shallow wells.

Characteristic	Range Point 6.1 (ppm)	Point 6.2 (ppm)	Point 6.3 (ppm)
Total nitrogen	0.4-2.2	–	–
NH_4^+-N	0.3-0.5	0.1-0.2	0.1-0.5
NO_3-N	0.2-0.2	0.0	0.1-0.2
NO_2-N	0.0	0.0	0.0
Organic nitrogen	–	0.9	–
Phosphate	0.02-1.96	0.01-0.05	.01-.07
Chloride	6.0-19.6	17.3-22.5	14.3-14.5
Fecal coliforms (MPN/100 ml)	0-0	0-100	0-640
BOD	1.8-7.2	1.2-3.7	0.3-0.6
Total solids	300-1462	–	324-453
Volatile solids	145-219	–	140-204
pH	7.5-7.6	7.6-8.5	7.7-8.1

Table IX. Yearly average concentration of effluent characteristics in the wastewater applied to the four test plots.

Constituents	No. of Samples	Concentration (ppm)
pH	24	8.3
BOD	19	6.6
Total phosphorus	24	0.24
Chlorides	21	80.3
NH_4^+-N	25	0.03
NO_2-N	25	0.005
NO_3-N	25	0.19
Total nitrogen	25	2.6
Organic nitrogen	25	2.3
Hardness (EDTA)	20	182
Fecal coliforms (MPN/100 ml)	20	44
Fecal streptococcus (MPN/100 ml)	19	341
Total solids	21	560
Total volatile solids	21	117
Suspended solids	21	162
Volatile suspended solids	16	37
Total soluble solids	21	381
Total volatile soluble solids	16	96

Fecal coliforms were found in the surface runoff (point 4) in the range of 30 to $> 5.5 \times 10^4$/100 ml (Table VII). The source of fecal coliforms (effluent or indigenous mammal population) could not be ascertained.

However, the effluent used in spray irrigation is not considered to be the source of this fecal coliform pollution, since fecal coliforms were detected only once in the chlorinated effluent, prior to application to the field plots (point 3, Table VII). Fecal coliforms were very seldom detected in the drainage water (point 5), shallow wells (points 6.1, 6.2, 6.3) or in the drainage water from different vegetation sites (Tables VII, VIII and IX). The range of coliform concentration in the surface runoff and drainage water was large. The high end of the range is represented by only one value, whereas the low portion of the range is represented by many values. The chloride concentration in the raw sewage was higher than expected at certain sampling times (Table VI). The presence of this chloride was probably due to the recharging of the water treatment system at the camping areas. Phosphate, BOD, chloride, nitrate-N, soluble salts and fecal coliforms in the tile drainage water from different points and shallow wells usually remained below that expected in natural waters and less than the limits proposed by EPA.

In 1976 the primary emphasis was placed on an experiment to assess the fate of aerosols from a spray application land treatment system so that the need of a buffer zone could more realistically be determined. The results from this study are being evaluated and will be published separately.[2] During the testing period the average concentrations of wastewater constituents being applied from the holding point (point 3) were carefully monitored by the Agronomy Department at Ohio State University.[3] The yearly averages are presented in Table IX. The data indicate that the stabilization and holding lagoon effectively remove the nitrogen and phosphorus in the wastewater prior to land application.

Seedlings

The number of seedlings remaining on test cell 1 were analyzed after the first two years of operation. Of the four species originally planted, sweetgum (*Liquidamber styraciflua*), sycamore (*Platamus occidentalis*), eastern white pine (*Pinus strobus*) and northern red oak (*Quercus rubra*), only the sweetgum and sycamore were growing well. The eastern white pine and northern red oak were chloritic and in very poor condition. Table X indicates the mortality of the trees planted at the site.

Generally, the seedling damage could be attributed to four factors: mechanical damage, transplanting, competition with other plant species, and soil and soil moisture conditions. The mechanical damage resulted from the management of the field plots. In certain instances the sycamores and the sweetgum seedlings had been damaged by the mower used to cut the grass between the trees. Most tree species such as pine and a majority of the hardwoods transplant easily with extremely good survival rates. Usually improper handling of the seedling stock or poor quality of the stock was the cause of

Table X. Tree mortality at the Deer Creek Lake land treatment of wastewater facility (1975-76).

Tree Species	Number of Trees First Planted	Mortality 1975	Trees Replanted	Mortality 1976	Mortality 2-yr
Red oak	640	20%	128	21%	34%
White pine	560	28%	155	29%	44%
Sycamore	560	14%	76	23%	32%
Sweetgum	560	29%	160	12%	31%

mortality. Most oak trees do not transplant well; therefore, this factor alone could account for the poor survival rate of the red oak.

Another important factor is the survival of the seedlings in competition with grass. If the grass height exceeds the height of the trees, it would affect the survival rate. Competition between grasses and seedlings for nutrients also plays an important role in survival. The red oak and white pines showed indications of stress as indicated by chlorosis and stunting of growth.

The moisture content exceeded the field capacity of the soils for a prolonged period during the growing season. Although all four species are tolerant to high moisture, not one can grow well under prolonged high moisture conditions. Sweetgum and sycamore are more tolerant to saturated soil conditions than the oak and pine seedlings; however, these species will ultimately be adversely affected by high moisture conditions.

Generally, the seedling species most suited to this site where low permeability and high pH of the soil is a critical factor are the bottom land hardwoods. Some bottom land conifer species could be considered, but the soil pH is marginal for adequate growth. Conifers should not be considered unless the species can grow well in neutral to slightly alkaline soils.

CONCLUSIONS

The Deer Creek Lake land treatment system has been operational since 1975 and has performed adequately to meet all point discharge requirements under PL 92-500. The crop management program has shown that grasses, alfalfa and soybeans can be grown when the application rate for these low permeable soils does not exceed one inch per week. The system removes nitrogen, phosphorus and BOD sufficiently to meet drinking water standards in the percolating water. The monitoring system has performed satisfactorily and a detailed data base for product water quality has been established.

REFERENCES

1. Lambert, D. J. and W. R. Dawson. "The Living Filter Comes to Deer Creek," *Water Spectrum* 7(1):1-8 (1975).
2. Schaub, S. *et al.* "Bacterial Aerosols From a Field Source During Sprinkler Irrigation with Wastewater" (to be published as a CRREL report, 1977).
3. Brar, S. S. *et al.* "Wastewater Demonstration Project at Deer Creek Lake," Status Report from the 1975 Research Program, Ohio State University, Department of Agronomy (1975).

7

PHYSICAL SITING OF WASTEWATER LAND TREATMENT INSTALLATIONS

W. J. Hartman, Jr.
U.S. Army Corps of Engineers
Washington, D.C.

INTRODUCTION

Urban growth places a sense of urgency on the matter of coming to grips with the wastewater treatment problem. In less than a century the population in the United States has shifted from a rural to urban nature. The 85% of our people who now live in urban surroundings occupy only 15% of our available land space.

At the close of the 18th century there were only about 20 cities in the world with a population of over 100,000. Even ancient Rome at the height of its power and glory numbered barely 350,000 inhabitants. Today there are at least 100 cities in the world with populations exceeding 1 million people. In the United States alone, based on the 1970 census data, there are 26 metropolitan areas with 1 million or more people. Thus, in a relatively short time, from the mid-19th century to the present, the large city has become the most characteristic form of modern social life. It is not surprising, therefore, that man is still learning how to supply wastewater treatment facilities and to live happily in large city communities.

Land treatment of wastewater has been advocated at various points in time. This method seems to surface most readily in times of crises. In Europe the method first became popular at the time of the cholera epidemics and again at the time of industrialization which caused widespread water pollution. In the U.S., land treatment emerged in areas where there were no large bodies of water to use for dilution and again in the environmental crisis of the 1970s.

PHYSICAL SITING OF TREATMENT FACILITIES

Engineering design parameters (soils, rate and quantity of application, soil moisture, geology and hydrology) control the design of a land treatment facility, but the design must be tempered by social, economic and environmental factors. Engineering parameters can be considered as controlling and the latter factors as constraining.

Siting of land treatment facilities has to be considered from a broad viewpoint; the delicate balances within the environment cannot be set aside. There can be many combinations of land uses to be considered. The people this treatment method serves often live close by and the comparably large amount of land required generally will be neighboring other land uses that may or may not be affected. How other uses can be affected by the land treatment process is very important. Thought has to be given to existing uses.

Social Components

In a relatively short history of the United States, urbanization has changed a nation which at birth was 95% rural to a country with 85% of its people living in urban areas. More than 150 metropolitan areas have populations between 100,000 and one million.

These changes were brought about by vast migrations of United States citizens from rural areas to metropolitan centers. Practically all of these people settled in suburban areas. At the same time many city dwellers sought a more pleasant environment in the same suburban sites, thus producing more and larger Standard Metropolitan Statistical Areas (SMSAs). Figure 1 shows the number and location of SMSAs as of January 1, 1974. There were 24 new SMSAs added by the Office of Management and Budget between 1970 and April 1973. Also, as of April 1973, the area of a large number (more than 50) of the then existing SMSAs was increased over the 1972 declared SMSAs. While central cities grew only about 1% in the last decade (1960-1970) suburban population increased by 28%, so that in 1970 more than half the people in metropolitan areas live outside the central cities. There are no strong deterrents in most states to stem this tidal spotty areal growth.

Assuming that the rural-to-urban movement persists, most of the United States population growth over the next decade (1970-1980) will be concentrated in the 12 largest urban regions. These 12 areas will contain over 70% of the population but will occupy only one-tenth of the land area. The result will be further decline in the vitality of small towns and rural areas so that farm populations may decline to only 2% of the total population by the year 2000. If the matter does not receive attention, smaller communities

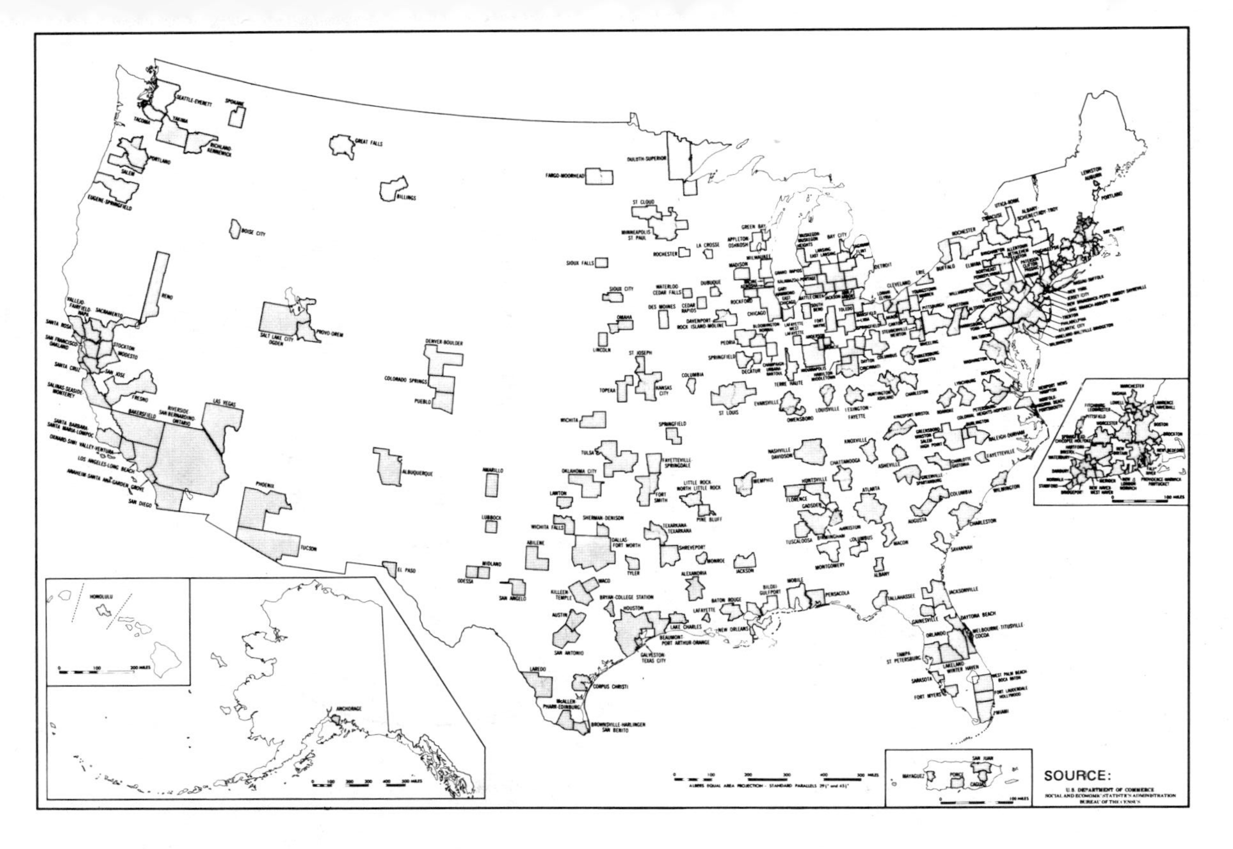

Figure 1. Standard Metropolitan Statistical Areas. Areas defined by Office of Management and Budget. January 1, 1974.

will be bypassed by the economic mainstream, while the vitality of central cities declines and growth proceeds in a disorderly, congested pattern at the edge of urban areas. This peripheral or suburban residential growth, much of it unplanned and sprawling, has been enormous and has affected every SMSA. The land area occupied by urbanities is growing faster than the number of urban dwellers.

Figure 2 shows the major land uses within the SMSAs of the U.S. in 1970. The largest category of land use was forest woodland, which accounted for almost one-third of the land area. One-quarter of the land was devoted to crops. In 1969, the last year for which data are available, 9% of U.S. wheat, 17% of our corn, 60% of our vegetables, 43% of our fruits and nuts, 16% of our hay, 15% of our soybeans and 22% of our cotton were produced within metropolitan areas. Figure 3 shows the principal cropland areas of the U.S. The impact of urbanization on cropland in the U.S. can be most vividly illustrated by comparing the SMSA map (Figure 1) with the principal cropland map (Figure 3).

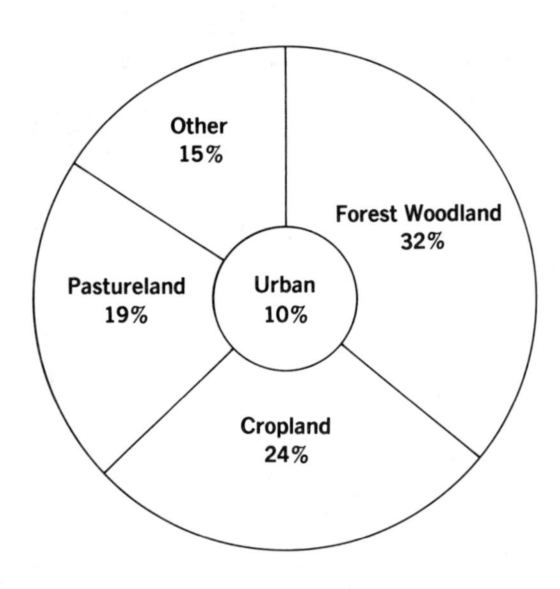

Figure 2. Land use within SMSAs 1970 for the 48 contiguous states.

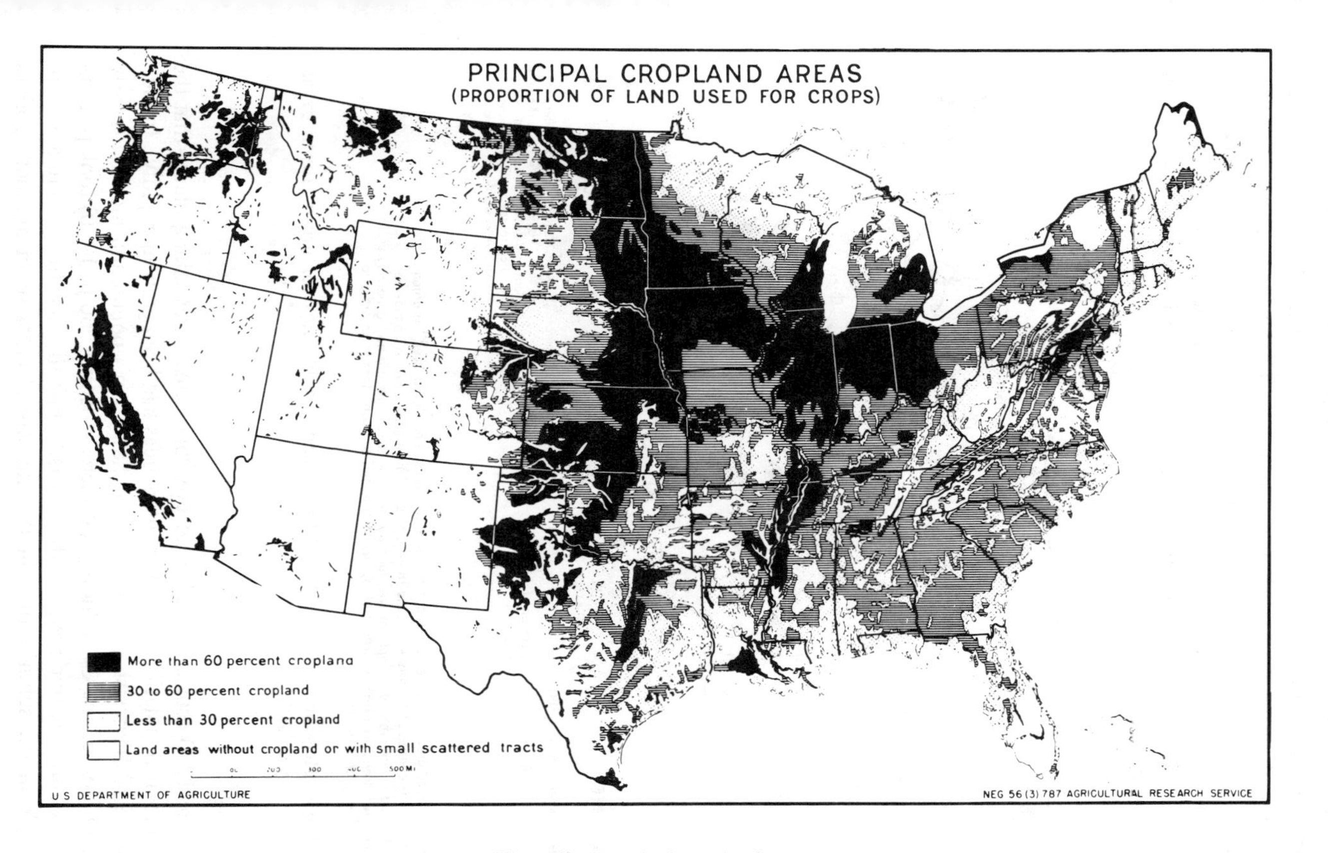

Figure 3. Principal cropland areas.

The various land uses within metropolitan areas do not arrange themselves into neat geographical patterns. Because of the way American cities have developed over the past 50 years, urban uses are widely interspersed with other kinds of uses. This intermixture of different uses has come to be known as urban sprawl. The pattern is not just a phenomenon of our newer cities. Figure 4 shows the results of this creature by establishment of the great megalopoli.

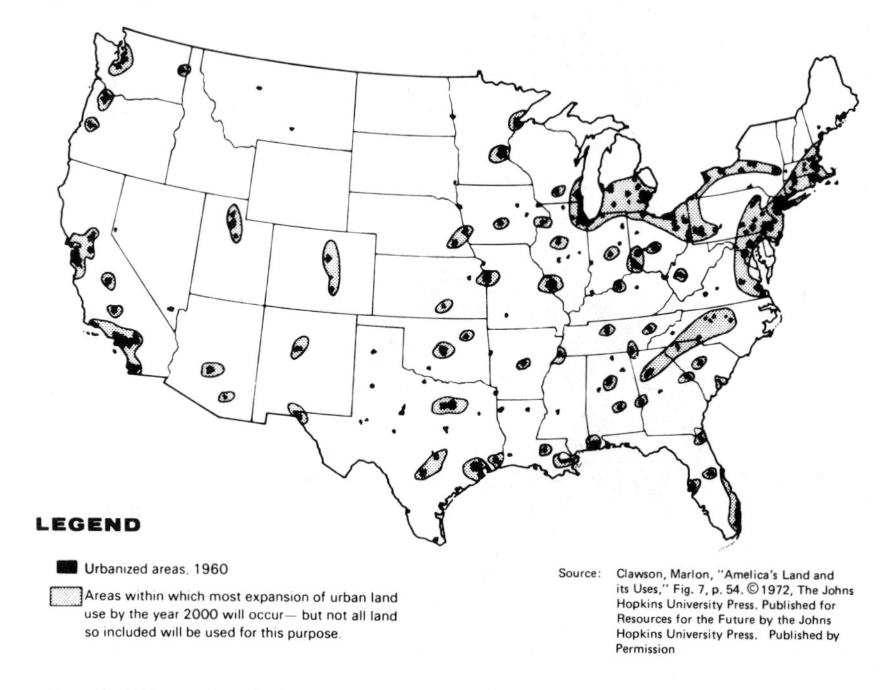

Note: Shaded areas indicate land areas that may be classed as "urbanized" if urban development continues to cluster around large cities. While some new cities may be built, most future urban development will take place in the areas that are now generally urban. All areas are slightly exaggerated, but perhaps not to equal degree, in order to show clearly on this scale.

Figure 4. Megalopoli of the United States and possible areas of urban growth by the year 2000.

The 1970 census proved what everyone sensed was taking place—that the suburban population in metropolitan areas had been growing rapidly. In fact, the 1970 figures show that most of the nation's population increase (nearly 17 million of the 24 million total increase from 1960 to 1970) took place in the suburbs of our metropolitan areas.[1] For the first time, there are more people in the suburbs than in the big central cities or in nonmetropolitan areas. The 1970 population totals were 75.6 million for suburbanites, 63.8 million for people in central cities, and 63.8 million for people in nonmetropolitan areas. Ten years earlier, the totals were 54.9 million for suburbanites, 58 million for those in central cities and 54.9 for those outside metropolitan areas.[1]

Translating this urbanization-suburbanization in terms of population density, the persons per square mile for the U.S. as a whole changed from 51 (132/sq km) in 1960 to 58 (150/sq km) in 1970. According to 1970 census data, New Jersey was found to be the most densely populated state, with 953.1 persons/sq mi (2469 persons/sq km). It took over from Rhode Island, which had been the most densely populated state ever since the first census in 1790.[1] The 1970 figure for Rhode Island was 905.4 persons/sq mi (2345 persons/sq km).

A very difficult situation for the application of the land treatment method will be in those places with the combination of the best cropland and high population density. This problem comes into focus by comparing the map of population density (Figure 5) with the map of principal cropland areas (Figure 3). It has been demonstrated, however, that land treatment facilities can be successful in areas with moderately high population density and on the fringe of highly productive cropland. There are three SMSAs in the state of California with 12 land treatment installations serving metropolitan areas. Some of these have been in operation for more than 50 years. The majority of these installations apply modern land treatment techniques and serve the public very well. Most of these are small installations under 1000 ac (405 ha) in area.

There are only a small number of cities with 1,000,000 or more people in the world that utilize land as a means of treating wastewater effluent. These include Berlin (in part), Paris (in part), Melbourne (99% part of year), Mexico City (in part), Delhi at Okhla, India (in part), Madras, India (in part), Ahmedabad, India (total), Hyderbad, India, Moscow, USSR (in part), and Johannesburg, South Africa (in part).

In the United States the largest city that utilizes land treatment, a combination of infiltration ponds and irrigation, of its pretreated wastewater effluent is Fresno, California (1970 population–165,972). The city has experienced only minor difficulties with the system and the city will continue using the system into the future at the same location.[1]

Land treatment is utilized by the greatest number of communities with a population of 500 to 10,000 people in the United States (Table I). In the period 1957 to 1968 the number of places with 500 to 10,000 people utilizing land for disposal of effluents from treatment plants decreased very rapidly and the overall application of the land for disposal declined while the stabilization pond method experienced exponential growth (Figure 6).

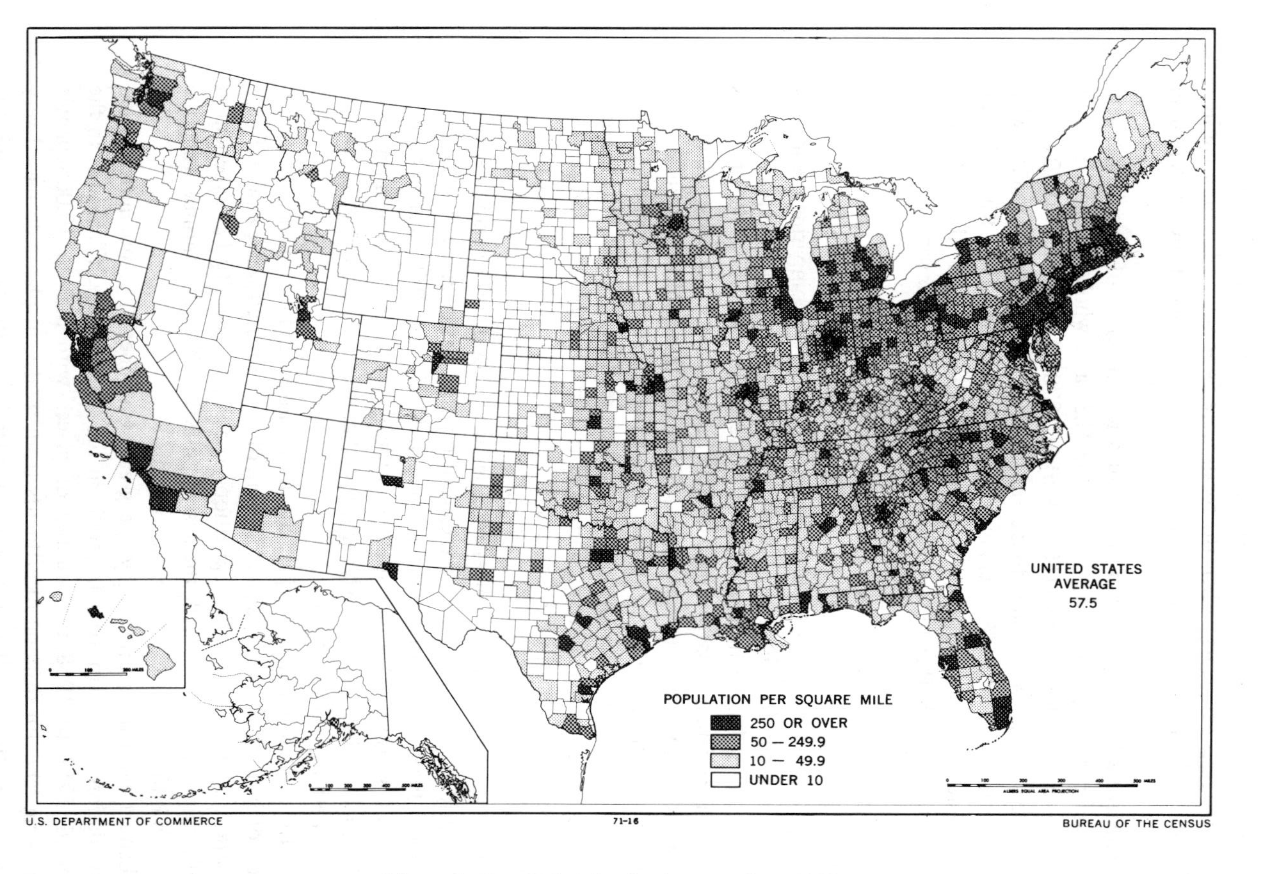

Figure 5. Population density by counties: 1970.

Table I. Number of places utilizing the application of wastewater effluent to land and total population served by population size for the United States.[a]

Population Size Group	1957 Number of Places	1957 Total Population Served	1962 Number of Places	1962 Total Population Served	1968 Number of Places	1968 Total Population Served
Under 500	76	17,916	51	14,519	36	10,580
500-1000	56	31,528	43	29,915	32	23,760
1000-5000	168	354,793	117	249,464	41	88,740
5000-10,000	27	148,670	30	167,575	8	52,335
10,000-25,000	7	124,485	15	140,875	5	52,560
25,000-50,000	4	175,000	5	81,410	2	88,685
50,000-100,000	3	144,000	4	157,000	3	96,000
200,000-250,000	-	-	2[b]	50,000	1	10
250,000-500,000	-	-	-	-	-	-
Over 500,000	-	-	-	-	-	-
Total	341	966,392	267	890,758	128	412,670

[a]Data from Statistical Summary-Municipal Waste Facilities in the United States, U.S. Department of the Interior.

[b]Population size groups changed in 1968 data; figures for 1962 were given for population over 100,000.

During the same period 1957-1968 the number of cities with populations of 25,000 to 100,000 utilizing land for effluent disposal did not change much.

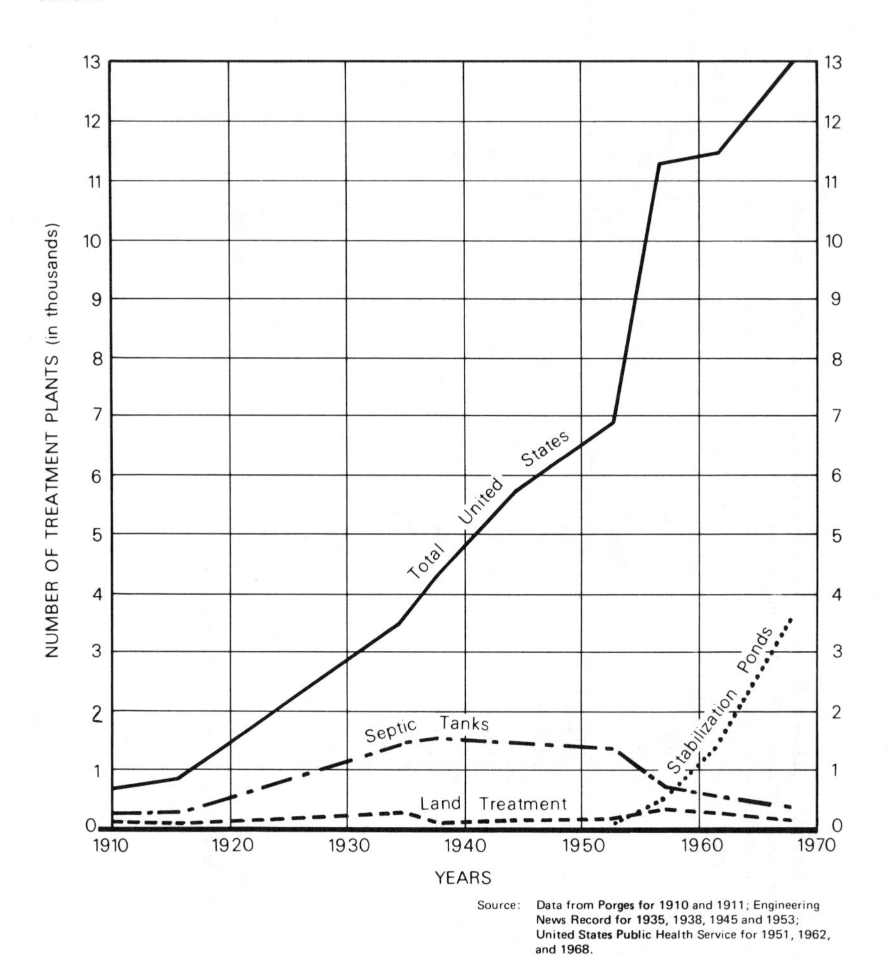

Figure 6. Use of land treatment method compared to septic tanks, stabilization ponds and to total treatment in the United States.

Studies of four characteristics of census tract data were made to determine the validity of the hypothesis that there is a significant difference in areas with land treatment facilities as compared to those without. The parameters used were taken from 1970 census data and are shown in Tables II and III. The characteristics of areas with treatment facilities showed that residents, on the average, had educational attainment and resident histories similar to people without such facilities. However, people near land treatment facilities have incomes and homes with higher dollar values. This deviation probably reflects the fact that land treatment facilities are generally located in suburban to rural settings where fewer houses exist and can have higher values. The results of these comparisons suggest that there is no significant difference in the characteristics of the populations residing in areas with or without land treatment facilities.

Table II. Comparison of census tracts with land treatment facilities to census tracts without such facilities in 1970.[a]

Area Characteristics	Tracts with Land Treatment Facilities	Tracts without Land Treatment Facilities
	N = 5	N = 27
Median Education	10.8	11.1
Median Income	$9,997	$8,550
Median Value of House	$22,000	$20,100
Average Year Moved into House	1966	1966

[a]Source: U.S. Bureau of the Census, 1970.

Table III. Comparison of census tracts with land treatment facilities to entire metropolitan area, 1970.[a]

Area Characteristics	Tracts with Land Treatment Facilities	Tracts without Land Treatment Facilities
	N = 5	N = 3
Median Education	10.8	12.1
Median Income	$9,997	$8,943
Median Value of House	$22,000	$15,800
Average Year Moved into House	1966	1966

[a]Source: U.S. Bureau of the Census, 1970.

There is also some support for this preliminary finding by way of visual examination of communities where land treatment is practiced. At Pleasanton, Palm Springs and Oildale, California, where there was no other land use (commercial, industrial or low-income housing) influencing the choice of construction site besides the wastewater land treatment facility, relatively expensive single-family housing developments have been constructed nearby in recent years. At Pleasanton, where overloaded primary treatment facilities were associated with land treatment, residents complained that they had made a mistake in establishing a home in the area. This was the only complaint registered by residents of four different communities visited. No new single-family housing developments had occurred in the vicinity of land treatment facilities at Bakersfield and Hanford, California, as of September 1973. In each of these cases other long-established land use (industry and low-income housing and commercial) establishments seemed to influence developers to use more attractive areas for additional housing developments.

Restraining Components

Besides the usual engineering data of geology, hydrology, soils and others, there are some very important restraining components that should be included in the selection of the site where land treatment of wastewater effluent can be carried on in a safe and economic manner. These components are restraining in that they primarily should influence site selection while having a minor impact on engineering design. The most apparent of these components are: (1) farm unit density and crop production; (2) existing and future land use patterns; (3) topography (landscape values); (4) existing vegetation and animal life; (5) meterology; (6) aesthetics; (7) public involvement. These studies should be used to determine whether a particular site can fulfill all of the requirements, present and future, beyond the strict engineering requirements (controlling components).

Farm Unit Density and Crop Production

Farm-unit density (farm population) within a given area in comparison to nonfarm population within the same area and the economic importance of the crops raised should be given consideration when the determinantion is made whether to buy the land needed for the land treatment development or to form a farmer-municipality cooperative. Where the land area under consideration for the land treatment plan is mainly farm-oriented, high-productive cropland, this should be strong persuasion toward the farmer-municipality cooperative method of providing the land needed for the treatment of wastewater. Low-yielding cropland intermingled with a high percent of nonfarm population may lead the planner toward the direction of acquisition into city ownership where perhaps chances would favor the municipality in increasing the yielding capacity of the land through municipality management.

Every effort should be made to avoid the reduction of good cropland at the benefit of lessening water pollution. This situation would be analogous to incineration of solid wastes where land pollution would be transformed into air pollution. Information should be obtained from state agricultural statisticians as to average crop yields. Only those lands with crop yields which fall below the national average for a given crop over the previous two to three cropping years should be included for further detailed investigation as a possible land treatment site. Some existing crops are specific as to soil requirements and should be kept in mind when land treatment sites are analyzed.

Nonfarm Land Uses

The farm density study should be followed with a land-use study of the surrounding area. Generally this study should encompass a 5-mile (8-km) wide area surrounding the potential treatment sites. A determination should be made of the location of residential, commercial, recreational and other highly sensitive uses, the direction and magnitude of future growth, and zoning restrictions of the land tentatively selected for treatment use and of the 5-mile (8-km) surrounding area. Generally county and regional planning agencies will be able to supply this information.

Topography

Detailed land form and feature studies are used to set out those landscape features that should be preserved and those features which are an asset or beneficial to the operation of the project. These features include wooded areas or knolls that might aid in breaking winds or that might conceal unsightly parts of the land treatment area, such as maintenance areas. These features could be set out with brief descriptions. Examples of special land forms are wooded landscape, rocks and ledges, brooks and ponds, open pasture, and ground surface undulations. Because spray irrigation methods can be used in designing land treatment systems, extensive grading of the site would not be needed. Above-ground sprinkler irrigation lines can be employed over rough terrain, thus avoiding destruction of natural land forms or the farm scenic values.

The planner of land treatment areas must recognize in the site those landscape values which are important to preserve. The planner must be able to apply and develop the land facility intelligently in studied relationship to the landscape features. The planned project must be ideally related to the best features of the site, and where developments must be within view of neighbors or casual visitors, it must not intrude on the visual senses of the viewer.

Vegetation and Animals

A complete inventory should be made of the vegetation and animals inside the study-site boundary. Trees, shrubs, vines, perennial, biennial and annual flowering plants, fungi and leek should be included as well as all animal life.

The listing should be made as to the places where each was found and relationships to other plants and microclimatic conditions as well as relationships to animal species found in the vicinity. These studies are necessary to determine the effects of the effluent on sensitive species and, if the site is used, to later determine changes that may occur in plant and animal communities due to the new use of the plant and animal habitat. The planner needs to know whether certain animal species and plant species can be displaced without endangering their existence. The increase in moisture will change the ecological balance, and this change should be taken into account in selecting the site.

Meteorology

Some knowledge of those conditions that affect the weather of the site should be set out. The planner needs to know expected wind velocities, direction and duration. The planner also needs to know something of the secondary micrometeorology of the region in which the study site is located, but more important is the need to know the details of the micrometeorology. A specialized area of micrometeorology is microclimatology and is defined[1] as the detailed study of the climate over a small area as determined by location and environment. Places only a few hundred feet apart may have radically different climates because of differences in factors such as elevation, soil and soil cover. Weather station records generally will not help in this study since the purpose of the weather bureau[1] is to determine representative air-mass characteristics as free as possible from the variations that result from the microclimate. On the other hand, the planner is interested in just those variations.

Some of the aspects of microclimatology and the reasons for climate variation in short distance are cold-air drainage and topographic effects on the wind. The nature of the earth's surface, *i.e.*, rough, smooth, bare of vegetation, grass cover, short crops, forest cover, etc., is of particular importance in determining the local vertical distribution of wind and temperature. Knowledge of wind variations and why they occur in a certain area can aid in preventing sprayed effluent from escaping beyond the site boundary.

Aesthetics

Siting land treatment developments should be approached from the standpoint of meeting more than just engineering requirements. It should involve more than a system to dispose of wastewater effluent. It should be as compatible with the environment as it is with the daily flow rates from the municipality served or the wastewater treatment plant. It should be ecologically restrained and humanly fitted in with the people it serves and protects. It should not encroach unduly on the rights and interests of others. Alternative studies of various sites would aid in this endeavor. If the proposed land treatment development takes these rights away from others (land owners–people not served) and gives them to people remotely related (those being served that live in another region or watershed), then some compensation or adjustment other than fair market value or the Uniform Relocation Act is in order. Perhaps a more expensive scheme would be required to maintain the balance of human rights and enjoyment under a system of government where these rights are guaranteed. Above all, the site should meet all of the requirements to renovate the wastewater to the degree that the water would be safe for use in the immediate area or at more distant points where the ground water with which it mingles surfaces as a stream frequented by wildlife or used by an industry which requires water with a high degree of purity.

Public Involvement

Early public involvement should be one of the very first studies made. Public relations cannot be overemphasized for the success of the land treatment facility. In order to achieve complete understanding by the general public, citizen organizations and state and local governmental elected and appointed public servants should be given complete information of what land treatment is all about. How this involvement is brought about and the extent of the involvement of the taxpayer should be a decision of the taxpayer himself. He should have available to him at all times, in a form he can understand, the complete technical information as it becomes available.

Implicit in this concept of the public decision-making process is the view that planning facilitates improved public actions by relating objectives to programs in a systematic way and by assessing costs and benefits in light of specific objectives. The specialist–the planner–does not decide what is best; instead, his expertise is utilized to determine the effects of alternative courses of action. The creativity of the planner will be reflected in his ability to develop programs from which the public will receive maximum benefit.

A formidable problem can arise when the planner attempts to convey to the public and its representatives a good understanding of what the effects of a given program will be. This problem has not been helped by those experts who have assumed that because they know a great deal about wastewater management, they also know what is best for society.

Another problem is that in every wastewater treatment situation an enormous number of alternatives might conceivably be considered. How can the choice be narrowed without the expert making the decision and substituting his value preferences for the value preferences of the individuals affected?

There are many things that need to be done to resolve these problems and to improve public decision making in the wastewater field. The ability of the expert to communicate with the public with regard to the technical aspects of wastewater projects should be improved. More attention should be given to techniques for reducing the alternatives to a small enough number for public consideration while maintaining a sufficient number to reflect the range of choice. Of utmost importance is the need to enrich the opportunities for constructive participation in public decisions by individuals and groups having differing objectives and value preferences.

Above all, it should be recognized that the public interest can be served only by a complete airing of the differing objectives and preferences that people hold. A decision-making framework that will quickly achieve consensus and avoid controversy is not the goal. Instead, a framework should be sought in which pressure groups and agencies reflect the full range of views and preferences which exist in the society and forcefully marshall and present their programs.

Siting Techniques

Planning for land treatment developments has to be accomplished by a group of scientists with diverse backgrounds. Areas of special knowledge and skills include agriculturists with expertise in agronomy, soils, agricultural engineering, economics and maintenance and operation; engineers with knowledge of geology, hydrology, wastewater treatment, soils, economics and mechanical engineering; planners; ecologists; landscape architects; microbiologists; public servants; land owners and the taxpayer.

None of these special skills should be left out in the planning phase of the land treatment development. To do so could mean disappointment and even failure sometime into the future as was experienced in so many of the areas no longer able to serve a most needed purpose.

The siting techniques used for land treatment facilities should combine traditional planning and design techniques for municipal wastewater treatment installations—comprehensive sewage plans—with studies of how to protect and preserve the amenities of those who will live nearby, *i.e.*, studies of land forms and features that should be preserved, land uses, aesthetics and ecological values to be protected. The combined information could then be applied to four siting techniques: (a) buffer area, where a certain size fringe area would be left undeveloped; (b) isolation, where the entire installation would be placed in the center of a large tract of land all of which would be under the control of the municipality or regional wastewater system manager—this technique would be similar to a watershed area; (c) natural resources preseve, an area which would be used jointly for regional or large metropolitan park purposes, wildlife preserve, forest preserve, flood plain preserve, industrial isolation preserve (atomic energy and other high-risk air pollution industries), open space preserve and land treatment of wastewater; and (d) a farm-municipality cooperative where farmers of an area would join a cooperative farm operation in which each farmer would agree to utilize a certain amount of wastewater effluent for supplemental irrigation of certain crops during the year.

Buffer Area

This siting concept can be planned so that the goals of protecting neighboring property and aesthetics of the surroundings can be accomplished over a long period of time. Even though there is little reason to believe that development will ever reach the site under consideration for land treatment of wastewater effluent, a plan for protection of the integrity of the existing landscape and the amenities of the area should be part of the development plan. First of all, there is no such site as a typical site. Each site will have to be treated individually. The concept of screening, however, is applicable to each and every site. Screening should provide protection and at the same time be attractive and add to the overall scenic character of the countryside. Screening of spray irrigation plots should be accomplished in the most casual manner and still provide a buffer against wind-carried spray from the site to areas beyond where the pathogens could be deposited indiscriminately onto human habitat or cause hazardous roadway conditions. Possible screening techniques are shown in Figures 7, 8 and 9.

The buffer-area technique offers a very versatile approach to the siting problem where lands are acquired at the inception of the plan for future expansion. In this technique, initial use of the site could take place in the center, particularly where holding ponds are provided on the site from which the irrigation pumping operation is begun. The distribution lines from the ponding area would encircle the pumping station and expansion could proceed in an orderly manner in an outward direction from the center of the site toward the site boundary. Thus the boundary area of the site would not come into use for some time after initial installation. This would allow for vegetation to grow in the buffer area as might be required prior to use of that part of the site according to the size of the site, the existing development beyond the site and the potential use of the area beyond the irrigation site boundary.

Planning for protection of the area beyond the site boundary should consider the possibility that residential development will occur and plans should be set forth accordingly when there is no definite land-use pattern established. Of course, when wooded areas exist along the site boundary, these should be retained and protected to perform the filtering and screening desired. Wooded areas near the site boundary should be used to the best possible advantage, such as placing the sprinkling lines on the windward side which will permit the sprayed effluent to be filtered through the foliage of the tree canopy, thus reducing or eliminating the amount of spray that might escape the land treatment site. Such vegetation areas should have a minimum width of 100 ft (30 m) for highways and should contain about 50% evergreen trees for effective winter season filtration. Height of tree stand should be a minimum of 20 ft (6 m). The filtering area should be 1000 ft (300 m) in width for health protection of housing developments of towns.

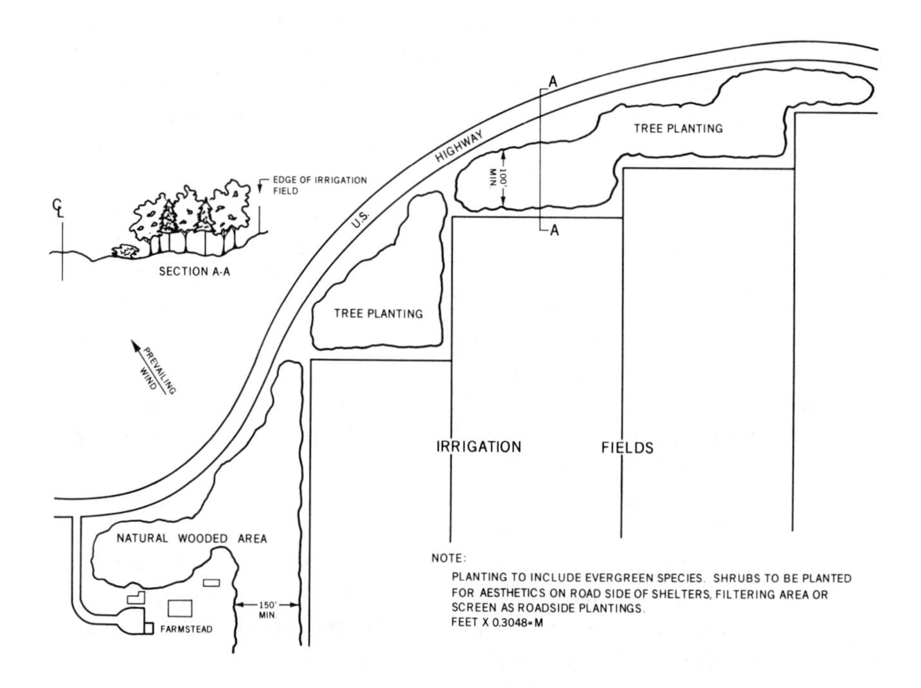

Figure 7. Screening technique involving a major highway and farmstead.

Isolation

As the term isolation implies, selection of an area that is far removed from present development as well as development potential for the future is the goal for this siting technique. Obviously, it would be most difficult to use this method in those areas of the United States with high population concentrations, such as the strip from Boston to Norfolk, Virginia, and from Milwaukee, Wisconsin, to Pittsburgh. This technique would be most applicable to small-size municipalities or regional systems where a small population is to be served. In this latter case, several sites would be used in order to achieve isolation. Only minor site work for aesthetics would be required for this method.

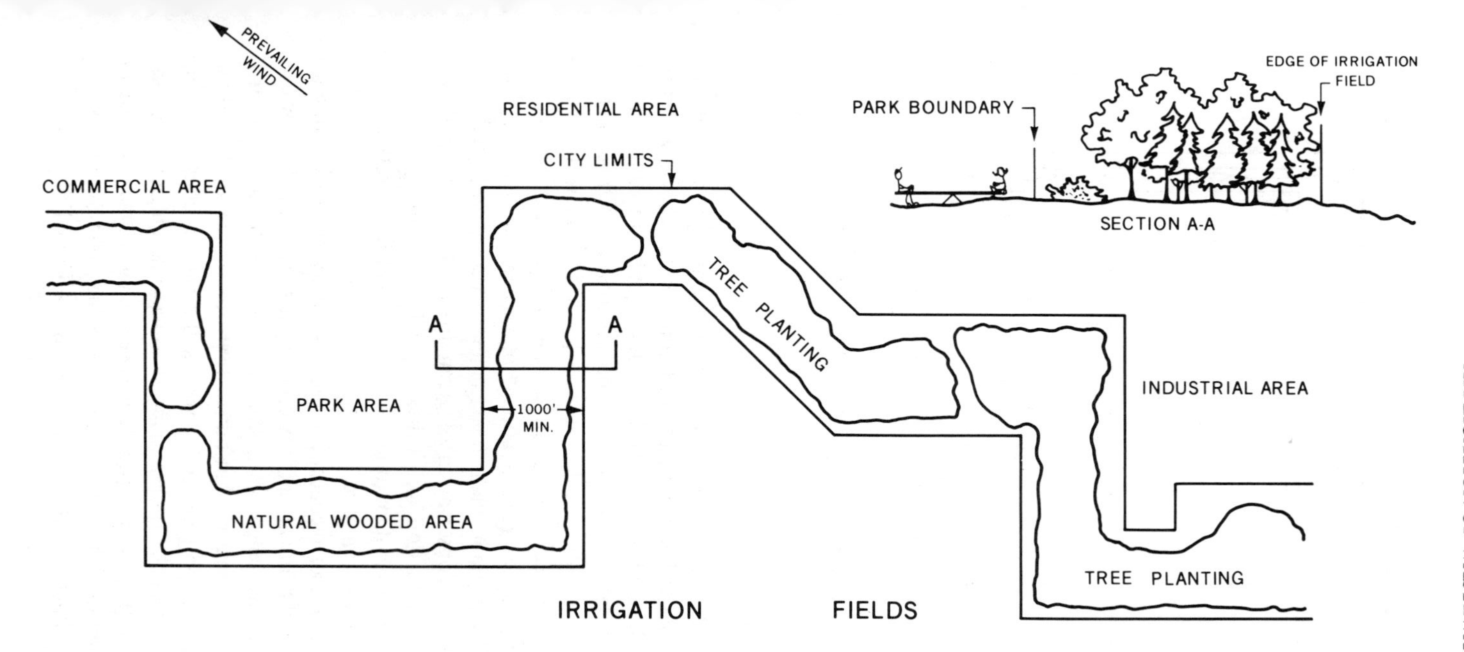

Figure 8. Screening technique involving town or city development.

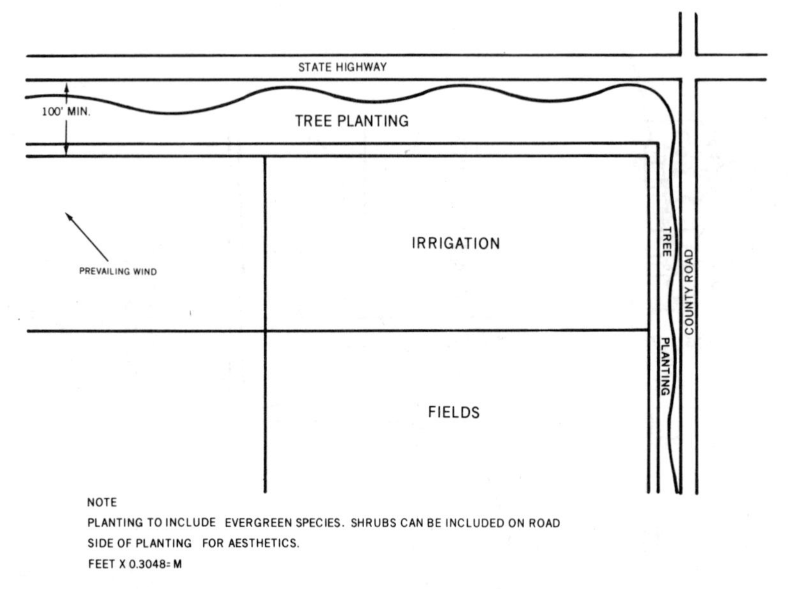

Figure 9. Screening technique for a state highway and county road.

Natural Resources Preserve

The most complex plan would be required for this technique involving land treatment of wastewater effluents on areas within the total site. Conservation practices would also be applied here. With the development of atomic energy for power production, large tracts of land close to concentrated populations would be required. The low population zone [2400 persons within a 3-mi (4.85 km) radius of the fence] required for atomic energy power plants could provide an excellent site for land treatment of wastewater effluent. This could be combined with flood plain management. Also to be included are solid waste landfills and conservation of green space. Heavy air pollution industries (aluminum reduction and others) could be included in this site together with the provision of outdoor recreation facilities in safe parts of this huge resource preserve complex.

The natural resource preserve concept is upon us. Many people are thinking about the concept[1] and many are already applying it to their needs.[1] Wasteful use of land is just as dangerous as careless use of water and air. It is simply a matter of which one we run out of first, for we cannot do without any one of these life support systems. We, as the total occupants of this limited earth, have played the foolish game of using something one time and disposing of it. It is the disposal part that is returning to haunt us, for we rapidly have poisoned our air, water and land at a high rate by our disposal technology. It is estimated that our land area in the United States is being converted to nonagriculture uses at the rate of 2 million ac each year, excluding surface mining, which in 1970 was using another 100,000 ac annually.[1] Of the 2 million ac, 580,000 ac are conveted to nonresource-protection use, such as urban development, airport and highways. We contaminate our air at the rate of 33.9 (sulfur oxides), 26.5 (particulates), 147.0 (carbon monoxide), 34.7 (hydrocarbons), 22.7 (nitrogen oxides) millions of tons per year as of 1970,[1] and the most haunting and visible destruction is that of our water supply, where we now have nearly six times as much pollution in our rivers, streams and lakes as 60 years ago.[1]

Farm-Municipality Cooperative

Not all of our farm land in the United States is high-yielding cropland. It is the low-yield, marginal soils where the cooperative technique could be most successful. Pasture land with well-drained soils could also be given a boost with irrigation of wastewater effluent. This technique requires that large blocks of land be within one border or nearly so for most efficient operation and maintenance. The size of these blocks of land is directly proportional to the amount of population to be served (size of municipality and daily wastewater flow). Soil, climate and type of crop have to be fitted in with any successful farm operation. Adjustment of the type of crop the individual farmer might grow in this cooperative adventure can result in an increase in net profit when the nutrients in the wastewater are properly utilized and taken advantage of. The buffer areas described above would apply to this siting technique.

REFERENCE

1. Hartman, Willis J., Jr. *An Evaluation of Land Treatment of Municipal Wastewater and Physical Siting of Facility Installations,* U.S. Department of the Army (May 1975).

SECTION III

INTERACTIONS OF SLUDGE, SOIL AND CROP PRODUCTION

8

NEW FEDERAL PERSPECTIVES ON THE LAND APPLICATION OF SLUDGE

E. Claussen
Office of Solid Waste
U.S. Environmental Protection Agency
Washington, D.C.

Sludge management is now entering a period of transition—transition from an era in which disposal practices were randomly selected and largely uncontrolled, to a period where environmental considerations and controls will play an increasingly important role in determining the ways in which sludge will be managed.

This trend toward more careful management can be seen in many state environmental agencies, where public officials are currently deliberating over what controls and techniques are acceptable for sludge disposal in their states. It can also be seen at the federal level, where two recent events are helping to set the direction for future U.S. Environmental Protection Agency (EPA) policy and programming on the land disposal of sludge. These are the approval of the EPA Residual Sludge Working Action Plan and, more important, the passage of the Resource Conservation and Recovery Act of 1976, PL 94-580.

PL 94-580, the new solid waste legislation, provides for the establishment of three major programs: (1) a hazardous waste program that provides a "cradle-to-grave" system for managing hazardous wastes; (2) a land disposal program that establishes "acceptable" practices for the land disposal of non-hazardous wastes; and (3) a resource conservation and recovery program. The first two of these programs are likely to affect wastewater treatment sludge land application and will therefore be the focus of this paper.

Before getting into specifics, however, it is important to look briefly at three of the definitions provided in the act:

1. Solid waste is defined as "any garbage, refuse, sludge from a waste treatment plant, water supply treatment plant, or air pollution control facility . . . "
2. Disposal is defined as "the discharge, deposit, injection, dumping, spilling, leaking, or placing of any solid waste or hazardous waste into or on any land or water so that such solid waste or hazardous waste or any constituent thereof may enter the environment or be emitted into the air or discharged into any waters, including ground waters."
3. Hazardous waste is defined as a solid waste that may "pose a substantial present or potential hazard to human health or the environment when improperly treated, stored, transported, or disposed of or otherwise managed."

These definitions clearly indicate that anything in the Act that refers to solid waste or solid waste management also refers to wastewater treatment sludge and wastewater treatment sludge management. Furthermore, the definition of disposal, which includes placing waste in or on any land, clearly encompasses both sludge land application and sludge disposal on a landfill. The definition of hazardous waste also leaves open the possibility that some sludges, like some solid wastes, may be hazardous and may, therefore, be covered under the hazardous waste control program of PL 94-580 (Subtitle C).

This hazardous waste disposal program requires that EPA (1) define what constitutes hazardous waste, (2) establish reporting requirements for those who generate and transport hazardous wastes, and (3) establish a permit system for those who treat, store or dispose of hazardous waste. The program is to be implemented at the state level, although EPA has enforcement authority in the event that a state does not provide for implementation.

The key to the program lies in the definition of what constitutes a hazardous waste. Section 3001(a) of the Act is somewhat helpful in this regard, stating that the criteria should take into account "toxicity, persistence, and degradability in nature, potential for accumulation in tissue . . . " If we assume that the definition will be based on toxic levels of various contaminants, the extent to which wastewater treatment sludge will be affected is entirely dependent on the extent to which a particular sludge contains the substances of concern. Most wastewater treatment sludges do contain some amount of such substances as cadmium, lead, mercury and PCB, and these substances are likely to be among the pollutants used in defining what constitutes a hazardous waste. However, it is presently anticipated that few wastewater treatment sludges will contain sufficient concentrations of these substances to render them hazardous. For those that would be considered hazardous, however, the following procedure would be envisioned:

1. The wastewater treatment plant would have to notify EPA or the state in which it is located that it generates a hazardous waste, and it would be required to keep records and provide reports on this waste.

2. If the waste is transported, the site would have to have a permit to treat, store or dispose of hazardous wastes and the shipment of the waste would be controlled by a manifest system.
3. If the treatment plant handles its own disposal, a permit for treatment, storage or disposal would be required.
4. Both landfilling and land application constitute disposal, and the permit conditions would be addressed to the ground-water protection, surface runoff, plant uptake for food crops and air emissions offered by the site.

It is obvious from the foregoing that any wastewater treatment sludge that is defined as a hazardous waste would place an additional burden on the plant operator. He would be very limited in his ability to turn the sludge over to another party for either landfilling or land application and would probably have to either clean up the sludge or provide for his own permitted treatment and disposal. However, it should be stressed again that the likelihood is that only a small percentage of municipal sludges would fall under the hazardous waste provisions of PL 94-580.

Most sludges would, on the other hand, fall under the Act's land disposal program, a program that deals with nonhazardous solid wastes. This program has three major facets: criteria for acceptable and unacceptable disposal facilities, guidelines for solid waste management, and state planning for solid waste management.

The criteria to be developed under section 4004 are the key to the land disposal program. These criteria will be used to determine which disposal facilities are acceptable and which are unacceptable. It is presently anticipated that the criteria will not be either waste or disposal-practice specific. Thus, "good" land application practices would also fall under the "acceptable" definition, and "bad" land application practices would be considered "unacceptable." These criteria are extremely important since any unacceptable solid waste practice is prohibited by the Act unless it is covered by a schedule for compliance that has been approved by EPA.

Implementation of these criteria would rest totally at the state level. The state agency may choose to institute a permit program to control the implementation process, and may require that records be kept at the treatment plant to insure that the criteria are met. Some testing and monitoring will likely be required, although this will depend upon the specific criteria that are to be developed.

The land disposal provisions of the Act also require EPA to publish guidelines for the development of state plans and the provision of financial assistance to state programs. To be approved for financial assistance for state planning, the Act requires that state plans must: (1) prohibit the establishment of unacceptable facilities, (2) provide for the closing or upgrading of existing unacceptable facilities and (3) provide a schedule for compliance with the criteria within a reasonable time period, not to exceed five years from the publication of an inventory of unacceptable disposal facilities that EPA will publish one year after the promulgation of the criteria.

EPA is also required to publish solid waste management guidelines, including those which describe disposal practices that can be used to meet the disposal facility criteria. These guidelines are not mandatory for states, and should therefore be considered as advisory only. They are, however, mandatory for all federal facilities and certain grantees under the Act.

What does all this mean in terms of the land application of wastewater treatment sludge? First, it is presently envisioned that the criteria will both establish "environmental limits" where possible, and require the use of best practicable technology where appropriate. For example, a limit on PCB concentrations in sludge to be spread on pasture land might be specified, or sludge application to vegetable crops that are to be eaten raw may not be permitted. Additional possibilities include specifications for soil pH and heavy metal loading rates. These are, of course, merely sample suggestions at this point, and are not EPA-approved criteria.

Descriptive operating methods will be covered in the guidelines to be developed under Section 1008. These guidelines will describe ways of protecting health and the environment, or best management practices, and they will be waste specific. Initially, we expect to write guidelines on municipal solid waste land disposal and municipal sludge land disposal. These are wastes on which information is presently available and where a definite and immediate need has been identified. Items such as recommended monitoring techniques and application methods may be included in these guidelines.

In closing the discussion on PL 94-580, it is important to point out that implementation at the federal level with regard to sludge land disposal is limited to: (1) regulations for hazardous waste disposal, (2) criteria for determining environmentally sound land disposal practices for nonhazardous wastes and (3) promulgation of guidelines and provision of technical assistance to states and localities in the criteria implementation process. The law assures enforcement for the hazardous waste program at either state or federal level. There are, however, no federal enforcement provisions for the land disposal program.

Fitting these new solid waste programs into a framework for sludge management within the entire Agency is, needless to say, not an easy task. The Agency is, however, fortunate in having formed the Residual Sludge Working Group over a year ago to assist the Office of Solid Waste in "coordinating the development of Agency policy, planning and guidance in the area of utilization and disposal of sludge."

The first major activity of the Residual Sludge Working Group was to prepare an Action Plan for Residual Sludge Management. This plan was designed to identify the constraints to Agency sludge management and to propose particular action items for the resolution of the problems identified. The plan was signed on October 19, 1976, by John Quarles, then Deputy Administrator of EPA, and work is now underway on the immediate action items proposed in the plan.

Perhaps the most important project to be developed is a Strategy Paper for Municipal Sludge. This is particularly significant in light of the passage of PL 94-580, a piece of legislation that provides for the development of several major sludge management programs. The purpose of the Strategy Paper is to provide a clear indication of the direction of all the Agency's programs for sludge management whether they impact oceans, air, water or land, to tie together all Agency efforts into a cohesive whole. The Strategy will thus alert the public to EPA's plans and programs in sludge management, indicating our overall objectives and implementation principles. It is presently anticipated that the paper, when complete, will contain the following:

1. A discussion of the nature of the sludge management problem.
2. A discussion of the legislative guidance on sludge management and existing controls at the federal and state level.
3. A set of goals, objectives and priorities.
4. A discussion of program management responsibilities, with emphasis on the roles and capabilities of the federal government, states, the local government and the public.
5. A discussion of major issues such as risk assessment and cost benefit relationship.
6. A series of recommendations for implementation of the strategy.

The strategy will be prepared by a working group composed of individuals within EPA. However, we will be sending drafts to a large number of interested parties so that we can obtain their comments prior to giving the general public a chance to comment. This review group will include members of public interest groups, agricultural organizations, states, localities, technical associations, universities and other interested federal agencies. Our current schedule for development of the Agency strategy includes a completed first draft by the end of June 1977, followed by a series of public hearings through September and a final publication date in October 1977.

As the foregoing doubtless indicates, EPA is now involved in getting its new sludge programs underway. The results of these activities will likely mean that some new directions will be set, and that some changes in current practices will be made. It is important to point out that decisions on how sludge should and will be managed will not be made in the dark. All ongoing sludge work, in EPA, other agencies, state and local governments and the research community will form the cornerstone of EPA's sludge policy and programs. If sludge is to be managed in an environmentally sound manner, we must rely and build on the expertise of those who have worked on these issues for many years. And we must expose our thinking to the general public, soliciting their comments on everything from our objectives to our conclusions. Having the authority to manage a problem is only one part of the equation. Working cooperatively with all concerned will yield the only viable solution.

9

GUIDANCE ON LAND APPLICATION OF MUNICIPAL SLUDGE–EPA CONSTRUCTION GRANTS PROGRAM

R. K. Bastian
U.S. Environmental Protection Agency
Office of Water Program Operations
Municipal Technology Branch
Washington, D. C.

INTRODUCTION

No matter what technologies are applied, there is nearly always something left over as a result of wastewater treatment. While treating sewage to acceptable discharge quality levels, various residuals or by-products are inevitably produced, be it sewage sludge from conventional physical/chemical and biological treatment plants, algae in pond systems or crops from land treatment systems.

The management of the conventional wastewater treatment process residuals (sewage sludges) is a twofold problem. The sludge must be disposed of or reused to complete the wastewater treatment efforts, and it must be done in an environmentally acceptable manner.

The requirements of the Federal Water Pollution Control Act Amendments of 1972 (PL 92-500) emphasize the need to employ cost-effective and environmentally sound waste management technology. At the same time its requirements for improved wastewater treatment will result in a nationwide increase in the production of sewage sludges from conventional sewage treatment processes. The current estimated 5 million dry tons of sewage sludge produced each year may more than double as a result of upgrading the nation's publicly owned treatment works to meet secondary treatment standards.

Although this increase in sludge volume to be managed will be faced by treatment plant operators across the country, the major problem areas are the large cities, especially those facing a phase-out of their current ocean disposal activities. The larger metropolitan areas simply have greater volumes of sludge to manage and in many cases run into problems obtaining sites and local approval for implementing land-based alternatives. We should not, however, disregard or underplay the sludge management problems being faced by the many smaller communities across the nation.

Current estimates indicate that as much as 50% of the municipal sewage sludges produced in the U.S. are applied to the land. While these estimates include a surprising volume of sludge that is simply stored in lagoons with no identified method of ultimate disposal, most of the land-applied sludges are utilized as soil amendments and supplemental fertilizer or are disposed of by landfilling. The scheduled phase-out of ocean disposal, increasing production of sludge with higher levels of wastewater treatment from conventional treatment processes and escalating fuel costs could change this picture dramatically over the next few years.

Since 30 to 50% of the construction costs of individual treatment plants may be clearly associated with sludge management facilities and EPA may provide as much as 75% of the capital funding required for these facilities under the Construction Grants Program, the Office of Water Program Operations (OWPO), which manages the EPA Construction Grants Program, has prepared the technical bulletin, "Municipal Sludge Management: Environmental Factors."* This bulletin, published in proposed form for public comment in the June 3, 1976, *Federal Register*, was prepared to assist the EPA Regional Administrators and their staffs in evaluating grant applications for construction of publicly owned sewage treatment works under Section 203(a) of PL 92-500. The document, which is now in the final agency approval process and expected to be published in approved form during June 1977, will also provide designers and municipal engineers with general information for selecting optimal sludge management options, but should not be construed to be a regulatory document.

HISTORY OF THE SLUDGE TECHNICAL BULLETIN

The sludge technical bulletin, as it is commonly referred to, has taken over four years to develop (involving numerous draft versions) by an agency

---

*By the time the proceedings of this conference are printed, single copies of the approved sludge technical bulletin should be available by writing to the General Services Administration (8 FFS), Centralized Mailing Lists Services, Building 41, Denver Federal Center, Denver, Colorado 80225. Please indicate the title of the publication and MCD-28 (EPA 430/9-76-004).

workgroup with substantial assistance provided by individuals from the CEQ, the USDA, the FDA and the Department of the Army. The original sludge workgroup was initially formed in 1973 to develop an agency "policy statement" on municipal sewage sludge disposal in response to regional requests related to implementing alternatives to ocean disposal. Several versions of draft agency policy statements were prepared by the workgroup but failed to gain agency management approval, mainly due to controversy over the position to take regarding agricultural uses of municipal sewage sludge. After thorough coordination with concerned agencies, the workgroup then prepared proposed "guidelines" on acceptable sludge management practices in an effort to clarify the controversial issues concerning land-application alternatives directly. Again, however, the proposed guidelines (in several draft versions) failed to gain agency management concurrence, this time due mainly to the arbitrary nature of recommended control measures, toxic substance concerns and the lack of specific guidance on regulation of operating facilities.

After several years of not achieving agreement on an acceptable agency "policy statement" or "guidelines," the EPA Office of Water Program Operations proposed to issue general guidance for the Construction Grants Program in the form of the technical bulletin, "Municipal Sludge Management: Environmental Factors," in an effort to highlight the major environmental factors to consider when reviewing grant proposals involving sludge management practices. Following extensive coordination with federal and state agencies and interested parties resulting in many versions, the proposed technical bulletin was published for public comment in the *Federal Register* on June 3, 1976. Written comments on the proposed document were invited from interested parties. The official comment period ended September 1, 1976, but was informally extended. Requests to submit comments were received as recently as December 1976. Over 150 individuals or groups submitted comments including several very thorough reviews and recommendations. In addition, over 920 form letters were received.

Comments were submitted by a variety of sources including federal and foreign agencies, state regulatory agencies, environmental groups, municipalities/counties/cities, consultants and special interest groups, university research and extension groups and the general public. The comments received indicated both support for and opposition to a number of points made in the proposed technical bulletin. There were comments strongly favoring the general site specific evaluation approach, while others indicated a desire for either more latitude or more restriction in the guidance. Conflicting points of view were often noted between comments received on a particular topic, even when received from the same type of reviewers. In many cases, however, agreement was indicated by numerous commenters who favored or supported a certain position. The EPA has carefully considered all comments received and modifications have been made to the proposed document wherever there was clear and strong support for change.

The principal comments received on the proposed technical bulletin and responded to in developing the final document fell into the following major categories:

a. requests for public hearings on sludge management issues;
b. need to clarify the roles of FDA and USDA in the development and review of facility plans addressing sludge management;
c. need for establishing allowable levels of heavy metals and other contaminants in sludges applied to the land;
d. exclusion of certain land application projects (those less than 1 mgd), "dedicated" publicly owned or leased sites, expansion of facilities producing bagged products and certain demonstration projects from conforming with each of the recommendations and monitoring requirements as "projects of minimum concern;"
e. need for additional information to be included in the bulletin covering each utilization and disposal option;
f. general tone of the proposed bulletin toward land application alternatives, especially agricultural uses;
g. monitoring requirements;
h. importance of adequate pretreatment requirements to control industrial discharges of contaminants into publicly owned sewage treatment works that may eventually create sludge management problems;
i. stabilization requirements;
j. determination of appropriate application rates for sludges applied to the land;
k. need to define, or better clarify, the meaning of terms such as stabilization, foodchain and nonfoodchain crops, utilization and disposal, agricultural and nonagricultural uses, high/average/low sludge quality and application rates;
l. need for clearer guidance on when additional pathogen reduction techniques are required and what pathogens need to be controlled;
m. restrictions recommended for the protection of food products and agricultural lands; and
n. need for regulations, criteria and detailed guidelines to provide better control of sludge disposal and utilization activities.

The proposed bulletin has been revised to take into account principal comments common to many reviewers. An explanation of these revisions and a response to the comments received is planned for publication in the *Federal Register* and will be made available upon request.

PURPOSE OF THE SLUDGE TECHNICAL BULLETIN

The final bulletin addresses only factors important to the environmental acceptability of the major sludge management options (land application, incineration, landfill and ocean disposal) and does so in a general manner to allow a maximum flexibility in its interpretation to meet varying regional needs and site-specific factors. Detailed information on costs and cost-effectiveness analysis procedures, pretreatment guidelines and regulations and sample collection/preservation/analysis procedures, as well as in-depth reviews of the somewhat controversial potential environmental impacts of land application, are or will be covered in additional supporting documents.

The sludge technical bulletin is based on current knowledge and will be modified from time to time as new regulations are developed and additional information becomes available from current and future research, development and demonstration projects. The document emphasizes land-application alternatives, since no agency guidance has been issued on this subject in the past, and contains existing agency guidance (and in some cases regulations) that is already available on the other major options–incineration, landfill and ocean disposal. The technical bulletin was not written as a regulatory document, although many reviewers of earlier versions have viewed it as the basis of EPA policy and regulatory actions since the agency has not issued a formal comprehensive sludge policy, and the document was derived from past efforts to develop such a policy statement.

The final bulletin will be issued in order to give the regions general guidance, calling for site-specific evaluations rather than nationwide implementation of arbitrary values that may be too limiting under some conditions and too liberal under others. The technical bulletin contains specific recommended numerical values only when they can be substantiated by an adequate experimental and operational data base.

TECHNICAL BULLETIN CONTENTS

The bulletin is divided into two distinct parts: one addressing methods in which the sludge is utilized as a resource and the second on those methods not utilizing the sludge in a beneficial manner. Appendices are also provided that cover the preparation of environmental impact statements, ground water requirements of BPT, guidance for the land disposal (landfilling) of solid wastes, incinerator emission and performance standards, criteria established for ocean dumping of municipal sewage sludges and typical sludge analyses, including levels of various contaminants in municipal sludges under different circumstances. Requirements for implementation plans to reduce toxic materials and interim continuation permits are discussed in light of established ocean-dumping criteria. Discussion of incineration alternatives includes information on pretreatment programs, new source performance standards, destruction of organic compounds, adequate ash disposal and monitoring plans. Discussion of sanitary landfill criteria covers information on stabilization, EPA guidance for federal facilities, ground-water protection and monitoring plans. Land application information includes discussions and suggestions for acceptable stabilization techniques, site selection, application methods, application rates, crop selection and monitoring plans, as well as precautions for protection of public health, ground water and surface waters.

During the development of the technical bulletin, considerable disagreement surfaced on the topic of utilizing sewage sludge by application to agricultural lands. Although utilization of sewage sludge as a resource to recover nutrients and other benefits has been encouraged by PL 92-500 and various advisory groups, the workgroup members, as well as reviewers, identified conflicting opinions concerning the overall merits vs potential risks associated with applying sludges to cropland.

Possible adverse effects upon the human foodchain (*e.g.*, the potential for increasing human cadmium intake) has remained a major concern expressed whenever this practice is considered. The relative risks of applying sewage sludges to croplands, when compared to other routes through which these contaminants enter the human diet, have yet to be determined. In addition, detailed trade-off analyses comparing environmental impacts of land application vs incineration vs landfill vs ocean disposal have rarely been undertaken in a manner that provided meaningful results.

The possible immediate and long-term effects of heavy metals and other contaminants in sewage sludge when used in agriculture and their translocation into crops that may eventually enter the human foodchain are currently being investigated. The fact that most of these same "materials of concern" currently enter the human foodchain through many routes and that these routes are not due to agricultural uses of sludge complicates the issue even further.

The EPA has encouraged the FDA to develop recommendations for acceptable tolerances or limits for heavy metals and other contaminants in human foods and animal feeds. Until the necessary human food and animal feed quality standards are established, strict regulation of crops produced on sludge-amended soils and design criteria for new projects will out of necessity have to be based upon rather arbitrary values. There are currently no acceptable tolerances or limits available for control of most contaminants in sludges for human foods and animal feeds that can be applied to crops grown on sludge-amended soils for human or animal use.

While not a regulatory document, the bulletin will apply some control to the design and proper management of applying municipal sludge to the land—at least to the extent that eligibility for capital funds from the Construction Grants Program is concerned. The bulletin was maintained as short and precise as possible while providing general guidance for the Construction Grants

Program on the environmental factors to consider regarding the major sludge disposal/utilization alternatives.

In an effort to provide the detailed data base needed in making site-specific evaluations, the EPA is planning a series of in-depth state-of-the-art information reports on major topics of interest concerning sewage sludge processing and disposal/utilization alternatives. The first of this series, a report titled "Application of Sewage Sludge to Cropland: Appraisal of Potential Hazards of the Heavy Metals to Plants and Animals " (EPA 430/9-76-013, November 1976), was recently published to provide an assessment of potential effects on agricultural crops and animals by heavy metals in sewage sludges applied to cropland, as well as some consideration of possible groundwater and surface-water contamination. Additional information reports are scheduled on various aspects of sludge management by land application and other alternatives. Planned reports include (1) sludge, soil and plant tissue sample collection/preservation/analysis procedures, (2) relative risks to human health from applying sewage sludge to the land, (3) updating the sludge processing and design manual, (4) a checklist for selecting a sludge management alternative, (5) sludge transport and management costs, (6) monitoring of land application systems and (7) pretreatment guidance.

OTHER EPA SLUDGE GUIDANCE ACTIVITIES

Although we anticipate issuing an approved sludge technical bulletin in June 1977, the development of an agency-wide sludge management policy statement is still desirable. Several efforts are currently underway that may eventually lead to an agency sludge policy statement.

An "Action Plan for Residual Sludge Management" has been developed by an agency-wide workgroup and was approved by the EPA Assistant Administrator in October 1976. The action plan calls for the preparation of a series of issue papers and other tasks that will help lead to the development of an agency residual sludge policy statement. This plan addresses both municipal and industrial sludges.

The action plan has only recently been approved. The immediate action items under the plan have been initialed. However, the creation of a sound agency policy statement will involve additional long-range actions that will take many months to complete.

The recently passed Solid Waste Act Amendments, the Resource Conservation and Recovery Act of 1976 (PL 94-580), also address municipal sewage sludge management by including municipal sludges in the definition of "solid waste." Guidelines and criteria are to be established by the EPA and utilized by states in developing and implementing all solid waste management programs. In addition, "hazardous waste" criteria are to be established, as well as a permit program for the control of these "hazardous" solid wastes. Whether the hazardous waste regulatory activities encompass some municipal sewage sludges depends on the criteria developed to identify hazardous wastes.

The future role of the newly enacted PL 94-580 in municipal sewage sludge management is yet uncertain. The problems of overlapping responsibilities with existing legislation [*e.g.*, PL 92-500, Safe Drinking Water Act (PL 93-532) and the Toxic Substances Control Act (PL 94-469)] have yet to be clarified, and it is unclear as to the impacts of this legislation upon the Construction Grants Program. However, criteria and guidelines for use in state-implemented solid waste or sludge management programs can be expected to be developed by the EPA under the implementation of the recently passed PL 94-580. Such efforts should help to address the concerns expressed by many commenters for better controls on operating sludge disposal and utilization activities.

MAJOR CONCERNS AND IMPEDIMENTS

Land application of sludge remains a controversial area, but most of the discussion centers around application to agricultural lands, especially for human food and animal feed crop production. However, the use of sludges on nonagricultural lands (stripmine reclamation, parks, construction sites, etc.) or in nonfoodchain crop production (thus allowing for sod production, etc.) can be used without much controversy to recover the nutrient and soil-building value of municipal sewage sludge.

Our experience shows that public acceptance and health effects concerns are two of the major impediments to the use of sludge in agriculture. Many of the concerns that have been expressed relate to possible problems or potential risks that might occur under adverse situations, rather than known dangers or hazards that can be dealt with directly. The fact remains that land application of sewage sludges has been an accepted and largely unregulated activity for many years—without known significant negative health impacts. While possible unnoticed problems associated with these practices are being investigated and questioned, well-managed systems can be expected to continue their operation into the future without significant health problems. Improved dissemination of available information to the public, consulting engineers and governmental officials may help improve the picture but certainly will not remove all opposition or concern to the practice.

One area of current comment is the need for increased pretreatment activities to improve the quality of many sludges that could be applied to the land. Several studies have indicated that industrial dischargers to publicly owned treatment works are significant contributors of heavy metals and other toxic chemicals that contaminate municipal sewage sludges. However, both residential wastes and stormwater runoff are also known to contribute toxics to municipal sewage and sewage sludges. It is also important to recognize that pretreatment of toxic substances is not equal to destruction of these materials. Toxic substances removed by industry must be disposed of in sludges resulting from pretreatment by those industries or recovered for reuse. In this regard, it is necessary to weigh the potential impacts of discharging toxics into the environment from the different waste streams (*i.e.*, in sludge or effluents from either municipalities or industries).

In accordance with the requirements of PL 92-500 and several recent court rulings, the EPA has embarked on an accelerated program to develop (1) pretreatment standards for the most significant polluting industries and (2) standards pertaining to the discharge of designated toxic pollutants. A concentrated effort has been initiated to implement an effective federal pretreatment program to achieve compliance with the provisions of PL 92-500. Additionally, the agency has revised and issued pretreatment guidelines to assist municipalities in developing local pretreatment requirements.

WORK NEEDED

Major technical needs to support the Construction Grants Program involvement in municipal sewage sludge management activities actually boil down to the development of (1) the best design criteria and cost information for the available technology and (2)innovative technologies for future implementation. With the current phase-out attitude toward ocean dumping, we are dealing with providing guidance to the regions on the best available land-based technologies for sludge management. The development of an EPA residual sludge policy and establishment of guidelines and criteria for use by states in developing and implementing all solid waste management programs, including sludge management, can also be expected.

From our viewpoint, the work most urgently needed in the municipal sludge management field includes:

a. resolution of health effects issues;
b. continued emphasis on innovative technology leading to beneficial use and resource recovery;
c. breakthrough in public acceptance; and
d. information dissemination to both design engineers/operators and elected government officials.

It is clear that the same sludge management alternative will not suit every community's needs. We must continue to improve upon the available technologies and perhaps even devise new approaches to sludge management so that a community will have a choice to come up with the best approach to meet their own local needs. However, we must keep in mind that all sludge disposal/utilization systems are going to cost money and that this expense is going to remain an integral part of the price we pay for ensuring clean water.

Further information concerning the municipal sludge management activities of OWPO are presented in "Municipal Sludge Management: EPA Construction Grants Program; An Overview of the Sludge Management Situation" (EPA 430/9-76-009) and in other documents which may be obtained from the Office of Water Program Operations, Municipal Technology Branch (WH-547), U. S. Environmental Protection Agency, Washington, D.C. 20460.

10

ENVIRONMENTAL ASSESSMENT OF MUNICIPAL SLUDGE UTILIZATION AT NINE LOCATIONS IN THE UNITED STATES

A. D. Otte
Office of Solid Waste
U.S. Environmental Protection Agency
Washington, D. C.

K. V. LaConde
SCS Engineers
Long Beach, California

INTRODUCTION

The disposal of municipal sewage sludge is a major environmental problem of our time. Increasing quantities of sewage sludge must be disposed of in an environmentally acceptable manner because:

1. the Federal Water Pollution Control Act of 1972 (PL 92-500) mandated higher wastewater treatment removal efficiencies of various pollutants;
2. the shortage of natural gas and oil has resulted in increased costs for energy, causing many communities to minimize use of sludge-conditioning or incineration facilities; and
3. the present restrictions on disposal of sludge to the oceans have created problems for major metropolitan areas along oceans.

Estimates indicate that about 25% of all municipal wastewater treatment sludge generated in the U.S. is disposed of by land application.

The purpose of this project was to assess the environmental impact from sludge land application at existing sites throughout the U.S. The assessment included impact on food and nonfood chain crops, public health implications, aesthetics, plant response, soils accumulation of heavy metals and uptake by plants and the cost of disposal at each site.

SUMMARY OF PROJECT

Nine landspreading sites representing a wide range of sludge application rates and characteristics, cropping practices, soils, climatological and geographical conditions and population densities were selected for study. Sludge characteristics, plant uptake and accumulation of heavy metals in soils, past and current operating procedures, public attitudes and landspreading costs were assessed at each location.

The study consisted of two separate phases. Phase I involved identification and preliminary screening of potential sites for inclusion in Phase II. Site visits were made to 16 sites meeting the desired selection criteria. Phase II involved a comprehensive and much more detailed onsite investigation at 9 locations selected from 16 Phase I sites. Data obtained during these investigations provided the basis for detailed and comprehensive site analyses.

Phase I

A potential candidate list using the selection criteria shown in Table I was prepared from a combination of related EPA studies, available literature and consultant and contractor experience. The intent was to have relatively unknown sites investigated in order to further the knowledge of soil amendment through sludge spreading.

Table I. Site selection criteria.

2 ha or larger in area	Control site available
Received sludge for at least 5 yr	Information available
Sludge applied at 11 tons/ha/yr or greater	Not previously studied
Used for agricultural purposes	

Initial calls were made to 95 sewage treatment plants and/or communities in 24 states to determine their conformance to the selection criteria. If an individual site appeared to meet some or all of the criteria, additional information was requested.

The 16 study sites ultimately selected met, to the maximum degree, the selection criteria and responded more favorably to the questions. Visits were made to each of the 16 sites at which time follow-up questions were asked. In addition, sludge, surface soil and plant samples were obtained and analyzed.

Phase II

Site characterizations, site analysis and case study development were prepared in Phase II. The information sought was similar to that identified in Phase I, but it was more comprehensive and was expanded to include public opinion of sludge management practices at the case study locations.

In addition to this information an extensive sludge, soil and plant sampling program was performed. Twice during the contract period composite raw and stabilized sludge samples were collected. Treated and control plots at all nine sites were each divided into five sections, from which surface and subsurface soil core samples were obtained at approximately 30-cm intervals to a depth of 122 cm.

Individual samples were digested with nitric-perchloric acid and analyzed for Cd, Cu, Ni and Zn. Composite samples were prepared from the individual soil samples for each depth and analyzed for other parameters.

Plants and edible grain samples were obtained from all treated and control plots. Individual sectional samples were analyzed for total Cd, Cu, Ni and Zn. Composites for each plot were prepared from the individual samples and analyzed for 14 other parameters.

DATA ANALYSIS

The data analysis is presented in two parts: (1) general evaluation, in which the site data are collectively examined and compared to data obtained during related studies, and (2) individual study site analyses in which the sludge, soil and plant findings from the field sampling program at the nine sites were summarized and analyzed. The four sites presented represent the range of conditions found during this study.

General Evaluation

A summary of information on the sewage treatment plants supplying sludges to the case study sites is presented in Table II. The size of the treatment plants ranged from small to moderately large. The largest served a population of 135,000 with an average flow of 75,700 m^3/day and the smallest a population of 9200 with an average flow of 6000 m^3/day. The chemical composition of the sludges obtained from the nine treatment plants (Table III) falls within the broad range of those normally reported for sludges.[1-3] Median concentration results reported are reasonably representative of median concentrations for sludges in general except for Cd and Se. This is due to high concentrations of these elements at one of the treatment plants.

Table II. Chemical composition of sludges from all study sites.

Parameter[a]	A	B	C	D	E	F	G	H	I	Comparative Sludge Data[b]	
										Range	Median
Vol. Solids (%)	36.900	74.700	52.400	59.200	56.400	41.100	48.300	36.100	55.000		
NO_3-N	7.800	0.410	13.700	51.400	13.000	10.200	13.400	4.050	60.000	2-4,900	140
NH_4-N (%)	1.100	0.315	1.010	4.180	2.120	0.540	0.739	0.276	1.270	0.005-6.76	0.092
Org-N (%)	3.450	4.520	2.800	4.300	2.460	1.830	2.300	1.480	2.300	0.1-10.84	3.208
P (%)	1.230	1.710	3.550	1.740	1.340	1.560	1.570	1.410	1.330	0.1-14.3	2.3
K (%)	0.220	0.157	0.211	0.535	0.222	0.131	0.224	0.526	0.318	0.02-2.64	0.3
Na (%)	0.388	0.200	0.200	0.798	0.668	0.195	0.618	0.651	0.611	0.01-3.07	0.24
Ca (%)	2.780	2.650	3.140	5.920	1.390	6.250	4.760	2.580	2.260	0.1-25.0	3.9
Mg (%)	1.210	0.490	0.716	0.634	0.524	0.359	0.546	1.230	0.585	0.03-1.97	0.45
SO_4	348.000	81.700	105.000	111.000	66.000	10.000	10.000	8.600	118.000	0.6-1.1	0.8
Cl (%)	0.103	0.148	0.275	0.495	1.090	0.025	0.120	0.674	0.320	0.05-1.02[c]	0.29[c]
Ag	1.580	3.450	2.420	0.950	5.920	1.160	1.950	10.200	1.230	5-150[d]	20[d]
As	2.210	4.130	1.130	6.210	2.520	1.140	3.060	5.110	2.190	6-230	10
B	72.000	71.600	37.500	40.500	42.500	32.100	43.600	43.800	38.700	4-760	33
Cd	11.900	11.100	11.200	56.100	7.030	7.600	1.500	50.000	5.700	3-3,410	16
Cr (%)	0.113	0.015	0.224	0.243	0.127	0.038	0.399	0.201	0.041	0.0010-9.90	0.089
Co	21.300	6.390	6.960	9.070	6.170	9.680	16.200	24.900	24.100	1-18	4.0
Cu (%)	0.096	0.095	0.041	0.082	0.137	0.055	0.714	0.073	0.039	0.0084-1.04	0.085
Fe (%)	2.360	0.692	1.360	1.350	1.330	0.746	2.140	4.810	1.020	0.1-15.3	1.1
Hg	16.400	49.000	1.930	10.600	24.100	9.820	4.410	3.080	2.600	0.5-10,600	5.0
Mn (%)	0.060	0.048	0.025	0.045	0.022	0.020	0.032	0.987	0.047	0.0018-0.71	0.026
Mo	14.500	10.200	45.500	43.400	6.600	11.100	6.770	7.540	21.100	5-39	30
Ni (%)	0.014	0.005	0.003	0.030	0.002	0.006	0.050	0.008	0.016	0.0002-0.352	0.0082
Pb (%)	0.061	0.004	0.111	0.015	0.010	0.009	0.025	0.748	0.010	0.0013-1.97	0.05
Se	13.700	7.020	4.230	8.170	7.600	6.360	5.680	17.200	6.640	1.7-8.7[c]	2.7[c]
Zn (%)	0.177	0.135	4.580	0.331	0.119	0.147	0.215	0.719	0.074	0.0101-2.78	0.1740
H_2O (%)	91.300	85.600	95.800	98.300	97.500	87.800	93.800	91.000	96.000		
pH	7.500	6.300	7.800	8.300	7.800	7.500	7.300	7.500	8.200		

[a]All units in μg/g (oven-dry weight basis) unless otherwise noted.
[b]Results from analyses of more than 250 sewage sludge samples from approximately 150 sewage treatment plants in the North Central and Eastern regions, U.S.;[2] sulfur values are for total sulfur, not SO_4.
[c]Data of 42 sewage sludges from locations in England and Wales.[3]
[d]Data of sewage sludges from 16 American cities.[4]

Table III. Chemical compositions of sludges for all sites.

Element	Range (%)	Mean (%)	Median (%)
Vol. Solids	36.1-74.7	51.1	52.4
NH_4-N	0.276-4.18	1.28	1.01
Organic-N	1.48-4.50	2.83	2.46
P	1.23-3.55	1.72	1.56
K	0.131-0.535	0.283	0.222
Na	0.195-0.798	0.481	0.611
Ca	1.39-6.25	3.53	2.78
Mg	0.359-1.23	0.700	0.585
Cl	0.25-1.09	0.36	0.275
Cr	0.015-0.399	0.156	0.127
Cu	0.039-0.714	0.148	0.082
Fe	0.692-4.81	1.76	1.35
Mn	0.020-0.987	0.143	0.045
Ni	0.002-0.05	0.015	0.008
Pb	0.004-0.748	0.110	0.015
Zn	0.074-4.58	0.722	0.177
H_2O	85.6-98.3	93.0	93.8
	(μg/g)	(μg/g)	(μg/g)
NO_3-N	0.41-60.0	19.3	13.0
SO_4	8.6-348	95.4	81.7
Ag	0.95-10.2	3.21	1.95
As	1.13-6.21	3.08	2.52
B	32.1-72.0	46.9	42.5
Cd	5.70-1500	185	11.2
Co	6.17-24.9	13.9	9.68
Hg	1.93-49.0	13.6	9.82
Mo	6.60-45.5	18.5	11.1
	4.23-17.2	8.51	7.02
pH	6.3-8.3	7.58	7.50
Cd/Zn	0.0002-0.698	0.084	0.007

Statistically significant differences at the 90% confidence level (treated vs control) were observed for surface and subsurface soils in terms of Cd, Cu, Ni and Zn concentrations. The significant differences at the 90% level for surface soils are as follows (treated vs control):

Cd – 5 sites showed increases
Cu – 8 sites showed increases
Ni – 7 sites showed increases
Zn – All sites showed increases

The subsurface soil data showed that metals applied have remained in the depth of incorporation with the exception of Site A where Zn has migrated to the 122-cm depth. This suggests that potential ground water contamination from surface sludge spreading operations at the 9 sites is remote except for Site A.

Small but significant differences at the 90% confidence level were observed in most plant Cd, Cu, Ni and Zn concentrations as follows (treated vs control):

Cd – 6 sites showed increases
Cu – 5 sites showed increases
Ni – 5 sites showed increases
Zn – 8 sites showed increases

In general, however, the concentrations of metals reported were within ranges not considered hazardous to health or phytotoxic to plants except for the following sites:

Site A – Zn concentration in cheatgrass ranged from 340-455 μg/g (mean = 380 μg/g), a level considered phytotoxic to some plant species.

Site G – Cd concentrations in wheat grains ranged from 0.6-1.9 μg/g (mean = 1.3 μg/g). Based on an average daily consumption of 100 g of wheat flour, this wheat may supply approximately twice the maximum amount of Cd as recommended by the World Health Organization for human consumption.

Individual Study Site Analyses

Comparative information on each of the nine sites is shown in Tables IV-IX.

Table IV. Comparative STP information.

STP	Sludge Generated/ Capita/Day (kg)	Type of Secondary Treatment	Sludge Digestion Process	Estimated Industrial Contribution (%)	Disposal Costs ($/dry m.ton)
A	0.03	Trickling Filter	2-stage anaerobic	30	7.98
B	0.15	Activated Sludge	aerobic	10	128.07
C	0.04	Activated Sludge	anaerobic	25	26.92
D	0.07	Activated Sludge w/Kraus modification	2-stage anaerobic	15	19.58
E	0.11	Activated Sludge	anaerobic	65	79.08
F	0.05	Trickling Filter	2-stage anaerobic	15	22.28
G	0.12	Trickling Filter	anaerobic	30	19.04
H	0.47	Trickling Filter	anaerobic	25	26.48
I	0.45	Activated Sludge	anaerobic	67	17.65

Table V. Comparative farm information.[a]

Site	Surface Soil Type[b]	Soil pH Treated/Control	Treated Plot Crop	Crop Use	Depth to Ground Water[c] (m)	Sludge Application: Total Sludge Treated Area (ha)	Sludge Application: Rates (m.tons/ha) Annual[d]	Sludge Application: Rates (m.tons/ha) Total[e]	Sludge Application: Years of Sludge Spreading[e]
A	sandy loam	3.7/4.1	cheatgrass	nondairy cattle feed	1.5	50.6	28[f]	308[f]	11
B	clay loam	6.5/7.1	ryegrass	nondairy cattle feed	213	7.3	50	149	7
C	silt loam	6.5/6.6	alfalfa	nondairy cattle feed	18	22.7	7	116	17
D	silt loam	7.0/6.9	fescue	nondairy cattle feed	41	12.6	16	237	15
E	sand	5.5/5.4	soybeans	open market	23	56.7	16[f]	80[f]	6
F	silt loam	6.2/5.2	fescue	nondairy cattle feed	varies	21	22[f]	16[f]	9
G	silt loam	6.7/6.0	wheat	flour mill	21	18.5	30[f]	360[f]	12
H	clay loam	6.4/6.6	alfalfa	nondairy cattle feed	30	40.5	20	81	13
I	sandy loam	6.6/6.4	corn	distillery	5	14.2	65	326	5

[a]Control plot crops from all sites were the same as the treated plot except Site A where control crop was oats.
[b]USDA classification.
[c]All ground water depths reported as estimated static levels.
[d]1975.
[e]Through December 31, 1975.
[f]Estimated.

Table VI. Annual Loading Rates of Sludge and Heavy Metals.

	Annual Sludge Application Rate– Dry Weight (m.tons/ha)	Annual Metal Loading Rate (kg/ha)					Total Sludge Application Rate– Dry Weight (m.tons/ha)
		Cd	Cu	Ni	Zn	Pb	
A	28[a]	0.3	27	3.9	50	17	308[a]
B	50[b]	0.6	48	2.5	68	2	149
C	7[a]	0.08	2.8	0.2	311	7.5	116[a]
D	16[b]	0.88	13	4.7	52	2.4	237
E	16[a]	0.11	22	0.3	19	1.6	80[a]
F	22[a]	0.17	12	1.5	32	2.0	66[a]
G	30[b]	45	214	15	65	7.5	360[a]
H	20[c]	1.0	14.4	1.6	142	147	81[a]
I	65[b]	0.37	25	10	48	7.8	326

[a]Estimated.

[b]Annual amount computed from multiyear averages.

[c]1975.

FINDINGS AND CONCLUSIONS

Results of observations and interviews at each of the nine study sites characterized indicate that:

- Sludge management practices were found to be uncontrolled and sometimes inadequate at all nine sites. Shortcomings included poor record keeping of sludge distribution and lack of preinvestigative work to determine sludge-soil-plant compatibility and phytotoxicity.

- There were no associated human health problems reported by any of the individuals interviewed.

- The sludge spreading program was favorably assessed at each of the nine sites in the opinion of those interviewed, including regulatory and health officials, newspaper editors, farmers and sewage treatment plant operators.

- An economic analysis of the sludge disposal costs at each of the nine sites showed a range of $8.00 to $130.00 per metric ton (dry weight basis).

Table VII. Metal concentrations of treated and control soil composites[a]

		Cd		Cu		Ni		Zn	
Site	Depth (cm)	Treated (μg/g)	Control (μg/g)	Treated (μg/g)	Control (μg/g)	Treated (μg/g)	Control (μg/g)	Treated (μg/g)	Control (μg/g)
A	Surface	5.89	0.76	267	11.9	67.6	19.1	475	27.7
	20-46	0.96	0.63	13.3	10.5	17.1	13.6	80.4	25.1
	46-61	0.47	0.51	10.1	6.00	9.44	12.3	39.0	15.5
	61-91	0.33	0.40	6.93	6.56	12.5	13.4	27.5	15.3
	91-122	0.99	0.95	19.6	20.4	24.4	24.0	88.5	52.4
B	Surface	3.58	3.77	45.4	27.7	52.9	48.0	120	98.1
	30-61	3.79	3.75	29.1	27.6	50.3	50.5	93.4	94.2
	61-91	3.41	3.22	28.5	28.0	49.5	50.2	93.8	90.2
	91-122	3.53	4.11	28.5	28.1	50.2	52.8	90.7	90.9
C	Surface	0.86	0.87	16.0	14.6	24.7	24.9	190	49.5
	18-30	0.89	0.91	19.0	17.6	32.3	31.4	62.5	57.3
	30-61	0.77	1.08	22.2	21.5	39.0	42.3	65.1	75.0
	61-91	1.00	1.08	22.8	23.4	44.3	46.2	66.0	74.2
	91-122	1.10	1.06	23.0	24.8	41.8	44.5	67.8	71.3
D	Surface	1.17	0.73	10.5	23.6	28.9	19.9	108	36.8
	18-30	0.77	0.68	13.1	14.3	28.2	22.7	41.8	34.8
	30-61	0.82	0.70	15.7	12.0	47.6	24.4	48.8	39.2
	61-81	0.83	0.82	20.0	13.2	38.7	28.8	52.9	45.7
	81-122	0.77	0.89	16.1	15.0	35.8	34.4	57.6	48.0
E	Surface	0.42	0.46	17.4	7.6	14.2	15.2	49.3	31.3
	20-46	0.38	0.34	10.5	7.6	16.7	14.3	29.4	24.2
	46-61	0.35	0.48	11.3	12.2	16.1	18.3	21.8	30.2
	61-91	0.31	0.31	13.0	15.7	14.6	19.6	22.1	27.8
	91-122	0.44	0.32	18.5	13.8	14.4	16.3	24.7	19.5
F	Surface	1.33	0.83	26.3	9.8	24.8	21.5	109	50.2
	15-30	0.95	0.77	17.7	13.7	27.0	25.6	68.0	62.3
	30-61	0.69	0.78	18.2	16.8	39.2	37.2	55.0	64.0
	61-91	0.84	1.02	15.4	15.4	26.3	26.8	60.4	71.6
	91-122	0.63	0.68	15.4	15.4	35.4	36.4	60.1	78.0
G	Surface	12.9	0.92	30.0	10.6	25.5	19.9	88.0	49.2
	15-40	2.17	0.75	15.1	15.5	24.5	32.6	61.6	54.0
	40-61	1.05	0.80	19.1	20.9	34.6	37.8	57.0	60.8
	61-91	0.92	0.79	19.1	22.0	34.3	42.0	56.1	57.2
	91-122	0.83	0.84	22.7	22.7	38.5	44.5	59.1	55.3
H	Surface	0.52	0.29	21.8	11.5	26.5	20.6	113	55.0
	20-30	0.16	0.22	19.6	14.6	37.3	28.2	64.8	60.3
	30-61	0.25	0.28	21.0	19.9	44.9	38.0	67.0	66.9
	61-91	0.46	0.55	21.3	19.4	44.6	40.6	65.7	61.9
	91-122	0.60	0.51	19.7	18.5	43.4	37.7	63.9	57.6
I	Surface	4.37	0.78	185	20.3	521	30.0	625	80.6
	30-46	0.39	0.49	19.7	18.8	41.1	34.5	70.0	75.0
	46-61	0.41	0.49	17.9	15.7	31.9	32.4	57.0	61.7
	61-91	1.46	1.57	21.7	9.65	38.4	32.8	57.0	35.5
	91-122	1.75	2.00	19.4	5.91	42.0	32.1	51.3	22.3

[a] HNO_3-$HClO_4$ digestion, all data expressed on oven-dry weight basis.

Table VIII. Mean metal concentrations[a] in leaf tissue.[b]

Site	Plant	Cd (μg/g)	Cu (μg/g)	Ni (μg/g)	Zn (μg/g)
A	Cheatgrass	0.71*[a]	21.98*	17.40*	383.40*
	Oats	0.31	5.61	3.63	31.20
B	Ryegrass	0.67	11.36	9.01*	58.50
		1.41*	10.53	5.08	61.07
C	Alfalfa	0.88	8.24	7.37	40.54*
		0.85	9.22*	8.27	32.40
D	Fescue	0.34	4.63	4.38*	31.28*
		0.43	5.25*	3.51	24.66
E	Soybeans	1.38*	13.32*	13.34	108.68*
		0.83	10.46	10.56	57.30
F	Fescue	0.27*	7.29*	4.28	37.58*
		0.18	4.39	3.71	20.52
G	Wheat	0.92*	6.75*	3.38	29.76
		0.28	5.90	4.15	27.92
H	Alfalfa	0.74*	9.31	7.02	83.04*
		0.60	8.46	6.70	36.48
I	Corn	2.31*	12.26*	9.22*	58.98*
		1.43	10.87	5.78	40.04

[a]For each metal, the first value is from the treated plot; the second, from the control plot. An asterisk indicates a significant difference between the treated and the control with 90% confidence level.

[b]HNO_3-$HClO_4$ digestion, all data expressed on oven-dry weight basis.

Table IX. Mean metal concentrations[a] in harvested grains.[b]

Site	Plant	Cd (μg/g)	Cu (μg/g)	Ni (μg/g)	Zn (μg/g)
E	Soybeans	0.13*[a]	19.18*	4.06*	96.82*
		0.07	11.14	1.82	69.56
G	Wheat	1.26*	4.65	1.24	45.08
		0.20	6.17*	1.98*	57.52*
I	Corn	0.43	2.36	4.93	29.20
		—	—	—	—

[a]For each metal, the first value is from the treated plot; the second, from the control plot. An asterisk indicates a significant difference between the treated and the control with 90% confidence level.

[b]HNO_3-$HClO_4$ digestion; data on oven-dry basis.

- The microbiological findings were typical of those reported in other similar studies and revealed no finding out of the ordinary.

- Low levels of DDT and Dieldrin (200 ppb) were found in 32% of the sludge samples. Very low levels were found in 47% of the surface soils samples and in only 10% of the plant samples.

- PCBs were detected in significant quantities (5100 ppb) in 74% of the sludge samples. Similarly, 44% of the surface soil samples were found to contain PCBs in levels of one magnitude lower (300 ppb). Only 13% of the plant samples were found to contain detectable PCB levels.

- Repeated sludge applications have resulted in increased surface accumulations of Cd, Cu, Ni and Zn in the sludge-treated fields.

- Total and DTPA-extractable metal analyses suggested that metals have not moved beyond the depth of incorporation at most sites.

- Zn was the most mobile of the heavy metals studied.

- The possibility of heavy metals contamination of ground water as a result of sludge spreading at eight of the nine sites is remote.

- When the soil pH was maintained at 6.5 or above and the total metal additions of Cd, Cu, Ni and Zn were limited to the USDA Agricultural Research Service recommended additions (Table X), the increase of the metal in the leaf tissue or grain was minimal.

- A sludge having a low heavy metal content can have adverse environmental impacts if proper management practices are not followed.

Table X. Maximum cumulative sludge metal applications for privately owned land.

	Soil Cation Exchange Capacity (meq/100 g)[a]		
	0-5	5-15	15
Metal	(Maximum Metal Addition, kg/ha)		
Zn	250	500	1000
Cu	125	250	500
Ni	50	100	200
Cd	5	10	20
Pb	500	1000	2000

[a]Determined on unsludged soil using the method utilizing pH 7 ammonium acetate for a weighted average to a depth of 50 cm.

ACKNOWLEDGMENT

This study was supported through EPA Contract No. 68-01-3265 and was conducted by SCS Engineers, Long Beach, California.

REFERENCES

1. Page, A. L. "Fate and Effects of Trace Elements in Sewage Sludge When Applied to Agricultural Lands: A Literature Review Study," National Environmental Research Center, EPA, Cincinnati, Ohio (January 1974).
2. Sommers, L. E. "Chemical Composition of Sewage Sludges and Analysis of Their Potential Use as Fertilizers," Contribution of the Purdue University Agricultural Experiment Station, Journal Paper No. 6420 *J. Environ. Qual.* 6 (1977).
3. Birrow, M. L. and J. Webber. "Trace Elements in Sewage Sludges," *J. Sci. Food Agric.* 23:93-100 (1972).
4. Furr, A. K. *et al.* "Multi-Element in Chlorated Hydrocarbon Analyses of Municipal Sewage Sludges of American Cities," *Environ. Sci. Technol.* 10:683-687 (1976).

11

LAND CULTIVATION OF INDUSTRIAL WASTEWATERS AND SLUDGES

D. H. Bauer, D. E. Ross and E. T. Conrad
SCS Engineers
Long Beach, California and
Reston, Virginia

INTRODUCTION

Disposal of industrial wastes presents problems to the waste generator and/or disposal site operator, especially since these wastes are very complex and are generated in large volumes. In addition, industrial waste disposal problems are expected to increase. As a result, there has been intensified interest in application of wastewaters and sludges to land and the development of alternative land-disposal techniques.

Industrial wastes are typically disposed of in sanitary landfills or deep wells or by incineration.[1] Some industrial wastes are recycled, stockpiled, stored or simply dumped into the ocean.

The land cultivation method of industrial waste disposal is being more widely practiced as one alternative to conventional or unacceptable techniques. However, very few published data are available on the land cultivation method and its potential environmental impacts. Accordingly, the U.S. Environmental Protection Agency, Solid and Hazardous Waste Research Division, Municipal Environmental Research Laboratory, is sponsoring a comprehensive study of the land cultivation practice. Basic project objectives include:

> gathering and assessing available information relating to past, existing and planned disposal activities entailing land cultivation or industrial wastewaters and sludges;

- evaluating pertinent technical, operational, economic and environmental factors;
- determining the chemical composition and accumulation of heavy metals and toxic constituents by soils and plants on selected disposal sites where land cultivation is practiced; and
- preparing a land cultivation site conceptual design.

The project is currently underway and is scheduled for completion by the end of this year (1977). This chapter presents a progress report reflecting findings of the literature review, personal interviews and field investigations to date. Investigations at the five land cultivation case study sites are not reported here, since data are incomplete. This chapter is concerned mainly with industrial waste land cultivation. A companion paper on land cultivation of municipal solid waste was presented in March 1977 in St. Louis.[2]

LAND CULTIVATION–A DISPOSAL ALTERNATIVE

Soil is a natural environment for the deactivation and degradation of many waste materials through complex physical, physiochemical, chemical and microbiological processes.[3] Land cultivation of waste is a disposal technique by which wastes are mixed with the surface soil to promote these processes, particularly microbial decomposition of the organic fraction. This technique is also known by other names such as landspreading, soil incorporation and land farming. If managed properly, the process could be carried out repeatedly on the surface of a disposal site, thereby "recycling" the land used for disposal. Proponents of land cultivation claim that under the most ideal conditions the site could be returned to any other land use, including agriculture, after cessation of disposal activities.

Although ideal conditions are rarely met in the field, and despite the fact that literature on the environmental impacts, regulatory controls and waste types and characteristics in relation to land cultivation is scarce, the practice of land cultivation has enough promise and has had enough preliminary success that many industries are already land cultivating their wastes or planning to do so.

The suitability of an industrial waste for land cultivation depends on such characteristics as concentrations of chemical elements in soluble as well as insoluble forms, bulk densities of waste solids, pH, sodium and soluble salt contents, as well as inflammability and volatility.[4,5] Local climatic conditions can influence the viability of this disposal practice. The presence of vegetative cover (primarily forages) can also significantly aid the disposal practice through the uptake of nutrients and water. To some extent, the Irrigation Water Quality Criteria[6] and existing proposed guidelines of heavy metal loading for sewage sludges can be used to determine the suitability of industrial wastewaters and sludge for land cultivation.[5]

Land Cultivation of Industrial Wastewaters

Land cultivation of wastewaters from food processing, pulp and paper, textile, tannery, wood preserving and pharmaceutical industries has been practiced on a limited scale.[7] At most locations, the practice is used primarily as a wastewater treatment method, and little or no attention has been paid to soil incorporation.

Three application methods are generally used; *i.e.*, overland flow, slow infiltration and rapid infiltration.[8] It appears that of the three methods, slow infiltration (spray irrigation) offers the highest degree of disposal reliability and potential for long-term site usage. Overland flow may require operational manipulations to realize the same useful site life as slow infiltration. Rapid infiltration (ground-water recharge) will require the most extensive and thorough site investigations to ensure that favorable conditions exist.

Land Cultivation of Industrial Sludges

The information gathered indicates that industrial sludges applicable to land cultivation have been either organic (*e.g.*, oil refinery, paper and pulp and fermentation residues) or treated inorganic (*e.g.*, steel mill sludge) wastes containing insignificant levels of extractable heavy metals. When the waste is applied to agricultural land, it is generally used as a soil amendment to improve soil characteristics and/or as a low-analysis nitrogen fertilizer.

Among the industrial sludges, oil refinery wastes have been most extensively disposed of by land cultivation.[9,10] Oil degradation rates vary, depending on climate, oil content in the soil and fertilization. The types of oily wastes that are disposed of by land cultivation include cleanings from crude oil, slop emulsion, API separator bottoms, drilling muds and other cleaning residues.[10] The sludge is spread to a depth of about 10 to 20 cm by a track-dozer and then disced into the soil. At existing sites, mixing intervals vary from once per week over several weeks to twice per year. The practice is strictly for disposal; no crops or vegetation are purposely grown at the sites.

Some industries and disposal firms are practicing land cultivation of hazardous industrial sludges on a trial basis. In most instances, the sludges have been pretreated to inactivate or remove the hazardous constituents in the waste. Hazardous waste disposal by land cultivation is not widely practiced; thus, as expected, no published information is available.

Limited data are available from greenhouse and field investigations conducted to evaluate the potential adverse effects of application of some industrial sludges on the yield and quality of crops. DeRoo[11] evaluated the use of mycelial sludges produced by the pharmaceutical industry in Connecticut as a nitrogen fertilizer and organic soil amendment. He concluded that if the mycelial sludge is applied repeatedly at high rates (222 m.ton/ha) to the same field, the soluble salt concentration and high zinc content in the sludge may be injurious to plants. Studies with similar objectives have been conducted using lagooned paper pulp sludge,[12] cannery fruit sludge[13] and nylon sludge.[14] Results indicated that these sludges would have value as a low-analysis nitrogen fertilizer and that no adverse effects were observed in crops and soils.

In field studies, the effects of a refractory metal sludge[2] and steel mill sludge[15] on the yield and chemical composition of forage and grain crops were evaluated. No adverse effects from heavy metals were observed. In the case of steel mill sludge, the growth was stunted at high waste application rates (sludge about 20 cm thick), which was attributed to nitrogen and phosphorus deficiencies and poor aeration from soil compaction.

In actual field disposal operations, sludges are either hauled directly from the wastewater treatment plant or from a lagoon to the disposal site. The sludges are applied to the land by spraying, spreading or subsurface injection. The field is then disced or plowed by conventional farm cultivation equipment. Loading rates generally depend on the waste's BOD, total dissolved solids and soluble salt contents, and on the soil's texture and drainage characteristics. Most state regulations do not presently limit the concentration of heavy metals or other potentially toxic constituents that are present at varying amounts in the sludges, although increased regulation is likely to occur if land cultivation becomes more common.

It should be noted that although soil can often serve as an effective disposal sink for industrial organic wastewater and sludge, if a specific soil cannot assimilate the applied quantity of organic sludge the soil will become anaerobic, resulting in nuisance conditions and failure of the system. Moreover, unless the waste materials are detoxified or decomposed by the soil or weather to nondeleterious products, the upper soil zone receiving the wastes eventually will become loaded to its ultimate capacity. As a result, disposal activities at the site will have to be terminated.

ENVIRONMENTAL ASSESSMENT

Factual data on environmental contamination resulting from land cultivation of industrial wastewaters and sludges are very scarce. It is recognized that improper land disposal of industrial wastes often goes unnoticed in the short term because the impacts are chronic rather than acute.[16] It takes decades, in some cases, for hazardous compounds which have been buried in the land to leach through soil into surface and ground water supplies.[16]

When the waste material is incorporated into the surface soil, it triggers a series of physical, chemical and biological processes. Phillips and Nathwani[3] have reviewed the various mechanisms involved in soil-waste interactions. Of significance are the migration of heavy metals and toxic organic compounds into the ground water, surface runoff, air pollution and potential hazards to the foodchain.

Ground Water Quality

Soils exhibit tremendous chemical and biological attenuation and degradation capacity due to their high reactive surface area, the available varied microbial populations and dilution effects. However, if a waste has a degree of solubility and is potentially mobile in any way, there is always a risk that it will affect the quality of subsurface waters. Adriano *et al.*[17] presented data showing that the nitrate and phosphate levels in subsurface waters exceeded public health standards and environmental guidelines, respectively, from long-term land treatment of food processing wastewaters.

The downward movement of heavy metals, oils and pesticides in soils is often restricted, due to low water solubility and high retention and degradation by various soil processes.[10,18,19] Other chlorinated hydrocarbons, phenols and detergent components may be present in varying amounts in the wastes, depending to a large degree on the type of generator industry. It is believed that these organic compounds will be eventually decomposed by soil microorganisms, and unless the soil is overloaded with wastes containing large amounts of these substances, land cultivation is not likely to pose a serious threat to the ground water quality.

Air Emissions

Like municipal effluents and sludges, industrial wastewaters and sludges also emanate odors when exposed to the atmosphere, impairing air quality of the disposal area. If the waste contains volatile components, land cultivation practices could increase evaporation of these components. Additionally, during soil incorporation, dust could present a health hazard to the personnel on the site. Subsurface injection of the waste and/or mixing with soil as soon as practical after deposition can alleviate, but not always eliminate, odor and evaporation problems.

Hazards to the Food Chain

In the ongoing instances of land treatment of industrial wastewaters, a cover crop is generally regarded as an integral part of the system, useful to improve water infiltration, remove nutrients and increase the hydraulic loading the site can accept. In disposing of industrial sludges by land cultivation, existing vegetation is usually removed from the site before waste application. Weeds and small bushes will become established in the disposal plot only if the plot is left untilled for some time. Available information indicates that crops grown on agricultural soils treated with selected industrial wastewaters and sludges do not accumulate toxic metals in sufficient quantities to affect plant growth adversely.[12,14,15,17] However, long-term effects of land cultivation on crop quality and the food chain are not known.

Land cultivation of waste material can pose hazards to the food chain through surface contamination and plant uptake. For example, pesticide residues have been shown to accumulate in various crops by these mechanisms,[20] rendering the crops unsafe for consumption.

The effects of toxic elements in sewage sludges and effluents on the food chain have been discussed in the literature. Cd, Cu and Zn are the elements commonly posing significant potential hazards to the food chain through plant accumulation. However, it should be noted that under improper site management (overloading, low pH, etc.), elements such as Pb, Hg, As, Se, Mo and Ni, if present in significant quantities in the wastes that are land cultivated, could also pose serious hazards to man and animals consuming the crops.

SOIL IMPACTS

As part of this study, samples of surface soil and typical native vegetation were taken from two oily waste disposal sites. Samples were also collected from the corresponding control sites which had similar soil and vegetation, but had received no waste. The objective of the sampling program was to determine the effect of land cultivation of oily sludges on soil chemical properties and elemental uptake. Pertinent information of the two sites studied is presented in Table I. Note that the site age, soil and sludge

Table I. Pertinent information on study sites used in the investigation of vegetative impacts of land cultivation.

	Site A–Southern California	Site B–Southeastern Texas
Area of Site	14 ha (35 ac)	8.2 ha (20 ac)
Waste Type	Drilling muds Tank bottoms	API oil/water separator sludges
Waste Volume	1000 to 1200 bbl/day	Periodic disposal, 185,000 bbl/yr
Soil Characteristics	Sandy, well-drained, slightly acidic	Clayey, poorly drained alkaline
Depth of Soil/ Waste Mixture	0.9 to 1.2 m (3 to 4 ft)	15 to 30 cm (6 to 12 in.)
Site Age	22 yr	5 yr
Vegetative Type	Weeds and shrubs are growing along the perimeter of the disposal area	Same as Site A
Vegetation Sampled	Tall grass, golden bush, ragweed, and ice plant	Nut grass leaves and cocklebur seeds

characteristics from the two sites are markedly different. Site A is managed by a disposal company, whereas Site B is located within and managed by an oil refinery plant.

Results of soil analyses (Table II) show that land cultivation of oily wastes at Site A resulted in increased pH, soluble salt content (expressed as EC) and levels of total Kjeldahl nitrogen (TKN) and organic carbon in the sandy soil.

Table II. Chemical characteristics of the surface soils from control and oil-treated plots.

	Site A		Site B	
Constituents[a]	Control	Treated	Control	Treated
pH	6.04	7.65	7.41	7.40
EC (mmhos/cm)	0.40	4.46	2.21	3.91
Oil (%)		2.28		2.06
TKN (%)	0.006	0.079	0.080	0.134
Organic C (%)	0.16	2.53	2.10	5.10
	(ppm)		(ppm)	
P	410	230	17.5	17.5
Na	110	280	185	375
B	0.2	2.28	0.2	0.22
Mn	35.4	55.0	65	71.6
Ni	1.5	2.5	4.8	5.3
Zn	7.5	40.7	53.5	71.5
Se	0.022	0.09	0.01	0.0
Mo	1.3	1.1	0.6	0.55
Cd	0.14	0.06	0.06	0.06
Pb	4.2	5.4	212	242

[a]Electrical Conductivity (EC) and B were measured in the saturation extracts; other elements in ppm were determined in 0.1 *N* NCl extracts.

Likewise, water-soluble B and acid-extractable Na, Mn, Ni, Zn, Se and Pb also increased. At Site B, similar increases were noted for soluble salts, TKN, organic carbon and acid-extractable Na, Mn, Zn, Se and Pb. Except for Pb concentrations at Site B, the elements analyzed had concentrations which are within the range typically found in soils.

The background levels of heavy metals in the clayey soil from Site B are higher than those in the sandy soil from Site A. In view of the coarse texture and low organic carbon content of the sandy soil at Site A, land cultivation would likely have a greater impact on the soil and plants grown at Site A than on those at Site B. The soluble salt concentration of the treated plot from Site A may be injurious to salt-sensitive plants such as green beans, celery and most common fruit crops.

Due to the limited scope of this study, a thorough survey and chemical analysis of all types of vegetation at the land cultivation sites was not made. Instead, some typical and prevalent species were sampled and analyzed.

Results from plant analyses indicate differences in plant uptake due to differences in site conditions, plant species and waste oil treatment methods. At Site A, ragweed and ice plant grown on the oil-treated plot contained higher concentrations of Na, Mn, Zn, Se and Pb than those from the control plot. These observations related to the trends in soil data. The plant N contents were reduced by the land cultivation treatment, probably due to immobilization of available N by the soil microorganisms. This suggests that heavy application of nitrogen fertilizer may be necessary to an oily waste disposal area if subsequent growth is planned.

Nut grass and cocklebur grown on the oil-treated plot from Site B contained higher concentrations of Zn, Mo and Pb than those grown on the control plot. Overall, concentrations of the elements analyzed were lower in the cocklebur seed than the concentrations in nut grass. The levels of Mo and Pb in both plant species from the oil-treated plot are worthy of notice since they have approached the undesirable level ($\geqslant$ 10 ppm) for animal consumption. The concentrations of Pb were exceedingly high, comparable to the Pb concentrations in pasture grasses in a lead-contaminated area near Antioch, California.[21] The data suggest that nut grass and cocklebur seed have accumulated significant amounts of Pb, particularly from the oil-treated plot.

SUMMARY AND CONCLUSIONS

Wastewaters and sludges that have been land cultivated are primarily from food processing, oil refinery, wood preserving, textile and pharmaceutical industries. The wastes are composed mainly of organic material and, thus, are biodegradable. It appears that some industrial sludges are applicable to land cultivation if the wastes are pretreated to remove or recycle the toxic constituents, thus alleviating the potential hazards of surface ground water and food chain contamination.

Land cultivation of industrial wastewaters and sludges has received very little attention, probably due to the lack of data on the economics, productive uses of the site after cessation of disposal activities and associated environmental problems. Preliminary results from field investigations indicate that land cultivation of oily sludges has resulted in increased soluble salt and organic carbon contents in soils and accumulation of heavy metals in soils and plants.

Based on available information and field study results on management and disposal of industrial wastewaters and sludges, the following conclusions can be drawn:

1. Land cultivation as a disposal alternative is presently practiced only by a few industries and on a limited scale in the United States. The trend indicates that there will be some slight increased use of land cultivation in the future.

2. Land cultivation is viable only where soil, climate, waste characteristics and environmental conditions permit. Depending on specific conditions, the disposal program can either be related to agriculture or can be solely a disposal practice.

3. Few comprehensive environmental monitoring programs are underway at ongoing land cultivation sites. A vast majority of states do not have statutes regulating land cultivation; thus, each instance is addressed on a case-by-case basis.

REFERENCES

1. Powers, P. W. "How to Dispose of Toxic Substances and Industrial Wastes," Noyes Data Corporation, Park Ridge, New Jersey.
2. Phung, T., D. Ross and R. Landreth. "Land Cultivation of Municipal Solid Waste," presented at Third Annual Research Symposium: Management of Gas and Leachate in Landfills, St. Louis, Missouri, March 14-16, 1977 (in press).
3. Phillips, C. R. and J. Nathwani. "Soil-Waste Interaction: A State-of-the-Art Review," Solid Waste Management Report EPS 3-EC-76-14, Environment Canada (October 1976).
4. Carlile, R. L. and J. A. Phillips. "Evaluation of Soil Systems for Land Disposal of Industrial and Municipal Effluents," Report No. 118, Water Resources Research Institute, University of North Carolina (July 1976).
5. Epstein, E. and R. L. Chaney. "Land Disposal of Industrial Waste," Proceedings of the National Conference on Management and Disposal of Residues from the Treatment of Industrial Wastewaters, Washington, D.C. (February 1975), pp. 241-246.
6. National Academy of Sciences–National Academy of Engineering. "Water Quality Criteria, 1972," (Washington, D.C.: U.S. Government Printing Office, 1974).
7. Wallace, A. T. "Land Disposal of Liquid Industrial Wastes," in *Land Treatment and Disposal of Municipal and Industrial Wastewater*, R. L. Sanks and T. Asano, Eds. (Ann Arbor, Michigan: Ann Arbor Science Publishers, Inc., 1976), pp. 147-162.
8. Hunt, P. G., L. C. Glide and N. R. Francingues. "Land Treatment and Disposal of Food Processing Wastes," Conference on Land Application of Waste Materials, Soil Conservation Society of America, Ankeny, Iowa (1975), pp. 112-135.
9. Kincannon, C. B. "Oily Waste Disposal by Soil Cultivation Process," EPA-R2-72-100, U.S. Environmental Protection Agency (December 1972).
10. Lewis, R. S. "Sludge Farming of Refinery Wastes as Practiced at Exxon's Bayway Refinery and Chemical Plant," presented at the National Conference on Disposal of Residues on Land, St. Louis, Missouri, September 13-15, 1976 (in press).

11. DeRoo, H. C. "Agricultural and Horticultural Utilization of Fermentation Residues," Bulletin No. 750, Connecticut Agricultural Experiment Station (1975).
12. Jacobs, L. W., Soil and Crop Science Department, Michigan State University, Lansing, Michigan, personal communication (1976).
13. Noodharmcho, A. and W. J. Flocker. "Marginal Land as an Acceptor for Cannery Waste," *J. Am. Soc. Hort. Sci.* 100:682-685 (1975).
14. Cotnoir, L., Plant Science Department, University of Delaware, Newark, Delaware, personal communication (1976).
15. Nelson, D. W., Agronomy Department, Purdue University, Lafayette, Indiana, personal communication (1976).
16. Lazar, E. C. "Summary of Damage Incidents from Improper Land Disposal," Proceedings National Conference on Management and Disposal of Residues from the Treatment of Industrial Wastewaters, Washington, D.C. (February 1975), pp. 253-257.
17. Adriano, D. C., L. T. Novak, A. E. Erickson, A. R. Wolcott and B. G. Ellis. "Effect of Long-Term Land Disposal by Spray Irrigation of Food Processing Wastes on Chemical Properties of the Soil and Subsurface Water," *J. Environ. Qual.* 4:242-248 (1975).
18. Page, A. L. "Fate and Effects of Trace Elements in Sewage Sludge When Applied to Agricultural Lands," EPA-670/2-74-005, U.S. Environmental Protection Agency (January 1974).
19. Letey, J. and W. J. Farmer. "Movement of Pesticides in Soils," *Pesticides in Soil and Water,* W. D. Guenzi, Ed. (Madison, Wisconsin: Soil Science Society of America, Inc., 1974), pp. 67-97.
20. Nash, R. G. "Plant Uptake of Insecticides, Fungicides, and Fumigants from Soils," *Pesticides in Soil and Water,* W. D. Guenzi, Ed. (Madison, Wisconsin: Soil Science Society of America, Inc., 1974), pp. 257-299.
21. Ganje, T. J., Soil Science and Agricultural Engineering Department, University of California, Riverside, California, personal communication (1976).

12

IMPACT ON FARM COSTS AND RETURNS FROM SLUDGE APPLICATION IN FORAGE PRODUCTION

G. C. Reisner and R. L. Christensen
Department of Food and Resource Economics
University of Massachusetts
Amherst, Massachusetts

INTRODUCTION

An economic analysis of sludge application on agricultural land should consider more than the costs and benefits to the agricultural sector. A complete analysis would also consider those alternatives and decisions with respect to waste generation and handling which are the economic responsibility of the municipality. However, such an analysis involves fairly complex techniques and is beyond the scope of this study.[1]

As Carlson and Young[2] suggest, costs and benefits from public services, such as waste disposal, are difficult to specify because, in some instances, units of output and/or prices are ambiguous. Both the farm and the waste-generating sectors (primarily municipalities, although the agricultural sector is a source of wastes as well) are faced with several goals or objectives, some of which are complimentary and some of which are in competition. Further, these objectives are often noncommensurable–dollar costs and social impacts cannot always be evaluated in the same terms. Further, since each sector has multiple alternatives for achieving these objectives, the joint product nature of agricultural land application of sludge suggests that the *opportunity cost* concept is paramount in a thorough economic analysis of the problem.

The concept can be illustrated for sludge disposal/use alternatives. For example, a relevant cost of incineration of sludge as compared with land application for agricultural purposes is the foregone value of the sludge in producing a crop. Another opportunity cost of incineration is the foregone value of the acquisition of knowledge about the relevant processes in agricultural use of sludge which may enable better decisions in the future. A third opportunity cost in selecting the incineration alternative might be the effects on regional income and employment which the land application alternative would have provided (see Seitz[3] and Seitz and Swanson[4] for some empirical evidence of this effect).

Finally, these opportunity costs (*e.g.*, dollar values and environmental effects) are often not expressible in the same terms or units. What often results, then, is that effects, other than monetary, are mentioned but not explicitly incorporated into the process of choosing a system.

Therefore, a complete analysis of the economic consequences of agricultural use of urban sludge would be couched in the broader framework implied by these considerations. However, a component in such a comprehensive approach is the microanalysis of impact on the farm sector. The study reported here focuses only on those economic consequences specific to the farm firm.

The following sections present a brief review of selected studies of the economics of sludge application on agricultural land, budgeted costs of sludge application for corn forage production on representative dairy farm situations, and finally, an overall summary.

REVIEW OF SELECTED STUDIES

One of the notable studies of economic aspects of the application of municipal wastes to agricultural land was reported by Seitz and Swanson[4] and Seitz.[3] It was based on the Chicago-Fulton County, Illinois, project. As background, the project involved transporting over 900 dry tons of anaerobically digested sludge per day 170 mi southeast to Fulton County for agricultural land application purposes. Over 50,000 ac of this land had been strip-mined and there was public pressure to return at least some of this land to agricultural use. The researchers developed an economic simulation model and found that the following variables most influenced the economic performance of the system: sludge transportation costs, site preparation costs, the nitrogen budget and cropping system used.

Their analysis suggested that the discounted present value of net costs of the project to Chicago would be about $68/dry ton using existing transportation methods (*i.e.*, railroad) and about $39/ton if a pipeline were built. If local labor and contractors were used, the project should generate economic activity in the county at the rate of $600 to $1200/ac. Their conclusion, then, was that a project of this sort is economically feasible. If the pipeline were constructed, Chicago could apply significant quantities of sludge at lower costs than for the alternative disposal methods previously used. Likewise, Fulton County benefits in that significant amounts of land are reclaimed and important new income and employment sources are generated within the county.

More recently, Forster *et al.*[5] surveyed the 50 to 60 communities in Ohio that are currently applying treated sewage sludge to agricultural lands as their primary method of disposal and arrived at estimates of landspreading costs and economic benefits accruing in the form of a nutrient supply. The surveyed communities were asked to provide information on: (1) capital investment in sludge disposal equipment and storage facilites, (2) average age of these capital investments, and (3) annual operating costs of sludge disposal. Adjustments were made to capital items so that all investments are in current dollar values.

The data indicate that the average unit sludge disposal cost declines as system size increases up to approximately a 25-mgd level. The average community in the sample operated at a plant capacity of 8 mgd with a cost of approximately $31/dry ton. For smaller plants, disposal costs increased rather rapidly. The median plant size of 4 mgd, for example, had an average sludge disposal cost of $43/dry ton.

A further objective of these researchers was to approximate the economic benefits of sludge application by the value of sludge as a substitute for commercial fertilizer applications–primarily the nitrogen, phosphate and potash required by crops. It was difficult to estimate these benefits due to the variable chemical composition of sludges as well as the fluctuation of nutrient prices. On the average, however, a dry ton of sludge was reported to contain roughly 100 lb each of nitrogen and phosphate and 5 lb of potash. The researchers note that this combination is not suitable for most crops, indicating that supplemental fertilization would be required. Assuming acceptable concentrations of trace elements and usage of sludge as a supplemental source of nutrients, Forster *et al.* reported estimates of the value of nutrients in sludge under six alternative assumptions regarding nutrient content of sludge and commercial fertilizer price. The range was reported to be from $10 to $49/dry ton.

For the communities surveyed, sludge disposal costs were slightly higher than the value of the sludge as a substitute for commercial fertilizer. Assuming the average nutrient content and current Ohio prices, the benefits were estimated at $23/dry ton. This compares, for example, to the $31/dry ton cost of spreading sludge on the land for the average community surveyed. Thus, the net cost per dry ton was $8 for agricultural land disposal.

The conclusion drawn was that the closeness of sludge application benefits and costs makes land application an extremely attractive alternative. The other primary alternatives are incineration and landfill, and their costs are generally higher (incineration costs were estimated in the range of $50 to $75/dry ton).

While the above studies were concerned with the broad issue of land application of sewage sludge, a detailed marginal analysis of the impact on the individual farm appears lacking. Economic analysis which focuses on the individual farm situation, involving consideration of labor and machinery use, is necessary to assess adequately this dimension of the agricultural use alternative.

ECONOMIC IMPACT

Introduction

The interest in the use of sludge on agricultural land in Massachusetts is motivated by (1) the knowledge that sludge contains basic plant nutrients and (2) the needs of Massachusetts farmers for fertilizer materials for crop production. The economic feasibility of sludge use in agriculture is dependent on the relative costs and benefits associated with this alternative as compared to others, with full recognition of the opportunity cost concept.

Nutrient Value of Sludge

The potential value of sludge lies in its nutrient content and in its characteristics as a "soil conditioner." The nutrient value is conventionally expressed in terms of the content of nitrogen, phosphorus and potassium contained in the sludge, although some minor elements such as boron or magnesium also contained in sludge, in relatively small amounts, may have economic value in crop production. The value of sludge as a soil conditioner is due to the fact that it is organic matter. Organic matter in the soil enhances soil texture, promotes aeration and increases moisture holding capacity. All of these characteristics may lead to increased crop production. If soils have been "run down" to the point where organic matter content is low, then application of sludge could have a significant effect. If, on the other hand, the soil has been well managed prior to sludge application, little effect may occur. Similarly, in years with good rainfall the moisture retention effect may not be significant, while in dry years it may be important. With this uncertainty relating to the value of sludge as a soil conditioner, one may either assume no difference or make some arbitrary adjustment to represent the effect over a period of years.

Table I contains the assumptions relative to the value of sludge. The content of nitrogen, phosphorus and potash from representative sludges is used. The price assumed for these nutrients is derived from the price of commercial fertilizing materials. It is assumed that nutrients in the sludge have "availability" comparable to the commercial fertilizers.

Table I. Nutrient content and estimated value by sludge type.

Sludge Type	Nutrient Content (%)[a]	Nutrient Price	Nutrient Value/Wet Ton Sludge[b]
5% Solids	N = 5.6	$0.255	$1.43
	P = 2.9	0.203	0.59
	K = 0.4	0.089	0.03
		Total	$2.05
15% Solids	N = 16.8	$0.255	$4.28
	P = 8.7	0.203	1.77
	K = 1.2	0.089	0.12
		Total	$6.17

[a]Nutrient content assumed: 5.6% N, 2.9% P, 0.4% K, all expressed as a percentage of solids.

[b]Value per dry ton is $41.

Dairy Farms as Potential Sludge Receivers

Dairy farms are hypothesized as the most logical potential receiver of sludge for several reasons. First, the physical handling of manures and sludge can be accomplished by essentially similar equipment already owned by the farmer. Secondly, the dairy farm uses manure in forage crop production and could use sludge for the same purpose. Thirdly, the crops on which sludge would be applied are not used for human consumption but for animal feed and there is scant evidence to show disease transmission through the food chain from animals fed on sludge-grown forage. Fourth, the nutrients contained in sludge can be substituted for commercial fertilizer use.

Some technical difficulties are apparent. Sludges with low solids contents are amenable for use only on farms with liquid manure handling systems. Sludges with high solids content can be handled with conventional spreading systems. But unless water is added to create a slurry, such filter cake sludges cannot be handled with liquid systems.

Use of sludge in crop production on a commercial farm must be keyed to the farm calendar. That is, unless an irrigation-type system is used, sludge cannot be applied during the crop growing season. For corn, this means from May through September, while for hay the prohibited period would be somewhat shorter, depending on the number of cuttings. It appears advisable to avoid spreading sludge on foliage of crops that will be consumed by livestock.

If sludge is to be immediately incorporated in the soil, a further restriction is imposed by climate. In Massachusetts, the soil is generally frozen from mid-November to March which would preclude tillage operations or soil injection.

In summary, these restrictions would indicate three possibilities: (1) that for small sewage treatment plants the frequency of sludge removal may conceivably be keyed to the spring and fall periods when land application is most feasible, (2) that sludge storage capability be provided either at the plant or at the farm to hold the sludge until field disposal is possible, or (3) that the municipality dispose of sludge to farmers during those seasons when possible and dispose in landfills or other means during other seasons.

The second alternative above may present difficulties if the sludge held in storage becomes anaerobic. The result would be highly objectionable odors during handling and application. The third alternative may be difficult to justify if a substantial investment is associated with the alternative disposal method. If all operating and fixed costs for the alternative must be paid even though some volume of sludge is diverted to land application, it will be difficult to convince municipalities of the economic wisdom of such an alternative.

Provided the type of sludge produced is compatible with the manure disposal equipment of the farm, costs of sludge application on farmland can be regarded as "marginal costs." Marginal costs are only those additional costs attributable to sludge application. This means that since the farmer already has the necessary equipment, the ownership costs (amortization, depreciation, etc.) would not be attributable to sludge application. Costs attributable to sludge application then become only the variable costs (fuel, repairs, labor, etc.).

Adjustments may be made in this concept to account for minor equipment requirements or the purchase of larger machinery to deal efficiently with higher volumes of material to be distributed on the land. It is also possible, where excess capacity exists, that the added volume of sludge handled will lower the unit cost of disposal for the farm manure handled.

Storage

If it is possible to schedule sludge disposal from the treatment facility with the farm operations schedule, as described previously, there would be no need for a storage facility. However, it may be necessary to provide short-term (up to four months) storage for sludge. For purposes of analysis, it was assumed that two types of on-farm storage might be used: (1) a bunker-type storage area for high-solids "dry" sludges or (2) a lagoon storage for liquid or slurry-type sludges. The capacity of each storage was assumed to be approximately equal at 1500 cubic yards.

Costs of constructing these on-farm storages vary considerably depending on construction details, site conditions, and availability of farm labor and equipment for construction. From reported costs of similar storages constructed by farmers in the state, it may not be unreasonable to assume a construction cost of $5000 for either storage. Assuming a 10-yr life, an 8% interest on investment and a $10/1000 tax rate, annual ownership costs could be estimated as approximately $750.

Sludge Application Systems

In evaluating the economies of sludge application on farmland, it is necessary to specify the alternative systems that might be used. Figure 1 illustrates some of the alternative technologies that are possible. Each of these systems will entail different investment and operating costs depending on the equipment complements and other facilities. Two systems will be considered: conventional manure handling and liquid manure systems.

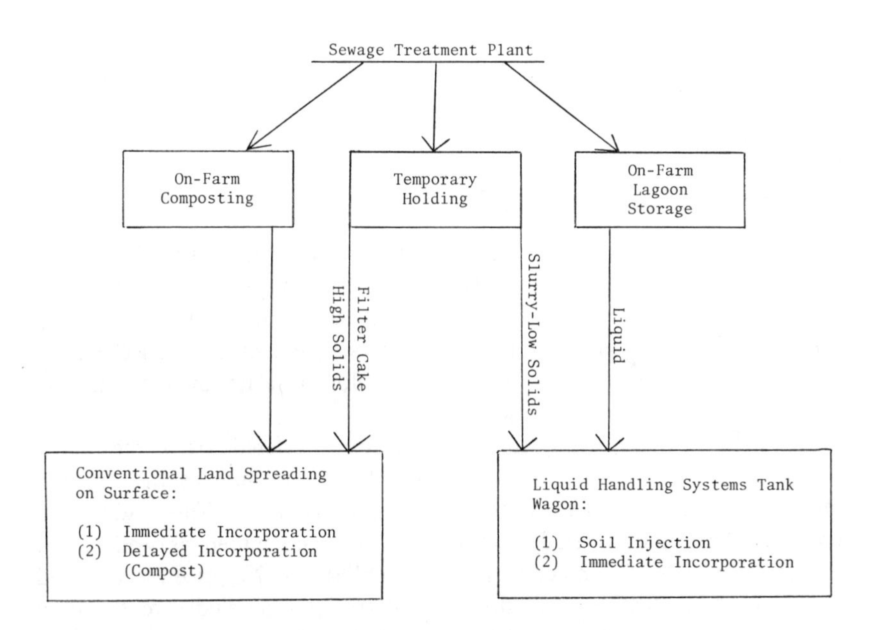

Figure 1. Sludge handling systems–on-farm.

It is assumed that the evaluation of costs will be those incurred on the farm with capital investment and ownership costs being borne by the farmer. Thus, transportation of sludge from the treatment plant to the farm will not be considered in this section.

On arrival at the farm, sludges would be deposited into one of three potential receiving systems:

1. An on-farm composting area. A composting pad with associated equipment would be necessary.

2. A temporary holding area. Several possibilities exist here. If a high-solids filter cake is involved, a dumping pad or bunker storage may be all that is necessary. If a semiliquid, low-solids sludge is the case, a "nurse tank" or small lagoon may be dictated.
3. A long-term holding lagoon or bunker-type storage. If it is necessary to provide storage for three to four months, a lagoon of substantial capacity may be needed.

The solids content of the sludge will dictate the method of on-farm transport and land application. High-solids filter cake can be transported and applied using conventional box-type manure spreaders. Low-solids liquid sludges must be transported and applied using a tank-type system. Conventional liquid manure handling equipment appears to be well adapted to sludge handling and application. Liquid-type sludges would appear to be best applied by subsurface injection, or, if surface spread, immediate incorporation is advisable. High-solids sludges should also be incorporated in the soil as soon as possible. Composted sludges, however, need not be incorporated into the soil immediately.

Estimates of On-Farm Sludge Application Costs

The model farm situation was specified as a dairy farm with 150 head of cattle and 200 ac of tillable land. The initial situation specified production of the corn silage needs for the herd, assuming the fertilizer nutrients for crop production were supplied by farm manures supplemented by purchases of commercial fertilizer. The cost data developed relate only to manure spreading and fertilizer application.

The situation involving sludge application differed from the initial situation in that sludge was used to supplement manure as the nitrogen nutrient source. No differences in yield were assumed. Those additional costs incurred due to sludge application were computed as well as the reduction in costs of commercial fertilizer application. No price was included for the sludge. Neither were the costs of transporting sludge to the farm included. Thus, it was assumed to be supplied to the farm as a free resource. Some supplemental applications of commercial fertilizer were found to be necessary to balance the total nutrient application rate.

The net costs/benefits to the farm situation were determined from considering the added costs of sludge application as compared with the reduced costs of commercial fertilizer.

Sludge Application Using Conventional Manure Handling Equipment (Sludge with 15% Solids)

Assuming that corn silage is to be grown on 200 ac with yield target of 20 tons/ac using farm-produced manure with supplementation from sludge and commercial fertilizer, the figures in Table II are derived. These figures assume that manure will be applied on 90 ac and sludge on 110 ac.

Table II. Sludge application using conventional manure handling equipment (sludge with 15% solids).

	N	P	K
Total nutrient requirement (lb)	30,000	25,000	36,000
Nutrients from manures (lb)	13,500	8,100	13,500
Nutrient deficit	16,500	16,900	22,500
Sludge (295 tons solids)	16,500	17,089[a]	2,357
Supplemental fertilizer required	0	3,150	20,143
Partial Budget			
Added Costs Associated With Sludge Application			$1,150
Reduced Fertilizer Expenses			7,210
Net Benefits Accruing to Sludge Application			$6,060
Net Farm Value Per Ton Sludge Solids			$20.54
Net Farm Value Per Ton of "Wet" Sludge			$ 3.08

[a]Acreage on which sludge only is applied will receive P in excess of needs. However, acreage on which manure is applied will require supplemental fertilizer application.

Sludge Application Using Liquid Manure Handling Equipment (Sludge with 5% Solids)

All assumptions and basic nutrient balance information will be identical with the situation described in the above example. However, three times as much total tonnage must be handled. In addition, the liquid manure handling equipment is somewhat more costly than conventional. Balancing these factors to some extent is increased load capacity.

The net benefits accruing to sludge application for low-solids sludge were estimated to be $4910. The net farm value per ton of solids (295 tons) was, therefore, $16.64. The net farm value per ton of "wet" sludge was then $0.83.

SUMMARY

It is difficult to reach and make general conclusions in a situation where agricultural conditions vary widely and sludge-generating conditions even more so. However, the following points seem clear. First, there are significant benefits in the form of foregone costs of commercial fertilizer associated with the land application option for sludge disposal which should be factored into a serious decision framework. The magnitude of these benefits depends heavily on sludge composition and price of commercial fertilizers, as well as cropping systems. To the extent that these commercial fertilizer costs are projected to increase relative to other prices and costs, the estimations of value of these benefits given earlier should be regarded as a lower bound on future benefits.

The literature also suggests that under a rather wide range of conditions, land application is the least expensive method of disposal (and use) of municipal sewage sludge, particularly when viewed from this broader perspective. Perhaps the most critical variable in determining the economic feasibility of the land application alternative is the distance (and cost) necessary to transport the sludge. This is a key reason why the agricultural use alternative is likely to be more economically feasible for small- to medium-sized communities than for larger metropolitan areas. The transportation costs can increase rapidly with increasing distance. A municipality will limit the distance carried to a point where disposal costs are equal to other alternatives. A farm using the above budgeting processes can now determine the value of the sludge to the farm and be able to make a more informed decision if asked to defray long-distance transportation costs.

The benefits accruing to the farm are in the form of reduced expenditures for commercial fertilizer. Other benefits not explicitly studied are the organic conditioning to the soils. Also, some indirect costs were not dealt with. An implicit cost may be identified as the opportunity cost of labor and machinery employed in sludge application which could be productively employed in other activities. These areas will need further study.

Finally, while little experience with agricultural use of municipal sewage sludge has been acquired in Massachusetts, largely due perhaps to institutional reasons, there is little reason to believe the critical conditions discussed earlier are substantially different from many other areas in the country.

ACKNOWLEDGMENTS

This is University of Massachusetts Experiment Station paper No. 2147.

REFERENCES

1. Perlack, R., C. Willis and B. Lindsay. "Application of Multiobjective Optimization to Sludge Disposal Alternatives," paper presented at the Annual Meetings of the Operations Research Society of America, Miami, November,
2. Carlson, G. and C. Young. "Factors Affecting Adoption of Land Treatment of Municipal Wastewater," *Water Resources Res.* 11:5:616-620 (October 1975).

3. Seitz, W. "Strip-Mined Land Reclamation with Sewage Sludge: An Economic Simulation," *Am. J. Agric. Econ.* 56:4:499-804 (November 1974).
4. Seitz, W. and E. R. Swanson. "Economic Aspects of the Application of Municipal Wastes to Agricultural Land," in *Recycling Municipal Sludges and Effluents on Land,* Conference Proceedings, Champaign, Illinois, July 9-13, 1973.
5. Forster, D. L. *et al.* "State of the Art in Municipal Sewage Sludge Land Spreading," paper presented at Conference on Agricultural Waste Management, Cornell University, Ithaca, New York, April 1976.

ADDITIONAL REFERENCES

Christensen, L., L. Connor and L. Libby. "An Economic Analysis of the Utilization of Municipal Wastewater for Crop Production," Agricultural Economic Report No. 292, Michigan State University (1975).

Commission on Organic Waste Recycling. "The Feasibility of Application of Municipal Sewage Sludge on Agricultural Land in Massachusetts," Robert L. Christensen, Task Force Chairman (1976).

Ecol. Science, Inc. *Environmental Impact Statement: Proposed Sludge Management Plan*, Vol. 1 (Boston, Massachusetts: MDC, 1976).

Johnson, J. B. and L. J. Connor. "Economic and Regulatory Aspects of Land Application of Wastes to Agricultural Lands," paper presented at Conference on Agricultural Waste Management, Cornell University, Ithaca, New York, April 1976.

"Process Design Manual for Sludge Treatment and Disposal," U.S. Environmental Protection Agency, Technology Transfer, EPA 625/1-74-006 (October 1974).

Young, C. E. "Land Treatment Versus Conventional Advanced Treatment of Municipal Wastewater," *J. Water Poll. Control Fed.,* 47:11:2565-2573 (November 1975).

Young, C. E. "The Cost of Land Application of Wastewater: A Simulation Analysis," Economic Research Service Technical Bulletin No. 1555, U.S. Department of Agriculture (November 1976).

13

ECONOMIC ANALYSIS OF RECYCLING SEWAGE SLUDGE ON AGRICULTURAL LAND

S. L. Ott and D. L. Forster
Agricultural Economics and Rural Sociology
The Ohio State University
Columbus, Ohio

INTRODUCTION

Landspreading, landfilling, ocean dumping and incineration are alternative methods to dispose of treated municipal sewage sludge. Landspreading of sludge appears to be growing in popularity among small- to moderate-sized communities due to its disposal cost advantages. However, numerous landspreading systems exist and little economic information is available to give guidance to cities in choosing a system. The intent of this chapter is to partially fill this void.

Treated sewage sludge and the effluent are the by-products of most waste treatment processes. Typically, the treated effluent is returned to a waterway while the treated residual is in the form of sludge to be disposed.

Sludge composition allows for convenient and low-cost handling (*e.g.*, in tank trucks) and provides nutrients to crops. Generally, the sludge is composed of 2 to 8% solids. Chemical composition varies, but nitrogen averages about 3%, phosphate (P_2O_5) 5%, and potash (K_2O) 0.4% of total solids. There is the potential hazard that heavy metals (*e.g.*, Zn, Cu, Ni, Cd and Pb) applied to land over several periods may build up in the soil. With repeated, large applications, toxicity to crops may develop and increases in the heavy metal concentrations in the food chain may result. However, "the impact of heavy metals . . . may be limited by using rational management methods intelligently."[1]

Applying the effluent to land appears to be economically feasible in many areas of the country. In dry climates, crops may benefit from both the water and nutrients supplied by land application, but for communities in moisture surplus areas of the country, the advantages of land application of wastewater must rest with the nutrients supplied to crops. High capital costs, institutional barriers to obtaining adequate tracts of land, and low-cost alternative effluent treatment methods seem to present barriers to widespread adoption of effluent land application.

As a result of the favorable position of landspreading sludge as a disposal alternative, communities in the north central U.S. are increasing their interest in landspreading.[2] Furthermore, monetary and environmental considerations as well as increases in sewered populations and upgrading treatment plants could increase landspreading significantly.

OBJECTIVES

The first objective of this study was to design a model to select economically optimal sludge-disposal systems. The technique was to be generalized in order to accommodate communities with differing characteristics such as size, sludge composition, distance to landspreading sites and so forth.

The sludge-disposal alternatives needing to be considered in the model included landfilling or incineration, as well as the almost endless combination of components associated with landspreading. The possible components included sludge dewatering and subsequent spreading or storage of dried sludge, storage of liquid sludge, spreading of liquid sludge by tank truck or tank wagon, and transportation of liquid sludge by a "nurse" tank or by the spreading vehicle. Also, the crops to which the sludge is spread need to be considered as a model component. Typical crops like corn, wheat, soybeans and hay differ not only in terms of their return to sludge nutrients, but also in application rates and time intervals in which sludge can be spread. Hence, the crop combination affects the storage and application components.

The second objective of the study was to apply the model to four representative communities in order to determine the optimal components for each. The four communities are part of a demonstration project to show proper landspreading management practices. These four sites were chosen on the basis of a number of criteria—including physiographic location, size, sludge composition and crops in the area—to represent a broad range of landspreading conditions.

The third objective of the study was to test the sensitivity of the optimal system components to the levels of critical parameters. Input prices are parameterized to examine the range in conditions over which the optimal system holds.

The final objective was to draw some preliminary inferences concerning the benefits resulting from landspreading sewage sludge. These benefits need to be framed from two different perspectives: (1) What are the benefits for the unit of government responsible for sludge disposal? (2) What are the benefits for the community as a whole, including the unit of government disposing sludge as well as the landowners receiving sludge?

MODEL

A linear programming model was developed to minimize the net sludge disposal costs. The model specifies several methods of sludge disposal, including incineration, landfilling and a number of alternatives for landspreading. There are six time periods to spread sludge (mid-November to March, April, early May, mid-May to mid-July, mid-July to October, and early November). Sludge can be stored between two or more periods if crop scheduling or weather makes spreading impractical in a time period. Prior to storage and spreading sludge, dewatering may occur. Dewatered sludge is less bulky than liquid sludge and demands different storage, transportation and spreading technologies. Spreading on fields may be accomplished by a liquid tank truck, liquid tank wagon or a truck spreader (for dry sludge). Transportation from the treatment plant to the field is by a "nurse" truck or by the spreading technology.

The model chooses that combination of disposal components and crop activities which minimizes net system disposal costs. Average cost per ton of sludge is supplied to the model for each storage, transportation, spreading, and cropping alternative. Average total cost per ton includes both average variable costs (*e.g.*, labor and fuel) and average fixed costs (*e.g.*, depreciation, interest and repairs). Average variable costs for a ton of sludge landspread on cropland are increased as the distance from the treatment plant to the spreading site increases. Average fixed costs per ton of sludge are decreased as the quantity of sludge spread increases.

Crops allowed in the model depend on the soil resources found in the area surrounding the community. For each of the four communities, a predominance of soil suitable for row crops is found; thus, some corn, wheat, soybeans and oats are allowed as well as a grass hay or pasture.

The model has a spatial orientation since it distinguishes several concentric circles surrounding a city. For example, the model allows a city to spread on areas of land having different distances from the treatment plant. Transportation costs increase as the sludge is spread farther from the plant; thus, the model tends to spread as near as possible. However, limitations on application rates or high returns to distant crops may force spreading to move farther away.

Crops used in the model for each of the concentric circles around the cities reflect actual crop acres.[3] The model allows some adjustment in actual crop acreage in order to reflect the demands of the optimum sludge-spreading system.

Application rates on crops are determined by one of three restrictions: phosphate, nitrogen or heavy metals. The phosphate and nitrogen restrictions limit annual application rates of sludge to those which meet the annual nutrient needs of the alternative crops. The heavy-metal restriction assures that the soils in each of the concentric circles may be used for at least 20 years before metal buildup forces curtailment of sludge spreading. This 20-yr limitation assures that no permanent damage is done to any of the soils in the area since the heavy-metal restrictions are thought to be quite conservative estimates of the soils' ability to withstand metal accumulations.[4]

A schematic representation of the model is shown in Figure 1.

COMMUNITIES MODELLED

The purpose of examining sludge disposal in several communities was to test the applicability of landspreading to a range of conditions found in the north central area. Those conditions which are likely to affect the applicability include: (a) community size, (b) soil type surrounding the city, (c) distance to potential disposal sites, (d) characteristics of the sludge including percent solids and concentrations of nitrogen, phosphorous, lead, zinc, copper, nickel and cadmium, and (e) cropping practices in the area.

After reviewing these conditions in approximately 45 Ohio communities, 4 communities were selected as representative. These communities are Defiance, Medina County, Montgomery County and Zanesville.

The city of Defiance was selected to provide soils representative of the Lake Plain region in northwestern Ohio, northeastern Indiana, and southeastern Michigan. The soils are fine-textured and poorly drained and are nearly level to gently undulating. Due to the relatively small size of the community and the proximity of the treatment plant to farms, the city has available cropland within a short distance of the city. Using the criteria for safe application of heavy metals developed by the North Central Regional Research Committee (NC-118), the Defiance sludge is a sludge with relatively low levels of heavy metals.[4] The crops grown in the Defiance area are primarily corn, soybeans and wheat.

The Medina area was selected to provide soils representative of the low-line glacial till region of northeastern Ohio and western Pennsylvania. Soils are generally medium textured with some fine-textured soils. Topography varies from nearly level to steep upland areas. Only small quantities of digested sludge are available for landspreading each year. The location of the treatment plant allows disposal within short distances of the treatment plant.

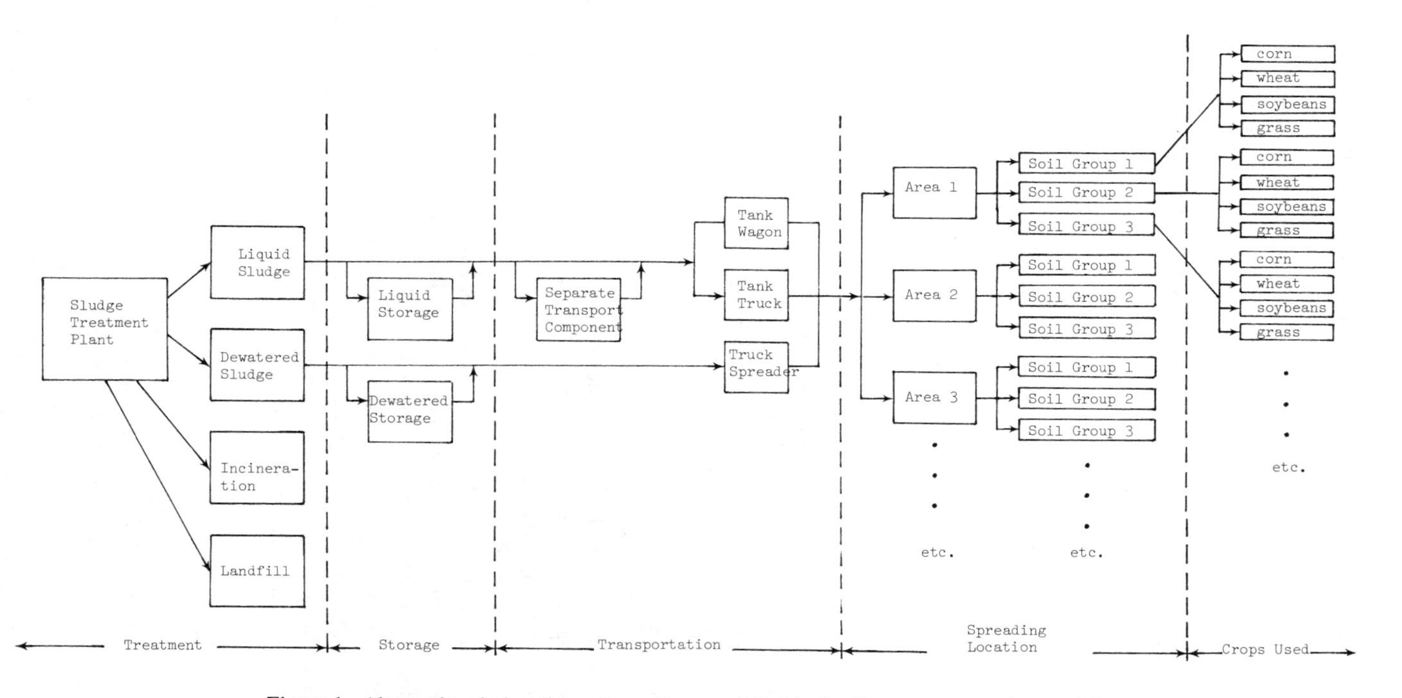

Figure 1. Alternative sludge disposal practices modelled in the linear programming model.

If the sludge is spread at rates to meet the nitrogen needs of a typical crop rotation in the area, NC-118 criteria allow continuous sludge application for 15 years. Thus, the sludge is relatively safe. The major enterprise in the area is dairy with primary crops being hay, pasture, corn, small grains and soybeans.

Montgomery County Sanitary District is located near medium-textured, high-lime, glacial till soils representative of many of the Corn Belt soils. The location of the treatment plants near the city of Dayton may require the sludge to be transported a distance of at least 10 miles. Furthermore, a relatively large amount of sludge needs to be hauled this distance. NC-118 criteria allow sludge, applied to meet the nitrogen needs of a corn-soybean rotation, to be spread at least 11 continuous years. Corn and soybeans are the predominant crops found in the area.

The Zanesville area is representative of the nonglaciated region of Ohio. The region is in the foothills of the Appalachian plateau and has topographic features ranging from nearly level to extremely steep. Although much of the area is not cropped, some productive soils are found on the ridges and plateaus. Pasture, hay and corn are the major enterprise found. Annual sludge production is at relatively small levels. Relatively high cadmium concentrations would limit continuous sludge applications, which meet crop nitrogen needs, to 1.5 yr on continuous corn and 2 yr on pasture-hay lands.

RESULTS

The choice of the optimal sludge-disposal system depends on the viewpoint taken concerning appropriate costs and benefits. One viewpoint is that of the city treatment plant or city utility. Their likely objective is to minimize disposal costs. The return from a farmer's crops is of only peripheral interest when the city receives none of the benefits. A second viewpoint is that of the community or society as a whole. Their objective includes not only the city's interests but also the farmer's.

Results are obtained for each city under these two different objectives. Under the first objective, labelled "city," only disposal costs are minimized. Under the second objective, labelled "society," crop returns less disposal costs (net economic benefits) are maximized.

Defiance

The city of Defiance, located in the Lake Plain region of northwestern Ohio, produces 970 dry tons of sludge/yr. The sludge contains 36.5 lb of available nitrogen, 42 lb of phosphorous and 12 lb of potassium per dry ton of sludge. The three major soils of the Defiance area are Paulding Clay (dark colored and very poorly drained), Latty clay (similar in properties to Paulding) and Roselin silty clay loam (light colored and somewhat poorly drained).

All soils have a cation exchange capacity of 15 or greater and can receive up to 40 dry tons of sludge before the heavy metals become limiting. Of the cropland, 85% is in either corn (20%), soybeans (45%), or wheat (20%). The waste treatment plant is located outside of town so that there is land close by for sludge disposal. The distance sludge is hauled is between 0 and 2 miles.

For the city its lowest cost method of sludge disposal is by tank truck at $20.27/ton. This is a savings of $29.73/ton over landfilling. Only minimal storage is provided under this least-cost solution for the city, and sludge is spread at an average rate of 1.5 tons/ac. If the "society" perspective is taken, the optimal solution calls for spreading sludge over many more acres (Table I).

Table I. Optimal sludge disposal systems for Defiance under two alternative objective functions.

	Objective Functions	
	City Perspective (Minimize Disposal Cost)	Society Perspective (Maximize Net Benefit)
Amount of Sludge (dry tons/yr)	970	970
Disposal Method	Landspreading	Landspreading
Disposal Technology	Tank Truck	Tank Truck
Required Storage (1000 gal)	Minimal	1,162
City's Disposal Cost ($/yr)	19,665	21,130
Net Economic Benefit of Landspreading		–
Over Landfilling ($/yr)	–	51,195
Average Annual Application Rate (dry tons/ac)		
Corn	4.1	0.5
Soybeans	0.4	0.4
Oats	1.2	0.5
Wheat-Summer Fallow	0.5	0.2
Wheat-Soybeans	0.7	0.5
Alfalfa	0.6	0.6
Grass Hay	3.8	0.7
Crops That Receive Sludge (ac)		
Corn	16.7	832
Soybeans	–	377
Oats	163.2	–
Wheat-Summer Fallow	279.7	203
Wheat-Soybeans	–	650
Grass Hay	189.7	–
Total	649.3	2,062

Also, this solution requires 1,162,000 gal of storage during the mid-May to mid-July period. While the city bears slightly higher costs than under its least-cost alternative, net returns from using sludge as a resource are substantial ($51,195). The city's higher costs are from the higher transport costs ($587) to travel the extra distance to dispose of sludge at the phosphorous rate and the added $878 storage costs.

Crops receiving sludge differ between the city perspective and society perspective. The minimize disposal costs, the city uses grass-hay and small-grain crops. To maximize returns to society, the crop acreage to which sludge is applied is corn, soybeans and wheat. Similarly, the application rates differ between the two perspectives. Under the city solution sludge is spread over 649 ac, but under that of society, sludge is spread over 2062 ac.

Medina County

The county of Medina is located in northeastern Ohio. The county waste system landspreads around 260 tons of sludge per year. The sludge contains 21.7 lb of available nitrogen, 89 lb of phosphorous and 6 lb of potassium per dry ton. The land for recycling sludge is 2 to 5 mi away from the sludge plants. Two representative soils of this region are Mahoning silt loam, a poorly drained soil that is formed in clayey glacial till and found in nearly level to gently sloping upland areas, and Ellsworth silt loam. The Ellsworth soil is a moderately well-drained soil also formed on clayey glacial till, and it occupies gently sloping to very steep upland areas. Both soils have a cation capacity of 15 or greater and they can receive up to 95 tons of sludge/ac before heavy metals become limiting. Cropland consists of 25% corn, 10% soybeans, 10% wheat, 10% oats and 30% hay.

From the city's viewpoint, the lowest cost method of disposal is landfilling. Landfilling is less expensive than tank-truck landspreading by $3010 ($11.58/ton). The main reason that landspreading was not the lowest cost method of disposal is that with only 260 tons the fixed cost per ton of landspreading becomes quite high.

For society the optimum solution is to landspread using a tank truck (Table II). This solution requires a total of 476,000 gal storage from the first of May to the middle of July. For the county to adapt to the optimal society solution it would incur an increased annual cost of $3370 ($12.96/ton): $3070 for using the tank truck and $300 storage cost. However, when society's net benefits are maximized, sludge is applied to corn and oats and realizes a fertilizer value of $45.57/ton. Annual application rates average only 0.2 tons/ac as the 260 tons are spread over 1300 ac. The net economic benefit to the community totals $11,804 by using landspreading rather than landfilling.

Table II. Optimal sludge disposal systems for Medina County under two alternative objective functions.

	Objective Functions	
	City Perspective (Minimize Disposal Cost)	**Society Perspective (Maximize Net Benefit)**
Amount of Sludge (dry tons/yr)	260	260
Disposal Method	Landfill	Landspreading
Disposal Technology	–	Tank Truck
Required Storage (1,000 gal)	Minimal	476
City's Disposal Cost ($/yr)	13,000	16,370
Net Economic Benefit of Landspreading Over Landfilling ($/yr)	–	11,804
Average Annual Application Rate (dry tons/ac)		
Corn		0.2
Soybeans		0.2
Oats	(No Landspreading)	0.2
Wheat-Summer Fallow		0.1
Wheat-Soybeans		0.2
Alfalfa		0.3
Grass Hay		0.3
Crops That Receive Sludge (ac)		
Corn		875
Soybeans		–
Oats		425
Wheat-Summer Fallow	(No Landspreading)	–
Wheat-Soybeans		–
Alfalfa		–
Grass Hay		–
Total		1300

Montgomery County

The Montgomery County Sanitary District produces 1300 tons of sludge annually. The sludge has the following characteristics: available nitrogen 34.6 lb, phosphorous 80 lb, and potassium 6 lb per dry ton. Furthermore, the sludge is only 1.5% solids. The main soils in the area are Miamian, Crosby, Celina and Brookston. Miamian soil is well drained and is found on slopes. The Celina soil is very similar to the Miamian in soil characteristics and in yield capacity but is found in more level areas than the Miamian. Brookston clay loam soil is poorly drained and occurs in level and depressed areas. Crosby silt loam is a light colored and somewhat poorly drained soil which is found on gently sloping low knolls. The Brookston soil has a cation exchange capacity of greater than 15 and can receive up to 31 tons of sludge/ac. The other soils have a cation exchange capacity of 10 to 15 and can only take up to 15.6 tons of sludge/ac before heavy metals become limiting. Corn comprises 45% of the cropland, soybeans 30% and wheat 10%.

The lowest cost method of sludge disposal is dewatering it and spreading it by the truck spreader. This method costs $38,415 ($29.55/ton), which is $20.45/ton cheaper than landfilling (Table III). In comparing costs between landspreading dewatered sludge and liquid sludge, the cost of dewatered sludge can increase by $41.52/ton before liquid sludge would be entered. Even if Montgomery County had liquid sludge with 3% solids, dewatered sludge could increase in cost by $22.37/ton before liquid sludge would be economical.

Table III. Optimal sludge disposal systems for Montgomery County under two alternative objective functions.

	Objective Functions	
	City Perspective (Minimize Disposal Cost)	Society Perspective (Maximize Net Benefit)
Amount of Sludge (dry tons/yr)	1,300	1,300
Disposal Method	Landspreading	Landspreading
Disposal Technology	Truck Spreader	Truck Spreader
Required Storage (1,000 gal)	Minimal	2,055
City's Disposal Cost ($/yr)	38,415	39,969
Net Economic Benefit of Landfilling Over Landspreading ($/yr)	–	48,259
Average Annual Application Rate (dry tons/ac)		
Corn	5.2	0.6
Soybeans	0.5	0.5
Oats	1.2	0.6
Wheat-Summer Fallow	0.6	0.3
Wheat-Soybeans	0.8	0.5
Alfalfa	0.7	0.7
Grass Hay	4.2	0.8
Crops That Receive Sludge (ac)		
Corn	–	1,390
Soybeans	–	26
Oats	–	–
Wheat-Summer Fallow	470	–
Wheat-Soybeans	–	823
Alfalfa	–	–
Grass Hay	255	–
Total	725	2,239

For society's interests, recycling sludge by the truck spreader is continued but at relatively low application rates. If just the city's interests are followed, sludge is spread on 725 ac of wheat and grass hay. If society's interests are followed, 2239 ac of corn, soybean and wheat are used at application rates averaging 0.58 tons/ac. Also, the society solution adds 2,055,000 gal storage during the mid-May to mid-July period and minimal storage during November. Under this method of landspreading, society's net returns are $48,259 ($37.12/ton) greater than under landfilling (Table III).

Zanesville

The representative city of nonglaciated southeastern Ohio is Zanesville. Sludge production is 600 tons of sludge/yr, and the sludge contains 29.6 lb of available nitrogen, 50 lb of phosphorus and 6 lb of potassium per dry ton. There is land available for sludge recycling 1 to 2 mi away from the waste treatment plant. There are two main soils, Muskingum silt loam and Zanesville silt loam. The Muskingum soil occupies smoothly rolling to somewhat hilly ridges and upper slopes and shoulders. It is a well-drained soil. The Zanesville soil is in areas of either level or undulating plateaus. Drainage is not a problem with the Zanesville soil. The cation exchange capacity for both soils is over 15. Due to the high amounts of cadmium in Zanesville's sludge only 11.1 tons of sludge/ac can be applied to the soil before it is no longer safe from heavy metals. Hay comprises 55% of the cropland and corn 30%.

Disposing sludge by landspreading with a tank-truck unit is the lowest cost method. Only minimal storage is provided, and application rates average 2.2 tons/ac (Table IV). Using this method costs the city of Zanesville $18,110 ($30.17/ton), (Table IV), which is a saving of $19.83 over landfilling.

The society solution also uses the tank truck, but application rates average only 0.45 tons/ac. While the city would prefer to spread sludge on grass hay and alfalfa to lower disposal costs, the society solution has landspreading occurring on corn and grass hay. Added storage must be built to accommodate 978,000 gal from mid-May to October. For the society solution the city would incur a cost of $616 in storage charges, but society would have net benefits of $28,328 above landfilling.

Sensitivity Analysis

For each of the four cities the costs used in the analysis are critical. Costs of the alternative disposal technologies largely influence the choice of the optimum systems estimated in the preceding section. Due to the difficulty in accurately specifying average annual costs of sludge disposal technologies, an analysis was made of the sensitivity of the communities' optimal solutions to critical cost parameters.

Table IV. Optimal sludge disposal systems for Zanesville under two alternative objective functions.

	Objective Functions	
	City Perspective (Minimize Disposal Cost)	**Society Perspective (Maximize Net Benefit)**
Amount of Sludge (dry tons/yr)	600	600
Disposal Method	Landspreading	Landspreading
Disposal Technology	Tank Truck	Tank Truck
Required Storage (1,000 gal)	Minimal	978
City's Disposal Cost ($/yr)	18,110	18,726
Net Economic Benefit of Landspreading Over Landfilling ($/yr)	–	28,328
Average Annual Application Rate (dry tons/ac)		
Corn	5.2	0.6
Soybeans	0.3	0.3
Oats	1.2	0.4
Wheat-Summer Fallow	0.7	0.2
Wheat-Soybeans	0.8	0.35
Alfalfa	0.5	0.5
Grass Hay	4.6	0.6
Crops That Receive Sludge (ac)		
Corn	56.7	598
Soybeans	–	–
Oats	–	–
Wheat-Summer Fallow	–	–
Wheat-Soybeans	–	–
Alfalfa	110.6	
Grass Hay	103.6	722
Total	270.9	1,320

For each of the cities, estimates of the community's disposal costs were made under five alternative disposal methods–tank truck, tank wagon, truck spreader, a hauling unit plus a tank truck and landfill (Table V). These costs vary with each community due to differences in size of community, distance to disposal site, soil type, crop receiving sludge and characteristics of the soil. For Defiance, the tank truck is the optimum technology with costs of $20.27/ton. However, if the present charge for tank wagon disposal ($34.41/ton) was reduced to less than $20.27/ton, it would be preferred. Similarly, the truck spreader using dewatering technology would enter as the best method if costs dropped from $28.51 to $20.27/ton.

Table V. Annual cost per ton for sludge disposal in four modelled communities.[a]

	Cities Modelled			
Technology	Defiance	Medina County	Montgomery County	Zanesville
Tank Truck	$20.27	$ 61.58	$121.09	$30.18
Tank Wagon	34.41	72.06	273.96	53.39
Truck Spreader[b]	28.51	63.43	29.55	36.44
Hauling Unit plus Tank Truck	40.75	119.59	71.07	58.65
Landfill[b]	50.00	50.00	50.00	50.00

[a]Annual costs are for disposal of sludge. No nutrient returns are assigned.
[b]Includes dewatering costs.

In the optimum solutions, Defiance, Montgomery County and Zanesville use landspreading. For two of the landspreading cities, Defiance and Zanesville, the tank truck with no sludge dewatering is used. On examination of Table V, slightly higher tank truck spreading costs cause the tank truck technology in Defiance and Zanesville to leave the solution and other spreading technologies to enter. Thus, landspreading technology is quite sensitive to the cost estimates; however, the sensitivity analysis shows that the cost of landspreading can greatly increase before landfilling is selected. In Montgomery County a more stable solution is found. Due to the distance hauling requirement (greater than 10 mi), the optimum plan calls for dewatering at the plant and spreading by truck spreader. This solution holds for a wide range of costs for all disposal technologies.

In Medina County, landfilling rather than landspreading is used in the "city's" plan due to the relatively small quantity of sludge and resulting high fixed costs for landspreading. However, this solution is sensitive to the cost estimate for landfilling. If landfilling costs increase from $50/ton to $61.58/ton, landfilling would be replaced by one of the landspreading technologies.

CONCLUSIONS

The increased interest in landspreading on the part of small- to moderate-sized communities is partly explained by this analysis. Landspreading of sludge is a low-cost alternative for the four modelled communities which represent a wide range in important characteristics such as community size, soil type, distance to disposal site, sludge characteristics and local cropping practices.

Conclusions regarding optimum sludge-spreading rates are significant in light of usual recommendations. Rather than spread at a rate corresponding to heavy metal restrictions or at rates corresponding to nitrogen needs for crops, a community using the "society" perspective would spread at rates corresponding to phosphorus needs of crops. These rates are low enough that heavy metal levels would prohibit landspreading with only extremely "dirty" sludge. From this viewpoint sludge is a resource to be allocated among competing crops. Crop nutrients are wasted if any of the important nutrients (nitrogen, phosphorus or potassium) are supplied in excess of crop needs. Thus, application rates are based on that nutrient in sludge which first satisfies the particular crop needs.

The low sludge application rates increase substantially the number of acres to which sludge is applied. However, for all communities modelled, this expanded acreage is well below the crop acreage in the area. Furthermore, the acres of individual crops actually grown in the areas could accommodate the sludge-spreading programs.

Sludge disposal costs increase only slightly if this "society" perspective is taken rather than the narrow "city" perspective. On the other hand, community benefits increase substantially when the broader perspective is used. A method of farmers reimbursing the city for the extra landspreading costs incurred in spreading at relatively low application rates may alleviate the problem of the "city's" desire to spread at high rates and "society's" desire to use low rates of application.

The kinds of crops receiving sludge differ considerably between the "city" and "societal" perspectives. Under the "city" objective of minimizing disposal costs, crops which provide nearly year-round spreading such as pasture, hay or small grains are used. With the "society" objective of maximizing net crop returns less disposal costs, sludge is spread on high-profit crops such as corn and wheat. These high-profit crops require the city to build added storage to accommodate the sludge during the growing season.

For communities producing relatively small quantities of sludge, high capital investments in spreading equipment may preclude landspreading of sludge. However, for communities producing above approximately 350 dry tons of sludge per year, landspreading appears to be a low-cost alternative.

Dewatering sludge is beneficial only when hauling distances are relatively long or percent solids in the sludge are low. For most small- to moderate-sized communities with solids of 3% or greater, spreading liquid sludge with the tank truck is the preferred technology for both spreading and transporting.

REFERENCES

1. Council for Agricultural Science and Technology (CAST). "Application of Sewage Sludge to Cropland: Appraisal of Potential Hazards of the Heavy Metals to Plants and Animals" Report No. 64 (November 22, 1976).

2. Forster, D. L., T. J. Logan, R. H. Miller and R. K. White. "State of the Art in Municipal Sewage Sludge Landspreading," *Land as a Waste Management Alternative,* R. C. Loehr, Ed. (Ann Arbor: Ann Arbor Science Publishers, 1977).
3. Crop Reporting Service. "Ohio Agricultural Statistics." 1975 Annual Report (May 1976).
4. Brown, R. E., *et al.* "Ohio Guide for Land Application of Sewage Sludge," Ohio Agricultural Research and Development Center, Research Bulletin 1079 (May 1976).

14

CROP YIELDS AND WATER QUALITY AFTER APPLICATION OF SEWAGE SLUDGE TO AN AGRICULTURAL WATERSHED

C. E. Clapp, D. R. Duncomb, W. E. Larson,
D. R. Linden, R. H. Dowdy and R. E. Larson
Agricultural Research Service,
U.S. Department of Agriculture, and
University of Minnesota
St. Paul, Minnesota

INTRODUCTION

Disposal of the products of municipal wastewater treatment plants has become one of our major environmental problems. Historically, land treatment methods have been concerned with sewage material disposal. Now, with greater concern for the environment and the need to conserve energy, a more modern approach to wastewater treatment by recycling on agricultural land seems desirable. Recent reviews by Carroll *et al.*,[1] Dowdy *et al.*[2] and Chaney *et al.*[3] have adequately covered potential hazards and problems as well as agricultural benefits of land treatment of sewage products. Many of the data in the United States have been gathered using small field plots and laboratory studies.

In this project it was proposed to develop a practical and complete farm management system for handling and using sewage sludge on agricultural land. Incorporated was a monitoring system for determining possible environmental effects.

The technology necessary for successful use of the sewage resource in agriculture needs development and demonstration under the soil and climatic conditions of the area in which it is used. The relatively short Minnesota summers offer an environment different from those where other major research is conducted on agricultural use of sewage sludge. Low soil temperatures result in relatively slow sludge decomposition, which will affect the permissible nitrogen loading rates. In addition, differences in soils and crops require a technology suitable to local conditions.

The overall objective of this study was to develop efficient, practical and environmentally safe methods for applying sewage sludge to land in harmony with agricultural usage. In this chapter we will present a summary of crop yields and water quality on an agricultural watershed that had received sewage sludge applications for three years.

EXPERIMENTAL

Site Selection and Construction

A 16-ha watershed at the Rosemount Agricultural Experiment Station of the University of Minnesota near Rosemount, Minnesota, was selected for study. The soils of the watershed have from 60 to 240 cm of silt loam overlying compact glacial till. Port Byron (Typic Hapludolls), Bold (Typic Udorthents) and Tallula (Typic Hapludolls) are the dominant soil types.

The watershed was terraced to give 13 treatment areas with separate surface tile inlets (Figure 1). Slopes of the treatment areas after terracing ranged from 2 to 10%. For construction details, see Larson *et al.*[4] The parallel terraces were built by a cut and fill technique, leaving a 2:1 backslope. The backslopes were seeded to Kentucky bluegrass. The terrace ridges, spaced 40 m apart, were designed to impound a maximum of 6.4 cm of runoff. The terrace channels were graded to the inlet pipes which were placed in natural depressions. Solid PVC pipe connected each inlet to sampling stations below the bottom treatment areas. At the sampling station, flow from runoff water was measured and sampled automatically. Water drainage from sampling stations was stored in a reservoir designed to handle runoff from a 100-yr storm (11.2 cm). Water stored in the runoff reservoir was periodically irrigated onto a 2-ha site in the northwest part of the watershed.

Sludge Storage and Application Method

Liquid digested sludge (approximately 2 to 5% solids) from four metropolitan wastewater treatment plants was transported to the watershed by tank truck and stored in two lagoons. The combined capacity of the lagoons was 11,400 m^3 (3 million gal) or approximately 6 months' storage capacity of sludge from the four treatment plants.

An underground pipeline extended from the sludge lagoons across the midline of the watershed. Riser pipes and outlet valves to which a traveling overhead sprinkler or subsurface injector could be connected by flexible hose were

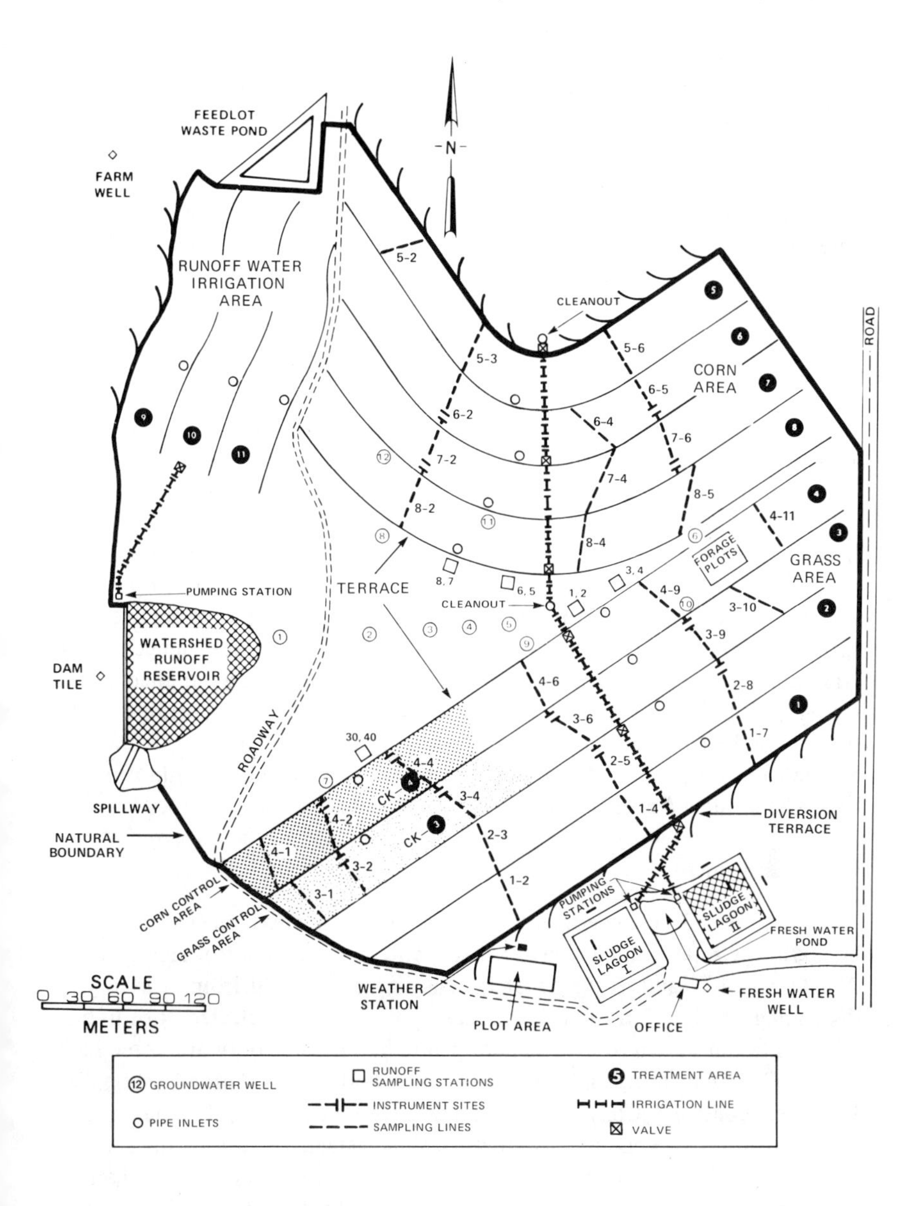

Figure 1. General plan showing engineering design and sample collection locations for the Rosemount sewage sludge watershed.

installed at the tops of treatment areas 1 and 5 and in the terrace channels of 2, 4, 6 and 8 (Figure 1). Before application, sludge in the lagoon was mixed by using a tractor-driven pump. Sludge was applied at rates of 0.6 to

1.1 ha-cm/hr to the grass areas in 1974-76 and to the corn areas in 1974-75 with a Nelson Big Gun[a] sprinkler attached to a Hydro Sprinkle Winch. Sludge was applied at the rate of 1.5 ha-cm/hr to the corn areas in 1975-76 with a Briscoe-Mathis subsurface injector.

During sludge applications, tile inlets were covered to prevent sludge from entering the runoff drainage system. To minimize movement of aerosols, the traveling overhead sprinkler was operated only when wind speed was less than 8 kmph (5 mph).

Design and Sludge Treatments

Ten separate treatment areas (terrace intervals), five cropped to corn (5, 6, 7, 8 and CK-4) and five to reed canarygrass (areas 1, 2, 3, 4 and CK-3), within the watershed were treated as replicated blocks in our experimental arrangement (Figure 1). One area of each crop was a control which received no sludge but did receive adequate amounts of commercial fertilizer. We believed that the sludge-treated corn areas would receive about 7.5 cm of sludge each year, half in spring and half in fall if applied by traveling overhead sprinkler, or all in fall if applied by subsurface injector. The sludge-treated grass areas would receive about 10 cm of sludge each year, applied in the spring and before significant regrowth after each of three cuttings in the summer and early fall. The runoff water irrigation site (areas 9, 10 and 11), seeded to reed canarygrass and quackgrass, received accumulated runoff water collected in the watershed reservoir but no sludge.

Crop Management

During the 3-yr experimental period, all corn areas received annually 66 kg/ha of K as commercial fertilizer, and the corn control area received 230 kg/ha of N and 20 kg/ha of P. Planting date varied from year to year depending on duration of spring sludge application. A selected corn hybrid was planted in 75-cm rows at 62,500 plants/ha. A corn rootworm insecticide was applied at planting and a postemergence herbicide was broadcast-applied for weed control. Corn was harvested about September 15 and ensiled.

The reed canarygrass areas were seeded and established during the summer of 1975. Before planting, all areas received 33 kg/ha of K, and the control area received 10 kg/ha of P and 94 kg/ha of N. In April 1976, all grass areas received 33 kg/ha of K, and the control area received 29 kg/ha of P and 134 kg/ha of N, applied half in April and half after the second cutting in

---

[a]Mention of trade products or companies in this chapter does not imply that they are recommended or endorsed by the Department of Agriculture over similar products of other companies not mentioned. Trade names are used here for convenience in reference only.

July. Three cuttings were harvested–in early June, mid-July and early September–and baled for livestock feed.

Sludge Application and Composition

Total sludge applications by seasons on the corn and grass areas are given in Table I. The differences in amounts applied result from variability in lagoon mixing and pumping. By fall 1975, conditions were satisfactory for adequate solids application.

Table I. Sludge applied by seasons for Rosemount liquid sewage sludge project (1974-1976).

Season	Appli- cations	Sludge (cm)	Solids (%)	Solids (m.ton/ha)	N (kg/ha)	P (kg/ha)	K (kg/ha)
Corn							
Spring 1974[a]	2	2.6	0.60	1.8	211	64	61
Fall 1974[b]	3	3.0	1.74	5.1	230	100	31
Spring 1975[b]	3	4.6	0.81	3.7	269	65	40
Fall 1975[c]	2	6.4	2.80	17.8	898	614	72
Total	10	16.6	–	28.4	1608	843	204
Grass							
Summer 1974[b]	4	5.1	1.46	7.4	522	164	95
Summer 1975[b]	2	2.0	1.19	2.4	93	52	14
Spring 1976[b]	1	1.9	3.03	5.7	327	214	22
Summer 1976[b]	2	3.9	1.22	4.9	314	96	30
Total	9	12.9	–	20.4	1256	526	161

[a] Application by tank wagon.
[b] Application by traveling overhead sprinkler irrigation.
[c] Application by subsurface injection.

Analyses of composite samples of sludge applied over the 3-yr period are given in Table II. High standard deviations illustrate the variations in mixing and spreading. For applications by traveling overhead sprinkler, sludge samples were collected in plastic pans at three sites on each of three sampling lines and composited for each treatment area. For subsurface injection applications, samples were taken from a manifold valve during injection over each of three sampling lines and composited for each corn treatment area.

Sample Collection

Water from runoff events on each area was collected by Isco model 1391 automatic samplers modified to begin collection when water flow started and at 1-hr intervals during runoff. Flow rates were measured by water stage

Table II. Analyses of sludge for Rosemount liquid sewage sludge project (1974-1976).

Characteristic	Mean[a] (%)	Characteristic	Mean[a]
Liquid			
Total Solids	1.82	Conductivity (μmhos/cm)	4260
Volatile Solids	0.80	pH	8.0
Total N	6.01 (1010 mg/l)		
NH_4-N	3.04 (482 mg/l)		
Freeze-dried[b]			(ppm)
OM	45.8	Cr	2660
C	23.0	Zn	895
P	2.76	Cu	665
K	0.82	Mn	346
Ca	4.09	Pb	280
Na	1.64	B	36
Mg	0.74	Ni	18
Al	0.98	Cd	7
Fe	0.82		

[a]Weighted means for 10 applications on corn areas and 9 applications on grass areas.
[b]Values based on 105°C weight as percentage of total solids.

recorder with slotted tube and stilling well.[5] Soil water was sampled at 3- to 4-week intervals by porous ceramic samplers installed at 60- and 150-cm depths at 24 sites (Figure 1). Duplicate samplers were installed at each depth. Two sampling sites were located in the channel and at midterrace interval on each of the two center corn and grass sludge-treated areas (four per treatment area). Four sampling sites were located on each of the control areas. Samples from 12 shallow ground-water monitoring wells within the watershed were collected monthly. Background samples from various water sources, both within and around the watershed, were taken bimonthly for the 1973 season before any sludge was applied on the project site and compared with water samples taken monthly during the 1974 to 1976 growing seasons.

Corn leaf samples (center 1/3 of leaf below and opposite the primary ear) were taken at the 75% silking stage along three sampling lines (Figure 1) of each corn area (15 plots). Corn was harvested usually at late dent stage. Grain was picked, then the rest of the plant was chopped as stover from 15 m of row (10 rows, 1.5 m long) on the same three sampling lines of each corn area. Reed canarygrass samples were harvested at three cutting times–early June, mid-July and early September. Plots of 0.9 x 6.1 m were located at three sites (in the terrace channel, at the midterrace interval and near the upper terrace interval) along three sampling lines of each grass area (45 plots).

Composite soil samples at 0- to 15- and 15- to 30-cm depths were taken in the fall of 1973 after the watershed was constructed, along sampling lines designated on each terrace interval. Single samples were also taken at specific sites in the channel and at the midterrace interval on the sampling lines. The same sites were resampled in the fall of 1975.

Analyses

Water and sludge samples were refrigerated immediately after collection. Analyses of organic components were performed within one week or the samples were acidified. Soil samples were oven-dried at 35°C; plant samples were dried at 65°C.

For sludge samples, total and volatile solids were determined at 105°C and 550°C, respectively. Total N by micro-Kjeldahl[6] and NH_4-N by distillation and titration were determined on liquid samples, but expressed both as percentage of total solids and milligrams per liter. Conductivity and pH were measured by standard procedures given in Taras *et al.*[7] Organic matter was estimated by heating in a muffle furnace 2 hr at 650°C. Total C was measured by Leco C determinator. Total P, K, Ca, Mg and heavy metals were determined by dry-ashing at 450°C for 24 hr. The ash was equilibrated with 1.1 *N* HCl and the extract analyzed by emission spectrography, or atomic adsorption spectroscopy using background correction (Pb, Zn, Cu, Cr, Cd and Ni).

For water samples, total N and total P (after digestion) and NH_4-N, NO_3-N and PO_4-P were determined by Technicon AutoAnalyzer. Analyses for Na, Ca, Mg and K were performed on a Perkin-Elmer 303 atomic adsorption spectrophotometer. Conductivity, pH and fecal coliform counts were determined by standard procedures.[7]

Plant sample analyses were run by the same procedures as those for sludge. Soil analyses included C by Leco C determinator and N by micro-Kjeldahl.[6]

RESULTS AND DISCUSSION

Crop Yield and Composition

Corn yields and total nutrients removed for the 3 years of the project are shown in Table III. Yields for 1974 and 1976 were comparable for control and sludge-treated areas, but the 1975 control area yields were reduced by corn rootworm and weed infestation as a result of the wet spring. Because corn was harvested at late dent stage, grain yields were somewhat lower than they would have been if corn had been grown to physiological maturity. Total nutrients removed reflect the yield differences and nutrient contents. Highest N removal (194 kg/ha) was in 1976 for the sludge-treated areas.

Table III. Corn yields and nutrients removed for the Rosemount liquid sewage sludge project (1974-1976).

		Yield			Total Nutrients Removed		
		Fodder[a]	Grain		N	P	K
Treatment	Year	(m.ton/ha)	(m.ton/ha)	(bu/ac)		(kg/ha)	
Control	1974[b]	15.1	6.7	106	172	30	171
	1975[c]	10.9	4.7	75	143	19	90
	1976[d]	15.2	7.8	125	177	34	127
	Mean	13.8	6.4	102	164	28	129
Sludge	1974[b]	14.0	6.2	99	144	29	136
	1975[c]	14.1	6.6	104	184	32	118
	1976[d]	15.4	7.7	123	194	33	126
	Mean	14.5	6.8	108	174	31	127

[a]Fodder = total dry matter = grain + cobs + stover.
[b]'Pioneer 3780' (105 day) 67,300 plants/ha. Means of 3 replicated plots on the control area; 12 replicated plots on the sludge-treated areas.
[c]'Northrup King PX476' (100 day) 62,900 plants/ha.
[d]'Northrup King PX448' (95 day) 52,800 plants/ha.

First-year reed canarygrass yields (Table IV) were higher for the sludge-treated areas than for the control area for all cuttings, with a total increase of 20% for the season. Total N removed by the reed canarygrass was quite high, averaging 33% higher than the N removed by corn in 1976.

Concentrations (Table V) of the major plant nutrients N, P and K in the corn leaf tissue of both control and sludge areas were within adequate limits for good corn growth,[8] except for an unexpected decrease in N content in 1976.

Table IV. Reed canarygrass yields and nutrients removed for the Rosemount liquid sewage sludge project (1976).

	Yield				Total Nutrients Removed		
	1st Cut	2nd Cut	3rd Cut	Total	N	P	K
Treatment		(m.ton/ha)[a]				(kg/ha)	
Control	3.9	2.5	1.4	7.8	242	27	311
Sludge	4.6	3.4	1.7	9.7	291	28	327

[a]'Rise' reed canarygrass. Means of 9 replicated plots on the control area; 36 replicated plots on the sludge-treated areas.

Table V. Elemental concentrations in corn and reed canarygrass tissue for Rosemount liquid sewage sludge project (1974-1976).

	1974		1975		1976	
	Control	**Sludge**	**Control**	**Sludge**	**Control**	**Sludge**
Element			(%)			
Corn leaf[a]						
N	3.11	2.96	3.22	3.25	2.80	2.64
P	0.33	0.34	0.32	0.34	0.28	0.30
K	2.42	2.45	2.01	2.04	2.23	1.96
			(ppm)			
Zn	32	29	23	23	25	41
Cu	8	10	7	10	8	11
B	6	7	6	4	10	12
Pb	2.4	2.5	1.3	1.2	1.0	0.7
Cr	0.8	0.7	0.5	0.4	0.5	0.7
Ni	0.3	<0.3	0.4	0.3	<0.3	<0.3
Cd	0.16	0.14	0.07	0.06	0.18	0.10
			(%)			
Corn grain[a]						
N	1.59	1.48	1.72	1.73	1.68	1.76
P	0.34	0.36	0.26	0.36	0.34	0.34
K	0.53	0.59	0.47	0.50	0.36	0.38
			(ppm)			
Zn	19	22	15	22	24	28
Cu	1.4	1.4	1.2	2.0	–	0.7
B	3.0	3.0	2.5	2.2	<1.0	<1.0
Pb	<0.3	<0.3	<0.3	<0.3	0.3	<0.3
Cr	<0.2	0.4	0.2	0.1	0.1	<0.1
Ni	<0.3	<0.3	0.3	0.3	0.3	<0.3
Cd	<0.06	<0.06	<0.06	<0.06	<0.06	<0.06
					(%)	
Reed Canarygrass[b]						
N					3.12	3.03
P					0.27	0.30
K					3.07	3.55
					(ppm)	
Zn					26	35
Cu					5.4	9.2
B					3.2	3.8
Pb					<0.3	0.8
Ni					1.8	1.9
Cd					<0.06	<0.06

[a]Means of 2 replicated plots on the control area; 8 replicated plots on the sludge-treated areas.

[b]Means of 6 replicated plots on the control area; 12 replicated plots on the sludge-treated areas.

This may be a varietal effect or related to sampling time variation during the dry 1976 season. Concentrations of N, P and K in corn grain tissue were normal for both control and sludge-treated areas. Heavy metal concentrations in corn leaf or grain tissue were comparable for control and sludge-treated areas. The reed canarygrass tissue analyses were normal for good quality grass and again showed no significant differences between sludge and control areas.

Soil Analyses

Analyses of composite and single-site soil samples taken in 1973 before sludge was applied and again in the fall of 1975 showed no measurable increases in soil C or N as a result of sludge addition. The quantities of sludge applied apparently were not great enough to determine by the usual procedures. Other possible changes in soil composition such as heavy metal buildup or cation imbalance are being studied further.

Water Quality

The 1975 season represented both typical and unusual features to serve as an example of runoff for the watershed. An average snow cover in the winter of 1974-75 equivalent to 7.0 cm of water produced significant runoff events in the spring. Runoff from the south-facing areas (corn, Figure 1) began earlier in the season, had greater intensity and lasted for a shorter time than did that from the north-facing areas (grass). Nutrient concentrations in snowmelt runoff were higher than those in rainfall runoff later in the season. Twelve rainfall events of more than 1.25 cm accounted for the remaining runoff. During snowmelt and early spring runoff, the corn areas yielded mean values of 0.02 and 0.03 m.ton/ha sediment, 1.8 and 22.4 kg NO_3-N/ha, 0.01 and 0.16 kg PO_4-P/ha and 0.1 and 0.1 kg K/ha for the control and sludge-treated areas, respectively. The season totals for the corresponding areas were 0.2 and 0.5 m.ton/ha sediment, 2.8 and 28.8 kg NO_3-N/ha, 0.05 and 0.23 kg PO_4-P/ha and 0.4 and 0.6 kg K/ha. Results from the reed canarygrass areas were incomplete in 1975 since the stand was just being established. Results of the heavy metal status during runoff are described by Larson and Dowdy.[9]

In April 1975, a record precipitation event (9.3 cm) caused erosion damage to some of the terrace structures. Recent backfilling of the underground irrigation line caused a portion of terrace ridges 1 and 6 to wash out. Some flooding and overtopping onto lower treatment areas resulted; consequently, data for this event were incomplete. The damaged terrace ridges were repaired shortly afterward, and no other problems were experienced for the remainder of the study. All treatment areas were bare during the winter of 1974 and spring of 1975. The 1976 season was dry, with no major runoff events other than snowmelt.

Soil water samples for chemical analyses were taken at the 60- and 150-cm depths for several midterrace interval and channel sites (Figure 1) on the corn areas during three growing seasons and on the reed canarygrass areas for two seasons. Seasonal means of nitrate-N and phosphate-P given in Table VI for the corn areas were lower for the control treatment and increased both with depth and moving downslope into the channel, as expected. Nitrate-N represented 95 to 100% of the total inorganic N; ammonium-N values never exceeded 0.1 mg/l. Nitrate-N has generally increased over the 3-yr period for both control and sludge areas, but values were still within reasonable limits for soil-water nitrogen in a well-fertilized corn system.[10] For the grass areas, nitrate-N levels in 1976, the first year of established sod, were lower than on comparable corn areas. Phosphate-P levels remained low for all treatments. Heavy metal concentrations (Cd, Zn, Cu, Ni, Pb and Cr) in soil water taken at both depths were not increased above background levels by a sludge application of about 10 m.ton/ha/yr.

Determination of surface and subsurface water quality in the watershed and surrounding area from sludge stored in the lagoons and applied on the treatment areas required a complete system of water sampling. During the three seasons, nitrate-N in the shallow ground water ranged from 1 to 19 mg/l; ammonium-N and phosphate-P were very low. The level of heavy metals in ground water was determined by analyzing composite samples collected between July and September from five of the wells below the areas receiving sewage sludge. In all cases, the concentration of heavy metals was less in the ground water samples than in the soil water samples. Background sampling from various water sources both within and around the watershed (Figure 1) were taken prior to the presence of any sludge on the project site in 1973. Analyses from several of the surface and subsurface water sources sampled in 1976 (Table VII) showed no movement of potentially polluting materials out of the watershed.

SUMMARY

Results for the three years of sewage sludge application to the terraced watershed showed that crop production was good. Surface and ground water quality remained high.

Total sludge applications through three growing seasons were 17 cm on the corn areas and 13 cm on the grass areas. These amounts were equivalent to 28 and 20 m.ton/ha total solids and 1610 and 1260 kg/ha total N, respectively.

Corn yield means for three seasons were 13.8 m.ton/ha fodder and 6.4 m.ton/ha grain (102 bu/ac) on the fertilized control area and 14.5 m.ton/ha fodder and 6.8 m.ton/ha grain (108 bu/ac) on the sludge-treated land. Reed canarygrass dry matter yields for one cropping season

Table VI. Nitrogen (NO_3-N) and phosphorus (PO_4-P) concentrations in soil water for Rosemount liquid sewage sludge project (1974-1976).

	Nitrate-N (mg/l)[a]			Phosphate-P (mg/l)[a]		
Treatment	1974	1975	1976	1974	1975	1976
Corn						
60-cm depth						
Midterrace						
Control	24	11	26	0.02	0.01	0.01
Sludge	32	72	95	0.08	0.01	0.06
Channel						
Control	26	18	40	0.03	0.01	0.11
Sludge	65	122	76	0.04	0.06	0.20
150-cm depth						
Midterrace						
Control	14	18	31	0.02	0.04	0.04
Sludge	28	48	78	0.06	0.02	0.01
Channel						
Control	25	24	54	0.01	0.01	0.01
Sludge	18	34	44	0.04	0.04	0.10
Grass						
60-cm depth						
Midterrace						
Control	–	44	22	–	0.07	0.02
Sludge	–	54	21	–	0.02	0.02
Channel						
Control	–	16	19	–	0.12	0.02
Sludge	–	36	5	–	0.02	0.04
150-cm depth						
Midterrace						
Control	–	35	21	–	0.01	0.01
Sludge	–	18	20	–	0.01	0.00
Channel						
Control	–	40	36	–	0.03	0.01
Sludge	–	28	21	–	0.01	0.02

[a]Means of 8 replicated samples on each corn and grass control area and 16 replicated samples on each corn and grass sludge-treated area taken by porous ceramic samplers at 2- to 4-week intervals between May and October.

were 7.8 m.ton/ha on the fertilized control area and 9.7 m.ton/ha on the sludge-treated areas. Plant tissue showed normal concentrations of N, P and K with no increase in heavy metal contents due to sludge application.

Table VII. Summary of water[a] analysis from various sources for Rosemount liquid sludge project (1973-1976).

	Watershed Reservoir		Dam Tile		Farm Well		Ground Water Well No. 1	
Characteristic	1973[b]	1976	1973[b]	1976	1973[b]	1976	1974[c]	1976
Total N (mg/l)	2.7	1.6	5.2	2.5	1.6	0.9	5.7	3.1
NO_3-N (mg/l)	2.1	0.0	5.0	2.0	0.8	0.8	3.4	2.6
NH_4-N (mg/l)	0.1	0.3	0.1	0.0	0.1	0.0	0.0	0.1
Total P (mg/l)	0.06	0.09	0.02	0.01	0.02	0.05	0.04	0.07
PO_4-P (mg/l)	0.06	0.02	0.01	0.00	0.01	0.01	0.04	0.02
Ca (mg/l)	21	29	74	88	70	79	85	95
Mg (mg/l)	10	15	30	32	24	25	30	32
Na (mg/l)	2	6	5	5	4	5	5	6
K (mg/l)	2	5	0	0	1	1	0	0
Conductivity (μmhos/cm)	180	310	720	670	560	560	700	690
pH (μmhos/cm)	7.8	8.3	8.1	8.1	8.3	8.0	8.0	8.0
Fecal coliform (MPN/100 ml)	0.50	21	0	0	0	0	0	0

[a]Water samples taken monthly from April to November.
[b]Samples from 1973 were taken before sludge was applied.
[c]No 1973 sampling.

Water runoff was highest in 1975; mean total sediment losses were 0.2 and 0.5 m.ton/ha for the corn control and sludge-treated areas, respectively. Total N (NO_3-N + NH_4-N), soluble P (PO_4-P) and K in runoff water averaged 2.8, 0.05 and 0.4 kg/ha for the control area and 29, 0.23 and 0.6 kg/ha for the sludge-treated areas, respectively.

Soil water samples showed some movement of nitrate-N to the 150-cm depth on the sludge-treated corn areas, but no increase over the control for the grass areas. Phosphate-P increased at the 60-cm depth for the corn areas only. Concentrations of heavy metals in soil water were not increased by sludge application.

ACKNOWLEDGMENTS

The authors acknowledge financial support for this project from the Metropolitan Waste Control Commission, St. Paul, Minnesota, and from the Municipal Environmental Research Laboratory, U. S. Environmental Protection Agency, Cincinnati, Ohio. This paper is a contribution from the Soil and Water Management Research Unit, ARS, USDA, St. Paul, Minnesota, in cooperation with the Departments of Soil Science and Agricultural Engineering, University of Minnesota, St. Paul, Minnesota 55108, and is Minnesota Agricultural Experiment Station Paper No. 1654, Scientific Journal Series.

REFERENCES

1. Carroll, T. E., D. L. Maase, J. M. Genco and C. N. Ifeadi. "Review of Landspreading of Liquid Municipal Sewage Sludge," Report No. EPA-670/2-75-049, U.S. Environmental Protection Agency, Washington, D. C. (1975).
2. Dowdy, R. H., R. E. Larson and E. Epstein. "Sewage Sludge and Effluent Use in Agriculture," *Land Application of Waste Materials* (Ankeny, Iowa: Soil Conservation Society of America, 1976), pp. 138-153.
3. Chaney, R. L., S. B. Hornick and P. W. Simon. "Heavy Metal Relationships During Land Utilization of Sewage Sludge in the Northeast," in *Land as a Waste Management Alternative,* R. C. Loehr, Ed. (Ann Arbor, Michigan: Ann Arbor Science Publishers, Inc., 1976), pp. 283-314.
4. Larson, R. E., J. A. Jeffery, W. E. Larson and D. R. Duncomb. "A Closed Watershed for Applying Municipal Sludge on Crops," *Trans. Am. Soc. Agric. Eng.* (in press, 1977).
5. Larson, C. L. and L. F. Hermsmeier. "Device for Measuring Pipe Effluent," *Agric. Eng.* 39:282-287 (1958).
6. Bremner, J. M. "Total Nitrogen," in *Methods of Soil Analysis*, C. A. Black *et al.*, Eds. *Agronomy* 9:1149-1178 (1965).
7. American Public Health Association. *Standard Methods for the Examination of Water and Wastewater*, 13th ed. (Washington, D. C., 1971).
8. Larson, W. E. and J. J. Hanway. "Corn Production," in *Corn and Corn Improvement*, American Society of Agronomy Monograph, Chapter 11 (in press, 1977).
9. Larson, W. E. and R. H. Dowdy. "Heavy Metals Contained in Runoff from Land Receiving Wastes," in *Proc. Natl. Conf. Disposal of Residues on Land,* Information Transfer, Rockville, Maryland (in press, 1977).
10. Clapp, C. E., D. R. Linden, W. E. Larson, G. C. Marten and J. R. Nylund. "Nitrogen Removal From Municipal Wastewater Effluent by a Crop Irrigation System," in *Land as a Waste Management Alternative*, R. C. Loehr, Ed. (Ann Arbor, Michigan: Ann Arbor Science Publishers, Inc., 1977), pp. 139-149.

15

THE POTENTIAL USE OF FOREST LAND AS A SLUDGE DISPOSAL SITE

R. C. Sidle
U.S. Department of Agriculture,
Agricultural Research Service
Morgantown, West Virginia

INTRODUCTION

The concept of sludge disposal in forested areas represents a relatively unique alternative in solid waste management. Current technologies for municipal sludge disposal include incineration, landfilling, ocean dumping and land application. Land application accounts for approximately 25% of the total disposed municipal sludge, with most of this received by cropland sites.[1] The land application alternative had been aided by the Federal Water Pollution Control Act Amendments of 1972, which allocate grant money for land to be used in disposal systems. Subsection 201(d) of the act specifically mentions the consideration of agricultural and silvicultural areas for recycling sewage wastes.

The basic concept behind land disposal of sludge is relatively simple. Constituents in sludge that would be pollutants if dispersed in a water environment become beneficial nutrients when applied to the land. One of the major problems with land disposal of sludge is the introduction of large amounts of heavy metals into the soil.

Heavy metals such as Cu, Zn and Ni can be toxic to vegetation if they are applied in excessive concentrations. Other metals, Cd, for instance, may accumulate in animals and humans, where they act as cumulative toxins. Another problem is the leaching of large quantities of soluble nitrates into ground-water reserves following mineralization of organic nitrogen in the surface-applied sludge. Almost all of the research dealing with the fate of heavy metals in sludge disposal sites has been in conjunction with productive agricultural cropland.[2-6] Sludge disposal in forested areas seems a viable alternative since vegetation is generally not harvested as an edible crop. The only means of foodchain transfer of toxic metals would be through wildlife feeding on vegetation in sludge-treated areas and through ground water recharge. Virtually no information is available concerning the distribution of environmental contaminants in forested waste disposal sites. Some research has been reported evaluating the movement of nitrogen and phosphorus in soil water percolate in forested sites irrigated with sewage effluent.[7,8] Sidle and Sopper[9] reported no increases in Cd levels in native vegetation and only surficial Cd increases in soils from two forested sites irrigated with sewage effluent and an effluent-sludge mixture. Urie[10] suggests the utilization of sewage sludge in intensive silvicultural practices to increase nitrogen removal over traditional forest land renovation potential. No existing forest sludge disposal areas are mentioned in this publication, however. The U.S. Forest Service position regarding the disposal of wastewater and sludge on national forest land has been cautious, with the present emphasis placed on treated wastewater irrigation rather than sludge disposal.[11]

AVAILABILITY OF FOREST LAND IN THE NORTHEAST

Availability of forest land is an important factor in determining the potential for its use as a sludge disposal site. Forest land represents 67.9% of the total land area in the northeast region and outranks cropland acreage by a ratio of almost 5 to 1. Statistics for forest and cropland are presented in Table I for all states in the northeast region.[12] Only the state of Delaware has a larger percentage of cropland than forested areas.

The amount of forest land in proximity to large urban-industrial areas is given on a county or area-wide basis in Table II.[13-18] These areas are some of the higher sludge-producing sectors in the northeast region. The New York City metropolitan area was not considered since its isolated location would make it very inaccessible to outlying forested areas. Also, the Camden, New Jersey, metropolitan area was not included because the only major forested area nearby is the pine barrens, which is an undesirable sludge disposal location due to very high water tables. Even though most of the urbanized regions listed in Table II have lower percentages of forested land than their respective state-wide averages, there is still abundant forest land available for sludge disposal. Of course, land economics will dictate the ultimate use of these forest resources. However, the availability of forested areas, even in proximity to some urban-industrial complexes, could make these areas

Table I. Percentage of total land area of northeastern states in forests and crops.

	Forested Land (%)	Cropland (%)
Connecticut	70.2	7.2
Delaware	30.8	39.0
Maine	89.6	4.5
Maryland	46.5	27.9
Massachusetts	70.2	5.0
New Hampshire	88.8	3.6
New Jersey	51.1	13.7
New York	56.7	19.0
Pennsylvania	61.9	19.3
Rhode Island	64.5	5.2
Vermont	74.0	12.8
West Virginia	79.0	5.7

Table II. Percentage of forested land in proximity to selected urban-industrial areas.

County or Area (Primary City)	Forested Land (%)
New Castle Co., Del. (Wilmington)	20.2
Baltimore Co., Md. (Baltimore)	36.9
Prince Georges Co., Md. (College Park)	52.9
Montgomery Co., Md. (Bethesda, Rockville)	32.4
Mercer Co., N.J. (Trenton)	21.6
Albany Co., N. Y. (Albany)	39.6
Monroe Co., N. Y. (Rochester)	15.8
Onondaga Co., N. Y. (Syracuse)	32.5
Allegheny Co., Pa. (Pittsburgh)	37.3
Southeastern Pa.[a]	25.7
Kanawha Co., W. Va. (Charleston)	83.6

[a]Includes cities of Philadelphia, Reading, Allentown, Pottstown, Lancaster and York.

potentially attractive sludge disposal sites. Certainly some smaller municipalities in more highly forested areas will consider the application of sludge to readily accessible forest land.

ANTICIPATED ENVIRONMENTAL BENEFITS AND PROBLEMS

As mentioned previously, the fact that no edible crops are harvested from forested areas would essentially eliminate foodchain transferral of heavy metals via vegetation. Wildlife feeding in forested sludge disposal areas would be the only means for accumulation of heavy metals in the foodchain. The possibility of a "dead-end" storage in the wood fiber for heavy metals makes this disposal method particularly attractive. There is evidence[19] that the high levels of organic matter and cation exchange capacity in the surface horizons of forest soils would be very important in binding surface-applied cationic heavy metals in sludges. However, the typically low pH levels of forest soils could increase heavy metal mobility and induce phytotoxicity at high loading rates of certain metals. Various types of macropores common in forest soil[20] could provide channels for deep percolation of heavy metals and nitrates into ground water reservoirs.

Fertility benefits from sludge applications would be especially realized in intensive silviculture production. However, forest sites are generally poorer nitrogen renovators than cropland, leading to a potential increase in NO_3^--N in the ground water. Evidence has been found[21] indicating that nitrogen mineralization rates in forest soils may be inhibited at very high heavy metal (especially Cu) loading rates.

Any sludge disposal operation involving surface spreading faces the prospect of undesirable odors, largely dependent on the characteristics of the waste. Since the use of sludge injectors as practiced in agricultural operations[22] is not likely to be adaptable to forested areas, odor problems would restrict forest land use at and near disposal sites.

The physical disposal of sludge in forested areas presents some problems. Under intensive silviculture practices, liquid sludge could be applied from a tanker pulled between rows of trees by a small tractor. Other mobile equipment such as the "rain gun" spray nozzle used by the Chicago Metropolitan Sanitary District[23] may be adaptable for liquid sludge disposal in forested areas. The use of a flexible hose that could be reeled from a tanker truck such as described by Ardern[24] could possibly be used to spray liquid sludge under pressure into relatively inaccessible forest areas. The application of dried sludge to forest areas does not hold much promise.

MIXED HARDWOOD RESEARCH SITE-BACKGROUND DATA

In light of the previous discussion, a field plot study was undertaken in conjunction with the Penn State Wastewater Renovation and Conservation Project to determine the fate of selected heavy metals and nutrients in a forested sludge disposal site. The disposal site consisted of eight 6.1 x 6.1 m plots randomly distributed in a mixed-hardwood forested area approximately 0.5 hectare (ha) in size. There were three plots each for the high and low sludge treatments and two control plots. The soil was a Hublersburg clay loam with an average pH of 5.0 in the 0- to 7.5-cm interval and 4.7 in the 7.5- to 15-cm interval. The cation exchange capacity (CEC) of the 0- to 7.5-cm depth was 18.4 millequivalents (meq) per 100 g of soil, while that of the 7.5- to 15-cm depth was 8.4 meq/100 g soil. The higher CEC in the surface depth of soil is largely due to its high organic matter content (10.6%), which was approximately twice that of the 7.5- to 15-cm depth (5.4%).

Anaerobically digested sewage sludge obtained from the Pennsylvania State University Sewage Treatment Plant was pumped from a holding pond to the forest plots where it was applied via a garden hose. Solids content of the liquid sludge varied from 0.102 to 3.094%. Sludge was applied to the treated areas from November 7 to December 12 in 1974 and again from April 22 to May 19 in 1975. Loading rates were based on the total solids content of the applied sludge, with the high treatment receiving approximately twice the loading from the two applications (26.96 m.tons/ha) as the low treatment (12.71 m.tons/ha). The control plots received no treatment. Loading rates for total solids and various chemical constituents for the fall 1974 and spring 1975 sludge applications are given in Table III for both treatments. Concentration ranges for some of the more toxic heavy metals in the sludge (dry weight basis) were as follows: Cu, 133 to 1278 μg/g; Zn, 307 to 1631 μg/g; Cr, 57 to 325 μg/g; Pb, 26 to 581 μg/g; Co, 7 to 274 μg/g; Cd, 3.2 to 28.3 μg/g; and Ni, 33 to 705 μg/g.

RESEARCH METHODOLOGY

Multiple applications of liquid sludge were necessary in all treated plots for both the fall 1974 and spring 1975 application periods in order to achieve the desired total solids loadings. Applications of sludge in the high treatment plots were limited to some extent due to surface clogging. A subsample of sludge from each application was taken for quality analysis including total solids, heavy metals and macronutrients. Two predominant tree species, red maple (*Acer rubrum* L.) and black oak (*Quercus velutina* L.), and a composite of ground vegetation were sampled from each plot for heavy metal analyses early in September 1974, prior to sludge application, and in September 1975, four months after the last sludge application. Soil samples were collected in October 1974, prior to any sludge applications, and again in November 1975, six months after the last sludge application. Sampling consisted of four sites in each plot at the following depths: 0 to 7.5 cm; 7.5 to 15 cm; 15 to 30 cm; 30 to 60 cm; and 60 to 120 cm. Soil samples were extracted with 0.1 *N* HCl and analyzed for heavy metals. Duplicate suction lysimeters at both the 15- and 120-cm depths were installed in all field plots to sample soil water percolate. Percolate samples were collected twice weekly when possible immediately following each sludge application period for approximately two months. After this time, sampling intervals were reduced to weekly, then monthly interims. Percolate samples were analyzed for selected heavy metals and NO_3^--N. Details on sampling methods and chemical analysis of all parameters are given by Sidle.[25]

Table III. Loadings of total solids and chemical constituents during sludge applications.

Treatment	Total Solids (m.ton/ha)	Mn (kg/ha)	Cu	Zn	Cr	Pb	Co	Cd	Ni	K	Ca	Mg	Na	NO_3^--N	NH_4^+-N	org-N	Total P
Low																	
Fall 1974	4.58	1.90	4.69	5.33	1.00	1.06	0.20	0.042	0.54	62.9	148.7	43.3	38.6	11.7	330.0	305.0	190.0
Spring 1975	8.13	3.89	6.64	7.27	1.98	2.02	0.18	0.068	0.46	41.1	266.8	42.4	26.6	9.2	364.0	397.0	169.0
Total Loadings	12.71	5.79	11.33	12.60	2.98	3.08	0.38	0.110	1.00	104.0	415.5	85.7	65.2	20.9	694.0	702.0	359.0
High																	
Fall 1974	12.70	5.38	12.97	15.88	2.88	3.55	0.52	0.135	1.26	134.6	405.0	123.1	88.3	12.5	969.0	729.0	534.0
Spring 1975	14.23	6.80	11.53	12.61	2.90	3.24	0.29	0.118	0.70	60.1	472.3	71.1	38.0	13.5	515.0	794.0	320.0
Total Loadings	26.96	12.18	24.50	28.49	5.78	6.79	0.81	0.253	1.96	194.7	877.3	194.2	126.3	26.0	1484.0	1523.0	854.0

RESULTS AND DISCUSSION

Vegetation

Total heavy metal concentrations in red maple, black oak and ground vegetation foliage are presented in Table IV. Mean separations are shown for differences between pretreatment (September 1974) and posttreatment (September 1975) samples of a given species within a particular treatment. No statistical analysis was performed for ground vegetation data since the composition of species in the treated plots changed drastically as the result of sludge application. For red maple, the low-sludge treatment increased Cd and Mn in the foliage and decreased Cu. The other metals, Zn, Cr, Pb, Co and Ni, were unchanged. In the high-sludge treatment, red maple foliage increased in Mn concentration and decreased in Cu concentration following sludge application. Other heavy metals remained unchanged in the high-treatment foliage. Levels of Cu, Zn, Cd and Pb in both pretreatment and posttreatment foliage were lower than reported[26] ranges for normal red maple foliage. Conversely, concentrations of Ni, Co and Cr were slightly higher than these reported ranges. Levels of Mg in the pretreatment foliage were within these ranges, while those in posttreatment samples were higher.

For black oak, the low-sludge treatment increased Ni in the foliage, while all other heavy metals remained unchanged. In the high-sludge treatment Ni levels also increased in black oak foliage following sludge treatments, while Co levels decreased. All other heavy metals remained unchanged in the high-treatment area. Levels of Cu and Zn in black oak foliage were similar to concentrations of these metals found in native foliage at a southern West Virginia site.[27] Levels of Mg were slightly higher than reported values.[27]

The ground vegetations consisted largely of Virginia creeper (*Parthenocissus quinquefolia* L.), wild hydrangea (*Hydrangea arborescens* L.) and wintergreen (*Pryola americana* Sweet) in 1974 (pretreatment), and clearweed (*Pilea pumila* L.) and pokeweed (*Phytolacca americana* L.) in 1975 (posttreatment). The dissimilarity of the two ground vegetation types confounds any interpretation of sludge effect on metal content.

Soils

Extractable (0.1 *N* HCl) levels of Cu, Zn, Cd, Cr, Pb, Co and Ni are presented in Table V for the various depth intervals in the high treatment sampled in October 1974, 1 month prior to sludge treatment and 6 months after the final sludge application (November 1975). Soils data for these heavy metals in the low-treatment and control plots have also been analyzed. Trends in metals data were very similar in the low treatment to those in the high treatment for both depth and year effects. Extractable Cu, Zn and Ni in the control plots exhibited some differences between years at selected depths.

Table IV. Heavy metal concentrations in foliar samples.

Species	Sampling Date	Cu	Zn	Cd	Cr	Pb	Co	Ni	Mn
		(µg/g)							
					Low Treatment				
Red Maple	Sept. 1974	7.00a†	56.32a	0.444b	1.28a	6.95a	3.27a	3.02a	734.0b
Red Maple	Sept. 1975	4.03b	56.70a	0.858a	1.00a	8.42a	3.16a	4.47a	2390.0a
Black Oak	Sept. 1974	9.67a	48.72a	0.197a	1.28a	6.58a	2.95a	6.00b	1319.0a
Black Oak	Sept. 1975	9.12a	40.57a	0.209a	1.36a	8.00a	3.14a	10.20a	1670.0a
Ground Vegetation*	Sept. 1974	10.77	52.74	0.647	2.82	13.08	4.38	4.69	365.0
Ground Vegetation**	Sept. 1975	14.59	58.95	0.275	2.50	13.07	3.97	7.48	859.0
					High Treatment				
Red Maple	Sept. 1974	8.19a	53.81a	0.358a	1.01a	7.58a	3.06a	4.55a	766.0b
Red Maple	Sept. 1975	4.67b	64.78a	0.474a	0.96a	9.21a	3.13a	4.69a	3534.0a
Black Oak	Sept. 1974	9.21a	47.11a	0.171a	1.38a	6.18a	2.59a	5.19b	1831.0a
Black Oak	Sept. 1975	9.53a	33.16a	0.199a	1.15a	6.66a	2.21b	8.20a	2161.0a
Ground Vegetation*	Sept. 1974	16.65	70.23	1.233	2.05	17.87	5.50	7.78	802.0
Ground Vegetation**	Sept. 1975	15.78	86.43	0.398	1.32	9.68	3.71	7.87	2287.0

* Primary species sampled were Virginia creeper, wild hydrangea and round-leaved American wintergreen.

**Primary species sampled were clearweed and pokeweed.

† Means of a particular metal for a given species and treatment followed by the same letter are not significantly different at the 0.05 level.

Table V. Concentrations of 0.1 *N* HCl extractable heavy metals in soils sampled in 1974 and 1975 at various depths in high treatment plots.

		High Treatment		
Element	Depth (cm)	1974 (Pretreatment) (μg/g)	1975 (Posttreatment) (μg/g)	Significance Between Years
Cu	0-7.5	1.76a*	8.30a	**
	7.5-15	0.79b	1.72b	**
	15-30	0.57c	1.33bc	**
	30-60	0.47c	0.85cd	**
	60-120	0.52c	0.79d	**
Zn	0-7.5	16.52a	27.23a	**
	7.5-15	4.95b	3.92b	-
	15-30	1.90c	1.63bc	-
	30-60	0.97c	0.87c	-
	60-120	0.75c	0.63c	-
Cd	0-7.5	0.188a	0.138a	-
	7.5-15	0.167b	0.038b	**
	15-30	0.031c	0.014c	**
	30-60	0.021c	0.009c	**
	60-120	0.011c	0.006c	**
Cr	0-7.5	0.258a	0.515a	**
	7.5-15	0.115b	0.134b	-
	15-30	0.090b	0.084bc	-
	30-60	0.069b	0.042c	-
	60-120	0.075b	0.037c	**
Pb	0-7.5	6.36a	6.29a	-
	7.5-15	2.95c	2.67c	-
	15-30	3.69c	4.06b	-
	30-60	5.72ab	5.64a	-
	60-120	5.35b	5.87a	-
Co	0-7.5	1.42a	1.08b	**
	7.5-15	0.76b	0.68c	-
	15-30	1.07b	1.10b	-
	30-60	1.76a	2.11a	-
	60-120	0.95b	1.25b	-
Ni	0-7.5	1.25a	2.37a	**
	7.5-15	0.51b	1.32b	**
	15-30	0.35c	0.90c	**
	30-60	0.31c	0.28d	-
	60-120	0.31c	0.28c	-

*Means of a particular metal for a given year and treatment followed by the same letter are not significantly different at the 0.05 level.

**Indicates a significant difference between years (0.05 significance level) for a particular metal at a given depth and treatment.

Extractable Cu, Zn, Cd, Cr and Ni levels generally decreased with soil depth, reflecting their native distribution in the soil profile. The native concentrations of Pb and Co in the soil showed no consistent gradient with depth. Maximum extractable Pb and Co concentrations were found in the 0- to 7.5-cm and 30- to 60-cm intervals, respectively.

Extractable soil Cu increased significantly at all depths sampled in the high-treatment plots following sludge applications. These increases were greatest in the upper soil depth and smaller in the lower depths, indicating that there was limited mobility of Cu within the soil matrix. Extractable soil Zn increased only in the upper 7.5 cm of the profile following sludge applications. No significant changes in extractable Cd levels were noted in the 0- to 7.5-cm depth; however, slight decreases in extractable Cd were found in the lower depths. Of the other heavy metals examined, only Ni showed any evidence of mobility in the soil matrix following sludge applications.

Soil Water Percolate

Mean concentrations of Cu, Zn and Cd in all percolate samples averaged over a 17-month period following the initial sludge treatment are presented in Table VI. For all three metals in both the 15-cm and 120-cm percolate samples, the control area had significantly lower concentrations than the low treatment, which in turn was significantly lower than the high treatment. Maximum monthly concentrations of Cu (37.5 ppb), Zn (576 ppb) and Cd (4.3 ppb) in the 15-cm percolate of the high treatment are still well below USPHS limitations for these metals in drinking water.[28] Maximum Zn and Cd levels in the 120-cm percolate lagged behind the corresponding maxima in the 15-cm depth by approximately 3 months. Figures 1 and 2 show Zn and Cd concentrations, respectively, in soil water percolate at the 15-cm depth over the entire 17-month period. At the bottom of each figure is a distribution graph for precipitation as well as arrows noting sludge application periods. Curves through percolate data of the high treatments represent the least squares third-degree equations of fit, while curves through control percolate data represent the least squares second-degree equations of fit.

Calculations of the total amounts of Cu, Zn and Cd moving out of the 120-cm soil depth in the high treatment indicated that only 0.3% of the applied Cu, 3.2% of the applied Zn and 6.6% of the applied Cd moved out of the 120-cm depth over the 17-month period. Percolate losses were calculated for each month based on mean monthly metals concentrations and mean monthly soil water velocities determined by a field water budget.[25]

The fact that neither extractable soil Zn nor Cd showed any increases below 7.5 cm following sludge applications (Table V) seems paradoxical to the concurrent increases in these metals in soil water percolate to a depth of 120 cm. One possible mechanism for the movement of these metals in the

Table VI. Mean concentrations of Cu, Zn and Cd in percolate collected during a 17-month period following the initial sludge treatment.

Treatment	Heavy Metal	15-cm Depth (ppb)	120-cm Depth (ppb)
Control	Cu	5.9c*	3.9c
Low	Cu	9.4b	6.7b
High	Cu	24.9a	8.6a
Control	Zn	65.9c	23.2c
Low	Zn	169.1b	52.4b
High	Zn	357.2a	94.0a
Control	Cd	0.3c	0.2c
Low	Cd	1.8b	0.6b
High	Cd	2.8a	1.8a

*Means of a particular metal for a given depth followed by the same letter are not significantly different at the 0.05 level.

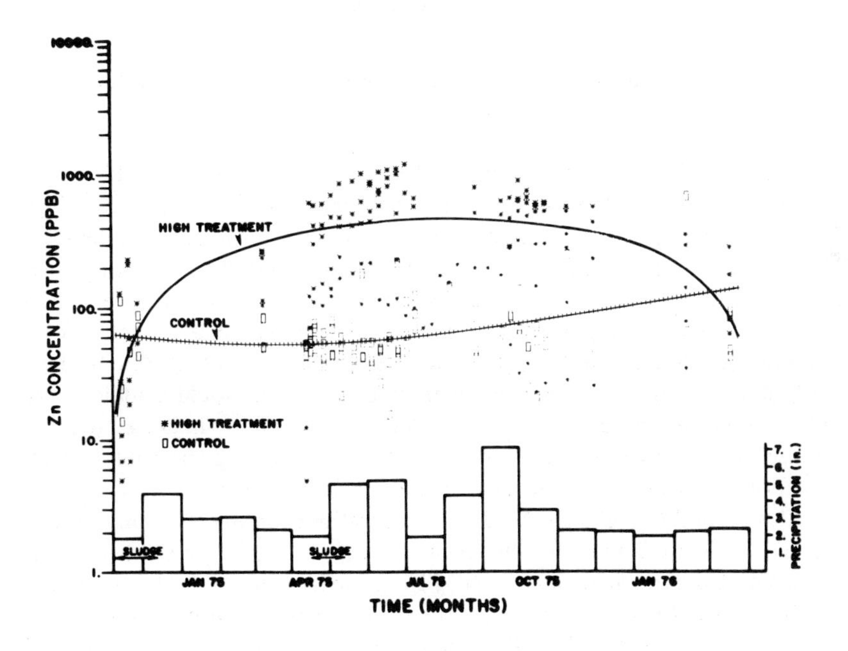

Figure 1. Percolate Zn concentrations in high-treatment and control plots at 15-cm depth.

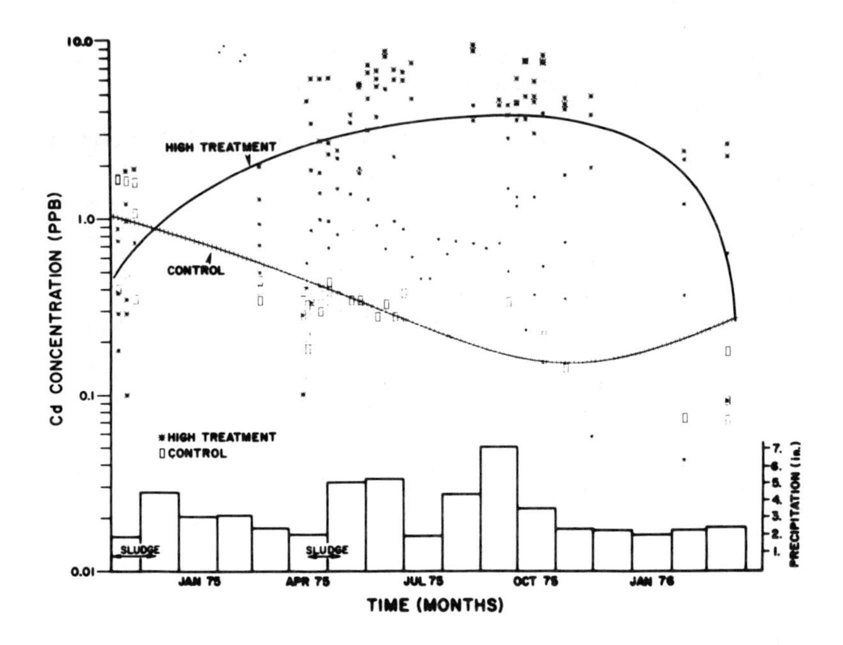

Figure 2. Percolate Cd concentrations in high-treatment and control plots at 15-cm depth.

percolate is through the transport of chelated metals which would not be readily absorbed by the soil. Another mechanism could be the "channeling" of metals through interconnected soil pores at times of high hydraulic loadings. Forest soils are particularly endowed with such water-conducting pores including root channels, macroinvertebrate passageways and freeze-thaw cracks, as well as interaggregate channels. When sampling percolate with suction lysimeters, water would be extracted preferentially from these channels of higher hydraulic conductivity and samples would contain heavy metals transported via these channels. A modeling study[29] indicated that the transport mechanism for these metals in the soil water was probably a combination of "channeling" and movement of chelated metals.

Mean monthly percolate NO_3^--N concentrations for all treatments are presented in Figures 3 and 4 for the 15-cm and 120-cm depths, respectively. It is apparent from these data that NO_3^--N in recharge water is the most site-limiting constituent monitored in this study. Nitrate concentrations at the 120-cm depth exceeded USPHS limitations for drinking water (10 mg/l NO_3^--N) from January 1975 throughout the study period. Nitrate levels in the 15-cm percolate declined to levels near 10 mg/l NO_3^--N by late spring 1976 in both sludge-treated areas. Maximum monthly NO_3^--N levels in the low and high treatments (15-cm depth) were 194 mg/l and 290 mg/l, respectively. Corresponding maxima at the 120-cm depth for the low (66 mg/l) and high (180 mg/l) treatments occurred approximately 6 months later.

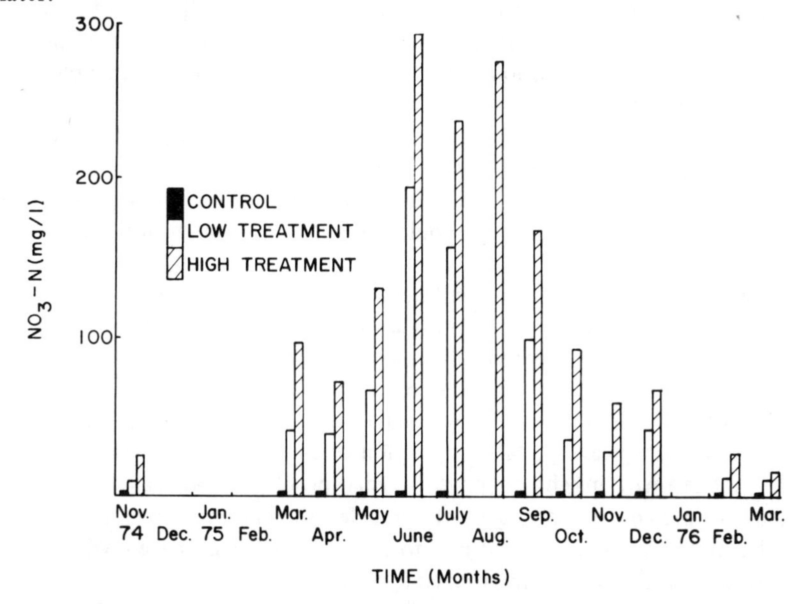

Figure 3. Mean monthly NO_3^--N concentrations in 15-cm percolate.

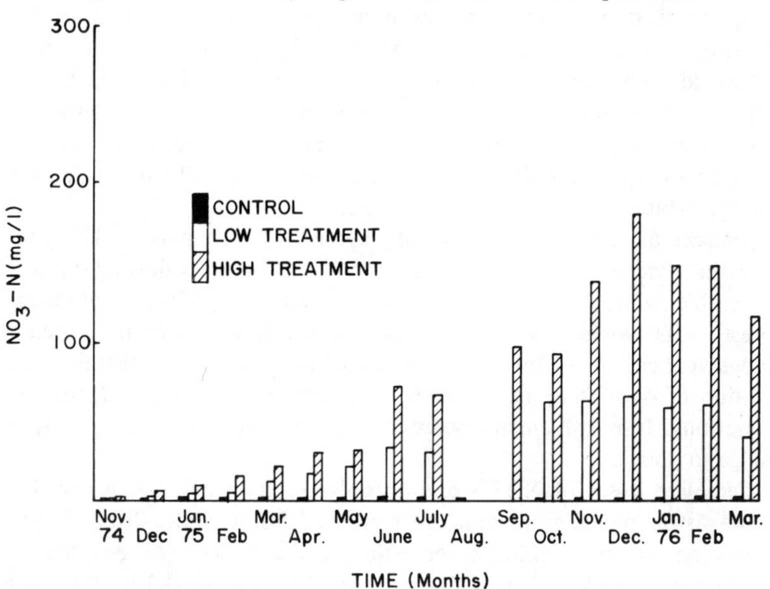

Figure 4. Mean monthly NO_3^--N concentrations in 120-cm percolate.

OVERVIEW

A case has been made establishing the availability of forest land in the Northeast, and it appears that some of this land will possibly be utilized for sewage sludge disposal in the near future. Data have been presented in this chapter showing the distribution of various potential environmental contaminants in a forest ecosystem treated with sludge. Since environmental data from land disposal systems are site-dependent, as well as influenced by sludge composition, it is difficult to make specific recommendations related to sludge disposal on forest land. However, some general observations have been made from this study that would be useful in planning and managing a forest sludge-disposal project.

None of the heavy metals monitored in various foliar samples were at concentrations abnormally higher than reported levels. Certain species tended to accumulate different heavy metals following sludge application. Red maple accumulated Mn and Cd (to a lesser extent), while black oak accumulated Ni. Knowledge of particular species accumulators of various metals together with the approximate heavy metal composition of the sludge to be applied could give pretreatment insight on phytotoxicity problems that may be encountered. As previously mentioned, the transferral of heavy metals through the foodchain is greatly minimized in a forest sludge disposal system when compared to the more commonly utilized agriculture systems.

The additions of nutrients, organic matter and water associated with the sludge applications effected a drastic change in ground vegetation. Native forest-floor vegetation was largely replaced by succulent annuals, such as pokeweed, following sludge applications. Similar ground vegetation changes were noted by Sopper and Kardos[30] in wastewater-irrigated forested areas. These vegetative changes should be taken into consideration in the overall planning of a forest sludge disposal area since they will have an impact on wildlife habitat.

Undesirable odors were prevalent, especially during summer 1975, in the immediate vicinity of the sludge-disposal plots. By the following summer no noticeable odors existed. Obviously, the surface application of sludge in forest areas would restrict other uses of the land immediately following sludge applications. However, since forested areas act as windbreaks, the dissipation of odors may be reduced compared to open-land sludge-spreading operations. It must be remembered that odor problems will vary greatly from sludge to sludge.

Percolate losses of NO_3^--N and possibly certain heavy metals may be the most restricting environmental factors in a forest sludge-disposal system, depending on the chemical composition of the sludge. Accelerated movement of these pollutants in soil-water percolate is enhanced by "channelization" in forest soils (or highly aggregated soils). Also, chelation of metals in sludge may increase the mobility of certain metals in percolate. Thus, even though increases in extractable metals in soils were not detected, the movement of certain metals (especially Zn and Cd) in percolate is not precluded. In this study, movement of metals in soil water was not at concentrations that would pollute recharge water. However, at higher metal loading rates, such as in acid soil, leaching of metals could present ground-water contamination problems, especially in high water-table areas. Deep percolation of soluble NO_3^--N may pose the greatest water quality problem. Climatic, plant uptake and soil physical data must be integrated with underlying aquifer characteristics, such as transmissibility and spatial extent, in order to determine acceptable N loading rates via sludge.

The possibility for increased N renovation under intensive silvicultural practices should be explored. Such operations could include Christmas tree farms, tree seedling nurseries and ornamentals. Liming certain forest soils may be considered in order to immobilize heavy metals associated with highly polluted sludges.

In conclusion, it is apparent that a complete chemical and physical characterization of sewage sludge to be applied to forest land is necessary. Levels of heavy metals and macronutrients in sewage sludges vary greatly even within sludges generated by the same treatment plant. Additional research needs associated with surface spreading of sludge in forested areas include surface clogging, health aspects (pathogens, etc.) and physical application systems. Answers to these questions together with reported data will provide information needed to better evaluate the potential use of forest land for sludge disposal on an individual site basis.

ACKNOWLEDGMENTS

The experimental research portion of this study was performed in conjunction with the Institute for Research on Land and Water Resources, The Pennsylvania State University, University Park, Pennsylvania. Manuscript preparation and selected data analysis were a contribution of the United States Department of Agriculture, Agricultural Research Service, Division of Plant Sciences, West Virginia University, Morgantown, West Virginia. The author wishes to thank William Haas, Robert Cowan and Sonja Kerr for their assistance in sample collection and chemical analysis.

REFERENCES

1. Bastian, R. K. "Municipal Sludge Management: EPA Construction Grants Program," in *Land as a Waste Management Alternative,* R. C. Loehr, Ed. (Ann Arbor, Michigan: Ann Arbor Science Publishers, Inc., 1976).

2. Boswell, F. C. "Municipal Sewage Sludge and Selected Element Applications to Soil: Effect on Soil and Fescue," *J. Environ. Qual.* 4:267-272 (1975).
3. Chaney, R. L., S. B. Hornick and P. W. Simon. "Heavy Metal Relationships During Land Utilization of Sewage Sludge in the Northeast," in *Land as a Waste Management Alternative,* R. C. Loehr, Ed. (Ann Arbor, Michigan: Ann Arbor Science Publishers, Inc., 1976).
4. Kirkham, M. B. "Trace Elements in Corn Grown on Long-Term Sludge Disposal Site," *Environ. Sci. Technol.* 9:765-768 (1975).
5. Page, A. L. "Fate and Effects of Trace Elements in Sewage Sludge When Applied to Agricultural Lands," Report No. EPA 670/2-74-005, U.S.EPA, Cincinnati, Ohio (1974).
6. Varanka, M. W., Z. M. Zablock and T. D. Hinesly. "The Effect of Digested Sludge on Soil Biological Activity," *J. Water Poll. Control Fed.* 48:1728-1740 (1976).
7. Sopper, W. E. "Effects of Trees and Forests in Neutralizing Waste," in *Trees and Forests in an Urbanizing Environment*, University of Massachusetts Cooperative Extension Service (1971), pp. 43-57.
8. Urie, D. H. "Phosphorus and Nitrate Levels in Ground Water as Related to Irrigation of Jack Pine with Sewage Effluent," Conf. on Recycling Treated Municipal Wastewater Through Forest and Croplands, Report No. EPA 660/2-74-003, Washington, D. C. (1974), pp. 157-164.
9. Sidle, R. C. and W. E. Sopper. "Cadmium Distribution in Forest Ecosystems Irrigated with Treated Municipal Waste Water and Sludge," *J. Environ. Qual.* 5:419-422 (1976).
10. Urie, D. H. "Nutrient and Water Control in Intensive Silviculture on Sewage Renovation Areas," *Iowa State J. Res.* 49:313-317 (1975).
11. Olson, O. C. and E. A. Johnson. "Forest Service Policy Related to the Use of National Forest Lands for Disposal of Wastewater and Sludge," Conf. on Recycling Treated Municipal Wastewater Through Forest and Cropland, Report No. EPA-660/2-74-003, Washington, D. C. (1974), pp. 414-419.
12. U.S. Department of Agriculture, Forest Service. "The Outlook for Timber in the United States," Forest Resource Report No. 20 (1973), p. 225.
13. Ferguson, R. H. "The Timber Resources of West Virginia," U.S. Forest Service Research Bulletin NE-2 (1964).
14. Ferguson, R. H. "The Timber Resources of Maryland," U.S. Forest Service Research Bulletin NE-7 (1967).
15. Ferguson, R. H. "The Timber Resources of Pennsylvania," U.S. Forest Service Research Bulletin NE-8 (1968).
16. Ferguson, R. H. and C. E. Mayer. "The Timber Resources of New York," U.S. Forest Service Bulletin NE-20 (1970).
17. Ferguson, R. H. and C. E. Mayer. "The Timber Resources of Delaware," U.S. Forest Service Bulletin NE-32 (1974).
18. Ferguson, R. H. and C. E. Mayer. "The Timber Resources of New Jersey," U.S. Forest Service Bulletin NE-34 (1974).
19. Sidle, R. C. and L. T. Kardos. "Adsorption of Copper, Zinc, and Cadmium by a Forest Soil," *J. Environ. Qual.* 6(3)(1977).

20. Aubertin, G. M. "Nature and Extent of Macropores in Forest Soils and Their Influence on Subsurface Water Movement," U.S. Department of Agriculture Forest Service Research Paper NE-192 (1971).
21. Tyler, G. "Heavy Metal Pollution and Mineralization of Nitrogen in Forest Soils," *Nature* 255:701-702 (1975).
22. Smith, J. L. and D. B. McWhorter. "Continuous Subsurface Injection of Liquid Organic Wastes," in *Land as a Waste Management Alternative,* R. C. Loehr, Ed. (Ann Arbor, Michigan: Ann Arbor Science Publishers, Inc., 1976).
23. Manson, R. J. and C. A. Merritt. "Land Application of Liquid Municipal Wastewater Sludges," *J. Water Poll. Control Fed.* 47:20-29 (1975).
24. Ardern, D. A. "The Agricultural Use of Municipal Sludge," in *Land as a Waste Management Alternative,* R. C. Loehr, Ed. (Ann Arbor, Michigan: Ann Arbor Science Publishers, Inc., 1976).
25. Sidle, R. D. "Transport of Heavy Metals in a Sludge-Treated Forested Area," Ph.D. Dissertation, Department of Agronomy, The Pennsylvania State University, University Park, Pennsylvania (1976).
26. Hanna, W. J. and C. L. Grant. "Spectrochemical Analysis of the Foliage of Certain Trees and Ornamentals for 23 Elements," *Torrey Bot. Club Bull.* 89:293-302 (1962).
27. Bennett, O. L., H. A. Menser and W. M. Winant. "Land Disposal of Leachate Water from a Municipal Sanitary Landfill," Proc. 2nd Nat. Conf. on Water Reuse, Chicago, Illinois (1975), pp. 789-800.
28. McKee, J. E. and H. W. Wolf. *Water Quality Criteria,* 2nd ed., California State Water Resources Control Board Publication 3-A, Sacramento, California (1963), pp. 169-171.
29. Sidle, R. C., L. T. Kardos and M. Th. van Genuchten. "Heavy Metals Transport Model in a Sludge-Treated Soil," *J. Environ. Qual.* (in press).
30. Sopper, W. E. and L. T. Kardos. "Vegetation Responses to Irrigation with Treated Municipal Wastewater," Conf. on Recycling Treated Municipal Wastewater Through Forest and Croplands, Report No. EPA-660/2-74-003, Washington, D. C. (1974), pp. 242-269.

16

FERTILIZER VALUE OF UNDIGESTED AND DIGESTED THERMORADIATION TREATED SEWAGE SLUDGE ON CALCAREOUS SOIL

B. D. McCaslin and P. S. Titman
Department of Agronomy
New Mexico State University
Las Cruces, New Mexico

INTRODUCTION

The use of nuclear reactor wastes as a radiation source to kill pathogens in sewage sludge is being studied at Sandia Laboratories, Albuquerque, New Mexico, and may add a new dimension to sewage sludge utilization. The use of ionizing radiation for the destruction of pathogens in sewage has been reviewed by Gerrard,[1] Ballentine[2] and Reynolds.[3] The recent findings by Sandia on the combined effects of ionizing radiation and heat (thermoradiation) and/or chemical treatments on destruction of pathogens in sewage sludge have been discussed by Sivinski[4,5] and Brandon *et al.*[6,7] The thermoradiation process being developed at Sandia using Cesium-137 from the nuclear power industry has cost/benefit potential outlined by Morris.[8]

The overall ongoing cooperative research program between Sandia Laboratories and New Mexico State University includes assessment of nutritive value of sewage solids when recycled directly as feedstuffs for ruminant animals and agronomic aspects related to usage of sludge products and/or wastewater on calcareous soils. The primary objective of the research presented herein was to examine the effects of thermoradiation on extractable and plant-available nutrients and heavy metals from sewage products collected at the Albuquerque sanitation district and thermoradiation treated at Sandia Laboratories.

METHODS

Characterization of Sludge Products

Approximately 400-gal liquid samples of both undigested (raw sewage as it entered the treatment plant) and anaerobically digested sewage sludge (after secondary sewage treatment) from the Albuquerque, New Mexico, sanitation district were collected and each thoroughly mixed and then divided into halves. One-half of each of the digested (DSS) and undigested (USS) sludge was thermoradiated (TRDSS and TRUSS, respectively); the other half of each sample of sludge was left untreated. The sludges were air dried and shipped to New Mexico State University.

Sewage sludges for chemical analysis were air dried in a forced air oven (25°C) and ground to pass a 10-mesh screen in a stainless steel Wiley Mill. The ground samples were analyzed for total nitrogen according to Bremner[9] and organic carbon content by the Walkley-Black method.[10] Sludges were prepared for total analysis by nitric + perchloric acid wet digestion[11] and for analysis of water-soluble elements by mixing deionized water with 50 g of sludge until an approximate saturation paste was obtained using 150% water by weight for digested sludge and 250% water by weight for undigested sludge. Saturation pastes were equilibrated for 24 hr and suction filtered. Sludges were prepared for DTPA-extractable metals determination by shaking 25 g of sludge with 150 ml DTPA extractant for 2 hr.[12] For 0.1 *N*, HCl extractable metal determination sludges were prepared by shaking 2 g sludge with 100 ml 0.1 *N* HCl for 2 hr. Saturation extracts, nitric-perchloric digests, HCl extracts and DTPA extracts were analyzed for Fe, Cu, Mn, Zn, Pb, Cr and Cd by atomic adsorption spectrophotometry and graphite furnace equipment where applicable. The phosphorus and potassium fertilizer value of the sludges were determined at the New Mexico Department of Agriculture, State Chemist, Laboratory according to methods in A.O.A.C.

Plant-Available Sludge Elements

A greenhouse experiment with eleven treatments (Table I) in a five-replication randomized block design was established. Each of the sludges was added to soil at rates to supply 500 and 1000 lb of elemental N/ac. An untreated soil and two commercial fertilizer treatments of 100 lb N + 40 lb P/ac and 200 lb N + 80 lb P/ac were included as checks in the experiment. Fertilizers or sludges were thoroughly mixed with 7.7 lb (3.5 kg) soil and placed in plastic 7-in. (17.8 cm) pots with drainage holes in the bottom. The soil used in the experiment was a surface [0-12 in. (30 cm)] sample of clay loam soil (Torrifluvent) from the New Mexico State University Plant Research Center (Table II). Eight sorghum seeds were planted in each pot

Table I. Treatment description and amount of fertilizer or sludge added in each treatment for 2-month greenhouse experiment.

Treatment Description	Amount of Sludge or Fertilizer Added to 3.5 kg soil	Comparable Amount of Sludge or Fertilizer Added per Land Area Basis
Check	0	0
100-40-0	0.389 g urea	222 lb/ac urea (249 kg/ha)
	0.152 g superphosphate	87 lb/ac superphosphate (97 kg/ha)
200-80-0	0.778 g urea	444 lb/ac urea (497 kg/ha)
	0.304 g superphosphate	174 lb/ac superphosphate (195 kg/ha)
USS - 500 lb N/ac (560 kg/ha)	32.423 g	9 ton/ac (20.2 m.ton/ha)
USS - 1000 lb N/ac (1120 kg/ha)	64.847 g	18 ton/ac (40.3 m.ton/ha)
TRUSS - 500 lb N/ac (560 kg/ha)	32.423 g	9 ton/ac (20.2 m.ton/ha)
TRUSS 1000 lb/ac (1120 kg/ha)	64.847 g	18 ton/ac (40.3 m.ton/ha)
DSS - 500 lb N/ac (560 kg/ha)	51.496 g	15 ton/ac (33.6 m.ton/ha)
DSS - 1000 lb N/ac (1120 kg/ha)	102.992 g	30 ton/ac (67.2 m.ton/ha)
TRDSS - 500 lb N/ac (560 kg/ha)	51.496 g	15 ton/ac (33.6 m.ton/ha)
TRDSS - 1000 lb N/ac (1120 kg/ha)	102.992 g	30 ton/ac (67.2 m.ton/ha)

Table II. Chemical properties of soil used in 2-month greenhouse experiment.

E.C. (μ mhos)	4.05	Cu (ppm)	3
pH	7.57	Mn (ppm)	7
Organic Matter (%)	0.53	Zn (ppm)	8
NO_3^- (ppm)	26	CEC (meq/100 g)	21
P (ppm)	19	Sand (%)	29
K (ppm)	448	Silt (%)	34
$CaCO_3$ (%)	7.5	Clay (%)	37
Fe (ppm)	7		

and thinned to 3 plants at 2 weeks. Plants were grown with natural illumination in a greenhouse for approximately 2 months, September 25 to November 19, 1976. Pots were watered by hand to approximately 90% of field capacity twice weekly. Plants were harvested from the soil surface, washed, dried at 70°C and weighed. Plant material was ground in a stainless steel grinder (60 mesh). Three replicate samples were analyzed for total N,[9] digested with nitric + perchloric acid and analyzed for P, K, Fe, Cu, Mn, Zn, Cr, Cd and Pb. After plant harvest, soil samples were taken from the surface to the bottom of the pot, air dried, sieved (2 mm) and analyzed for extractable P, K and exchangeable NH_4 and NO_3^-, pH, electrical conductivity and DTPA extractable Fe, Mn, Zn, Cu, Cd, Pb and Cr.

RESULTS AND DISCUSSION

Chemical Composition of Sludges

The total analysis of the dried sludges used in the study shows a difference in nutrient concentration in N, organic carbon, pH, electrical conductivity, C/N ratio, Pb, Cd, Cr, Fe, Zn, Cu and Mn between the undigested and the anaerobically digested sewage sludges (Table III). The thermoradiation process had little or no effect on total chemical analysis. The digested sewage sample did not come from the same undigested sewage sample before anaerobic digestion treatment. Since some variation in nutrient concentration occurs with sampling time, the variation in total chemical analysis between digested and undigested sludges is partially confounded by different sampling times.

Analysis of variance (F-Test at 5% level) for the triplicate extracts for heavy metals from both digested and undigested sewage materials indicated no significant increases in extractable metals resulting from the thermoradiation treatment for any of the extractants (Tables IV and V). However, it is interesting to note that the HCl-extractable Zn and Cd in the digested sludge was approximately 93% of the total Zn and Cd, compared to 69 and 72% of the total Zn and Cd, respectively, in the undigested sewage. The three extractants were selected to give three degrees of strength of extractability and do not necessarily represent plant-available heavy metals.

Effect of Thermoradiation on Plant Nutrient Availability

Significantly more sorghum dry matter was produced in all undigested sewage sludge treatments and in the 30 ton/ac rates of the digested sewage than in the check treatments (Table VI). The two high-rate (18 ton/ac) treatments of USS gave the highest yields, followed by the two high rates of DSS and the two low rates of USS. There was no significant difference

Table III. Total analysis of undigested (USS), thermoirradiated undigested (TRUSS), digested (DSS), and thermoirradiated digested (TRDSS) sewage sludges collected at the Albuquerque, New Mexico, sanitation district.

	Sludge Type			
	USS	TRUSS	DSS	TRDSS
Commercial Fertilizer				
N (%)	2.8	2.8	1.9	1.8
P_2O_5 (%)	2.0	2.3	2.4	2.5
K_2O (%)	0	0	0	0
Organic Carbon	30	30	15	15
pH[a]	5.9	5.8	6.5	6.3
E.C. (mmhos/cm)[a]	5.0	4.9	8.1	8.2
C/N Ratio	11/1	11/1	8/1	8/1
Pb (ppm)	626	608	678	676
Cd (ppm)	15	14	26	27
Cr (ppm)	194	194	379	395
Fe (ppm)	10771	10864	14671	14857
Zn (ppm)	1364	1333	1684	1676
Cu (ppm)	852	796	1132	1121
Mn (ppm)	145	143	236	234

[a]pH and E.C. are on approximate saturation extracts, 150% water by weight for digested sludge and 250% water by weight with undigested sludge.

Table IV. HCl-, DTPA- and H_2O-extractable Fe, Zn, Cu, Mn, Cr, Cd, Pb from digested dried sewage solids, half of which was T.R. treated and half untreated.

	Fe (ppm)	Zn (ppm)	Cu (ppm)	Mn (ppm)	Cr (ppm)	Cd (ppm)	Pb (ppm)
Extractant	Digested						
HCl	3389	1580	577	150	74.0	25	525
DTPA	156	437	145	14	0.6	8	72
H_2O	2	11	6	2	0	0	0.02
Total	14671	1684	1132	236	379	26	678
	T.R. Digested						
HCl	3204	1528	560	144	68.0	24	512
DTPA	166	421	161	14	0.6	8	74
H_2O	2	11	6	2	0	0	0.03
Total	14857	1676	1121	234	394	27	676

Table V. HCl-, DTPA- and H_2O-extractable Fe, Zn, Cu, Mn, Cr, Cd, Pb from undigested dried sewage solids, half of which was T.R. treated and half untreated.

	Fe (ppm)	Zn (ppm)	Cu (ppm)	Mn (ppm)	Cr (ppm)	Cd (ppm)	Pb (ppm)
Extractant				Undigested			
HCl	4155	975	374	89	35.0	10	397
DTPA	795	352	45	19	0.50	4	183
H_2O	9	16	3	4	0	0	0.1
Total	10771	1364	852	145	194	15	626
				T.R. Undigested			
HCl	4248	971	374	88	38.0	10	403
DTPA	809	310	40	19	0.50	3	184
H_2O	8	16	3	4	0	0	0.1
Total	10864	1333	796	143	194	14	608

Table VI. The average sorghum dry matter produced per sludge and fertilizer treatment after approximately 2-month growing period in greenhouse conditions.

Treatment	Yield (g/pot)
Check	0.159 c*
100-40-0	0.142 c
200-80-0	0.242 c
USS (9 ton/ac)	1.770 b
USS (18 ton/ac)	2.973 a
TRUSS (9 ton/ac)	1.782 b
TRUSS (18 ton/ac)	2.443 a
DSS (15 ton/ac)	0.554 c
DSS (30 ton/ac)	1.480 b
TRDSS (15 ton/ac)	0.426 c
TRDSS (30 ton/ac)	1.538 b

*Means followed by the same letter are not significantly different at the 5% level by the New Duncan's Multiple Range Test.

between the two low rates of DSS and the checks. Also, no significant growth differences occurred between any corresponding thermoradiation and untreated sludge treatments, indicating little or no effect from the thermoradiation treatment in respect to growth response for sorghum.

The sorghum tissues were then analyzed to see whether nutrient concentration varied, thus testing whether thermoradiation had significant effects on tissue elemental concentrations perhaps not displayed in plant growth differences (Table VII). No significant increases in plant tissue elemental concentrations resulted from corresponding TR treated and untreated sludges for the elements analyzed by orthogonal contrasts at the 5% level of significance. The total nutrient uptake (*i.e.*, elemental concentration in tissue multiplied by the total dry matter yield) does indicate the effect of sludge applications on total heavy metals removed from the soil (Table VIII). Since the application rate of each sludge was based on the nitrogen content of the sludge, the

Table VII. Sorghum tissue elemental concentrations of N, P, K, Fe, Zn, Mn and Cu after 2-month growing period in greenhouse.

Treatment	N	P	K	Fe	Zn	Mn	Cu
Check	3.00	0.59	3.48	262	62	63	31
Fertilizer Nutrient (lb/ac)							
100-40-0	2.83	0.57	3.28	204	68	69	37
200-80-0	2.48	0.49	3.37	176	55	62	20
Undigested Sludge (ton/ac)							
USS (9 ton/ac)	2.63	0.60	3.58	95	133	55	29
USS (18 ton/ac)	2.40	0.56	3.67	90	177	47	35
TRUSS (9 ton/ac)	2.33	0.59	3.55	96	131	63	26
TRUSS (18 ton/ac)	2.35	0.58	3.49	93	170	54	29
Digested Sludge (ton/ac)							
DSS (15 ton/ac)	2.38	0.59	3.74	192	128	72	24
DSS (30 ton/ac)	1.81	0.56	3.33	105	138	63	27
TRDSS (15 ton/ac)	2.22	0.44	3.52	147	118	67	33
TRDSS (30 ton/ac)	2.00	0.51	3.08	98	129	58	35

Table VIII. Plant nutrient uptake per pot in 2-month greenhouse experiment.

	N	P	K	Fe	Cu	Mn	Zn
Treatment		(mg/pot)			(μg/pot)		
Check	5	1	6	42	5	10	10
100-40-0	4	1	4	29	5	10	10
200-80-0	12	1	8	43	5	15	13
USS (9 ton/ac)	47	11	63	168	51	97	235
USS (18 ton/ac)	71	17	109	267	104	140	526
TRUSS (9 ton/ac)	42	11	63	172	46	113	233
TRUSS (18 ton/ac)	57	14	85	228	71	132	415
DSS (15 ton/ac)	13	3	21	106	13	40	71
DSS (30 ton/ac)	27	8	49	155	40	94	204
TRDSS (15 ton/ac)	9	2	15	63	14	29	50
TRDSS (30 ton/ac)	31	8	47	151	54	90	198

total amount of heavy metals added per pot varied with sludge type as well as with amount added (Table IX). It is obvious that more metals were added per pot in the DSS treatments.

To examine total metal uptake on a more equitable basis, an index termed "Plant Extractability" was calculated as follows:

[Total plant uptake of a specific metal in a sludge treatment minus total plant uptake of that metal in the check] divided by total metal added to the pot from the sludge treatment.

or

Plant extractability of element X =

$$\frac{\frac{\mu g\ X}{pot}(\text{treatment}) - \frac{\mu g\ X}{pot}(\text{check})}{\frac{\mu g \text{ added in sludge}}{pot}} \times 10^6$$

The plant extractability index depicts a larger difference between sludge types than between rates per sludge type (Table X).

At the end of the greenhouse experiment the soil in the greenhouse pots was again characterized (Table XI). Sludge treatment had no effect on soil tests for NH_4 and K at the end of the experiment compared to checks by orthogonal contrasts at the 5% level of significance. Extractable Cu, Cd, Pb, Fe, Zn, Mn, NO_3^- and P were significantly higher at the end of the experiment for the sludge treatments compared to checks by orthogonal contrasts at the 5% level of significance, but thermoradiation did not significantly increase these nutrients. The higher sludge rates did significantly increase the extractable Cu, Cd, Pb, Fe, Zn, Mn, NO_3^- and P, comparing high and low rates of the same sewage material analyzed by orthogonal contrasts (5% level of significance). The average tissue concentrations for N, Fe and Mn were decreased compared to checks by sludge treatments and Zn concentration increased (tested at 5% level by orthogonal contrasts). Extractable Cu, Cd, Pb, Fe, Zn, Mn and P were significantly higher for the DSS treatments than for the USS treatments, but it should be remembered that more of these elements were added initially to the soil in the DSS treatments.

SUMMARY AND CONCLUSIONS

Information has been given indicating that the "Thermoradiation" process of reducing pathogens in sewage products being developed by Sandia Laboratories, Albuquerque, New Mexico, does not significantly increase the chemical extractability and plant uptake of a broad range of nutrients and

Table IX. Amount of metals added per pot in sludge treatments in greenhouse experiment and comparable amount of metals added on a per acre basis.

Sludge Treatment	Fe		Cu		Zn		Mn		Cd		Cr		Pb	
	(mg/pot)	(lb/ac)	(mg/pot)	(lb/ac)	(mg/pot)	(lb/ac)	(mg/pot)	(lb/ac)	(mg/pot)	(lb/ac)	(mg/pot)	(lb/ac)	(mg/pot)	(lb/ac)
USS (9 ton/ac)	349	199	28	16	44	25	5	3	0.5	0.3	6	4	20	12
USS (18 ton/ac)	698	399	55	32	88	50	9	5	1	0.6	13	7	41	23
TRUSS (9 ton/ac)	352	201	26	15	43	25	5	3	0.5	0.3	6	4	20	11
TRUSS (18 ton/ac)	704	402	52	30	86	50	9	5	1	0.6	13	7	40	23
DSS (15 ton/ac)	755	432	58	33	87	50	12	7	1.3	0.8	20	11	35	20
DSS (30 ton/ac)	1511	863	117	66	173	100	24	14	2.6	1.6	39	22	70	40
TRDSS (15 ton/ac)	765	437	58	33	86	49	12	7	1.4	0.8	20	12	35	20
TRDSS (30 ton/ac)	1530	874	115	66	173	98	24	14	2.8	1.6	41	23	70	40

Table X. Plant extractability of sludge-borne metals in sludge-treated soil in 2-month greenhouse experiment.

Sludge Treatment	Fe	Cu	Mn	Zn
USS (9 ton/ac)	363	1658	18544	5126
USS (18 ton/ac)	323	1802	13852	5867
TRUSS (9 ton/ac)	370	1592	22365	5200
TRUSS (18 ton/ac)	264	1267	13070	4715
DSS (15 ton/ac)	86	144	2492	702
DSS (30 ton/ac)	75	258	3489	1124
TRDSS (15 ton/ac)	28	156	1535	468
TRDSS (30 ton/ac)	71	425	3326	1090

heavy metals. Therefore, results of experiments on using sludges on calcareous soils should not differ greatly whether the sewage products are thermoradiation treated or untreated.

However, thermoradiation treatment greatly facilitates handling sewage for experimentation because pathogen contamination precautions are eliminated and objectionable odors are almost completely eliminated. Rosopulo *et al.*[13] studied the effects of sludge irradiation on plant nutrient uptake and found no concentration increases agreeing with results presented herein.

Experimentation published to date is sufficient to forego recommending indiscriminate use of sewage sludge on all agricultural land, (CAST[14]). However, sewage products may have special potential for use on calcareous soils, such as in New Mexico. For instance, in New Mexico the lack of potassium in sewage products is not a problem because most New Mexico soils contain sufficient K for good crop growth, and K is not routinely added as a fertilizer. The naturally high pH (7.5 to 8.0 or higher) of New Mexico soil greatly reduces plant availability of many problem heavy metals.

Dramatic increases in yield over and above that expected for N, P inputs from sewage are typified by the greenhouse experimental results presented herein, especially for the known micronutrient-deficient soils of New Mexico. However, more research needs to be done before the economics of sludge application can be calculated, and more information is needed on heavy metals that might need to be controlled on calcareous high-pH soils.

ACKNOWLEDGMENTS

This research was conducted under contract #F (29-2)-3626 with U.S. Energy Research and Development Administration, Albuquerque Operation Office, Albuquerque, New Mexico, in cooperation with Sandia Laboratories,

Table XI. Soil properties at termination 2-month greenhouse experiment.

Treatments	EC_{SE}	pH_{SE}	NH_4^+-N	NO_3^--N	DTPA Extract (ppm)								
					K	P	Fe	Zn	Mn	Cu	Pb	Cd	Cr[a]
Check	4.30	7.77	15	85	1446	9	10	4	8	2	2	0.08	0
100-40-0	4.48	7.75	13	118	1448	13	10	4	8	2	2	0.08	0
200-80-0	5.23	7.68	16	184	1364	13	10	3	8	1	2	0.08	0
USS (9 ton/ac)	4.21	7.67	20	90	1424	17	16	9	11	4	5	0.16	0
USS (18 ton/ac)	4.73	7.51	11	124	1441	21	20	15	12	7	7	0.24	0
TRUSS (9 ton/ac)	4.41	7.57	15	100	1440	17	16	10	11	4	5	0.16	0
TRUSS (18 ton/ac)	4.76	7.51	19	119	1376	19	20	14	12	6	7	0.23	0
DSS (15 ton/ac)	4.87	7.68	14	113	1434	16	15	10	9	6	5	0.24	0
DSS (30 ton/ac)	4.89	7.60	12	118	1441	22	18	16	10	9	7	0.40	0
TRDSS (15 ton/ac)	4.61	7.66	16	99	1446	17	15	10	10	6	5	0.24	0
TRDSS (30 ton/ac)	4.54	7.59	18	116	1438	23	20	17	11	11	8	0.43	0

[a]Cr was not detectable in the soil DTPA extract by the atomic adsorption procedure used in this study.

Albuquerque, New Mexico. This paper is published as Paper 101 from the New Mexico State University, Las Cruces, New Mexico 88003.

REFERENCES

1. Gerrard, M. "Sewage and Wastewater Processing with Isotopic Radiation: Survey of the Literature," *Isotopes Radiation Technol.* 8(4):429 (1971).
2. Ballentine, D. S. "Potential Role of Radiation in Waste Water Treatment," *Isotopes Radiation Technol.* 8(4):415 (1971).
3. Reynolds, M. C., R. L. Hagendruber and A. C. Zuppere. "Thermoradiation Treatment of Sewage Sludge using Reactor Waste Fission Products," SAND 74-0001, Sandia Laboratories, Albuquerque, New Mexico (1974).
4. Sivinski, H. P. "Treatment of Sewage Sludge with Combinations of Heat and Ionizing Radiation (Thermoradiation)," *IAEA Symposium,* (Munich, 1975). (Albuquerque, New Mexico: Sandia Laboratories-5440, 1975).
5. Sivinski, H. D. "Progress Report: Waste Resources Utilization Program," SAND-75-0580, Sandia Laboratories-5440, Albuquerque, New Mexico (1976).
6. Brandon, J. R. and S. R. Langley. "Inactivation of Bacteria in Sewage Sludges by Ionizing Radiation, Heat and Thermoradiation," SAND-75-0168, Sandia Laboratories, Albuquerque, New Mexico (1976).
7. Brandon, J. R. and S. L. Langley. "Sludge Irradiation: Bacteriology and Parasitiology," Proc. JHU Conf. on Evaluation of Current Developments in Municipal Waste Treatment, Johns Hopkins University, Baltimore, Maryland, January 1977.
8. Morris, M. "Cost-Effectiveness Comparisons of Various Types of Sludge Irradiation and Sludge Pasteurization Treatments," Proc. JHU Conf. on Evaluation of Current Developments in Municipal Waste Treatment, Johns Hopkins University, Baltimore, Maryland, January 1977.
9. Bremner, J. M. "Total Nitrogen, Inorganic Forms of Nitrogen, and Organic Forms of Nitrogen," in *Methods of Soil Analysis, Part 2,* C. A. Black, Ed. *Agronomy* 9:1149-1254 (1965).
10. Allison, L. E. "Organic Carbon," in *Methods of Soil Analysis, Part 2,* C. A. Black, Ed. *Agronomy* 9:1372-1375 (1965).
11. Piper, C. S. *Soil and Plant Analysis* (New York: Interscience, 1944).
12. Lindsay, W. L. and W. A. Norvell. "Development of a DTPA Micro-Nutrient Test," *Agron. Abstr.* 84 (1969).
13. Rosopulo, A., I. Fiedler, H. Slaerk and A. Suess. "Experience with a Pilot Plant for the Irradiation of Sewage Sludge: Analytical Studies on Sewage Sludge and Plant Material," *IAEA Symposium* (Munich, 1975). (Albuquerque, New Mexico: Sandia Laboratories-5440, 1975).
14. Council for Agricultural Science and Technology. "Application of Sewage Sludge to Croplands: Appraisal of Potential Hazards of the Heavy Metals to Plants and Animals," Report No. 64, CAST, Iowa State University, Ames, Iowa (1976).

17

DIFFERENCES IN THE SUSCEPTIBILITY OF SOYBEAN VARIETIES TO SOIL CADMIUM

S. F. Boggess and D. E. Koeppe
Department of Agronomy
University of Illinois
Urbana, Illinois

INTRODUCTION

Trace elements, certain of which have ubiquitous low concentrations in soils, are being dispersed in substantial quantities in localized situations through the activities of man.[1-3] While certain of these trace elements are known to be essential for normal metabolic activities of many organisms (including higher plants),[4] others (including Cd) have no known essential role but are readily absorbed and accumulated.[5-10] Toxic effects due to Cd uptake have been observed in many species at relatively low soil and plant tissue concentrations.[5,7,8,10,11] Variations in uptake and effect are observed in different species,[5,7,8,10-12] and the influence of soil parameters including pH, CEC, phosphorus and interactions with other elements plays a substantial role in the ultimate effect a trace element has on the plant.[6,14-18]

Sewage sludge is known to contain high trace element concentrations,[19] but is being spread on land at rates which could produce potentially toxic trace element concentrations within a few years.[20,21] While the possibility of "poisoning" land to the growth of plants clearly exists from sludge disposal practices, numerous studies suggest that the buffering and binding capacity of soils, combined with limited uptake and accumulation patterns by plants, make such disposal practices acceptable over a limited time period.[20,22] Fewer problems are anticipated with crops grown mainly for grain, since accumulation studies show a substantial step-down pattern in trace element concentrations from roots through vegetative tissue to the grain.[19,20]

Studies utilizing soils amended with trace element inorganic salts and/or sewage sludge strongly point to Cd and/or Zn accumulation by plants as possibly the greatest potential trace element problem to emerge from sewage sludge amendments to agricultural land,[20,23] as Cd is toxic to growth of virtually all plant species tested regardless of the growth medium. The magnitude of this problem remains questionable at this time, however, since most of the more definitive studies of Cd toxicity have been conducted with Cd-salt-amended soils which may provide substantially different Cd availability patterns than those created by comparable Cd additions through sludge amendments.[24]

To our knowledge, no consideration has been given to varietal differences in soybeans in response to Cd. This is somewhat surprising in light of the conclusion by Millikan[25] that the efficiency of nutrient absorption and utilization by plants is often greater between varieties of the same species than those differences found between related species or genera. The University of Illinois-Urbana has two projects which are assessing the uptake and effect of Cd by different inbred lines of corn[20,23] and varieties of soybeans. A portion of the soybean varietal work is presented here. Soybean varieties from several maturity groups exhibit a range of susceptibility to Cd added to soil as an inorganic salt. Abbreviated data are also presented which show a significant correlation in the degree of Cd susceptibility of soybean varieties grown on Cd-salt-amended soil and sludge-amended soil.

MATERIALS AND METHODS

Soybean varieties (*Glycine max* L.) were grown in the glasshouse in 4-in. pots containing three soils: Flanagan, Bloomfield and Plainfield (Table I). Lighting was supplemented with irradiation from GE 1000 w metal halide lamps, with a photoperiod of 14 hr light, 10 hr dark. The Flanagan soil was used without fertilization, while the Bloomfield and Plainfield soils were

Table I. Designation, source and characteristics of soils used.

Soil Type and Source	pH	K	P	CEC
		(μg/g dry soil)		(meq/100/g)
Bloomfield (Psammentic Hapludalfs) Havana, Illinois	5.5	38	8	1.8
Plainfield (Typic Udipsamments) Hancock, Wisconsin	6.5	9	60	–
Flanagan (Aquic Argiudolls) Urbana, Illinois	7.3	231	18	23

treated with KNO_3 (reagent grade) to give N and K levels recommended for a yield of 50 bu/ac (Illinois Agronomy Handbook, 1976). Additions of KNO_3 and $CdCl_2$ were made to enough distilled water to moisten weighed quantities of dry soil. After thorough mixing, the soil was placed in pots and seeds planted. Treatments were set up in duplicate and plants thinned to two per pot subsequent to cotyledon emergence from the soil. Three weeks after initial watering visual symptoms of Cd toxicity were recorded and the shoots were harvested, air dried at 60°C and weighed. The dried samples were then dry-ashed at 490°C in a muffle oven, allowing four hours for the oven to come to ashing temperature. This was followed by cooling overnight before the samples were removed from the oven. The ash was dissolved in 3 *N* HCl (diluted from 6 *N* constant boiling distillate) and the Cd concentration was determined by atomic absorption spectrophotometry. The reliability of this procedure has been previously established.[9]

A composite index was assigned from the sum of a dry weight term [1 - (treated dry weight/control dry weight)] x 10, Cd tissue concentration $\cdot 10^{-1}$, and the visual index. The visual index was on a 0 to 5 scale, broken down as follows: 0—no visual symptoms; 1—reddening at the base of leaflets; 2—a spreading redness along the veins toward the tip and margin of the leaflets (no deformation of leaflets); 3—a slight downward curvature of the leaflet midvein accompanied by further reddening of the veins; 4—leaflets curved downward, puckered or rumpled, slightly chlorotic and smaller than the control (reddening of the veins still very prominent); and 5—leaves small, increasingly chlorotic, so curved downward as to appear clasped, brittle and broken off easily (reddening of veins prominent). Pictures in Figure 1 correspond to the following rankings: 1A = 0; 1B = 1 to 2; 1C = 2 to 3; 1D = 4 to 5. It should of course be noted that symptoms of Cd toxicity present themselves as a continuum and that our visual symptom ranking scale and the pictures in Figure 1 are arbitrary.

Several experiments have been completed with 2 to 3 levels of amended Cd in three soils. In each experiment, 20 to 30 varieties were tested. In addition, a preliminary experiment with 10 varieties selected from the Cd-salt-amended soil studies for a broad range of Cd tolerance were grown on Keomah soil (aeric ochraqualfs, Fulton County, Illinois) amended with two levels of sludge. The concentrations of Cd in the sludge-amended soil were 3.92 and 7.54 μg/g soil, while Zn levels were 121 and 170 μg/g soil. CEC values were 15.0 and 18.4 meq/100 g soil for the moderate and higher sludge applications, respectively.

Criteria used in choosing the tested varieties were: economic importance, performance in other screening tests,[26,27] diversity of genetic background and maturity group designation.

Spearman's coefficient of rank correlations were used in comparing the varietal rankings of the tested parameters from the various experiments.

Figure 1. Soybean trifoliate leaves showing various degrees of cadmium toxicity. Darkening along veins shows as red to red-brown coloration in living tissue. See Materials and Methods for details.

RESULTS AND DISCUSSION

Few characterizations of the visual symptoms of Cd toxicity are presented in the literature. Root *et al.*[9] have indicated the presence of a Cd-induced interveinal chlorosis that was correlated with a decrease in chlorophyll content in corn, and Haghiri[5] has noted a progressive browning of the veins culminating with chlorosis which correlated with increased Cd treatment in soybeans. No evidence, to our knowledge, has been presented previously to identify the curled-leaf configuration (Figure 1) with increased Cd concentration. The type of symptoms shown in Figure 1 have appeared in all soybean varieties in all soils in response to one or more of the Cd levels used. Figure 2 shows a typical range of Cd effects on six soybean varieties grown on Bloomfield soil amended with 2 μg Cd/g. Table II presents data from plants grown on the same soil with a Cd concentration (1 μg/g) giving less severe toxicity. These 10 varieties were selected for divergence in Cd toxicity ranking from a larger experiment in which 27 varieties were tested. Results of the Cd effect on dry weight, Cd concentration in the shoot and visual Cd toxicity symptoms are presented. The composite index is derived from the previous information (see Materials and Methods). Numbers in parentheses represent the rank of the variety with regard to the characteristic being considered, with (1) meaning the least effect of Cd (greatest tolerance). Using Spearman's coefficient of rank correlations, the rankings based on dry weight, plant Cd concentration and visual Cd toxicity symptoms were well correlated with each other and with the composite index, the ρ value being less than 0.03 with the exception of the correlation of the dry-weight and visual-index ranking. Here the dry-weight ranking is probably influenced by poor control growth of several varieties. The maturity group ranking, on the other hand, did not correlate with any of the measures of Cd toxicity (ρ values ranged from 0.21 to 0.97).

Table III shows the varieties tested on Bloomfield soil ranked according to the composite index. When all 27 varieties were considered, significant linear correlations ($r > 0.79$: $\rho = 0.0001$) were found between soil Cd concentration and plant dry weight (negative) and plant Cd concentration.

In another experiment, the varieties considered in Table II were grown in soil treated with sewage sludge at two levels (Table I). Table IV shows shoot Cd concentration after 3-weeks growth in the sewage sludge-amended soil, the ranking according to plant Cd concentration and a visual index for plants grown at the higher sewage sludge level. No visual symptoms were observed at the lower level of application, although some appeared in plants grown for longer periods. While the Cd concentrations of the plants grown on the sludge-amended soil are substantially less than those grown in the Cd-amended Bloomfield soil, a very good correlation was obtained between the ranking of the Cd concentration of the $CdCl_2$-soil-amended plants and the Cd

Figure 2. Variable response of soybean varieties to 2 μg/g Cd as $CdCl_2$ Bloomfield soil. The varieties from left are Clark 63, Hawkeye 63, Dunfield, Dare, Arksoy and Harosoy 63.

Table II. Dry weight, cadmium concentration, visual index and composite index of 3-week-old plants of 10 soybean varieties grown on Bloomfield soil amended with 1 μg/g Cd as $CdCl_2$.

Variety	Maturity Group	Composite Index	Dry Weight (treated/control)	Plant (Cd) (μg Cd/g dry wt)	Visual Index
Richland	II (4.5)	6.06 (1)	0.92 (1)	17.6 (5)	3.5 (6.5)
Jackson	VII (10)	6.31 (2)	0.72 (2)	15.1 (2)	2.0 (2.5)
Clark '63	IV (8)	6.55 (3)	0.62 (5.5)	12.5 (1)	1.5 (1)
Grant	O (1)	7.46 (4)	0.66 (4)	15.6 (4)	2.5 (4)
Mandarin	I (2)	7.54 (5)	0.70 (3)	15.4 (3)	3.0 (5)
Amsoy '71	II (4.5)	10.60 (6)	0.38 (9)	24.0 (7)	2.0 (2.5)
Corsoy	II (4.5)	10.80 (7)	0.49 (7)	22.0 (6)	3.5 (6.5)
Dunfield	III (7)	11.70 (8)	0.62 (5.5)	34.0 (8)	4.5 (8)
Harosoy '63	II (4.5)	15.04 (9)	0.47 (8)	47.4 (10)	5.0 (9.5)
Arksoy	VI (9)	16.09 (10)	0.28 (10)	47.0 (9)	5.0 (9.5)

concentration of the high-sludge-amended plants (ρ = 0.0114). A similar significant correlation was found for the ranking of the visual symptoms of the $CdCl_2$-soil-amended plants and the Cd concentration of the high-sludge-amended plants (ρ = 0.0088). Of probably greatest interest is that regardless of absolute rank, the 10 varieties generally sort out in all ranked categories into a susceptible group (Dunfield, Harosoy and Arksoy) and a less descript, less susceptible group. Dowdy and Ham[24]

Table III. Twenty-seven soybean varieties grown in Bloomfield soil amended with cadmium (1 μg/g) and ranked according to a composite index of susceptibility to cadmium based on dry weight, cadmium concentration and visual symptoms.

Variety	Maturity Group	Composite Index
Perry	IV	3.43
Richland	II	6.06
Jackson	VII	6.31
Clark '63	IV	6.55
Lee	VI	6.94
Capital	0	7.35
Grant	0	7.46
Mandarin	I	7.54
Roanoke	VII	7.75
Hawkeye '63	II	7.76
Mukden	II	8.35
Ford	III	8.97
Blackhawk	I	9.95
Bragg	VII	10.45
Amsoy '71	II	10.60
Corsoy	II	10.80
Illini	III	11.09
Lincoln	III	11.42
Dunfield	III	11.70
Beeson	II	12.05
Fiskeby	00	12.45
Scioto	IV	13.92
Flambeau	00	14.85
Chief	IV	14.90
Harosoy '63	II	15.04
Dare	V	15.85
Arksoy	VI	16.90

in a recent paper have concluded that "the uptake of sludge-borne metal cannot be predicted from data obtained in studies using inorganic salts as a metal source." There is no reason to question their conclusion as it relates to one soil, one soybean variety (Corsoy) and one defined sewage sludge. However, the latitude their statement would seem to cover is questioned. While the limited nature of our comparative rankings on Cd-salt-amended soils with the sludge-amended soils is recognized, the data reported here, and pertinent observations from experiments in progress, suggest that varietal rankings of Cd toxicity obtained on Cd-salt-amended soils will be accurate in predicting sludge Cd toxicity in soybean varieties from the highly susceptible grouping.

Knowledge of the wide-ranging soybean varietal differences in Cd uptake and/or tolerance should be useful in a number of ways. These include: (1)

Table IV. Cadmium concentration and visual symptoms in 10 soybean varieties grown on soil amended with sewage sludge.

Variety	Plant (Cd)[a] Moderate Application[c]	Plant (Cd) Higher Application[d]	Visual Index[b]
Richland	0.63 (3)	3.48 (7)	0
Jackson	0.92 (7)	3.22 (6)	0.5
Clark '63	0.65 (5)	1.36 (1)	0
Grant	0.86 (6)	2.91 (5)	0.5
Mandarin	0.57 (2)	1.98 (2)	0
Amsoy '71	0.53 (1)	2.36 (4)	0
Corsoy	0.64 (4)	2.17 (3)	0
Dunfield	1.57 (9)	4.25 (8)	1.0
Harosoy '63	2.08 (10)	5.97 (10)	1.0
Arksoy	1.55 (8)	4.79 (9)	1.0

[a]Number in parenthesis represents the rank.
[b]Visual symptoms were observed only at the higher application.
[c]3.92 μg Cd/g soil.
[d]7.54 μg Cd/g soil.

the reduced Cd contamination of agricultural products through careful varietal selection; (2) the use of varieties with an exceptionally high affinity for Cd as scavengers of Cd from contaminated soils; (3) the possible use of especially sensitive varieties, such as Dunfield, Harosoy or Arksoy, as visual indicator plants to warn of undesirably high Cd levels in soils without resorting to laboratory analyses (such an indicator would have the advantage of measuring available Cd, which, to our knowledge, is not presently available); and (4) the use of soybean varieties of varying Cd tolerance in studies of the mechanism of Cd toxicity, particularly in localizing the overall processes, metabolic steps, etc., that are most influenced by the metal.

SUMMARY

When a number of soybean varieties were grown on soil amended with $CdCl_2$ or sewage sludge, cadmium toxicity symptoms appeared as a continuum from slight effects observed as a red to red-brown coloration at the junction of the leaf blade and petiole, to severe leaf curling and extensive reddening of the leaf veins, chlorosis and finally a brittle condition, followed by abscission. Marked differences in the severity of symptoms appeared among the varieties, and a visual symptom index correlated well with indices based on Cd-induced dry-weight reduction and the concentration of Cd in the plant.

Using the above indices, a similar ranking of susceptibility of the experimental varieties to Cd was found in experiments with three $CdCl_2$-amended soils, although detailed data from only one soil are presented here. The susceptibility ranking also agrees for the most part with Cd concentrations found in a group of varieties grown in sewage sludge-treated soil. This suggests that data from plants grown on $CdCl_2$-amended soils can be used to predict the soybean varietal response to Cd on sludge-amended soils.

ACKNOWLEDGMENTS

This research was supported in part by NSF-RANN grant GI-31605 and the Illinois Agricultural Experiment Station. The authors wish to thank Susan Willavize, Roberta Steward and Mark Rubin for excellent technical assistance. Valuable discussions of the research with Drs. J. J. Hassett, T. Hymowitz and T. D. Hinesly are also gratefully acknowledged.

REFERENCES

1. Fleischer, M., A. F. Sarofim, D. W. Fassett, P. Hammond, H. T. Shacklette, I. C. T. Nisbet and S. Epstein. "Environmental Impact of Cadmium: A Review by the Panel on Hazardous Trace Substances," *Environ. Health Pers.* 7:253-323 (1974).
2. Fulkerson, W. and H. E. Goeller, Eds. "Cadmium the Dissipated Element," Oak Ridge National Laboratory Report NSF-EP-21 (1973).
3. Page, A. L. and F. T. Bingham. "Cadmium Residues in the Environment," *Residue Rev.* 48:1-44 (1973).
4. Gauch, H. G. *Inorganic Plant Nutrition* (Stroudsburg, Pennsylvania: Dowden, Hutchinson and Ross, Inc., 1972).
5. Haghiri, F. "Cadmium Uptake by Plants," *J. Environ. Qual.* 2:93-96 (1973).
6. Haghiri, F. "Plant Uptake of Cadmium as Influenced by Cation Exchange Capacity, Organic Matter, Zinc, and Soil Temperature," *J. Environ. Qual.* 3:180-183 (1974).
7. Jarvis, S. C., L. H. P. Jones and H. J. Hopper. "Cadmium Uptake from Solution by Plants and its Transport from Roots to Shoots," *Plant & Soil* 44:179-191 (1976).
8. Page, A. L., F. T. Bingham and C. Nelson. "Cadmium Absorption and Growth of Various Plant Species as Influenced by Solution Cadmium Concentration," *J. Environ. Qual.* 1:288-291 (1972).
9. Root, R. A., R. J. Miller and D. E. Koeppe. "Uptake of Cadmium–Its Toxicity and Effect on the Iron Ratio in Hydroponically Grown Corn," *J. Environ. Qual.* 4:473-476 (1975).
10. Turner, M. A. "Effect of Cadmium Treatment on Cadmium and Zinc Uptake by Selected Vegetable Species," *J. Environ. Qual.* 2:118-119 (1973).
11. John, M. K. "Cadmium Uptake by Eight Food Crops as Influenced by Various Soil Levels of Cadmium," *Environ. Poll.* 4:7-15 (1973).

12. Bingham, F. T., A. L. Page, R. J. Mahler and T. J. Ganje. "Growth and Cadmium Accumulation of Plants Grown on a Soil Treated with a Cadmium-Enriched Sewage Sludge," *J. Environ. Qual.* 4:207-211 (1975).
13. Hassett, J. J., J. E. Miller and D. E. Koeppe. "Interaction of Lead and Cadmium on Corn Root Growth and Uptake of Lead and Cadmium by Roots," *Environ. Poll.* 11:297-302 (1976).
14. John, M. K. "Interrelationships Between Plant Cadmium and Uptake of Some Other Elements from Culture Solutions by Oats and Lettuce," *Environ. Poll.* 11:85-95 (1976).
15. Koeppe, D. E. "The Uptake, Distribution and Effect of Cadmium and Lead in Plants," *Sci. Total Environ.* 6 (in press, 1977).
16. Latterell, J. J., R. H. Dowdy and G. E. Ham. "Sludge-Borne Metal Uptake by Soybeans as a Function of Soil Cation Exchange Capacity," *Commun. Soil Sci. Plant Anal.* 7:465-476 (1976).
17. Miller, J. E., J. J. Hassett and D. E. Koeppe. "Uptake of Cadmium by Soybeans as Influenced by Soil Cation Exchange Capacity, pH, and Available Phosphorus," *J. Environ. Qual.* 5:157-160 (1976).
18. Miller, J. E., J. J. Hassett and D. E. Koeppe. "Interactions of Pb and Cd on Metal Uptake and Growth of Corn Plants," *J. Environ. Qual.* 6:18-20 (1977).
19. Jones, R. L., T. D. Hinesly and E. L. Ziegler. "Cadmium Content of Soybeans Grown in Sewage Sludge-Amended Soil," *J. Environ. Qual.* 3:351-353 (1973).
20. Hinesly, T. D., R. L. Jones, J. J. Tyler and E. L. Ziegler. "Soybean Yield Responses and Assimilation of Zn and Cd from Sewage Sludge-Amended Soil," *J. Water Poll. Control Fed.* 48:2137-2152 (1976).
21. Webber, J. "Effects of Toxic Metals in Sewage on Crops," *Water Poll. Control Eng.* 71:404-406 (1972).
22. John, M. K. and C. J. Van Laerhoven. "Effects of Sewage Sludge Composition, Application Rate, and Lime Regime on Plant Availability of Heavy Metals," *J. Environ. Qual.* 5:246-251 (1976).
23. Hinesly, T. D., R. L. Jones, E. L. Ziegler and J. J. Tyler. "Effects of Annual and Accumulative Applications of Sewage Sludge on Assimilation of Zinc and Cadmium by Corn (*Zea mays* L.)," *Environ. Sci. Technol.* 11:182-188 (1977).
24. Dowdy, R. H. and G. E. Ham. "Soybean Growth and Elemental Content as Influenced by Soil Amendments of Sewage Sludge and Heavy Metals, Seedling Studies," *Agron. J.* 69:300-303 (1977).
25. Millikan, C. R. "Plant Varieties and Species in Relation to the Occurrence of Deficiencies and Excesses of Certain Nutrient Elements," *Aust. Inst. Agri. Sci.* 27:220-233 (1961).
26. Armiger, W. H., C. D. Foy, A. L. Fleming and B. E. Caldwell. "Differential Tolerance of Soybean Varieties to an Acid Soil High in Exchangeable Aluminum," *Agron. J.* 60:67-70 (1968).
27. Howell, R. W. and R. L. Bernard. "Phosphorus Response of Soybean Varieties," *Crop. Sci.* 1:311-313 (1960).

18

NUTRIENT USAGE AND HEAVY METALS UPTAKE BY SHEEP FED THERMORADIATED, UNDIGESTED SEWAGE SOLIDS

G. S. Smith and H. E. Kiesling
Department of Animal and Range Sciences
New Mexico Agricultural Experiment Station
New Mexico State University
Las Cruces, New Mexico

H. D. Sivinski
Environmental Programs and Waste Management Division
Sandia Laboratories
Albuquerque, New Mexico

INTRODUCTION

Recent concerns for scarcity of fossil-fuel energy and portents of impending world food shortage, together with staggering costs of present technology for sewage disposal, have given urgency to additional research and development of technology for recycling of nutrients in sewage solids. Domestic sewage represents a vast resource of nutrients that are potentially usable by direct refeeding to animals, including livestock. Refeeding of products from poultry and animal manures is being practiced on a large scale in present-day livestock production and is based on a firm foundation of research demonstrating important economic and ecologic benefits. The "state of the art" on current feedlot practice for recycling manures for nutrient recovery was summarized in reports at a seminar at Ada, Oklahoma, on April 7, 1977.[1] Abstracts of current research on feeding of animal manure, feathers, offal and other waste products to livestock have been compiled and reported.[2]

Refeeding of human wastes has been practiced in some primitive societies but is restricted in modern society because of known health hazards as well as cultural stigmata and low economic incentive. Over 20 years ago, Illinois researchers[3,4] demonstrated the nutritive value of sewage products for ruminant animals, and recent reappraisals of economic benefits[5] have renewed interest in direct refeeding to livestock.

Technology which promises favorable cost/benefit relationships[6] has been developed by Sandia Laboratories for destruction of pathogens in sewage products through combination of heat and *gamma*-radiation from "waste" radionuclides from the nuclear fuel cycle.[7-9] Using thermoradiated products representing undigested ("raw") sewage solids collected from the Albuquerque sanitation district, researchers at New Mexico State University have demonstrated nutritive energy and nitrogen in sewage solids which compare favorably with cottonseed meal when used as supplements to fibrous, low-nitrogen diets for sheep and cattle.[10-11] It appears at present that the primary constraint on direct refeeding of properly-processed sewage products is the potential hazard of chemical toxicants in sewage. Preliminary investigations at New Mexico State University[12-15] with rumen cultures, rats and sheep have suggested that chemical toxicants in Albuquerque sewage products pose little hazard of acute toxicity. It is recognized, however, that research on longer term effects from chemical toxicants, mutagenic agents and carcinogens is lacking and obviously will be required in assessing the possible usage of sewage products as feeds.

The research reported herein was designed to extend observations on the nutritive value of thermoradiated sewage products used as supplements in feeds for ruminants, and to assess the retention of heavy metals and certain trace elements by sheep fed sewage products and subsequently fed a conventional diet.

NUTRITIVE VALUE OF SEWAGE SOLIDS

Ruminant Digestive Physiology

It has long been recognized that ruminant animals (cattle, sheep, goats, deer, etc.) exist in symbiotic relationship with vast populations of microorganisms which inhabit the fore-stomach compartments of the digestive tract. These microbes "predigest" the feeds consumed by the host, utilizing the products of anaerobic fermentation for growth of microbial cells and releasing fermentation products for absorption and usage as nutrients by the host. A major supply of nutrients for ruminant animals is the mass of microbial cells and fermented feeds which passes into the lower digestive tract, and a major by-product of microbial fermentation is methane gas which is eructated from the fore-stomach. This microbial fermentation enables ruminant animals to utilize feedstuffs that are indigestible by or sometimes toxic to nonruminants. Ruminants thus utilize energy from fibrous, cellulosic feeds and utilize microbial proteins and amino acids synthesized from dietary nonprotein nitrogen. Likewise, most (sometimes all) of the B-vitamins needed by ruminants can be derived from microbial synthesis in the digestive tract. Moreover–and quite importantly–the microbial fermentation effectively detoxifies numerous organic and inorganic compounds such as ammonia, cyanide, nitrate and sulfide when ingested in small amounts and when "active fermentation" is sustained by suitable management practices.[16] Such detoxification has important implications for the potential recycling of wastes such as animal manures, sewage products and contaminated feeds.

The NMSU-Sandia Research Program

A cooperative research program involving the New Mexico Agricultural Experiment Station at New Mexico State University, Las Cruces (NMSU) and Sandia Laboratories, Albuquerque (SANDIA) was initiated in 1975 under contract with the U.S. Energy Research and Development Administration (Albuquerque Operations Office) and is projected to continue through 1981. SANDIA has developed technology for destruction of pathogens by thermoradiation and has provided NMSU with thermoradiated sewage solids for assessment of nutritive value. Although the NMSU program to date has not been primarily "toxicological" in design, considerable attention has been given to measurement of potential toxicants in the sewage products fed and to assessment of apparent toxicity in experimental animals.

Sewage products were analyzed chemically and then studied in cultures of rumen microorganisms to assess effects on microbial degradation of fibrous substrates. In these studies, sewage products were compared with cottonseed meal and urea, when provided in isonitrogenous amounts. When assayed in terms of fiber degradation, the solids from undigested ("raw") sewage supported greater fermentative activity ($P < .05$) than either cottonseed meal or urea, suggesting nutritive benefits in addition to the nitrogen provided.[10] In these studies, microscopic examinations of the microbial populations showed normal numbers of typical rumen protozoa, suggesting that the sewage solids were relatively nontoxic to these sensitive organisms. Likewise, rather large amounts of sewage solids (80 to 220 g daily) have been placed directly into the fore-stomachs (via rumen cannulae) of rumen-fistulated sheep fed conventional and experimental diets for extended periods (20 to 50 days) without apparent adverse effects on feed intake, animal appearance and behavior, or presence of typical protozoa in the rumen fluid.

Prior to, or concurrent with, feeding experiments with sheep and cattle, the sewage products evaluated have been tested in long-term feeding experiments with rats, whereby male and female rats were fed diets comprised of commercial chow with up to 50% by weight of added sewage product from weaning through young adulthood and one or two phases of reproduction. Although the products used provided little nutritive value to rats (as expected), the growth, appearance and reproductive performance of rats fed thermoradiated sewage solids suggested little, if any, adverse effect other than dilution of dietary nutrients. Balance trials with such diets indicated that most of the dietary heavy metals and (selected) trace elements with potential for toxicity were excreted in the feces; however, some enlargement of livers was noted, and tissue retention (livers and kidneys) of certain elements (notably Fe and Cu) suggested possible adverse effects from long-term feeding of products tested. Data from typical experiments with rats have been reported.[10]

Nutritive Evaluation with Sheep

In a series of short-term digestibility trials with sheep fed basal, fibrous, "poor quality" (low nitrogen) diets in comparison with equal intakes of basal components plus supplemental cottonseed meal (CSM) *vs* thermoradiated, undigested sewage solids, digestibility of basal components was apparently improved by supplements, and sewage solids compared favorably with CSM. Apparent digestibility of energy in the supplements (calculated by difference in total energy digested for basal *vs* supplemented diets) varied from 40 to 60% for sewage solids in comparison with 58 to 65% for CSM. Likewise, apparent digestibility of supplement nitrogen (calculated by difference) ranged from 50 to 60% for sewage solids as compared with 67 to 73% for CSM. In the most definitive of these trials, the calculated true digestibility of supplement nitrogen (calculated by difference and corrected for metabolic fecal nitrogen based on literature values), was 82.8 ± 5.7% for sewage solids *vs* 82.2 ± 1.7% for CSM ($\bar{x}$ ± S.D., three lambs each diet). These data were reported earlier.[10]

More recent experiments, reported in part originally herein, evaluated sewage solids collected from the Albuquerque sanitation district and processed by centrifugation (to enrich solids from about 5% to about 20 to 25%) prior to thermoradiation and drying by forced-air ventilation in vats lined and covered with plastic. This product (TRUSC) was sampled, analyzed chemically, processed through a hammer mill and incorporated into experimental diets at levels comprising 10, 20 or 30% of diet dry matter. Comparable diets were formulated to contain 5, 10 or 15% cottonseed meal (CSM), and a control ("BASAL") diet was formulated to be somewhat similar but without either CSM or TRUSC. All diets were pelleted at a local feed mill. The ingredient composition and crude protein content of these diets are shown in Table I. Chemical composition of the TRUSC is shown in Table II.

Commercial crossbred lambs were group-fed these diets (with salt and water *ad libitum*) for 29 days during which time feed intake and body weight changes were recorded. Thereafter, all lambs were provided a commercial, pelleted finishing diet for feeder lambs, and feeding performance was evaluated during 100 days. The results, reported elsewhere,[17] showed poor intake of diets with TRUSC during the first two weeks but acceptable intakes thereafter. Some of these lambs were maintained on 30% TRUSC for 69 days and achieved intakes approaching 4% of body weight. Feeding of TRUSC did not adversely affect subsequent feeding performance.

Concurrent with the feeding trial, selected lambs in each of three diet groups (Basal *vs* 10% CSM *vs* 20% TRUSC) were fitted with fecal collection bags and used in a 10-day digestion trial while consuming the diets *ad libitum* under group-feeding conditions. Total fecal collections were accomplished and dietary intakes calculated from fecal lignin output and dietary lignin content. Energy and nitrogen intakes and digestibility were calculated. These data are summarized in Table III. Digestibility of both energy and nitrogen was improved by CSM or TRUSC over the basal diet, and the amounts of digested energy and digested nitrogen were greater for lambs fed 20% TRUSC than for those fed 10% CSM. The data in Table III are presented in terms of nutrients consumed and nutrients digested per unit of metabolic body size, $(\text{kg body weight})^{0.75}$ in order to facilitate comparisons with similar data for cattle, presented in Table IV and described below.

The same diets studied with sheep were also fed to steers penned separately and offered diets in amounts calculated at 2 to 3% of body weight. Diets were fed during a 10-day adjustment period and a 5-day collection period during which time total feces were collected by fitting the animals with fecal collection bags. Dietary intakes and fecal outputs were measured directly and digested nutrients calculated by difference. The data (Table IV) show that CSM and TRUSC tended to increase the amounts of digested energy and nitrogen beyond the levels for the basal diet and tended to improve digestibility of dietary energy and nitrogen over the basal diet. TRUSC at 20% of the diets provided slightly more digested energy and digested nitrogen than CSM at 10% of the diet. Although the differences between diets are not statistically significant ($P > .05$) because the numbers of individuals are small and individual variations are relatively large, the trends suggest nutritive value of sewage products which has biological and economic importance.

Table I. Ingredients and protein content of diets fed to sheep and cattle.[a]

Ingredients (%)	Diets						
	BASAL	5% CSM	10% CSM	15% CSM	10% TRUSC	15% TRUSC	30% TRUSC
Sorghum stover	40.0	45.0	40.0	35.0	45.0	40.0	35.0
Alfalfa hay	30.0	27.0	24.0	21.0	27.0	24.0	21.0
Sorghum grain	24.7	18.2	21.6	24.8	13.3	11.5	9.8
Cane molasses	5.0	4.5	4.0	3.5	4.5	4.0	3.5
Cottonseed meal (CSM)	none	5.0	10.0	15.0	none	none	none
TRUSC[b]	none	none	none	none	10.0	20.0	30.0
Phosphorus supplement[c]	0.3	0.3	0.4	0.7	0.2	0.5	0.7
Protein content (%)	11.0	11.2	12.8	14.0	10.9	11.8	12.0

[a]NMSU experiments, summer 1976.

[b]TRUSC represents undigested sewage solids from Albuquerque, New Mexico, treated by thermoradiation at Sandia Laboratories, Albuquerque, New Mexico.

[c]"X-P-4,"® commercial polyphosphate supplement.

Table II. Chemical composition of thermoradiated, undigested sewage solids fed to sheep.[a]

Components	Content
	(% of dry matter)
Ash (550°C)	29.6
Crude silica	14
Ether extract	13.1
Nitrogen (Kjeldahl; HgO catalyst)	2.89
Crude protein (N x 6.25)	18.1
Acid-detergent fiber	33[b]
Acid-detergent lignin	8[b]
Heat of combustion	4.65 kcal/g
Calcium, as Ca	1.85
Phosphorus, as P	0.71
Magnesium, as Mg	0.251
Sodium, as Na	0.143
Potassium, as K	0.209
Trace Elements[c]	(ppm)
Ag	25
Cd	12
Co	5
Cr	192
Cu	636
Fe	5040
Hg	4
Mn	107
Pb	99
Zn	1147

[a]Used in NMSU experiments with sheep and cattle, summer, 1976.

[b]Data expressed on ash-free basis.

[c]Values represent total content, as measured by atomic absorption spectrophotometry using acidic solutions from wet-ashed samples.

TISSUE RETENTION OF HEAVY METALS AND TRACE ELEMENTS

Attention has been given to potential hazards of heavy metals and organic pollutant in the food chain arising from land application of sewage sludges.[18,19] Sheffner *et al.*[20] discussed toxicological aspects in the nutritional evaluation of activated sludge and presented data on growth, reproductive performance and teratology of rats and rabbits, and egg production by chickens fed sewage products. Adverse effects were noted in terms of production when sewage products were used at 10% or more of diets, but levels of 5% or 10% of the diet did not result in gestational or teratogenic

Table III. Digestibility of energy and nitrogen in fibrous diets for lambs as affected by cottonseed meal (CSM) or thermoradiated undigested sewage solids (TRUSC).[a]

Items	Diets		
	Basal	10% CSM	20% TRUSC
Lambs per group	4	4	4
Energy intake [Kcal/(kg BW)$^{0.75}$]	375 ± 76[b]	378 ± 43	400 ± 63
Energy digested [Kcal/(kg BW)$^{0.75}$]	191 ± 57	191 ± 17	214 ± 27
Energy digestibility (%)	50 ± 6	51 ± 2	54 ± 3
Nitrogen intake [g/(kg BW)$^{0.75}$]	1.69 ± 0.34	2.04 ± 0.23	1.96 ± 0.31
Nitrogen digested [g/kg BW)$^{0.75}$]	0.68 ± 0.24	0.91 ± 0.09	0.93 ± 0.18
Nitrogen digestibility (%)	39 ± 7	45 ± 1	47 ± 3

[a]New Mexico State University experiments, summer, 1976. TRUSC represents undigested sewage solids from Albuquerque, New Mexico, treated by thermoradiation at Sandia Laboratories, Albuquerque, New Mexico.

[b]Values shown are means ± standard deviations.

Table IV. Digestibility of energy and nitrogen in fibrous diets for cattle as affected by cottonseed meal (CSM) or thermoradiated undigested sewage solids (TRUSC).[a]

Items	Diets		
	Basal	10% CSM	20% TRUSC
Steers per group	4	4	4
Energy intake [Kcal/(kg BW)$^{0.75}$]	280 ± 9[b]	284 ± 9	316 ± 24
Energy digested [Kcal/(kg BW)$^{0.75}$]	127 ± 30	152 ± 26	171 ± 25
Energy digestibility (%)	45 ± 10	53 ± 7	54 ± 9
Nitrogen intake [g/(kg BW)$^{0.75}$]	1.31 ± 0.05	1.53 ± 0.05	1.55 ± 0.12
Nitrogen digested [g/kg BW)$^{0.75}$]	0.46 ± 0.20	0.76 ± 0.15	0.80 ± 0.09
Nitrogen digestibility (%)	35 ± 15	59 ± 8·	50 ± 8

[a]New Mexico State University experiments, summer, 1976. TRUSC represents undigested sewage solids from Albuquerque, New Mexico, treated by thermoradiation at Sandia Laboratories, Albuquerque, New Mexico.

[b]Values shown are means ± standard deviations.

abnormality. Cheeke and Myer[21] evaluated activated sewage sludge as a protein source for rats and Japanese quail, but reported no data on tissue composition. Schönborn[5] cited reports, mostly from European literature concerning nutritive value of sewage products for rats, poultry and swine, and discussed potential hazards of chemical toxicants. Recently, Kinzell, Cheeke and Chen[22] reported growth, organ size, tissue heavy metals and trace elements, reproductive performance and pentobarbital sleeping times of rats as affected by dietary activated sewage sludge. Although growth and reproductive performance were adversely affected, livers and kidneys were enlarged, and pentobarbital sleeping times suggested liver metabolism of some toxic constituents in the product fed, heavy metals (Cd, Hg and Pb) were not accumulated "beyond acceptable levels" in the muscle tissue, even when sludge comprised 50% of the diet. Metals contents (Cd, Cu, Fe, Hg, Mn, Pb and Zn) in livers, kidneys, hearts, spleens, testes, muscles, brains, etc. from rats fed activated sewage sludge at 50% of the diet were reported.

Although it is generally recognized that ruminants are the animals with greatest potential for direct refeeding of sewage solids, and that data from nonruminants are often unapplicable for ruminants, there are very few reports in the American literature on tissue retention of heavy metals, trace elements and organic toxicants by ruminants fed sewage products. Kienholz, Ward and Johnson[23] presented preliminary results from studies of heavy metals and organic toxicants in tissues of cattle grazing forages from sludge-treated soils and of cattle fed digested sludge in feedlot. These Colorado researchers reported further results, indicating measurable tissue levels of certain metals, pesticide residues and chlorinated hydrocarbons, at the Third National Conference on Sludge Management, Disposal and Utilization.*

Data have been reported on apparent absorption and retention of heavy metals and certain trace elements by sheep fed thermoradiated undigested sewage solids.[10] Those data, based on analysis of feeds, feces and urine from short-term balance trials, showed that absorption of Ag, Cd, Co, Cr, Cu, Fe, Hg, Mg, Mn, Pb and Zn from sewage solids was low. These elements from ingested sewage products were excreted mainly in the feces. However, positive balances were recorded in some cases, although amounts detected in livers from six sheep fed sewage products at levels of 20 to 30% of diet dry matter for about three months were not appreciably elevated over comparable values for experimental control animals. Likewise, blood samples from sheep fed sewage solids at 20 to 30% of diet showed no appreciable difference in content of heavy metals or trace elements when compared to experimental controls. These same trends have been confirmed and reported in subsequent studies.[24]

---

*Miami, Florida, December 1976, *Proceedings* in press (Rockville, Maryland: Information Transfer, Inc.).

In more definitive experiments (reported originally herein), samples of blood were taken from sheep fed the experimental diets shown in Table II after 29 days feeding, and other samples were obtained from sheep fed two of these diets (15% CSM and 30% TRUSC) after an additional 40 days. Contents of heavy metals and trace elements in these samples are shown in Tables V and VI. These data show that blood levels of the heavy metals and trace elements measured were not appreciably affected by dietary sewage solids.

Samples of livers and kidneys were collected from sheep fed the basal diet or 30% TRUSC for 69 days and then fed a pelleted, commercially prepared conventional sheep diet for 61 days. Likewise, samples of kidneys were collected from sheep fed the 30% TRUSC diet for 29 days and then the conventional diet for 104 days. Elemental composition of these livers and kidneys is shown in Table VII. The results show that ingestion of TRUSC at 30% of the diet for 29 or 69 days, followed by a conventional diet for 104 or 61 days, failed to significantly affect liver and kidney contents of Ca, P, Na, K, Fe, Cu, Mg, Mn, Zn, Co and Cr. In the cases of Ag, Cd, Hg and Pb, the means for values from animals fed TRUSC were slightly higher than comparable means for control animals, suggesting slight increases in content of these elements as a result of ingested TRUSC. The values (in ppb) for Hg in livers (< 14 for "controls" vs 31 ± 10 for "experimentals") and for Pb in kidneys (148 ± 13 for "controls" vs 281 ± 24 and 226 ± 35 for "experimentals") indicate significant increases ($P < 0.05$) in these elements as a result of earlier ingestion of sewage solids, although the biological importance of these differences is questionable.

Heavy metals and potentially toxic trace elements tend to accumulate in livers and kidneys at levels higher than in muscle tissue and other typically edible meat products, and therefore, levels of these elements in livers and kidneys can be regarded as indicators of potential hazards to the human food-chain. On this basis, the results presented herein suggest that feeding of TRUSC, followed by prolonged feeding of a conventional diet, poses little if any hazard from the elements assayed to consumers of meat products. It should be noted, however, that livers and kidneys are not necessarily the most important tissues to assay for possible hazards *to the animals* consuming contaminated feedstuffs, since chronic toxicity may relate more closely to levels accumulated in other tissues such as brain, bone, lung, spleen, etc. These tissues have not been analyzed to date in our studies.

The data in Table VII provide two notable observations, one of which has been widely recognized: copper accumulates in livers of sheep at levels much higher than in kidneys, whereas cadmium accumulates at higher levels in kidneys. Less frequently discussed in literature we have seen in the observation that silver (Ag) accumulated in livers at 800 to 1300 ppb but was undetected in kidneys of the same animals. The biological importance of this finding is not known at present.

Table V. Elemental composition of whole blood in lams as affected by diets with and without sewage solids.[a]

Elements	Diets						
	BASAL	5% CSM[b]	10% CSM	15% CSM	10% TRUSC[b]	20% TRUSC	30% TRUSC
			mg/100 ml				
Ca	6.2 ± 2.7	4.3 ± 0.6	4.1 ± 0.6	3.7 ± 1.1	4.3 ± 0.9	4.8 ± 0.8	5.3 ± 0.9
P	17.9 ± 1.5	18.4 ± 1.6	19.4 ± 1.7	22.1 ± 1.8	20.1 ± 1.3	19.2 ± 2.2	18.4 ± 1.0
Na	201 ± 6	203 ± 2	203 ± 2	207 ± 1	204 ± 2	206 ± 5	208 ± 5
K	44 ± 11	31 ± 4	35 ± 2	37 ± 6	31 ± 3	36 ± 13	34 ± 5
Fe	38 ± 4	38 ± 5	38 ± 2	34 ± 5	45 ± 12	38 ± 5	35 ± 7
Mg	2.5 ± 0.3	2.4 ± 0.3	2.5 ± 0.1	2.5 ± 0.2	2.3 ± 0.2	2.6 ± 0.1	2.5 ± 0.1
			μg/100 ml				
Ag			< 3.6*				
Cd			< 2*				
Co			< 1.4*				
Cr			< 7.8*				
Cu	93 ± 9	102 ± 8	94 ± 16	102 ± 6	98 ± 0	115 ± 13	96 ± 4
Hg	65 ± 13	54 ± 18	70 ± 16	45 ± 7	75 ± 13	66 ± 9	62 ± 10
Mn			< 4*				
Pb	24 ± 2	23 ± 5	22 ± 1	23 ± 0	29 ± 3	27 ± 3	26 ± 2
Zn	572 ± 183	522 ± 143	531 ± 135	480 ± 122	535 ± 155	645 ± 75	562 ± 210

[a]NMSU experiments, summer 1976; diets had been fed for 29 days; samples were wet-ashed and analyzed by atomic absorption spectrophotometry. Values shown are means ± standard deviations representing 4 lambs for each group.

[b]CSM is cottonseed meal; TRUSC is thermoradiated undigested sewage solids from Albuquerque, New Mexico.

*Values represent limits of detection under analytical conditions used.

Table VI. Elemental composition of whole blood in lambs as affected by diets with cottonseed meal (CSM) or thermoradiated sewage solids (TRUSC).[a]

Elements	Diets	
	15% CSM	30% TRUSC
	(mg/100 ml)	
Ca	5.7 ± 0.6	5.6 ± 1.3
Na	260 ± 30	265 ± 10
K	37 ± 10	39 ± 6
Fe	45 ± 9	45 ± 7
Mg	1.2 ± 0.3	1.2 ± 0.4
	(μg/100 ml)	
Ag all < 3.6*		
Cd	6.0 ± 1.4	6.2 ± 0.7
Co all < 1.4*		
Cr all < 7.8*		
Cu	164 ± 7	162 ± 8
Hg	48 ± 20	37 ± 10
Mn all < 4*		
Pb	49 ± 4	34 ± 1
Zn	259 ± 17	245 ± 40

[a]New Mexico State University experiments, summer, 1976; diets had been fed for 69 days; samples were wet-ashed and analyzed by atomic absorption spectrophotometry. Values shown are means ± standard deviations representing 6 lambs per group.

*Values represent limits of detection under analytical conditions used.

In our experiments, potential organic toxicants such as pesticides, chlorinated hydrocarbons, polychlorinated biphenyls, aflatoxins, etc., have not been measured in the sewage products fed or in animal tissues. Such investigation is obviously of importance and will be conducted with samples from experiments reported herein, although results from our bioassays with rats indicated no apparent adverse effects on reproductive performance from ingestion of sewage products at levels up to 50% of diets from weaning through adulthood and one breeding cycle. A very recent report[25] of neoplastic skin lesions in neotenic tiger salamanders from a sewage lagoon suggests chemical etiology and provides further incentive for continued, rigorous research in assessing the hazards of recycled sewage products.

Table VII. Elemental composition of livers and kidneys from lambs fed diets with and without thermoradiated undigested sewage solids (TRUSC).[a]

Dietary Regimen	Livers		Kidneys		
	Basal, 69 days	30% TRUSC, 69 days	Basal, 69 days	30% TRUSC, 69 days	30% TRUSC, 29 days
	Conventional Diet, 61 days		Conventional Diet, 61 days		Conventional Diet, 104 days
Number of lambs	5	5	4	5	6
Elements			[mg/kg (ppm), Fresh Tissue]		
Ca	52 ± 14	58 ± 4	75 ± 14	86 ± 11	79 ± 15
P	3824 ± 236	3646 ± 128	6163 ± 267	5799 ± 356	6089 ± 516
Na	783 ± 42	748 ± 57	1903 ± 228	1664 ± 151	1703 ± 102
K	2819 ± 174	2679 ± 138	2641 ± 141	2481 ± 208	2565 ± 283
Fe	63 ± 10	87 ± 13	59 ± 16	52 ± 11	54 ± 18
Cu	95 ± 18	93 ± 22	3.3 ± 0.2	3.3 ± 0.3	3.4 ± 0.3
Mg	183 ± 9	177 ± 1	182 ± 10	172 ± 7	181 ± 17
Mn	3.1 ± 0.4	3.0 ± 0.3	1.2 ± 0.1	1.2 ± 0.2	1.2 ± 0.1
Zn	35 ± 2	35 ± 2	19 ± 1	24 ± 2	23 ± 3
			[μg/kg (ppb), Fresh Tissue]		
Ag	950 ± 119	1182 ± 146	< 25*	< 25*	< 25*
Cd	78 ± 34	90 ± 13	131 ± 104	125 ± 13	81 ± 8
Co	< 50*	< 50*	< 50*	< 50*	< 50*
Cr	< 60*	< 60*	< 60*	< 60*	< 60*
Hg	< 14*[b]	31 ± 10[c]	< 14*	< 14*	< 14*
Pb	226 ± 44	285 ± 66	148 ± 13[b]	281 ± 24[c]	226 ± 35[c]

[a]New Mexico State University experiments, summer, 1976; samples were wet-ashed and analyzed by atomic absorption spectrophotometry; values shown are means ± standard deviations.

*Values represent limits of detection under analytical conditions used.

[b,c]Values with different superscript letters are significantly different ($P < 0.05$).

DISCUSSION

It is recognized that the experimentation to date is insufficient to provide conclusions regarding potential hazards of recycling sewage solids as feedstuffs. Wide variations in composition of sewages, even from the same source from day to day, preclude generalizations. Nevertheless, results reported support the conclusion that economically important amounts of nutrients are involved and usable by ruminant animals. At current prices for conventional protein supplements, the feeding value of "raw" solids from domestic sewages projects easily into the range of $100 or more per ton. The production of cattle and sheep on marginal croplands and uncultivatable rangelands is a multibillion-dollar industry in the U.S. Ruminant animals are destined to become ever more important in meeting the nutritive needs of the growing world population:

> Because of . . . natural variability in forage quality, supplementary feeding may be necessary during nongrowing seasons in some regions to avoid ruminant mortality. Supplementary feeding is often necessary to maintain reproductive level and is generally necessary to maintain milk production. However, supplementary feeding is not always profitable. Hence, research to improve technology for ruminant feeding continues in the use of hay, silage, feed grains, cassova and other crops. Supplementation of grazing with molasses, urea, crop residues, millings and industrial by-products and wastes is especially important.[26]

Perhaps it is not premature to suggest that sometime in the future, for some cities (assuredly not all), the development of a "feedstuff industry" may well replace the typical sanitation district and thus provide the incentive to "decontaminate" the products of domestic waste. At current prices, most cities of 100,000 persons are presently "disposing," at great cost, potential feedstuffs worth almost half a million dollars yearly.

ACKNOWLEDGMENTS

Research conducted under contract No. E (29-2)-3626 with the U.S. Energy Research and Development Administration, Albuquerque Operations Office, Albuquerque, New Mexico, in cooperation with Sandia Laboratories, Albuquerque, New Mexico. Published as Publ. S-100 from the New Mexico Agricultural Experiment Station.

The authors wish to acknowledge technical assistance by Mr. Leroy Ben. Bruce, Mr. Joe Cadle, Mr. Charlie Staples and Ms. Phyllis Walters.

REFERENCES

1. U.S. Environmental Protection Agency. "Seminar on Feedlot Manure Recycling for Nutrient Recovery," East Central University, Ada, Oklahoma, April 7, 1977. Sponsored by Great Plains Extension Committees in cooperation with Robert S. Kerr Environmental Research Laboratory, U.S. EPA, Ada, Oklahoma.
2. Smithsonian Science Information Exchange, Inc. "Feeding of Manure, Feathers, Offal and Waste Products to Livestock," Publ. No. EKIOD, SSIE, Rm. 300, 1730 M Street N.W., Washington, D.C. 20036 (1977).
3. Hackler, L. R., A. L. Neumann and B. Johnson. "Feed from Sewage. III. Dried Activated Sewage Sludge as a Nitrogen Source for Sheep," *J. Animal Sci.* 16:125 (1957).
4. Hackler, L.R. "Dried Activated Sewage Sludge as a Nitrogen Source for Ruminants," Ph.D. Dissertation, University of Illinois, Urbana, Illinois (1958).
5. Schönborn, W. "On the Use of Sewage Sludge as Fodder," *Proc. IAEA Symposium on Use of High Level Radiation in Waste Treatment–Status and Prospects, Munich, 1975,* IAEA-Sm-194/701 (Frankfurt Am Main, West Germany: Batelle Institute, 1975).
6 Morris, M. "Cost-Effectiveness Comparisons of Various Types of Sludge Irradiation and Sludge Pasteurization Treatments," Proc. JHU Conf. on Evaluation of Current Developments in Municipal Waste Treatment, Johns Hopkins University, Baltimore, Maryland, January 1977.
7. Sivinski, H. D. "Treatment of Sewage Sludge with Combinations of Heat and Ionizing Radiation (Thermoradiation)," *Proc. IAEA Symposium on Use of High Level Radiation in Waste Treatment–Status and Prospects, Munich, 1975* (copies available from Sandia Laboratories-5440, Albuquerque, New Mexico 87115).
8. Brandon, J. R. and S. R. Langley. "Inactivation of Bacteria in Sewage Sludges by Ionizing Radiation, Heat and Thermoradiation," SAND-75-0168, Sandia Laboratories, Albuquerque, New Mexico (1976).
9. Brandon, J. R. and S. R. Langley. "Sludge Irradiation: Bacteriology and Parasitology," Proc. JHU Conf. on Evaluation of Current Developments in Municipal Waste Treatment, Johns Hopkins University, Baltimore, Maryland, January, 1977.
10. Smith, G. S., H. E. Kiesling, J. M. Cadle, C. Staples, L. B. Bruce and H. D. Sivinski. "Recycling Sewage Solids as Feedstuffs for Livestock," *Proc. Third Natl. Conf. on Sludge Management, Disposal and Utilization, Miami, Florida, December, 1976* (Rockville, Maryland: Information Transfer, Inc., 1977).
11. Smith, G. S. and B. D. McCaslin. "Agronomic and Animal Feeding Evaluations of Thermoradiated Sewage Solids in New Mexico," Proc. JHU Conf. on Evaluation of Current Developments in Municipal Waste Treatment, Johns Hopkins University, Baltimore, Maryland, January, 1977.
12. Hoffman, M. and G. Smith. "Recycling Sewage Solids as Feedstuffs: Bioassays," *J. Ariz. Acad. Sci.* 11:146 (1976).
13. Smith, G. S., J. M. Cadle, P. Walters and H. E. Kiesling. "Sewage Solids as Feedstuffs for Ruminants," abstr. *J. Animal Sci.* 43:467 (1976).
14. Smith, G. and C. Staples. "Heavy Metals in Rats Fed Sewage Solids," abstr. *J. Animal Sci.* 43:233 (1976).

15. Smith, G. S., B. D. McCaslin and H. E. Kiesling. "Recycling Sewage Solids as Feeds and Fertilizer," abstr. *J. Nutr.* 106:xxxi (1976).
16. Phillipson, A. T. *Physiology of Digestion and Metabolism in the Ruminant* (Newcastle upon Tyne, England: Oriel Press, 1969).
17. Cadle, J. M., H. E. Kiesling and G. Smith. "Feedlot Performance of Lambs Fed Sewage Solids," *Proc. West. Sec. Am. Soc. Anim. Sci.* 28 (in press, 1977).
18. Dean, B. "Hazards from Metals and Organic Pollutants in Sludge from Municipal Treatment Plants," paper presented at International Water Conservancy Exhibition, September, 1975, Advanced Waste Treatment Research Laboratory, U.S. EPA, Cincinnati, Ohio.
19. Council for Agricultural Science and Technology. "Application of Sewage Sludge to Croplands: Appraisal of Potential Hazards of the Heavy Metals to Plants and Animals," Report No. 64, CAST, Iowa State University, Ames, Iowa (1976).
20. Sheffner, A. L., J. W. Keating, A. L. Palanker, M. S. Weinberg and R. Dean. "Toxicological Aspects in the Nutritional Evaluation of Activated Sludge," paper presented at the Am. Chem. Soc. meeting, September, 1974. (Copy available from R. Dean, EPA, Cincinnati, Ohio.)
21. Cheeke, P. R. and R. O. Myer. "Evaluation of the Nutritive Value of Activated Sewage Sludge with Rats and Japanese Quail," *Nutr. Reports International* 8(6):385 (1973).
22. Kinzell, J. H., P. R. Cheeke and R. W. Chen. "Nutritive Value of Activated Sewage Sludge for Rats: Growth, Tissue Heavy Metals and Reproductive Performance," *Proc. West. Sec. Am. Soc. Anim. Sci.* 27:142 (1976).
23. Kienholz, E. W., G. M. Ward and D. E. Johnson. "Sewage Sludge Metals in Cattle Tissues," abstr. *J. Animal Sci.* 43:230 (1976).
24. Bruce, L. B., G. S. Smith and C. Staples. "Silicate Effects on Minerals in Lambs Fed Sewage Solids," *Proc. West Sec. Am. Soc. Anim. Sci.* 28 (in press, 1977).
25. Rose, F. L. and J. C. Harshbarger. "Neoplastic and Possibly Related Skin Lesions in Neotenic Tiger Salamanders from a Sewage Lagoon," *Science* 196:315 (1977).
26. Byerly, T. C. "Ruminant Livestock Research and Development," *Science* 195:450 (1977).

SECTION IV

NUTRIENT MANAGEMENT

19

AVAILABILITY AND TRANSFORMATION OF SEWAGE SLUDGE NITROGEN

B. R. Sabey
Department of Agronomy
Colorado State University
Fort Collins, Colorado

INTRODUCTION

Sewage sludge accumulations in some of our major metropolitan areas have posed serious disposal problems for those cities. With the society-imposed restraints on incineration, land filling and/or ocean dumping, using land as a terminal treatment and recycling system has gained tremendous momentum in the United States. Many municipalities are moving rapidly to land application of their sewage sludges. Several potential problems exist, however, among which are heavy metal accumulation and entrance into the food chain,[1] pathogen survival and disease spread[2-5] and phosphorus and nitrogen contamination of soils and waters.[6-8] With the increased cost of commercial fertilizers (from the standpoint of both money and energy), using sewage sludge as a fertilizer looks increasingly attractive to many.

Several authors and guideline bulletins have advocated nitrogen as being one of the more important limiting factors for sewage sludge application to land.[9-11] They reason that if application rates could be limited to those quantities supplying adequate nitrogen for plant growth, then the possibility of nitrogen accumulation or nitrate leaching would be diminished. Additionally, those application rates would increase the period of time before other problems would occur. However, making accurate sludge application recommendations based on nitrogen supply and availability presupposes a knowledge of organic nitrogen decomposition, nitrogen transformation and plant uptake rates. A generalized diagram showing possible transformations of nitrogen added to the soils in anaerobically digested sewage sludge is illustrated in Figure 1. The rates of these transformations are dependent upon many environmental conditions that influence biological, chemical and physical activities. Nitrogen availability depends on method and duration of

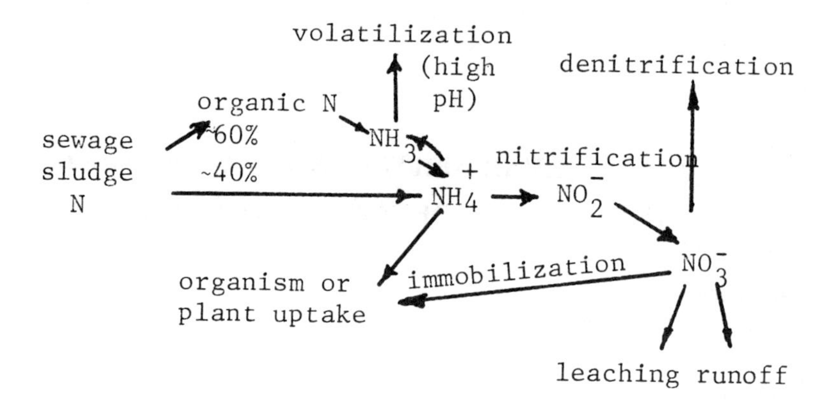

Figure 1. Generalized transformations of nitrogen added to soils in anaerobically digested sewage sludge.

sludge treatment prior to field application, amount of total and ammonium nitrogen, method and time of field application, soil properties, climatic conditions and numerous other factors. Therefore, management of the land application operation will probably have as much effect on the amount of nitrogen that is or becomes available for plant growth as any other factor. One of the main reasons that management (after digestion) is so important is that ammonium can make up from 20 to 60% of the total nitrogen of anaerobically digested sewage sludge. Much of that ammonium can be volatilized and lost if the sludge is applied and dried on a sand drying bed or on a soil surface, especially at a high pH. If the ammonium is not lost, much of it will be immediately available for plant growth. If the ammonium is lost, the remaining nitrogen will be largely in the organic form which must be mineralized before becoming available for plant growth. The first inorganic compound formed via mineralization is ammonia which unites with water to form ammonium. This also can be volatilized or nitrified (Figure 1). Nitrification is quite rapid in most soils with ambient temperatures and moisture conditions; therefore, nitrate accumulates in the soil unless it is utilized by organisms or plants or lost by leaching, runoff and/or denitrification. It if is lost by denitrification, potential pollution of air could occur. If it is lost by leaching and runoff, possible water pollution could result. These should be avoided if possible in attempting to manage for maximum benefit and utilization of this resource. One of the pressing objectives of sludge research should be to gain a more complete understanding of the mechanisms and rates of transformations of the nitrogen added to soils so that recommended amounts of sludge will be more accurate and management systems can be designed for optimum recycling of this valuable resource.

An excellent review of the nitrogen cycle in soils amended with sewage sludge has been prepared by Kelling.[12] Nitrogen transformations are very complex because of the many factors and variables that affect losses and additions. Volatilization, denitrification and plant removal have been considered the accepted, desirable removal mechanisms. Some scientists, however, are now wondering about the desirability of volatilization and denitrification because of possible effects in the atmosphere or stratosphere.[13] Each can remove appreciable quantities of the total nitrogen depending on such management factors as methods, rate and time of application, as well as other important factors.

The most desirable mechanism for removal or utilization of the nitrogen is through crop or plant growth in a recycling system. An excellent corn crop can remove from 150 to 250 kg/ha per year. Some grasses according to King and Morris[14] can use more nitrogen than this. They report uptake of 360 to 800 kg of nitrogen per hectare in a two-year period. Coker[15] from England applied lower rates ranging from 3.5 to 9 m.ton/ha for four years and obtained 32 and 24% nitrogen recovery by grass. Equivalent amounts of ammonium nitrate fertilizer resulted in recovery values of 35 and 32%, respectively. In another study by Coker[16] about 46% of the applied nitrogen was recovered by plants from a mixture of sewage sludge and fertilizer when low application rates were added. Of the added nitrogen, 60% was in the ammonium form. Most of these kinds of studies have shown that as the application rate of sewage sludge increases the percent recovery of nitrogen during the first year decreases. These studies have been concerned about nitrogen recovery by plant growth from plots receiving sewage sludge.

Other studies have attempted to determine the amount of the organic nitrogen or carbon in sewage sludge that is decomposed or mineralized with time. Estimates have ranged from 2 to 3% to greater than 50% of the organic nitrogen becoming available in one growing season. Miller[17] determined that 17 to 20% of the organic carbon was mineralized in a 6-month laboratory incubation study. These values were similar to those of Rothwell and Hortenstine.[18] Larson, Clapp and Dowdy,[19] however, estimated a 6% annual mineralization of organic nitrogen that was added as sewage sludge. In a laboratory column study Miller[20] estimated that 3.3 to 3.4% of the sludge organic nitrogen was mineralized and leached out of the column after a 6-month incubation period. Ryan[21] found 4 to 48% of the organic nitrogen was apparently mineralized in their incubation study that lasted 16 weeks at a constant temperature. King[22] in a laboratory study found that nitrate accumulation was 22% of the applied nitrogen when 2.5 cm of liquid sludge was applied to the soil surface and 38% when it was incorporated into the soil. Ammonium accounted for only 20% of the applied nitrogen initially. From 16 to 36% of the initial applied nitrogen was accounted for at the end of the incubation period for incorporated and surface applied sewage sludge, respectively. King attributed most of these losses to denitrification. He also indicated that about 60% of the initial nitrogen added was mineralized at the end of the 16-week incubation period. Pratt, Broadbent and Martin[23] have suggested a decay series of 35, 10 and 5% of the organic nitrogen in sewage sludge becoming mineralized for the first three years after application to soil. They emphasized however, that the series is based on the best estimate of the three authors rather than on hard data. Probably no one is certain just how rapidly the organic nitrogen in sewage sludge is mineralized in the field as yet, though it is certain that it depends on many factors which vary greatly from site to site.

The pursuit of information on this topic should be vigorous since in the absence of solid data one can only estimate the nitrogen supply from sewage sludge. Error in the estimates can either cause unnecessary loss of nitrogen and possible environmental pollution or cause a nitrogen deficiency in the plants that are being fertilized.

The study upon which the following data are based has been primarily concerned with determining the value of sewage sludge compared to inorganic fertilizer for plant growth and nitrogen uptake.

METHODS AND MATERIALS

A greenhouse study was set up with the following treatments: five application rates, including 0, 22.4, 56, 112 and 224 m.ton/ha of sewage sludge; and 4 application rates of inorganic nitrogen fertilizer, including 0, 67, 134 and 269 kg/ha of ammonium nitrogen. These treatments were added to two kg of air-dry samples of the A-one horizon of a Nunn clay loam (aridic Argiustoll) that was crushed and sieved through a 6.35-mm sieve. The sludge was an air-dried anaerobically digested municipal sewage sludge from the Fort Collins, Colorado, East Drake Plant, that had been screened through a 2-mm sieve. The nitrogen analyses of the sludge and soil are found in Table I. Each treatment was replicated five times.

Table I. Nitrogen analyses of the sewage sludge and soil used in this study.

	N Concentration (ppm)	
N Determination	Sludge	Soil
Total N	36,700	1,050
NH_4^+-N	10,200	10
NO_3^--N	23	47

Water was added to reach field capacity and the soil-sludge mixtures were allowed to incubate two weeks, after which the top 13 mm of soil were removed. Thirty-six seeds of Wichita wheat (*Triticum aestivum L. em Thell*) were evenly spread over the surface of the exposed soil, then the top 13 mm of soil were replaced and firmed around the seeds. After germination and emergence, the plants were thinned to three average-sized plants in each pot. The surviving three plants were allowed to grow for about seven more weeks before harvesting. The plants were harvested by clipping the aboveground growth, were oven dried at 60°C for 24 hr, then were weighed to determine the dry matter production. These dried samples were ground in a Wiley mill with a 20-mesh screen in preparation for total nitrogen determination.

After harvest the soils were screened through an 8-mesh sieve and mixed well, then subsamples of each pot of soil were dried in preparation for determining total nitrogren, nitrate and ammonium nitrogen. The procedures for determining total nitrogen analyses are those outlined by Bremner,[24] using salicylic acid to include nitrate. The nitrates and ammonium were determined by the steam distillation method of Bremner.[25] Total nitrogen for the plant material was determined by the same method used for total nitrogen in the soil.

RESULTS AND DISCUSSION

Table II shows the effect of various application rates of sewage sludge as well as rates of inorganic fertilizer only on plant growth and on nitrogen

Table II. Plant growth and nitrogen uptake in relation to sludge and inorganic nitrogen additions.

Sewage Sludge Application Rate (m.ton/ha)	Plant Growth (g/pot) ↓	Nitrogen Uptake by Plants (mg/pots)
	1.8	45
22.4	4.6	212
56	8.9	341
112	8.3	335
224	6.0	265
Inorganic Fertilizer Only (kg/ha)		
0	1.8	45
67	3.0	65
134	3.7	100
269	5.3	170

uptake in the aboveground tissue (reflecting nitrogen uptake). Maximum plant growth for this study occurred with sludge at 56 m.ton/ha. This resulted in 8.9 g of dry matter production. Higher sludge applications caused a decreased wheat growth. None of the inorganic nitrogen rates produced wheat growth as great as 56 m.ton/ha rate of sludge. It took 269 kg/ha of inorganic fertilizer nitrogen to result in a plant growth value beyond the 22.4 m.ton/ha rate of sewage sludge addition.

The highest nitrogen concentration in plant tissue occurred also with the 56 m.ton/ha sludge-addition rate. None of the inorganic nitrogen additions caused as much nitrogen uptake as the 22.4 m.ton/ha sludge addition.

In an attempt to evaluate the relative value of sludge compared to inorganic fertilizer nitrogen in the effect on plant growth and nitrogen uptake, the data in Table II were used to develop Figures 2 and 3. Figure 2 shows the relative effect of the four rates of sewage sludge on wheat growth compared to the inorganic fertilizer. This was done by plotting the wheat growth obtained with various additions of inorganic fertilizer nitrogen (the

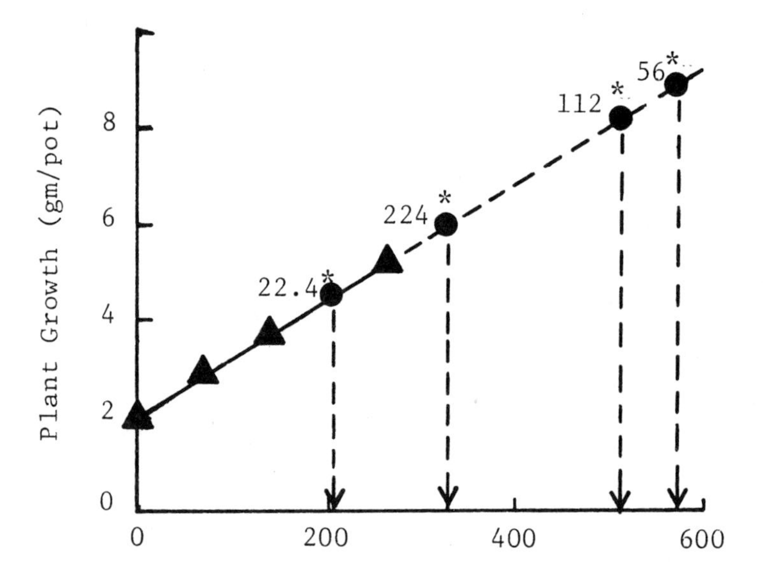

Figure 2. Estimation of inorganic fertilizer nitrogen equivalent of sewage sludge added to soil at various rates based on plant growth equivalents. *Sludge application rates (m.ton/ha).

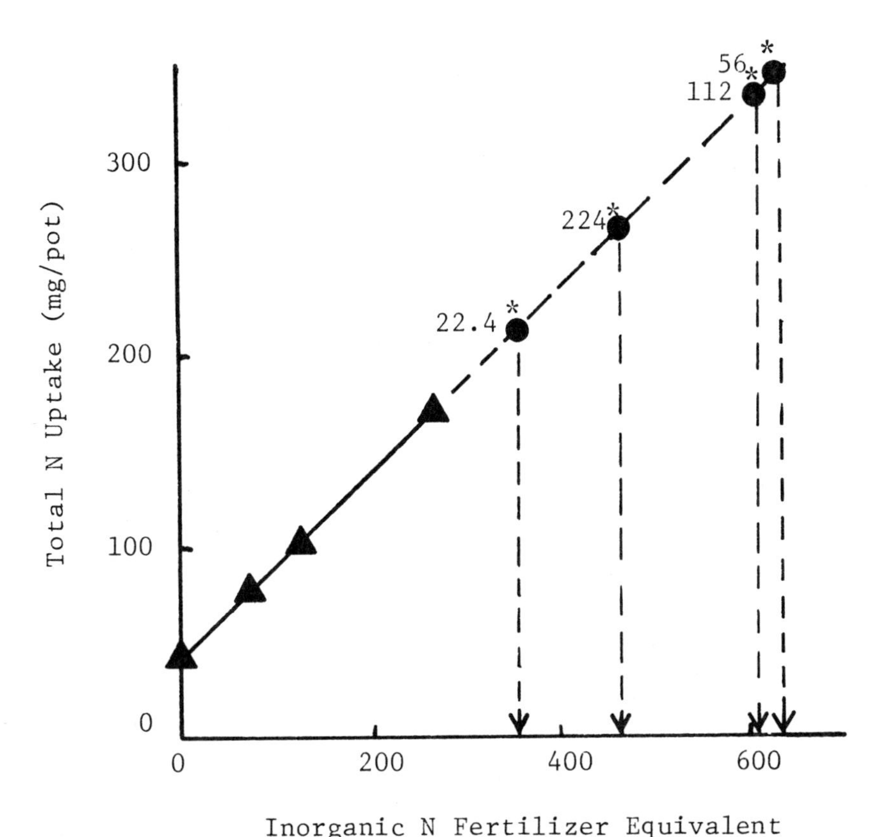

Figure 3. Estimation of inorganic fertilizer nitrogen equivalent of sewage sludge added to soil at various rates based on plant uptake of nitrogen. *Sludge application rates (m.ton/ha).

triangle symbols), then extrapolating the line beyond the point of highest plant growth value for the inorganic fertilizer addition (dashed part of the line). The plant growth values for each application rate of sewage sludge were then located on the line as shown by the solid black dots. By drawing a line from the plant growth values noted by the black dots parallel to the ordinate down to the abscissa, one can estimate the inorganic nitrogen fertilizer equivalent of each rate of sludge addition. These values were 210, 575, 515 and 330 kg/ha for 22.4, 56, 112 and 224 m.ton/ha of sewage sludge, respectively, as noted by the dashed arrows on Figure 2 and on line 5 of Table III. It is obvious that the line could not be straight indefinitely because in a typical curve showing plant growth response to fertilizer addition, the

Table III. Parameters measured and calculated in the greenhouse study.

Line No.	Parameter Measured	Sewage Sludge Application Rate (m.ton/ha)				
		0	22.4	56	112	224
1	Total N concentration in soil after harvest (mg/pot)	1886	2544	3326	5098	7028
2	Total N in pots initially (mg/pot) (soil N plus sludge N)	2100	2834	3935	5720	9440
3	Total N recovered in plant tissue and in the soil after harvest (mg/pot)	1931	2756	3667	5433	7293
4	Total N recovered after harvest (%)	91.9	97.3	93.2	94.1	77.2
5	Equivalent fertilizer N value based on plant growth (kg/ha)	0	210	575	515	330
6	Apparent N availability (%)	–	26	28	13	4
7	Equivalent fertilizer N value based on N uptake (kg/ha)	0	360	630	615	470
8	Apparent N availability (%)	–	44	31	15	6
9	N use efficiency (%)	–	29	19	9	4
10	N uptake due to sludge addition only (mg/pot)	0	167	296	290	220
11	N in the original pots minus N in the original soil (mg/pot)	0	734	1835	3670	7340
12	N use efficiency from treatment only (%)	0	22.7	16.1	7.9	5.0
13	Overall N use efficiency (%)	2.1	7.5	8.7	5.8	2.8

line tends to level off when the fertilizer rates are increased greatly. However, the data obtained from the graph will be in error on the low side, since if the line tended to level off as most yield curves do, the inorganic nitrogen fertilizer equivalent values would be larger than those noted in Figure 2. The inorganic fertilizer equivalent of sewage sludge is at least as much as indicated and may be more. It was felt, therefore, that this approach was valid, especially for a greenhouse experiment wherein greater quantities of fertilizer per unit weight of soil are required for optimum growth than in the field, due to the confined root volume in the pots.

Figure 3 indicates the relative effect of the four rates of sewage sludge on nitrogen uptake by the wheat plants compared to inorganic nitrogen fertilizer. This was done in a similar way by plotting the nitrogen content in the aboveground portion of the wheat plants with increasing inorganic fertilizer nitrogen (triangle symbols), then extrapolating the line beyond the point of the highest nitrogen concentration for the highest inorganic fertilizer addition (dashed portion of the line). The values for nitrogen content in the wheat plants were located on the line (black dots), then a line from these values parallel to the ordinate down to the abscissa indicated the inorganic nitrogen fertilizer values for each rate of sludge addition. These values were 360, 630, 615 and 470 kg/ha, or 22.4, 56, 112 and 224 m.tons/ha of sewage sludge, respectively, as noted by the dashed arrows in Figure 3 and on line 7 of Table III. There were considerable differences in the equivalent nitrogen fertilizer values derived by the two approaches, but the trends were quite similar. The similarity of these trends could have been predicted, since there was a close correlation between plant growth and total nitrogen content of the plant tissue per pot. These data could be used to recommend application rates of sewage sludge to land where nitrogen supply to plants is the basis for such rates. Care must be taken, however, when extrapolating from a greenhouse study to the field. Generally, the amount of nitrogen needed per unit weight of soil for optimum plant growth in the greenhouse will be considerably more than in the field because of the confined rooting volume in greenhouse pots. Therefore, the fertilizer value of sewage sludge in the field might be quite different from that in the greenhouse.

Table III shows some of the important parameters that were measured or calculated in the study. Line 4 of Table III shows the percentage of the total nitrogen accounted for at harvest time. These figures were derived by dividing the values on line 3 by those on line 2 and multiplying by 100. The initial nitrogen recovered at harvest time ranged from 77.2 up to 94.1%. These values are typical for studies of this nature. Less than 100% recovery may have been due to denitrification, ammonia volatilization and/or experimental error.

The apparent nitrogen availability of the sewage sludge nitrogen that was added was calculated by dividing the values in lines 5 and 7 of Table III by the amount of total nitrogen initially added with the sewage sludge to the soil in the pots. These values for the two approaches described above are indicated on lines 6 and 8 of Table III. The apparent nitrogen availability by the two methods are quite similar except at the 22.4-m.ton/ha rate, in which case the nitrogen uptake method value is almost double that of the plant growth method value.

Another method of evaluating the nitrogen supplied to plants by the added sewage sludge or the percent efficiency of nitrogen use is to divide the total nitrogen in the plant tissue by the total nitrogen added with the sewage sludge. These values are shown on line 9 of Table III. They are somewhat lower than the apparent nitrogen availability percentages. This approach considers only that nitrogen added by the sewage sludge and disregards that nitrogen supplied to the plant by the soil. If the 45 mg per pot taken up by the wheat plants in the untreated soil is subtracted from the nitrogen uptake figures (Table II) of the plants grown with the four rates of sewage sludge added, the values on line 10, Table III, are obtained. If the 2100 mg of nitrogen per pot contained in the soil are subtracted from the total nitrogen in the initial pots (including that in the soil and the sludge–line 2, Table III), values on line 11, Table III, are obtained. By dividing the values on line 10 by those on line 11, the nitrogen use efficiency figures on line 12, Table III, are derived. Even this, however, ignores the so-called "priming effect" wherein the added sewage sludge causes the soil to supply a greater amount of nitrogen due to decomposition of the native organic matter in the soil. In an attempt to take this into account, the nitrogen uptake by the wheat plants was divided by the total nitrogen in the pots initially (including that in the soil and the sludge). These nitrogen use efficiency values are shown on line 13 of Table III. These data are considerably lower than the previous approaches and are undoubtedly too low to reflect the actual percentage of the sludge nitrogen that became available for plant uptake, since it included the extremely low availability of the unamended soil organic matter. Only about 2% of the nitrogen contained in the control soil was taken up by the wheat plants. It is concluded that the actual nitrogen use efficiency of the sewage sludge ranges somewhere between the figures on lines 12 and 13 of Table III for the conditions of this greenhouse study. The values simply reflect the percentage of the amount of the total nitrogen that was contained in the sludge, in the soil or in both, that was taken up and utilized by the aboveground portions of the wheat plant. The best approach of these outlined depends on the use one plans to make of the data, but in any case they should be used with a clear understanding of their origin.

SUMMARY

Numerous authors and guideline bulletins have advocated that available nitrogen from sludge be the initial limiting factor in governing sewage sludge applications to soils. If the sludge application adds or supplies only enough nitrogen for optimum plant growth, then nitrate accumulation and leaching will be limited. Other potential problems would also be diminished compared to heavier application rates.

Making accurate application recommendations requires a knowledge of not only the nitrogen species in the sludge but also the rate of nitrogen transformations as influenced by environmental conditions and management.

Most anaerobically digested sludges have considerable ammonium nitrogen, ranging from 20 to 60% of the total nitrogen. This is immediately available for plant growth unless the sludge is surface applied, wherein significant amounts may be lost through volatilization. The organic nitrogen must be mineralized before plant utilization. Estimates of the amount of organic nitrogen that becomes available in one season have ranged from 2 or 3% to greater than 50%.

This study attempted to determine the fertilizer value of several rates of sludge addition to soils. This included the total available nitrogen for plant growth either from the ammonium initially present or from mineralization of organic nitrogen. Wheat plants were grown in the greenhouse for a 50-day period in soil treated with sewage sludge at rates of 0, 22.4, 56, 112 and 224 m.tons/ha. This growth was compared to wheat plants grown in soils to which the equivalents of 0, 67, 134 and 269 kg/ha of inorganic nitrogen were added. Also, the total nitrogen uptake by wheat plants grown on sludge-treated soils and fertilizer-treated soils was compared. By a comparison of growth and nitrogen uptake it was concluded that the equivalent fertilizer values of the four rates of sludge addition were 210 to 360, 575 to 630, 515 to 615, and 330 to 470 kg N/ha, respectively. These values indicated that the apparent total nitrogen availability was 26 to 44, 28 to 31, 13 to 15, and 4 to 6%, respectively, for the four rates of sludge addition.

The actual nitrogen use efficiencies (or the nitrogen that was utilized by the plant divided by the total nitrogen added in the sludge treatment), were 23, 16, 8 and 3% for 22.4, 56, 112 and 224 m.tons of sludge/ha, respectively. These data indicate that the amounts of inorganic nitrogen taken up by plants originating from the organic nitrogen of the sludge were less than the above values.

REFERENCES

1. Chaney, R. L. "Crop and Food Chain Effects of Toxic Elements in Sludges and Effluents," in *Recycling Municipal Sludges and Effluents on Land,* Proc. Joint Conf. of U.S.EPA, USDA and Nat. Assoc. of State Univ. and Land Grant Coll. (1973), pp. 129-146.
2. Dunlop, S. G. "Survival of Pathogens and Related Disease Hazards," Proc. Symposium on Municipal Sewage Effluent for Irrigation, Louisiana Polytechnic Institute (1968), pp. 107-122.
3. Krone, R. B. "The Movement of Disease-Producing Organisms Through Soils," Proc. Symposium on Municipal Sewage Effluent for Irrigation, Louisiana Polytechnic Institute (1968), pp. 75-104.
4. Malina, J. F., Jr. and B. P. Sagik. "Virus Survival in Water and Waste-water Systems," Water Resource Symposium No. 7, Center for Research in Water Resources, The University of Texas at Austin, Texas (1974).
5. Rudolfs, W. L., L. L. Falk and R. A. Ragotzkie. "Literature Review of the Occurrence and Survival of Enteric, Pathogenic and Relative Organisms in Soil Water, Sewage Sludges and on Vegetation," *Sew. Ind. Waste* 22:1261-1281 (1950).
6. Frink, C. R. "Plant Nutrients and Water Quality," *Agric. Sci. Rev.* (second quarter, 1971), pp. 11-25.

7. Hauck, R. D. "Quantitative Estimates of Nitrogen Cycle Processes. Concepts and Review," in *Nitrogen-15 in Soil-Plant Studies* (Vienna, Austria: International Atomic Energy Agency, 1971), pp. 65-80.
8. *Nitrogen and Phosphorus–Food Production, Waste and the Environment*, K. Porter, Ed. (Ann Arbor, Michigan: Ann Arbor Science Publishers, Inc., 1975).
9. Keeney, D. R., K. W. Lee and L. M. Walsh. "Guidelines for Application of Wastewater Sludge to Agricultural Land in Wisconsin," Technical Bulletin No. 88, Department of Natural Resources, Madison, Wisconsin (1975).
10. Ohio Task Force on Land Application of Sewage Sludge. "Ohio Guide for Land Application of Sewage Sludge," Research Bulletin 1079, Ohio Agricultural Research and Development Center (1975).
11. Illinois Environmental Protection Agency. "Design Criteria for Municipal Sludge Utilization on Agricultural Land," Technical Policy WPC-3.
12. Kelling, K. A. "The Effect of Field Applications of Liquid Digested Sewage Sludge on Two Soils in South Central Wisconsin," Ph.D. Thesis, Soils Department, University of Wisconsin, Madison, Wisconsin (1974).
13. Crutzen, P. J. "Estimates of Possible Variations in Total Ozone Due to Natural Causes and Human Activities," *Ambio* 3:201-210 (1974).
14. King, L. D. and H. D. Morris. "Land Disposal of Liquid Sewage Sludge: I. The Effect on Yield *in vivo* Digestibility and Chemical Composition of Coastal Bermudagrass (*Cynodon dactylon* L. Pers.)," *J. Environ. Qual.* 1:325-329 (1972).
15. Coker, E. G. "The Value of Liquid Digested Sewage Sludge. I. The Effect of Liquid Digested Sludge on Growth and Composition of Grass-Clover Swards in Southeast England," *J. Agric. Sci. Camb.* 67:91-97 (1966).
16. Coker, E. G. "The Value of Liquid Digested Sewage Sludge. II. Experiments on Rye-Grass in Southeast England, Comparing Sludge With Fertilizers Supplying Equivalent Nitrogen Phosphorus, Potassium and Water," *J. Agric. Sci. Camb.* 67:99-103 (1966).
17. Miller, R. H. and D. D. Zaebst. "Factors Affecting the Rate of Sewage Sludge Decomposition in Soils," *Agron. Abstr.* (1972), p. 98.
18. Rothwell, D. F. and C. C. Hortenstine. "Composted Municipal Refuse: Its Effect on Carbon Dioxide, Nitrate, Fungi, and Bacteria in Arendonda Fine Sand," *Agron. J.* 61:837-840 (1969).
19. Larson, W. E., C. E. Clapp and R. H. Dowdy. "Interim Report on the Agricultural Value of Sewage Sludge," U.S. Department of Agriculture, Agricultural Research Service and the Department of Soil Science, University of Minnesota, St. Paul, Minnesota (1972).
20. Miller, R. H. "The Microbiology of Sewage Sludge Decomposition in Soil," U.S. Environmental Protection Agency Report (1973).
21. Ryan, J. A., D. R. Keeney and L. M. Walsh. "Nitrogen Transformations and Availability of an Anaerobically Digested Sewage Sludge in Soil," *J. Environ. Qual.* 2:489-492 (1973).
22. King, L. D. "Mineralization and Gaseous Loss of Nitrogen in Soil-Applied Liquid Sewage Sludge," *J. Environ. Qual.* 2:356-358 (1973).
23. Pratt, P. F., F. E. Broadbent and J. P. Martin. "Using Organic Wastes as Nitrogen Fertilizers," *Calif. Agric.* (January 1973), pp. 10-13.

24. Bremner, J. M. "Total Nitrogen," in *Methods of Soil Analysis, Part 2,* C. A. Black, Ed., *Agronomy* 9:1162-1164 (1965).
25. Bremner, J. M. "Inorganic Forms of Nitrogen," in *Methods of Soil Analysis, Part 2,* C. A. Black, Ed., *Agronomy* 9:1195-1206 (1965).

20

MANAGEMENT OF FERTILIZER NITROGEN FOR POTATOES CONSISTENT WITH OPTIMUM PROFIT AND MAINTENANCE OF GROUND WATER QUALITY

D. R. Bouldin
Department of Agronomy
Cornell University
Ithaca, New York

G. W. Selleck
Long Island Vegetable Research Station
Riverhead, New York

INTRODUCTION

The population of Long Island, which lies to the east of New York City, is in excess of 3 million people. The most convenient sources of fresh water for this population are the aquifers under the island. Currently some of the aquifers contain in excess of 10 mg/l of nitrate nitrogen. Hence control of loading of nitrate appears essential if the quality of water in the aquifers is to be maintained at acceptable levels.

In many areas of Long Island, the nitrate contamination is associated with suburban pollution such as septic tanks and lawn and turf fertilization. However, as illustrated in Figure 1, some of the well water in the predominantly agricultural areas north of Riverhead contains more than 10 mg/l of nitrate nitrogen. Much of this land is devoted to potato production, and the most likely source of the nitrate in this area is the fertilizers added for the potatoes.

The objective of the research reported here was to devise schedules of fertilizer nitrogen application for potatoes on Long Island which would maintain producer profits and at the same time keep the nitrate nitrogen content of the ground water at less than 10 mg/l. Although some aspects of the results are specific for potatoes on Long Island, the general principles have

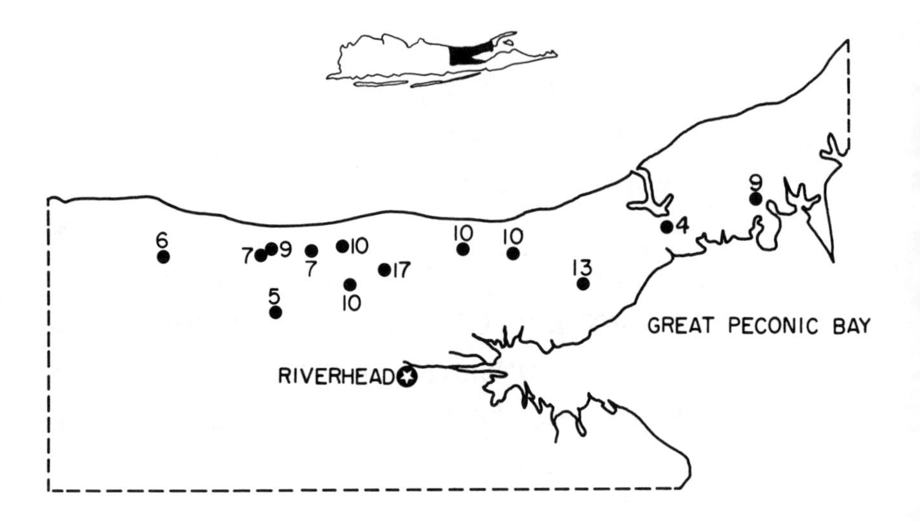

Figure 1. Location of sampled wells in Suffolk County; numbers are mg/l of nitrate nitrogen.

applications for other crops and other areas. Our objective in this chapter is to present an overall summary of several years of work by a number of people. The details of the data and individual experiments will be presented in several publications now in preparation.

BACKGROUND STUDIES

A detailed analysis of nitrogen in the Long Island potato system has been carried out.[1] This study found that several simplifying approximations were justified for the budget on Long Island. For instance, the denitrification losses of N on these acid sandy soils were balanced by N inputs in the rainfall, irrigation and seed potatoes. Similarly, since these soils have been cropped for a long period of time under the same system (continuous potatoes with a cover crop of rye seeded in the fall), the changes in soil organic matter are small, which means that the nitrogen mineralized from soil organic nitrogen is compensated for by the nitrogen returned in crop residues (potato residue plus rye cover crop). The remaining portions of the nitrogen budget (runoff losses, mineral weathering, etc.) were shown to be very small. The end result of this analysis is that nitrogen leaching losses can be approximated by the difference between fertilizer additions and the nitrogen contained in the harvested tubers. The estimated average recharge (express of precipitation over evapotranspiration) is 55 cm per year.

Presently farmers on Long Island are applying about 225 kg N/ha (200 lb N/ac) on potatoes. The removal of nitrogen in the harvested tubers is about 120 kg N/ha (100 lb N/ac) and hence on the order of 100 kg N/ha is the maximum amount of nitrogen leached to the ground water based on the foregoing analysis. When this is dissolved in 55 cm of recharge, the resulting expected concentration in the ground water is 18 mg/l of nitrate nitrogen. The results shown in Figure 1 are in reasonable agreement with this estimate; the lower concentrations in the wells are probably the result of dilution by water from unused land and sampling of aquifers too deep to be contaminated by current agricultural practices.

In the period 1975-1976, the nitrate concentration in the shallow ground water under the fields of four representative farmers in Suffolk county was measured. The sampling was performed by installing tube wells screened about 50 cm below the ground water level. These samples, which are representative of current recharge, averaged 18 mg/l nitrate nitrogen. This is the expected nitrate concentration based on the input-output analysis described above.

The foregoing analysis demonstrates that the amount of nitrate leached from potato fields will be approximately equal to fertilizer inputs minus tuber removal, subject to the condition of continuous potatoes fertilized such that yields are not reduced appreciably from their present levels. Furthermore, the concentration in the ground water under potato fields will be approximately equal to fertilizer inputs minus tuber removal dissolved in 55 cm of recharge. Thus, if nitrate nitrogen concentrations are to be maintained at 10 mg/l, fertilizer inputs minus tuber removal must not exceed 55 kg N/ha (50 lb N/ac) per year on Long Island.

The current farmer practice is to apply all of the nitrogen at planting in April. Usually 4 to 6 weeks pass before large amounts of nitrogen are taken up by the plants; during this interval, an appreciable fraction of the fertilizer nitrogen is leached beyond the rooting zone in some years. Thus farmers apply an amount of fertilizer which will compensate for this anticipated loss of nitrogen.

MANAGEMENT OF FERTILIZER NITROGEN

The objective of the research reported here was to devise methods of applying fertilizer nitrogen such that plants would take up most of the nitrogen before it was leached. Thus the quantity of fertilizer nitrogen applied could be reduced and yet yields maintained at current levels.

After several preliminary experiments, two fairly comprehensive experiments were carried out in the period 1973-1976 in which yields were measured under increasing rates of fertilizer nitrogen split between about one-third added at planting and the remainder sidedressed when the plants were

5 to 15 cm tall. In the preliminary experiments, it was found that sidedressed applications should be made fairly early since potatoes take up large amounts of nitrogen during the vegetative stage of growth, which occurs in the 6-week period following emergence.

The experimental area was located on the Vegetable Research Station near Riverhead, New York, on soils of the Haven-Riverhead association, which are well-drained loams and sandy loams containing about 2% organic matter. Experiment I consisted of treatments (a) with different rates of nitrogen all applied at planting and (b) with different rates of nitrogen split between planting and when the plants were about 5 cm tall. Katahdin variety of potatoes were grown with the same treatments applied to the same plots in the period 1973 to 1975. In Experiment II, four rates of nitrogen were applied to each of four varieties of potatoes (Katahdin, Cascade, Hudson and Superior), with 35 or 50 kg N/ha applied at planting and the remainder sidedressed when plants were approximately 5 to 15 cm high. The same treatments were applied to the same plots for each of the years 1974 to 1976.

Phosphorus, potassium and magnesium were applied at rates judged sufficient to eliminate these nutrients as factors limiting yields. Marketable yields and nitrogen contents of the tubers were determined.

At the highest rate of application (224 kg N/ha) in Experiment I, there were no differences in yields between methods of applying nitrogen (all at planting or split), indicating that the split applications as applied here had no adverse effects on yields. In the earlier experiments, similar results were obtained, indicating that so long as the sidedressed nitrogen was applied early enough, there were no differences in limiting yields between split applications and all at planting.

The regression of yield of tubers on fertilizer nitrogen applied is illustrated in Figure 2A. The standard error of the estimated mean is about ± m.ton/ha. The important point to note is that maximum yields were obtained when about 170 kg N/ha (150 lb N/ac) is applied, one-third at planting and the remainder when plants are 5 to 15 cm tall.

The nitrogen balance is illustrated in Figure 2B, where the shaded area between the fertilizer N and tuber N line represents the approximate amount of nitrogen leached to the ground water. One of the conditions of the nitrogen balance analysis was that yields remain at their present levels (essentially those yields obtained when nitrogen is not limiting). This means that the analysis is valid only when at least 125 to 140 kg N/ha is applied.

The analysis is carried one step further in Figure 2C, where the amount of nitrogen leached to the ground water is plotted against fertilizer nitrogen added. The scale on the left is in kg N/ha, and the scale on the right is in mg/l of nitrate nitrogen if this nitrogen is dissolved in 55 cm of recharge.

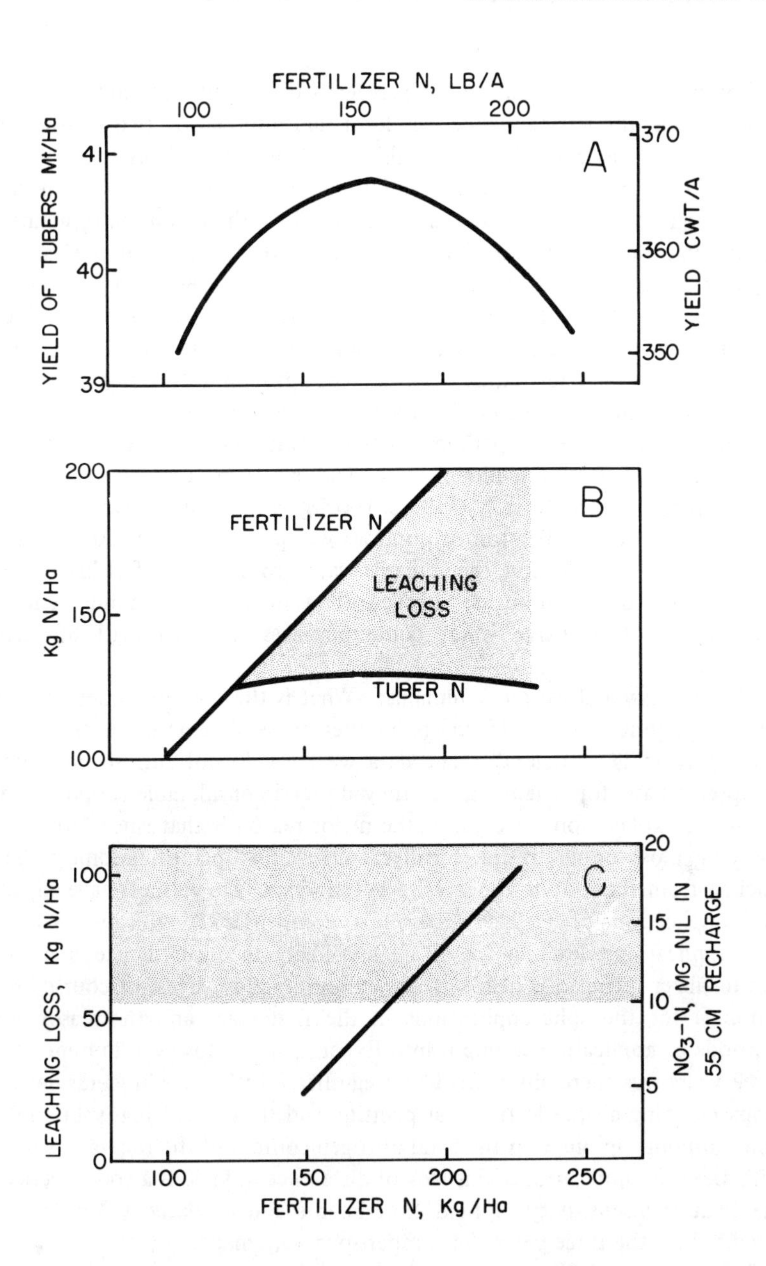

Figure 2A. Regression of yield of tubers on fertilizer nitrogen additions.

B. Nitrogen content of fertilizer and tubers plotted against fertilizer nitrogen added; shaded portion represents expected leaching losses.

C. Leaching losses (left scale) or nitrate nitrogen expected in recharge water (right scale) plotted against fertilizer nitrogen added.

The net result of the above analysis leads to a small "slit" in the fertilizer N scale. On the left-hand side, the limits are imposed by two factors: the limitation of the conditions of high yields on the analysis itself and the farmer's economic interest in obtaining about 99% of the maximum yield. On the right-hand side, the limits are imposed by the condition of maintaining ground-water nitrate content at less than 10 mg/l. Thus the range of application rates should be between about 135 and 180 kg N/ha. At the lower limit, yields will be maintained at about 98 to 99% of their value in the presence of optimum nitrogen fertilizer and, at the higher level, yields should be approximately equal to those with optimum nitrogen and yet maintain ground-water nitrate content at acceptable levels.

The results of these experiments demonstrate that maximum yields were obtained when fertilizer nitrogen applications were about 170 kg N/ha (150 lb/ac). Approximately 150 to 160 kg N/ha would maximize profit, since on the order of 0.1 ton of potatoes will purchase about 10 kg of fertilizer nitrogen. In addition, the recharge water from fields so fertilized should average about 8 mg/l of nitrate nitrogen. Thus maximum economic yields are consistent with recharge water containing less than 10 mg/l of nitrate nitrogen.

One unanswered question remains: What is the reason farmers are currently applying about 225 kg N/ha, yet these data indicate that about 170 kg N/ha is sufficient? The data we have do not furnish an entirely acceptable basis for judgment, but they do lend considerable support to the following explanation. We judge the major reason is that since farmers are applying most or all of the fertilizer nitrogen at planting, some is being leached from the rooting zone in the wetter years. The farmers have adjusted their applications of fertilizer nitrogen to compensate for this loss. With the split applications, leaching losses are less likely to occur; hence, somewhat less fertilizer nitrogen is needed for maximum yields. This, of course, is the rationale for the split applications in the first place and the reason both methods of application were put into Experiment I. However, in none of the three years was there any statistically significant difference in yields between crops receiving all the fertilizer at planting and those receiving split applications, although in one year the total nitrogen content of the tubers was higher with the split application. This lack of difference in yield is a consequence of the relative insensitivity of field experiments and a relatively low leaching potential for the three years the comparisons were made.

In other experiments conducted over the last 25 years, there is considerable evidence that nitrogen applied at planting is not as effective in increasing yields as nitrogen applied after plants have emerged in some years, but there was no marked difference in effectiveness in other years. However, these earlier experiments were not designed to measure the relative effectiveness of all nitrogen at planting vs nitrogen split in the manner used in Experiments I and II. On the basis of the foregoing evidence and arguments, we conclude that with split applications of fertilizer nitrogen, most of the fertilizer is taken up by the plant soon after application; hence, no appreciable leaching losses occur. When all nitrogen is applied at planting, some leaching losses occur some years; hence, larger amounts of fertilizer nitrogen are required to get maximum yields.

The results of these experiments are another illustration of what appears to be a fairly general and important principle: by properly timing fertilizer nitrogen applications and adjusting the quantity applied to proper levels, a large fraction of the applied nitrogen can be recovered in the plant. This principle was illustrated in previous reviews of response of corn to nitrogen fertilizer in the U.S.[2,3] It has also been applied to corn in Puerto Rico.[4]

SUMMARY

With potatoes on Long Island, the evidence indicates that optimum economic yields can be obtained and yet the nitrate nitrogen content of the ground water can be kept at less than 10 mg/l. If in a drier climate recharge were one-half of that on Long Island, for example, this might not be possible. To maintain less than 10 mg/l of nitrate nitrogen in one-half as much recharge would mean that fertilizer applications would have to be reduced to about 150 kg N/ha (Figure 2C). This quantity of fertilizer nitrogen is likely to be slightly below the economic optimum, but the reduction in profits would not be catastrophic. Thus, these results suggest that in many areas and with many crops nitrate loading can be controlled in a fairly inexpensive manner by proper timing of the proper amount of fertilizer nitrogen.

ACKNOWLEDGMENTS

This is paper No. 707 from the Department of Vegetable Crops and the Department of Agronomy, Cornell University, Ithaca, New York. This manuscript was prepared in consultation with Dr. J. J. Meisinger, Soil Scientist, Agricultural Research Service, U.S. Department of Agriculture, Beltsville, Maryland, formerly Saltonstall Scientist, McCain Food Ltd., New Brunswick, Canada, formerly Research Associate, Department of Vegetable Crops, Cornell University; and Chang-Chi Chu, Department of Vegetable Crops, Cornell University, Ithaca, New York. In addition, the following individuals made invaluable contributions to the project: Stewart Dallyn, P. A. Schipper, Gary Rathburn and R. Kossack, Department of Vegetable Crops; and Gilbert Levine, Department of Agricultural Engineering. This project was financed by the College of Agriculture and Life

Sciences, Cornell University, the Rockefeller Foundation and the Suffolk County Legislature.

REFERENCES

1. Meisinger, J. J. "Nitrogen Application Rates Consistent with Environmental Constraints for Potatoes on Long Island," *Search* 6(7), Cornell University, Ithaca, New York (1976).
2. Lathwell, D. J., D. R. Bouldin and W. S. Reid. "Effects of Nitrogen Fertilizer Application in Agriculture," in *Relationship of Agriculture to Soil and Water Pollution,* Proc. Cornell Waste Management Conf., Ithaca, New York (1970).
3. Bouldin, D. R., W. S. Reid and D. J. Lathwell. "Fertilizer Practices Which Minimize Nutrient Loss," in *Agricultural Wastes, Principles and Guidelines for Practical Solutions,* Proc. Cornell Waste Management Conf., Ithaca, New York (1971).
4. Fox, R. H., H. Talleyrand and D. R. Bouldin. "Nitrogen Fertilization of Corn and Sorghum Grown in Oxisols and Ultisols in Puerto Rico," *Agron. J.* 66:534-539 (1974).

21

USE AND RELATIVE ENVIRONMENTAL EFFECTS OF FERTILIZERS APPLIED TO CROPLAND AND TURF IN A MIXED RURAL AND SUBURBAN AREA

K. S. Porter
Department of Agricultural Engineering
Cornell University
Ithaca, New York

L. B. Baskin and D. H. Zaeh
Department of Environmental Engineering
Cornell University
Ithaca, New York

INTRODUCTION

This chapter briefly outlines a regional study of the use of nitrogenous fertilizers and an evaluation of their impact on ground water relative to other major sources of nitrogen. The study was undertaken in Nassau and Suffolk Counties, a region which represents in microcosm many of the environmental problems encountered in other rapidly developing areas in the United States.

Since World War II, Nassau and Suffolk Counties have experienced remarkably rapid urbanization, in large part due to encroachment from New York City. In consequence, highly fertile farm lands have been transformed into housing tracts leaving a residue of farming located primarily in Eastern Suffolk County. Nevertheless, through intensive cultivation, the approximately 50,000 acres which remain devoted to farming produce crops which exceed in value those of any other county in New York State.

Associated with transformation in land uses has come an expansion in recreational and tourist facilities. As a result there is an increasing source of nitrogen in the region due to the use of nitrogenous fertilizers on golf courses, parks and highways. When use of fertilizers on lawns is also considered, there is in sum a source of nitrogen of potentially greater impact than that from agriculture.

Although data have been lacking, previous studies indicate that fertilizers used on Long Island may constitute one of the major sources of nitrogen which have contributed to the deterioration in the quality of the region's ground waters. Such an hypothesis has considerable significance for management policies designed to protect the aquifers. For example, the replacement of septic tanks by sewerage and wastewater treatment plants may be an inadequate remedy if the use of fertilizers constitutes a continuing source of nitrogen.

OBJECTIVES

The work described in this chapter attempts to confirm or refute the above hypothesis by:

1. quantifying the use of fertilizers by major users,
2. investigating the fate of nitrogen within the soil after application, and
3. assessing the relative significance of fertilizers as a source of nitrogen by the construction and application of an overall hydrological nitrogen model for the region.

METHODS

To estimate the use of nitrogenous fertilizers, field surveys were carried out during 1976. A particular effort was made to estimate use of fertilizers on garden lawns. A pilot survey was undertaken in Riverhead, Suffolk County, to test the survey procedures and questionnaires. Following this preliminary work, seven areas were selected. Several criteria were applied in the selection of the sites, the most important being population density. Since it was believed that there may be a relation between income levels and the care of gardens, household income was also taken into account. The selected areas are depicted in Figure 1. Within each area the houses chosen for inclusion in the survey were determined by a random selection procedure.

By statistical analyses it was further determined that 60 houses per site would give an acceptable level of reliability for the estimates obtained. Hence, approximately 500 houses in total were included in the survey.

A summary of preliminary results from the survey is given in Table I. The figures shown represent the amount of fertilizers used by householders in 1976 up to the time of the survey. A supplementary questionnaire was mailed to estimate the amount of fertilizers used during the remainder of 1976. From returned questionnaires it was estimated that the average additional amount applied was about 50% of that already applied. Assuming

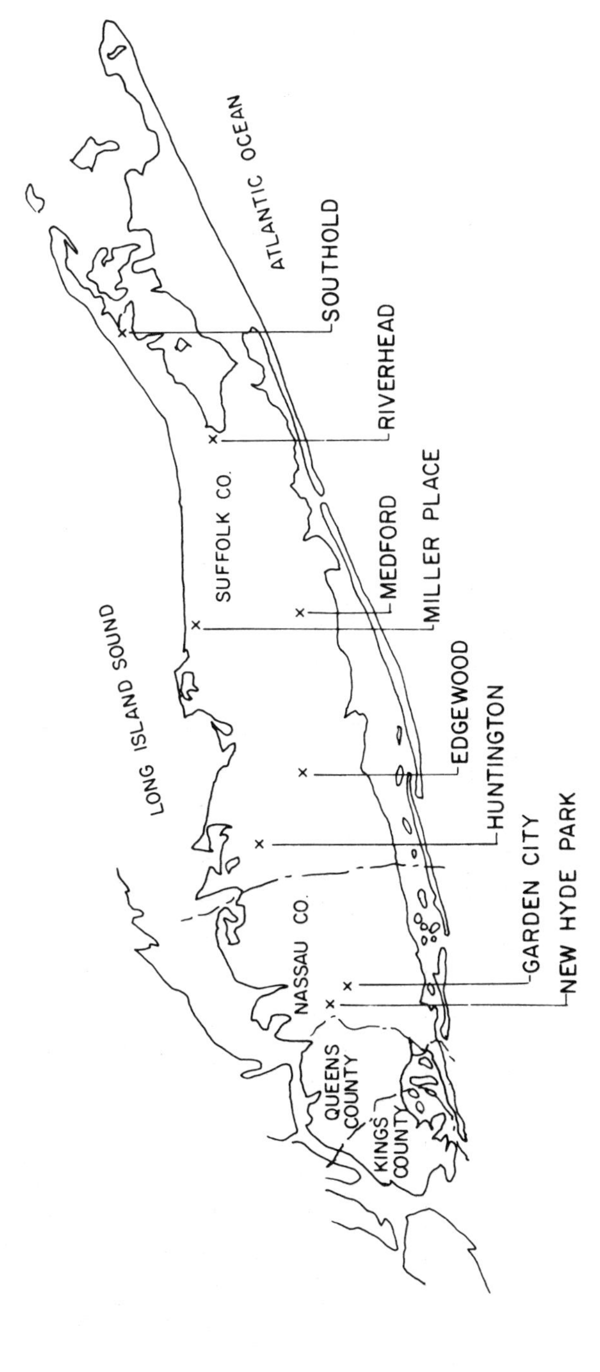

Figure 1. Field survey areas to estimate household use of pesticides and fertilizers.

Table I. Preliminary summary of results of household survey of nitrogenous fertilizers applied to lawns. Figures apply only to date of field survey completion.

Location	Housing Density (persons/ac)	Income Level ($ Thousands/ family/yr)	Number of Questionnaires	Mean Nitrogen Input to Date (lb/1000 ft^2)	Standard Deviation	Range	Mean Turf Area (1000 ft^2)
Garden City	8.01	21.8	61	2.5	1.1	0-4.3	6.2
New Hyde Park	10.90	15.0	65	2.0	1.1	0-5.0	4.1
Huntington	5.51	14.7	52	2.2	2.3	0-13.3	10.9
Miller Place	0.40	10.7	50	1.1	2.0	0-11.4	21.9
Southold	0.34	10.4	63	1.2	1.6	0-8.3	21.1
Medford–12 Pines	1.26	9.1	109	1.5	1.3	0-8.0	8.7
Brentwood	4.21	8.7	60	1.2	1.9	0-7.7	6.6

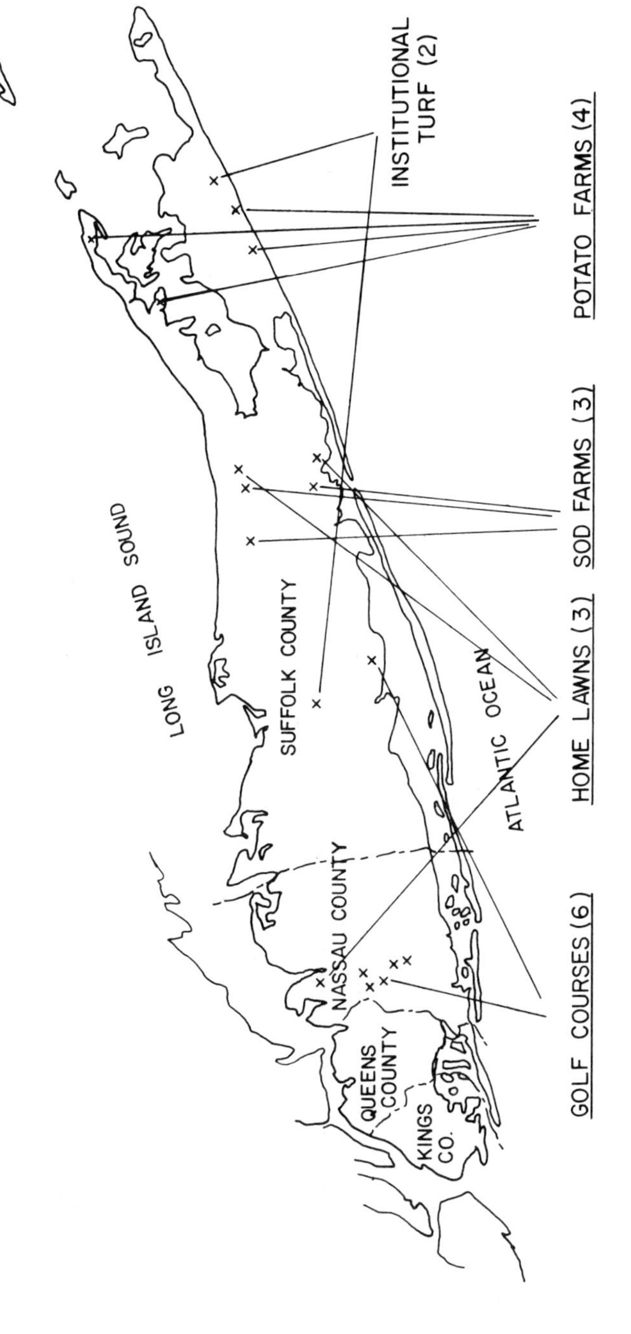

Figure 2. Sites of soil nitrogen field survey on Long Island.

this and extrapolating to the whole bicounty region, the total use of nitrogenous fertilizers by householders has been provisionally estimated to be approximately 7000 tons.

Parallel to the household survey, field work was undertaken to assess simultaneously the use and fate of nitrogen applied to various types of turf and potatoes grown commercially. The turfed areas included six golf courses, three home lawns, three sod farms, a highway and a school ground (institutional turf) and four commercial potato farms. Figure 2 shows the location of the sites in Nassau and Suffolk Counties.

Each site was sampled every two to three weeks for soil nitrogen and soil water levels. Measurements were taken at various depths to determine the soil profiles of nitrogen and water within and immediately below the root zone. It was assumed that any nitrogen detected below the roots would eventually leach to ground water. Where possible, concentrations of nitrogen in the underlying ground were were obtained.

An example of such measurements is shown in Figure 3. The measurements were taken in one of the potato fields included in the study conducted by the Long Island Vegetable Research Farm (Bouldin and Selleck, 1977). The field was divided into two equal parts, each part being managed differently. Figure 3 shows the levels of nitrogen in the soil fertilized according to the farmer's normal practice. Fertilizer was applied twice; on April 7 nearly

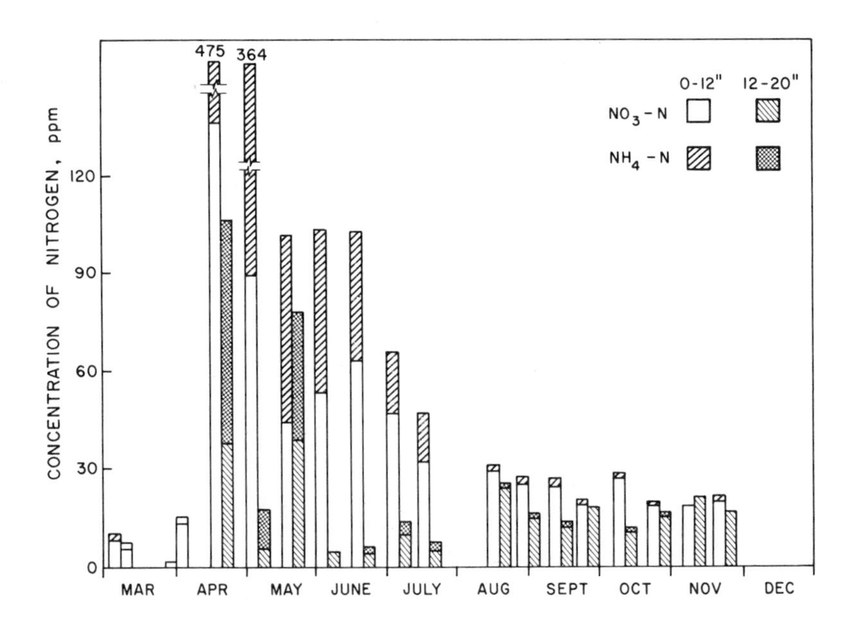

Figure 3. Inorganic soil N in potato field (Farmer's Current Management, 1976).

192 lb of nitrogen per ac were applied, and on May 5 a further 100 lb were applied. The concentration measured in the 0- to 12-in. root zone is depicted on the left, and below the root zone to the right. As can be seen, the soil cores taken through the fertilizer band indicate that there were high levels of nitrogen both in and below the root zone. By the end of the year, concentrations were returning to background levels observed in the early spring.

Figure 4 shows the results of measurements taken in part of the field fertilized according to new recommendations developed at the Long Island Vegetable Research Farm. In mid-March about 50 lb N/ac were applied, followed by another 100 lb at the end of May. Although less is applied in this way than is the farmers' normal practice, by judicious timing sufficient nitrogen may be available for crop requirements.

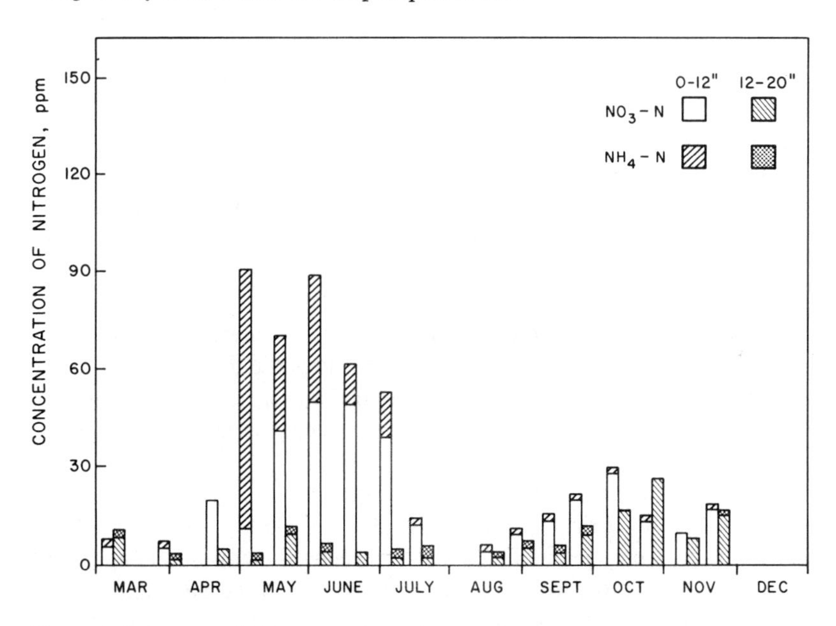

Figure 4. Inorganic soil N in potato field (Cornell's Experimental Management, 1976).

It is evident from the figure that measured levels of nitrogen are substantially lower, especially immediately below the root zone. The rise in levels later in the year would possibly be due to mineralization or organic nitrogen from crop residues. Concentrations of nitrate nitrogen in the ground water underlying the two parts of the field during 1976 indicated much higher levels under that half of the field to which the higher applications had been applied. Parallel results were obtained from the other three potato farms, each of which managed a field in the same way.

As indicated earlier, however, the major crop grown in Nassau and Suffolk is grass. Initially, fifteen turf sites representing different uses were regularly monitored (Figure 2). In some cases the sites were subdivided and fertilized at different rates. As expected, levels of nitrogen in the grass generally were directly related to amounts applied.

At the outset of the study, the principal question with respect to turf was, What was the fate of applied nitrogen? Ornamental grass is not cropped in the conventional sense. Even when clippings are removed, their disposal elsewhere does not necessarily remove the nitrogen from the overall system. It may be argued that, in the long run, the nitrogen added to turf may be lost to the atmosphere as a result of ammonia volatilization or denitrification, there may be a net increase of organic nitrogen in the soil or it may be removed by water. The following argument (Bouldin, 1976) outlines the various possibilities.

Under mature turf evidence indicates that soil organic nitrogen is in equilibrium with the soil. Thus prior to maturity there is an increase in soil organic nitrogen followed by no net changes as illustrated hypothetically in Figure 5B. It follows that inputs of nitrogen to the turf must equal outputs (A = E in Figure 5A). Since the soils on Long Island are generally light and sandy, well aerated and with low pH, it may be argued that conditions do not favor gaseous losses of nitrogen. Also, Long Island is flat typographically, and there is little runoff. Under such conditions, therefore, additions of nitrogen must eventually leach to ground water. The Long Island Vegetable Research Farm is currently undertaking long-term experiments to investigate this hypothesis. Within the program of study discussed in this chapter, some short-term field experiments were made to measure the leaching of nitrate following application to turf.

Several plots of mature turf, each being approximately 100 ft^2 in area, were fertilized and irrigated at various rates and times. Levels of nitrogen and soil water were measured daily to determine the corresponding profiles in the soil. Figure 6 shows the measured levels of total inorganic nitrogen in five 4-in. layers of soil following an application of 4 lb KNO_3-N/1000 $ft.^2$ One inch of water was applied twice daily. About 0.06 lb N/1000 ft^2 were applied in each inch of water. As is evident from Figure 6, there was rapid movement of the nitrogen downwards throughout the period of the experiment.

The result of a similar experiment is shown in Figure 7. In this example, 2 lb $NaNO_3$-N/1000 ft^2 and only 1 in. water/day was applied. However, the nitrogen profile obtained is similar to the experiment illustrated in Figure 6.

Previously it was indicated that the fate of nitrogen contained in clippings represented an uncertainty. In particular, it could be argued that ammonia might volatilize from freshly cut clippings, in which case that fraction of the nitrogen added to the grass which so volatilized would clearly not leach.

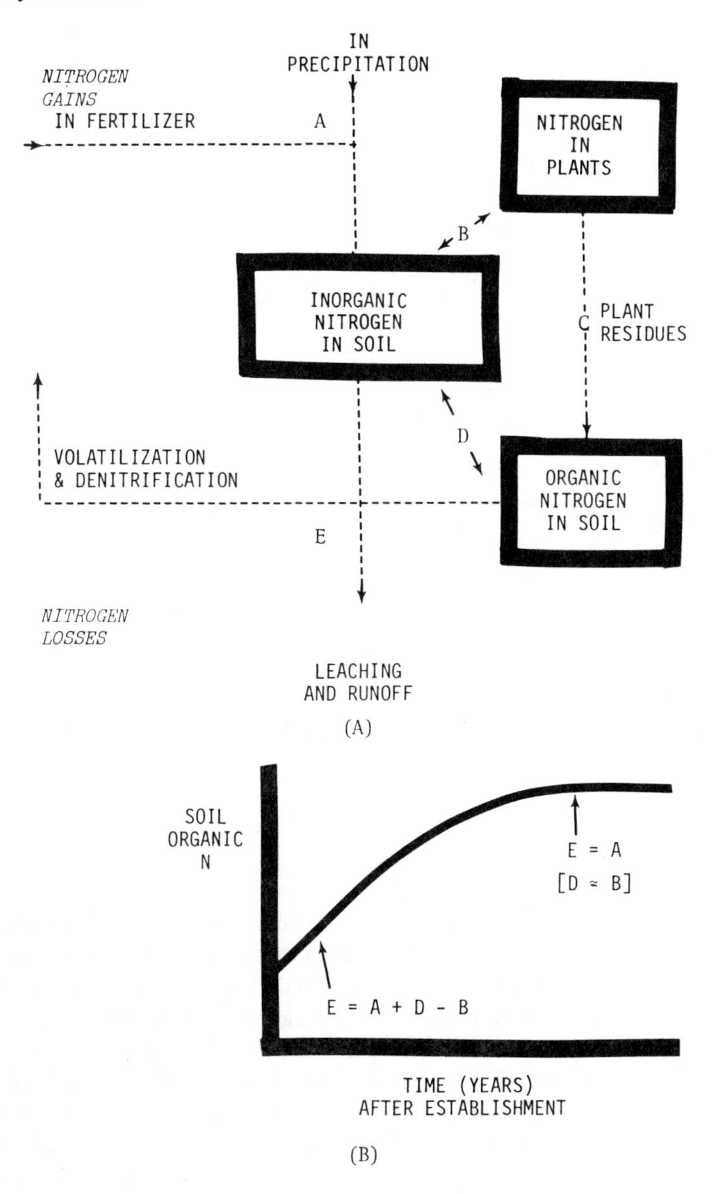

Figure 5. Soil nitrogen balance under turf and its expected disposition toward equilibrium (Bouldin, 1976).

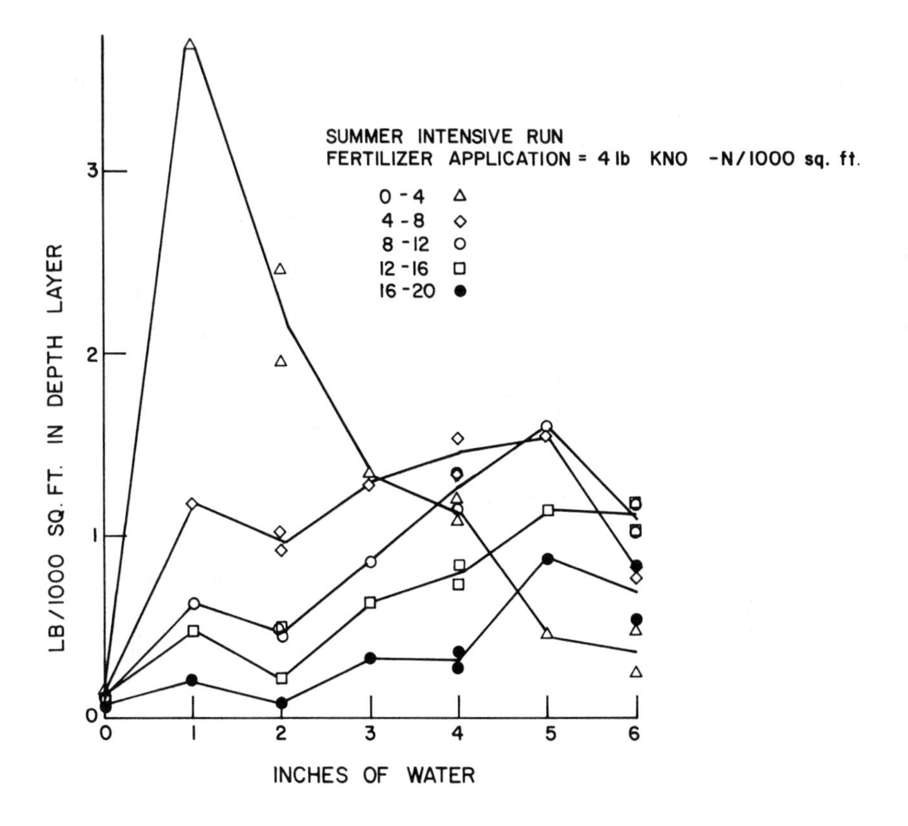

Figure 6. Total inorganic nitrogen in each depth layer versus amount of water applied.

To investigate this hypothesis the following simple experiment was performed. Twenty samples of freshly cut grass clippings were each measured for the percentage of nitrogen they contained. Each sample was divided into four equal parts. The N level in the first part was determined immediately following clipping, and the second part was oven-dried prior to determination. Then the third and fourth parts were allowed to air-dry for three days before being measured for N in the same way, *i.e.,* one of the parts being oven-dried prior to the N determination. The results are summarized in Table II. The greatest loss of N, 12.8%, occurred in the samples which were immediately oven-dried. It may therefore be provisionally concluded that losses of nitrogen via volatilization from clippings are relatively small.

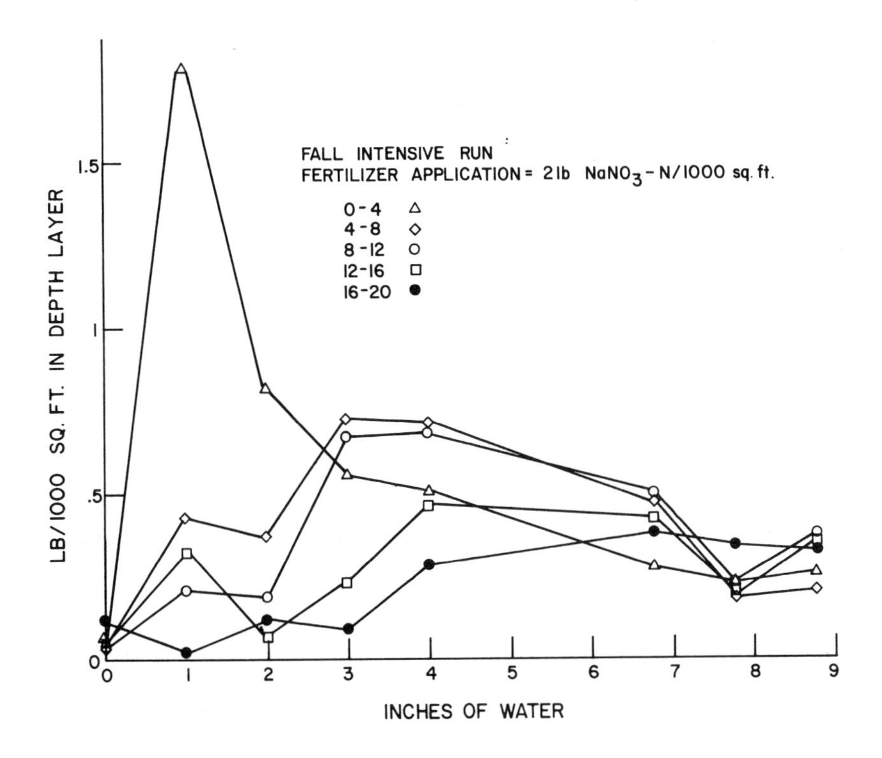

Figure 7. Total inorganic nitrogen in each depth layer versus amount of water applied.

Table II. Determination of loss of nitrogen from freshly cut grass clippings treated four ways.

		Mean % Nitrogen in Grass Samples (Dry Weight Basis)		
		Not Oven-Dried	Oven-Dried	% Loss in Oven-Drying
Time of Air-Drying	0 Days	4.68	4.08	12.8
	3 Days	4.25	4.02	5.4
	Apparent % Loss in Air-Drying	8.8	1.5	

ANALYSIS OF FIELD DATA

A primary factor in the leaching of nitrate nitrogen and enrichment of ground water is the flow of soil water, both as the medium of transport and as that which is enriched. The analysis of the field data discussed above is therefore now proceeding in two main phases: (1) a macrohydrological budget has been constructed to enable the estimation of the percolation of soil water throughout the bicounty region, and (2) a corresponding nitrogen-loading macromodel is now being superimposed onto the hydrological model to estimate the mass of nitrate which is leached. Both models are being calibrated and validated in a micromode.

HYDROLOGICAL MODEL

Both the nitrogen model and the hydrological model are based on a mass balance. The calculations are performed for each of the 762 cells specified by a grid system defined by the Nassau-Suffolk Regional Planning Board. Each cell is 2.25 mi^2 in area.

The simplest form of the mass balance equation for each cell is:

$$S_i(t+1) - S_i(t) = I_i(t) - O_i(t) \tag{1}$$

where $S_i(t)$ = storage of water in element i at the beginning of month t
$I_i(t)$ = mass flow of water into element i during month t
$O_i(t)$ = mass flow of water out of element i during month t

For the unsaturated zone, the mass flow into a cell i consists of the sum of precipitation $[P_i(t)]$, irrigation $[W_i(b)]$, and net runoff leaving the cell i $[R_i(t)]$:

$$I_i(t) = P_i(t) + W_i(t) - R_i(t) \tag{2}$$

Similarly, the mass flow from the cell is the sum of evapotranspiration, $E_i(t)$, and net subsurface flow out of the cell $[L_i(t)]$ down to ground water.

Thus by substituting and rearranging terms in Equation 1, one obtains an equation for subsurface flow $[L_i(t)]$ in cell i during month t:

$$L_i(t) = S_i(t) - S_i(t+1) + P_i(t) + W_i(t) - E_i(t) - R_i(t) \tag{3}$$

Clearly one can thereby sum over space and time and obtain an estimate of total recharge for the region in the specified time:

$$L = \sum_i \sum_i L_i(t) \tag{4}$$

A simplified flow chart of the model whereby Equation 3 is computed is shown in Figure 8. As indicated, the computation of the components in Equation 3 requires the assumption of underlying parameters such as land use characteristics in each cell, or the computation of auxiliary factors upon which one or more components depends such as available moisture capacity of the soil. A brief outline of the computation of Equation 3 follows. More details may be found in Baskin .[3]

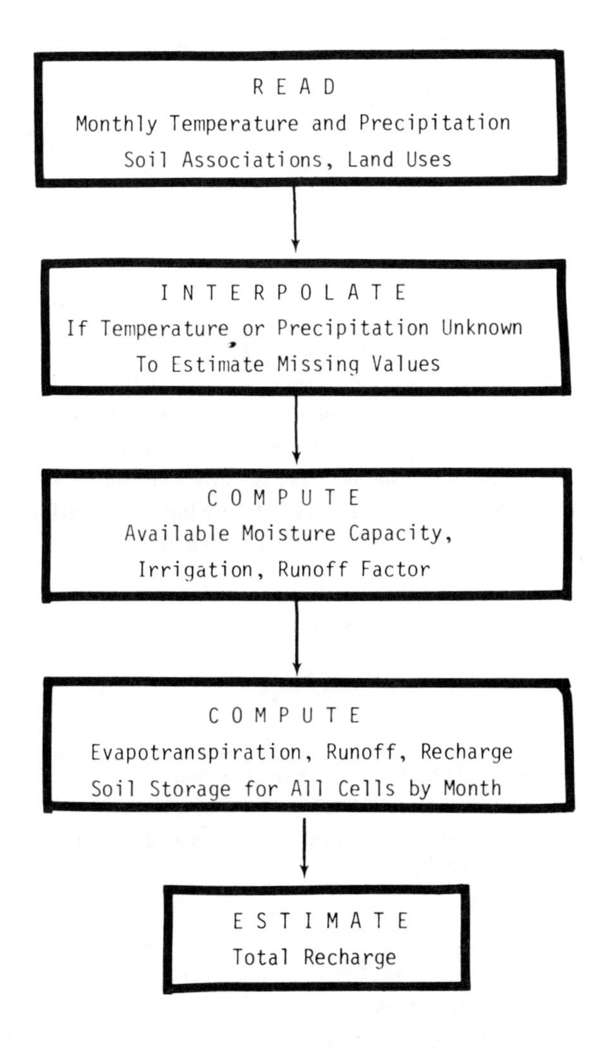

Figure 8. Simple flow chart of recharge model.

Change in Soil Storage

In the model it is assumed that whenever the soil moisture increases above field capacity, the excess water will percolate below the root zone and thereby constitute recharge. Evapotranspiration will deplete the available water in the soil below field capacity, hence creating a "water deficit." This deficit must be eliminated by further precipitation or irrigation before further leaching is assumed to occur in the model.

Considerable effort was made to obtain soil measurements at various sites which would permit the model to be calibrated and validated. Field capacities at the various sites, available moisture and wilting points were all estimated. The model was operated in a mode specific to several sites and calibrated. Validation was then accomplished by comparing the simulations of the model for other sites with actual soil moisture measurements at those sites. Very close agreement between the predicted and observed values was thus obtained.

Precipitation

In 1976, precipitation was measured at thirty reasonably well distributed points in Nassau and Suffolk. For the purposes of the model, an estimate of precipitation was required for each cell, thereby necessitating the use of interpolation for 732 cells in which precipitation had not been measured. The usual methods of interpolation were reviewed, but none were considered to be entirely satisfactory. Therefore, the following computation was devised in which the precipitation P_{it} in cell i at time t is estimated by:

$$P_{it} = \frac{\sum_{k=1}^{N} \alpha_{ik}(P_{kt})}{\sum_{k=1}^{N} \alpha_{ik}^{m}} \qquad i \neq k \tag{3}$$

where P_{kt} = estimated precipitation at point k at time t
N = number of recording stations
α_{ik} = a weighting factor defined as the inverse of the distance between points i and k

This procedure was tested by omitting each of the thirty points in turn at which precipitation was measured and determining the accuracy with which the point so omitted was estimated for different powers of m.

Irrigation

Considerable quantities of irrigation are usually applied during the summer on Long Island on both cropland and turf. During the summer of 1976, the Soil Conservation Service cooperated in this study by attempting to estimate the quantities of water so applied. However, the results obtained are considered to be speculative, especially in the case of house lawns, where accurate estimates are extremely difficult to obtain.

Evapotranspiration

There are several commonly used methods of computing evapotranspiration. The methods may roughly be classed as being based either on mass transfer theory, the energy balance, or on empirically derived relations. Each has advantages, but the most readily usable are empirical methods since generally the data requirements for these are most easily satisfied. For this reason, the empirical formula derived by C. W. Thornthwaite[4] was chosen for this study. Thornthwaite specifically developed his formula for estimating the potential evapotranspiration in the humid areas of eastern North America. Previous estimates of evapotranspiration on Long Island have generally agreed that approximately 21 to 22 in. of precipitation are returned to the atmosphere by evapotranspiration.[5] The Thornthwaite method was found to give good agreement with these estimates.

Runoff

Soils on Long Island are sandy and generally the potential rates of infiltration are higher than the rates of precipitation. During the field surveys made in 1976, rates of infiltration, after the soils had achieved saturation, varied from 0.2 in./hr on a cultivated field, to 18 in./hr on a mature loam. A typical example of the results obtained is shown in Figure 9. Also, the land has little or no gradient throughout most of the two counties, hence conditions do not favor runoff.

On impervious surfaces in developed areas, direct runoff to surface water may not occur. The cumulative effects of such areas on surface hydrology have yet to be fully quantified.[6] Holzmacher[7] has proposed that runoff in urban areas is equal to the amount of precipitation falling on land covered by impervious surfaces.

Throughout Nassau and Suffolk, surface runoff is collected. Where the ground-water table is high and where there is adjacent surface water, the runoff is usually conveyed and discharged to the river or marine bay as the case may be. The greater part of runoff, however, is collected into recharge basins, of which there are several thousand in the region, through which the water percolates down to ground water.

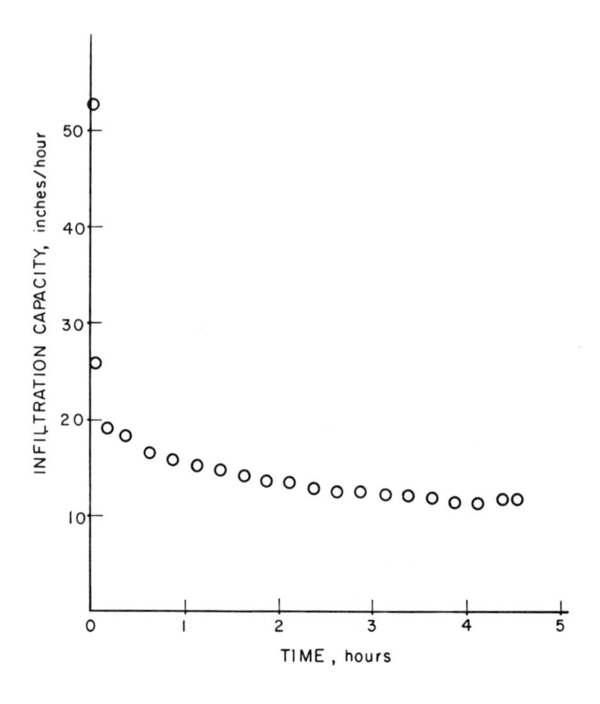

Figure 9. Example of determination of infiltration capacity in mature turf.

For the purposes of the model it was therefore assumed that:

$$R_i(t) = 0 \text{ if cell i does not contain a surface water body}$$
$$= \sum_{j=1}^{N} \lambda_j L_j [P_i(t)] \text{ otherwise,}$$

where λ_j is the fraction of land use j which is impervious and L_j is land use type j.

A summary of the results for the cells summed over a year is given in Table III.

An obvious question arises concerning how the results of the model compare with those of earlier studies. No previous work has estimates for the whole bicounty region. However, several investigators considered part of the two counties, and it is possible to compare their results with those obtained by the model for the same areas. This comparison is shown in Table IV. Estimates of recharge obtained from the previous work are in surprisingly close agreement with those obtained by the model. This is not the case with the estimates of runoff. Estimates of runoff provided by the model

Table III. Summary of results obtained by the recharge model (in./yr).

	Max.[a]	Mean	Min.[a]
Precipitation	50.1	43.0	36.7
Runoff	24.2	4.5	0
Evapotranspiration	24.9	17.8	8.8
Recharge	37.0	20.7	5.6

[a]The estimated maximum and minimum yearly values are for individual cells.

are consistently much higher than those from the previous studies. This is not surprising, however, because the model attempts to estimate the total runoff which occurs, including both diffuse and stream flows. With the exception of Seaburn, the other investigators based their estimates primarily on measured stream flows only.

Table IV. Comparison of previous work to model results.

Site	Area (sq mi)	Source	Estimated Quantity (mgd/sq mi)	Cornell/CES Estimated (mgd/sq mi)
Net Recharge				
Huntington, Smithtown	146	Lubke[8]	1.01	1.00
Nassau, W & C Suffolk	760	Franke & McClymonds[9]	1.03	1.12
Babylon, Islip	190	Plukowski & Kantrowitz[5]	1.13	1.05
Nassau, W & C Suffolk	760	Cohen *et al.*[10]	1.08	1.12
Nassau	291	Greely & Hansen[11]	0.69	0.99
N & S	1166			0.91
Runoff			**Quantity (in./yr)**	
Nassau, W & C Suffolk	760	Franke & McClymonds[9]	0.55	3.09
Nassau, W & C Suffolk	760	Cohen *et al.*[10]	0.55	3.09
Babylon, Islip	190	Plukowski & Kantrowitz[5]	1.0	5.29
East Meadow	5.6	Seaburn[6]	9.8	
N & S	1166			4.54

THE NITROGEN MODEL

The nitrogen model is currently being developed in a way which parallels the hydrological model just described. Further details may be found in Zaeh.[12]

Although the model is not fully operational, it is possible to estimate, very provisionally, the total loads of nitrogen from major sources on the Island. Major inputs of nitrogen to Long Island are shown in Table V, which also shows the corresponding figures for the nation for purposes of comparison.

Table V. Provisional estimates of annual loads of nitrogen from indicated sources.

	Long Island (Nassau & Suffolk)		Nation[a]	
Source	Thousand Tons	%	Million Tons	%
Fertilizer	11.5	68	6.8	60
Fixation	3.7	22	3.0	27
Precipitation	1.6	10	1.5	13
Totals	16.8	100	11.3	100

[a]Based on Wadleigh.[13]

The corresponding load of nitrogen from the population is estimated to be approximately 20,000 tons. If allowance is made for losses of nitrogen to the atmosphere or the ocean from the wastewater during treatment and disposal, the actual losses of nitrogen to ground water from fertilizers and the population may be of comparable magnitude.

ACKNOWLEDGMENTS

A very large number of persons and agencies have cooperated in the work outlined in this paper. In particular, without the help and facilities at the Long Island Vegetable Research Farm, this research could not have been carried out. The assistance of Cooperative Extension personnel on Long Island and individuals at Cornell must also be very gratefully acknowledged.

The work was performed primarily under a contract with the Nassau-Suffolk Regional Planning Board pursuant to Section 208, Federal Water Pollution Control Act Amendments of 1972 (PL 92-500). This Project has been financed in part with federal funds from the United States Environmental Protection Agency under grant number P002103-01-0. The contents do not necessarily reflect the views and policies of the United States Environmental Protection Agency.

REFERENCES

1. Bouldin, D. R. and G. W. Selleck. "Management of Fertilizer Nitrogen for Potatoes Consistent with Optimum Profit and Maintenance of Ground Water Quality," presented at 9th Cornell University Waste Management Conference, Syracuse, New York, April 1977.
2. Bouldin, D. R. Department of Agronomy, Cornell University (personal communication), 1976.
3. Baskin, L. "A Recharge Model for Nassau and Suffolk Counties, New York," Masters Thesis, Department of Environmental Engineering, Cornell University (1977).
4. Thornthwaite, B. W. "Report of the Committee on Transpiration Evaporation, 1943-1944," *Trans. Am. Geophysical Union* 25 (1944).
5. Pluhowski, E. J. and I. H. Kantrowitz. "Hydrology of the Babylon-Islip Area, Suffolk County, New York," United States Geological Survey, Water Supply Paper 1768 (1964).
6. Seaburn, G. E. and D. A. Aronson. "Influence of Recharge Basins on the Hydrology of Nassau and Suffolk Counties, Long Island, New York," U.S. Geological Survey, Water Supply Paper 2031 (1974).
7. Holzmacher, McLendon and Murrell. "Comprehensive Public Water Supply Study, Suffolk County, New York," Holzmacher, McLendon and Murrell—Consulting Engineers, Melville, New York (1970).
8. Lubke, E. R. "Hydrogeology of the Huntington-Smithtown Area, Suffolk County, New York," United States Geological Survey, Water Supply Paper 1669-D (1964).
9. Franke, O. L. and N. E. McClymonds. "Summary of the Hydrologic Situation on Long Island, New York, as a Guide to Water Management Alternatives," United States Geologic Survey, Professional Paper 627-F (1972).
10. Cohen, P., O. L. Franke and B. L. Foxworthy. "An Atlas of Long Island's Water Resources," New York Water Resources Commission Bulletin 62 (1968).
11. Greeley and Hansen. "Comprehensive Public Water Supply Study," County of Nassau, State of New York CPWS-60 (June 1971).
12. Zaeh, D. H. Masters Thesis, manuscript in preparation. Department of Environmental Engineering, Cornell University (1977).
13. Wadleigh, C. H. "Wastes in Relation to Agriculture and Forestry," Misc. Publ. No. 1065, U.S. Department of Agriculture, Washington, D.C. (1968).

22

THE ECONOMIC TRADE-OFFS OF COMMERCIAL NITROGEN FERTILIZERS, LEGUMES AND ANIMAL WASTES IN MIDWEST AGRICULTURE

T.J. Considine
Department of Agricultural Economics

R.E. Muller, Jr.
Department of Agricultural Engineering

R.M. Peart
Department of Agricultural Engineering

O.C. Doering III
Department of Agricultural Economics
Purdue University
West Lafayette, Indiana

INTRODUCTION

Declining domestic reserves of petroleum and natural gas have recently resulted in shortages and significant price increases for these resources. Given the present rates of energy consumption, significant economic and technological adjustments appear to be necessary over the next decade. The economic and political consequences related to dependence on foreign sources of fossil-fuel energy have led many people to advocate energy conservation as one way to reduce this dependence and to extend our remaining supplies.

U.S. agricultural production currently invests large amounts of fossil-fuel energy in collecting and storing solar energy in plant and animal products.[1] The application of technology and scientific research over the past thirty years has facilitated the rapid expansion of agricultural output with less labor and land requirements. This development reflects not only the advancement of science and education but also changing relative prices for purchased farm inputs.[2] The utilization of relatively abundant supplies of natural gas and petroleum by industries producing fertilizers, pesticides and farm machinery in part permitted their rapid expansion. Current farming practices utilize calories of fuel energy in tillage, fertilizers, irrigation, harvesting and processing to help crops convert calories of sunlight into calories of food energy for man and animals.[3] Presently, fossil-fuel energy is the major input in maintaining the high level of agricultural production.

Fossil-fuel energy consumed in agricultural production can be divided into four components. Fuel is used off the farm to manufacture farm inputs: natural gas for nitrogen fertilizers and petroleum for herbicides and insecticides. In addition, fuel is "directly" used on the farm in applying these inputs: by tractors during planting, cultivation, pest control and harvesting; by crop dryers; and by irrigation pumps. Energy is also expended in transporting purchased inputs and agricultural products. Finally, energy is used in manufacturing farm capital such as farm machinery and equipment, agricultural steel and the construction of farm buildings. These components and their percentage of the total U.S. energy budget are presented as follows:[4]

Indirect	Direct	Transportation	Capital	Total
1.1	1.0	0.4	0.4	2.9

Notice that the indirect component is more than the direct component. The fertilizer industry consumes the largest amount of energy in the indirect component, comprising 0.7% of total U.S. energy consumption.[4] This figure is probably underestimated, since it does not include heat and power used in processing nitrogenous fertilizers.

Corn is the predominant crop in Midwest agriculture, and nitrogen fertilizer is the single largest energy input into corn production. Depending upon the application rate, nitrogen fertilizer accounts for around 50% of the total energy input. This chapter focuses on the technical and economic feasibility of cropping technologies which economize on nitrogen fertilizer. Two cropping technologies are evaluated. The first involves the inclusion of nitrogen-fixing legumes into cropping rotations. The second technology considered is the application of animal wastes to replace commercial fertilizers.

The objective of this chapter is twofold: (1) to provide estimates of the energy inputs and efficiencies of these alternatives and (2) to examine their economics. In particular, the total resource use and profitability of these technologies are studied. Secondly, the adoption of these technologies as a function of rising nitrogen fertilizer prices and relative crop prices is examined.

METHOD OF ANALYSIS

A technical and economic analysis of the above alternatives must consider the interactive impacts that change in production technology and relative input and output prices would have upon product quantity and quality and resource use. The principal concern of the analysis is to focus on the production process itself and the interactions within that process. Weather is considered an important variable in the analysis, not only as a contributing factor to the profitability of an enterprise, but also as one of the main determinants of the actual energy efficiency of a given cropping system.[5]

Three analytical models were linked together for the analysis: an agricultural energy input-output simulator, a linear programming farm management model and a livestock feed ration model. The main objective of the simulator is to determine how alternative practices would affect the productivity of a farm as well as measure changes in the energy inputs to the farm. Practices that reduce energy usage in one component of crop production could possibly increase energy usage in another component or significantly reduce crop yields and wipe out any gains in energy savings. The large number of interactive effects in crop production makes a systems approach a necessity. A simplified flow diagram of the simulator is presented in Figure 1.[6]

A major feature of the simulation model is that it goes through the necessary field operations on the basis of actual weather data for a specific year. Crop yields are calculated by plant growth models based upon date of planting, the level of inputs and weather conditions. The corn yield model developed by Dale and Hodges is the most sophisticated, using daily solar radiation, leaf area index and a moisture stress function.[7] This is in contrast to a soybean yield regression model based on monthly weather data.[8]

The simulator's results represent a technical evaluation of a cropping system and provide an accounting of the inputs and outputs. The results of the cropping simulation, that is, yields, input use, the timing and rate of field operations and so on, are then used as basic inputs into the economic model. This model is an extended version of the Purdue Model B-9 Linear Programming Farm Management Model. The model is designed to find a farm plan which maximizes net income from crop operations in one year.[9] The cropping activities are constrained by the availability of land, labor, machine time and good working days.

The input-output simulator does not explicitly consider animal production activities. However, animal wastes can be included as a source of plant nutrition. The handling and spreading of manure would simply be considered as another field operation subject to the same weather, machinery and labor constraints. Since a livestock operation may be an important input supplier and product user of the cropping enterprise, a livestock model was also linked

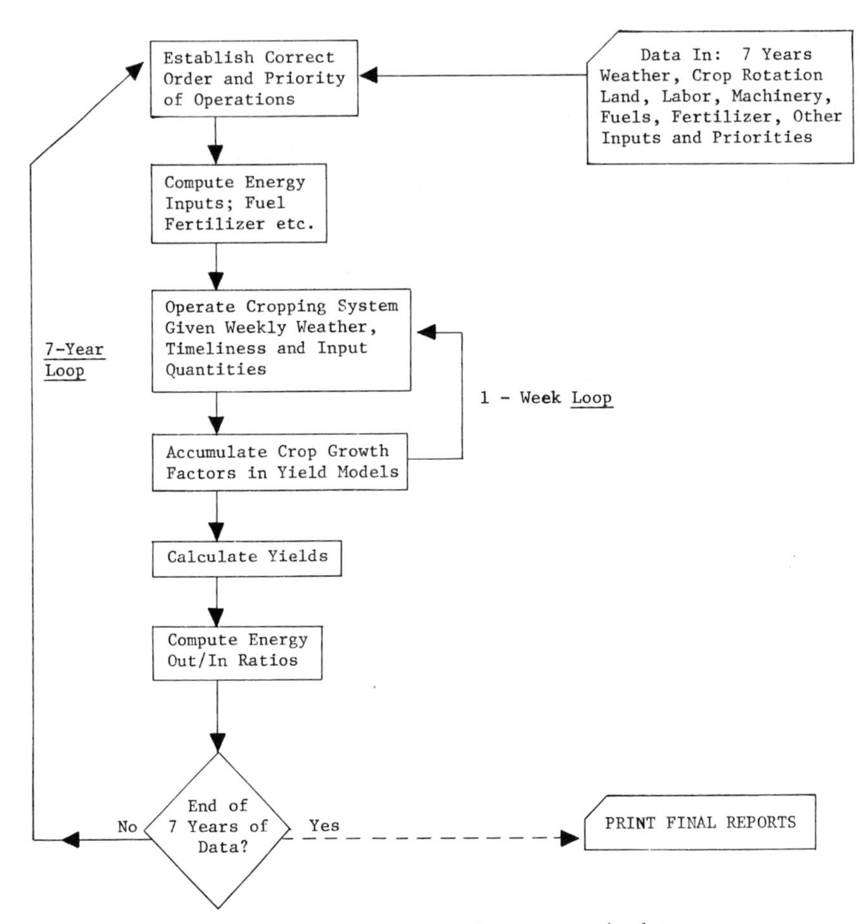

Figure 1. Agricultural energy input-output simulator.

to the analysis.[10] This model essentially finds a balanced feed ration among a number of feed rations available, given the nutritive value of these feeds.

The overall system of analysis integrating these three models is represented in Figure 2. The analysis starts with a search for base data concerning land, labor, machinery, weather, management resources and price data for inputs and outputs. The simulator then evaluates the cropping pattern considering the resource constraints and provides input to the linear programming model in the form of a technically evaluated cropping plan. When evaluating an integrated agricultural production operation, the livestock model is interwoven in the analysis framework via direct input and output flows and budgeting operations.

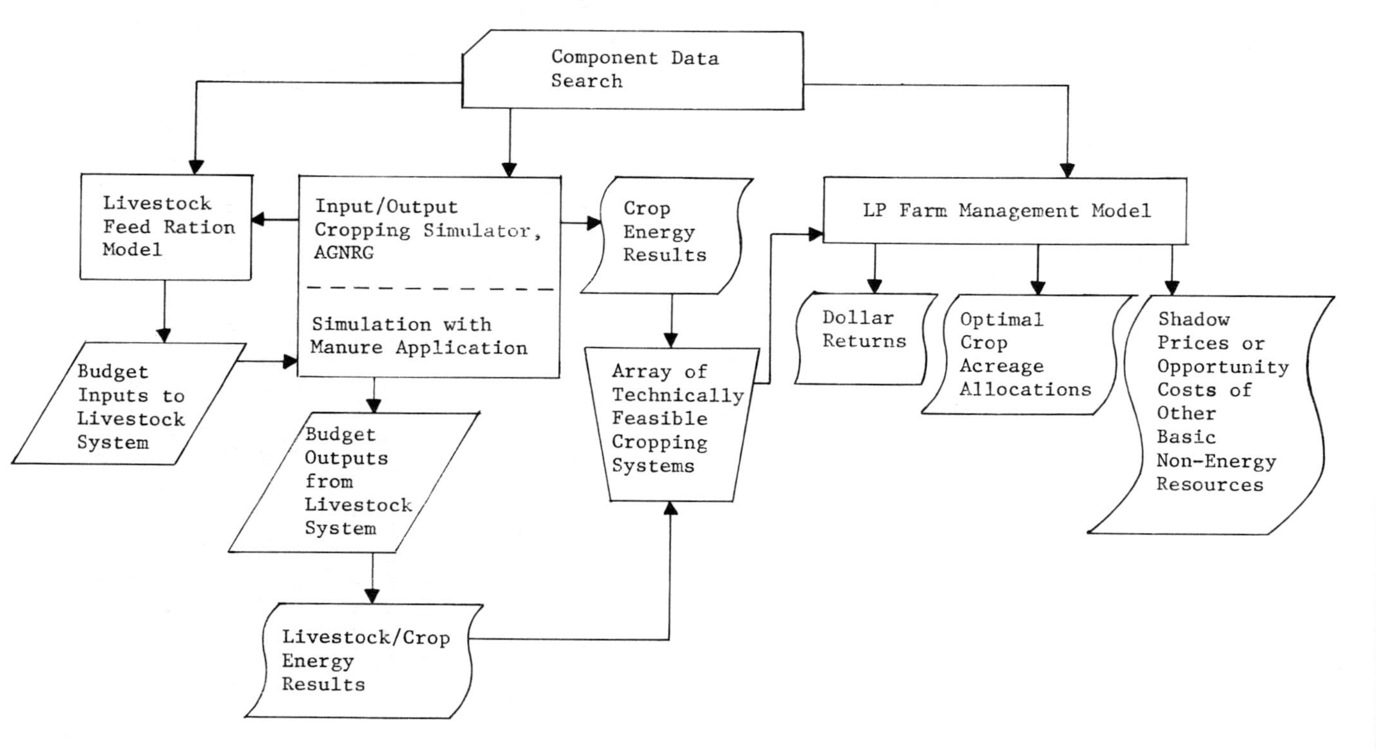

Figure 2. Framework of analysis.

THE CASE EXAMPLES

This system of analysis was used to examine four different cropping systems. The systems are based upon similar resource endowments primarily to examine the relative efficiencies of different production methods. The weather data are from central Indiana during the period 1968 to 1974. The soils are well drained Class I or Class II or the Crosby, Miami or Brookston series which sustain high levels of productivity. All systems are based upon 600 acres of land.

The systems are described as follows:

System 1: corn-corn-soybeans (400 and 200 ac in rotation);

System 2: corn-corn soybeans (400 and 200 ac in rotation, with swine manure providing the entire nutrient requirements of the corn);

System 3: corn-soybeans-winter vetch (300 ac corn, 300 ac beans, followed by winter vetch);

System 4: corn-soybeans-wheat-alfalfa (200 ac corn; 200 ac soybeans; 200 ac of combined alfalfa including 67 ac of wheat/alfalfa, 67 ac of second-year alfalfa and 66 ac of third-year alfalfa).

Timeliness is an important concern of any farm producer, and special attention was paid to this factor in the simulation modeling. The model explicitly takes into account planting date and adjusts yields accordingly. The machinery complement chosen for each of these systems was sufficient for good timeliness. The basic equipment remained the same for each of the systems, with new equipment being added as new crops and operations were considered. This provided a uniform basis for comparison of the timeliness of these systems.

System 1 is a conventional system that would be representative of a typical cropping rotation used by Corn-Belt farmers. The second system is essentially the same except for the fact that swine manure, rather than commercial fertilizer, provides the nutrients for the corn crop. Problems related to off-farm transportation and the regional distribution of manure have been ignored to concentrate on the farm production aspects of using manure as a complete fertilizer. The main concern here is to examine the timeliness of this system and the potential yield effects of this practice. To allow for maximum flexibility, it was assumed that the manure was knifed down in the fall (with N-serve)* or in the spring. The assumptions relating to the nutrient composition of the manure, manure production and application rate are presented in Table I.

*This is not a proven technology; however, it is considered feasible. Research on this is underway in the Agronomy Department at Purdue University.

Table I. Summary of assumptions concerning manure supply for System 2 (swine).

Annual Fertilizer Content (lb/1000 lb animal weight):[11]

Handling and Disposal Method	N	P_2O_5	K_2O
Manure Pit Knifing	124	111	119

Manure Production:[11]

Animal	Size	Gallons	Total Manure Production/yr gal/1000 lb	Water/1000 lb	gal Total
Swine	200 lb	547.5	2737.5	91.25	2828.75

Animal Nutrient Content and Application Rate

Nutrient (lb)	Per 1000 gal/Manure	4000 gal/ac
N	43.75	175
P_2O_5	39.25	157
K_2O	42.00	168

Field Application Rate: 1.96 ac/hr

System 3 is an unconventional cropping system and involves the incorporation of a nitrogen-fixing legume, hairy vetch (*Vicia Villosa*) into a cash grain rotation. This crop is seeded along with wheat to give good winter cover and prevent erosion. Both the wheat and the vetch are plowed under in the spring before corn is planted, thus providing nitrogen and organic matter for the corn. Once again, the concern is to examine the timeliness of this system.

The fourth system is unique due to its high proportion of alfalfa. The alfalfa fixes nitrogen and can be used as a forage feed. A dairy operation was included with this cropping system, utilizing the entire quantity of alfalfa produced as the basic feed for the herd (121.6 milking cows plus an equal number of replacements). The livestock feed ration model was used to find a complete, balanced ration. The quantity of manure produced was then calculated, given the herd size and handling and storage methods (Table II). The manure was applied to the land going into corn production. Less manure was required for the 67 ac of corn following alfalfa because of the carry-over nitrogen from the alfalfa crop (Table II).

SIMULATION RESULTS

Table III summarizes the average simulated performance of each of the four systems over seven years of actual weather. Production is expressed in terms of 100,000-lb units and in terms of digestible protein and energy. The energy inputs and associated output/input ratios are also presented.

Table II. Summary of assumptions concerning manure supply for System 4 (dairy).

Annual Fertilizer Content (lb/1000 lb animal weight):[11]

Handling and Disposal Method	N	P_2O_5	K_2O
Manure Pit Knifing	114	54	107

Manure Production:[11] **(Assume equivalent manure to 186.3 cows @ 1400 lb each)**

Animal	Size	Gallons @ 12.7 Solids	Gallons @ 8% Solids	Liquid Handling System (gal)
Dairy	1000 lb	3613.5	5700	5700

Annual Nutrient Supply and Application Rates

Nutrient (lb)	Per 1000 gal	133 ac @ 8650 gal/ac	67 ac @ 5000 gal/ac
N	20	173	100
P_2O_5	9.5	82.2	47.5
K_2O	18.8	163	94

Field Application Rates:

133 ac @ 1.22 ac/hr
67 ac @ 1.73 ac/hr

Table III. Summary of four cropping system simulations.

	System 1	System 2	System 3	System 4
Crop (ac)				
Corn	400	400	300	200
Soybeans	200	200	300	200
Wheat	-	-	-	67
Alfalfa	-	-	-	200
Winter Vetch	-	-	300	-
Crop Produced (lb x 10^5)				
Corn	31.55	31.51	22.80	15.89
Soybeans	5.738	5.702	8.391	5.732
Wheat	-	-	-	2.152
Alfalfa	-	-	-	17.44
Energy Inputs (Btu x 10^9)	3.678	1.852	2.552	2.781
Outputs				
Digestible Energy (Btu x 10^9)	24.68	24.63	20.94	22.93
Digestible Protein (Btu x 10^5)	4.019	4.005	4.221	5.107
Ratio of Outputs/Energy Inputs				
Digestible Energy (Btu/Btu)	6.71	13.30	8.20	8.25
Digestible Protein (10^{-5} lb/Btu)	10.92	21.62	16.54	18.37

For System 2, virtually the same level of output was maintained as in System 1. Approximately 83% of the manure available to System 2 was applied in the fall, leading to no yield reductions due to delays in planting. Energy inputs were reduced roughly 50% from System 1, and the ratio of digestible energy output to energy input increased 98%. If the manure application were constrained to the spring, delays would most certainly occur. This point and variation in machinery capacity require further analysis.

System 3 illustrates a good example of a timeliness problem. Planting delays occurred in the spring due to competition between land preparation and planting for the corn crop and the plow-down operation for the vetch. As a result, corn yields were reduced to 137 bu/ac, down from the average of 140 bu/ac for Systems 1 and 2. Energy inputs were reduced about 31% from System 1. The energy efficiency as measured by digestible energy output per unit of energy input increased to 8.20 from the System 1 average of 6.71 (Table III).

Simulation results are also presented for System 4. Timeliness was no real problem in terms of contributing to yield reductions. However, competition did exist between corn and soybean postplanting cultivations and the first cutting of alfalfa. Again, machinery capacity was not a limiting factor in this analysis. Variation of this resource to examine possible output effects due to timing warrants further examination.

As noted above, a dairy operation was connected to System 4. Energy budgeting was then undertaken to account for the energy flows into and out of the livestock system. Table IV presents the results of this budgeting. Good estimates for the digestible protein and digestible energy of the meat sold were not available; hence, the respective ratios are underestimated.

ECONOMIC ANALYSIS

Each of the above cropping systems and System 4 with dairy operation were evaluated by the linear programming model. This is a static equilibrium model in that product and input prices are fixed and assumed to be given. One of the main features of the model is that field operations are constrained by the availability of good working days, machinery and labor.

The base data—field working rates, machinery complement and input use—are the same as those used by the simulator. In addition, the average weather, yields and timing of the simulator were used as basic data for the LP model. Input prices were the average prices paid by Indiana farmers during 1976. The corn price selected was $2.36, which was the forward contract bid for October-November-December delivery as of March 1, 1977. Soybean price ($5.71/bu) and wheat price ($2.86/bu) were derived from the above corn price based upon the mean of crop-year price ratios from 1964 to 1976.

Table IV. Summary of System 4 with dairy operation.

	Crop			
Category	**Corn**	**Soybeans**	**Wheat**	**Alfalfa**
Produced (ac)	200	200	67	200
Produced (10^5 lb)	15.86	5.733	2.146	17.41
Fed (10^5 lb)	8.071	0	0	17.41
Sold (10^5 lb)	7.789	5.733	2.146	0

Dairy Inputs	**Energy (10^9 Btu)**
Crop production	2.847
Gasoline (1820 gal)	0.226
Electricity (49,500 kWh)	0.169
LP gas (953 gal)	0.089
Vitamin A (2048 lb)	0.051
Dicalcium phosphate (5341 lb)	0.016
Soybean oil meal (6804 lb)	0.007
	3.405
Fertilizer saved by use of manure	- 0.842
Net Inputs	2.563

Final Products	**Digestible Protein (10^5 lb)**	**Digestible Energy (10^9 Btu)**
Crops sold	2.547	10.57
Milk sold	0.596	2.13
Meat sold	-	-
Total output	3.143	12.70
Total output/Energy input	12.26	4.96

The alfalfa price and milk price used were \$60/ton and \$10.70/cwt, respectively.

The net income generated by each of the above systems, given this specific set of input and product price data, is as follows:

	System 1	System 2	System 3	System 4	System 4 (with dairy)
Net Income (\$/yr)	45,644	62,713	38,714	45,716	88,789

These results on net returns to management can be used only for the sake of comparison and should not be used to make any normative conclusions as to the selection of a particular cropping system. Selection of cropping systems depends on a multitude of factors, primarily managerial abilities, risk aversion preferences and unique production characteristics of various farms.

The data in Table V illustrate the variations in variable costs for the different cropping systems. Total variable costs for System 2 are 33% less than for

System 1 because of the large reduction in fertilizer cost. As a consequence, net income for System 2 is much higher. Nitrogen costs for System 3 declined relative to System 1 due to the reduction in corn acreage and the nitrogen contribution from the vetch. However, the savings from lower nitrogen costs were reduced by increased expenditures on seed and potash for the vetch and higher fuel costs due to the greater number of field operations. The lower corn yields of System 3 result in a decline in net income relative to System 1.

System 4 had the highest phosphate, potash, insecticide and fuel costs of the four cropping systems because of its high proportion of alfalfa. Net income for System 4 is slightly higher than for System 1. Inclusion of a dairy operation with System 4 resulted in much higher net income than the other systems. The relatively high milk price and high productivity of the milking cows assumed in the analysis may be introducing an upward bias on the net income of this system.

Another feature of the LP model is that it provides a tabulation of seasonal labor use (Table VI). Such labor accounting may be valuable in evaluating labor bottlenecks. System 2 provides a good example of a labor bottleneck in the period September 27 to October 10, during which corn is harvested and manure is knifed down. A labor bottleneck also occurs in System 3 from April 26 to May 9 due to competition between plow-down of the vetch and corn planting. The labor use for System 4 is much greater than for the other systems; however, it is more evenly distributed.

Overall, the main resource substitution taking place among Systems 1 through 4 appears to be energy and labor. As energy inputs are reduced, labor hours used increase. However, the relatively high labor input of System 4 may also be associated with high levels of protein output. Thus, some key factors in the selection of more energy-efficient systems may be the opportunity cost of labor of the farm operator and hired labor wage rates. In this analysis the opportunity cost of the farm operator's time was set equal to zero, while the part-time wage rate was set at $3.50/hr. Variation of these parameters may be an interesting area for further research.

SENSITIVITY ANALYSIS

Technological change can come about either from new discoveries in scientific knowledge or with the adoption of known alternative production techniques. The cropping systems discussed above would be of this second type of technological change. The dynamics of the decision process that lead to technological change has been described as the interaction of "coercion" resulting from declining profits and the search for higher profits.[12] The cause of these declining profits is changing relative factor and product prices.

Table V. Summary of total variable and fixed costs.

	System 1	System 2	System 3	System 4	System 4 (w/dairy)
Variable Cost ($)					
Nitrogen	11,900	0	4,634	6,178	1,082
Phosphate	6,800	2,000	6,600	8,193	5,961
Potash	3,600	1,080	5,265	6,924	5,664
Seed	6,000	6,000	9,000	6,606	6,606
Herbicide	7,000	7,000	7,200	4,800	4,800
Insecticide	2,800	2,800	2,100	3,590	3,590
N-serve	0	1,472	0	0	736
Credit & Misc.	4,400	4,400	4,200	4,140	4,140
Fuel & Repairs	5,346	5,302	5,726	6,710	6,692
Drying	3,264	3,566	2,247	1,931	1,931
Part-Time Labor	866	981	505	2,156	2,156
Total	51,976	34,601	47,477	51,228	43,358
Livestock	-	-	-	-	23,243
Fixed Cost ($)					
Machinery	25,904	27,299	26,533	27,428	28,832
Land	55,800	55,800	55,800	55,800	55,800
Labor	8,000	8,000	8,000	8,000	16,585[a]
Animals	-	-	-	-	17,097
Livestock Bldg. & Land	-	-	-	-	14,725
Equipment	-	-	-	-	14,938
Total	89,704	91,099	90,333	91,228	147,968

[a]Includes the labor requirements of the dairy operation.

Table VI. Total farm usage of labor hours.

Time Period	System 1	System 2	System 3	System 4
March 29-April 11	50	50	50	–
April 12-April 25	20	5	51	62
April 26-May 9	207	145	206	90
May 10-May 23	100	–	8	100
May 24-June 6	–	72	79	208
June 7-June 20	85	56	41	19
June 21-July 4	–	–	40	37
July 5-July 18	–	28	–	231
July 19-August 1	–	–	–	–
August 2-August 15	–	–	–	–
August 16-August 29	–	–	–	286
August 30-September 12	–	–	–	–
September 13-September 26	72	72	108	74
September 27-October 10	228	319	114	139
October 11-October 24	108	108	108	247
October 25-November 7	–	106	143	106
November 8-November 21	–	83	12	83
November 22-December 5	–	58	–	65
December 6-December 19	–	–	–	22
December 20-January 3	–	–	–	–
January 4-January 17	–	–	–	–
January 18-January 31	–	–	–	–
February 1-February 14	–	–	–	–
February 15-February 28	–	–	–	–
March 1-March 14	–	–	–	10
March 15-March 28	–	–	–	6
Total	870	1102	960	1769

In the short run, reductions in the energy consumption of the agricultural sector would probably come about through price-related changes rather than supply-related ones.[13] Although restricted supplies of LP gas and anhydrous ammonia have been experienced, scarcities in the factor markets in the last two years have been reflected in prices. This trend is assumed to continue.

Since the above cropping systems represent nitrogen fertilizer-saving crop technologies, the nitrogen fertilizer price would be a key variable in examining their profitability. In addition, the relative price of corn to soybeans is an important variable. To examine the impact these variables have on the profitability of these systems, three farms were hypothesized, each with an energy-saving cropping option. They are defined as follows:

Farm A: Systems 1 and 2;
Farm B: Systems 1 and 3;
Farm C: System 4 with the option of utilizing the dairy waste for corn fertilization.

The linear programming model was modified so that the most profitable "cropping system" was selected.

Three nitrogen fertilizer prices and three different corn/soybean price ratios were used in the sensitivity analysis. Table VII summarizes the results. Given our assumptions concerning timing and application rates, System 2 is profitable under current nitrogen prices. Timing and machinery constraints need to be varied to examine yield effects and changes in net income. Also, different types of animal wastes as fertilizer sources need to be evaluated.

Table VII. Summary of LP runs: Farm A.

	Soybean Price		
Corn Price = $2.36/bu	**($5.17/bu)**	**($5.71/bu)**	**($6.56/bu)**
Nitrogen price = 17¢/lb:			
Crop mix (ac)			
Conventional corn	200	178	400
Corn with manure	400	400	200
Soybeans	–	21	–
Net income ($/yr) (optimum plan)	63,327	62,726	70,873
Manure shadow price ($/10^3 gal)	11.20	10.50	5.41
Nitrogen price = 34¢/lb:			
Crop mix (ac)			
Corn with manure	400	400	400
Soybeans	200	200	200
Net income ($/yr) (optimum plan)	57,529	62,713	70,873
Manure shadow price ($/10^3 gal)	17.46	14.22	9.12
Nitrogen price = 51¢/lb:			
Crop mix (ac)			
Corn with manure	400	400	400
Soybeans	200	200	200
Net income ($/yr) (optimum plan)	57,529	62,713	70,873
Manure shadow price ($/10^3 gal)	21.18	17.94	12.84

With the options available to Farm A, conventional corn (System 1) did not enter the optimal plan for nitrogen prices above 34¢/lb.

A constraint concerning the availability of the swine waste to Farm A was added to the model to estimate an imputed value for the manure. Figure 3 charts the various shadow prices for manure as a function of nitrogen fertilizer prices and relative soybean prices. As nitrogen prices increase, the shadow price for manure also increases. This is consistent with economic theory.

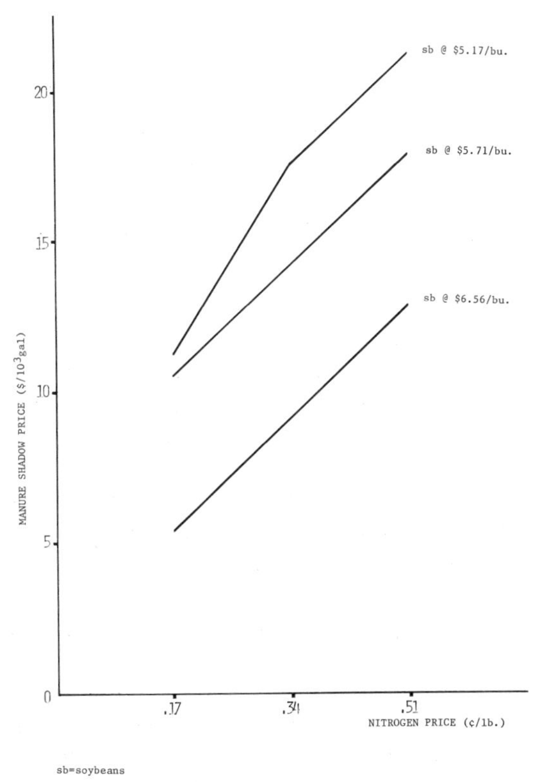

Figure 3. Manure shadow price ($\$/10^3$ gal).

For these experiments, manure availability was restricted to an amount great enough to cover 400 ac at 4000 gal/ac. In an additional experiment, manure availability was changed to estimate a derived demand for the swine waste (Figure 4). This relationship was estimated under current nitrogen prices, a corn price of $2.36/bu and a soybean price of $5.71/bu. The maximum allowable acreage for corn grown with manure was set at 400 ac to be consistent with the technical simulation. The shadow price is an imputed

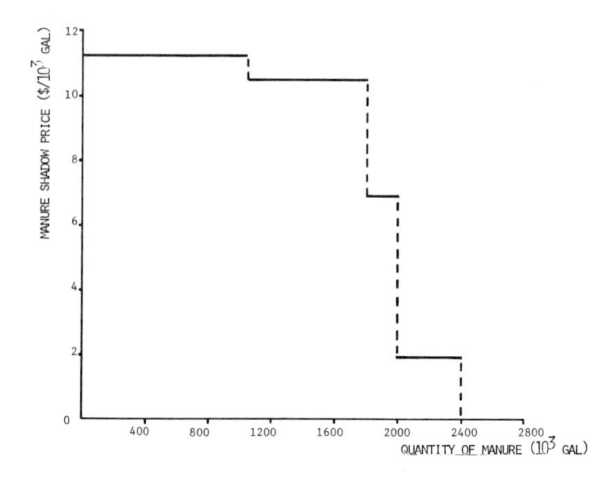

Figure 4. Manure shadow price ($\$/10^3$ gal).

value and was derived for quantities of manure between zero and 2.4×10^6 gal. For quantities of swine waste between zero and 1 million gal, the shadow price was $11.20/1000 gal. The shadow price was zero for quantities greater than 2.4×10^6 gal.

Farm B was also examined for changes in net income and crop mix under different nitrogen prices and relative crop prices. Table VIII summarizes the results. Corn grown with vetch (System 3) enters the optimal plan under all three relative crop price ratios if nitrogen price is 51¢/lb. This is primarily because of the corn yield reductions incurred when using vetch as a source of nitrogen.

A sensitivity analysis was also performed on Farm C. Only nitrogen prices were varied in this experiment. The primary objective was to examine the competition between manured crops and legumes under rising nitrogen prices. The results are presented in Table IX. Alfalfa acreage was reduced in the optimum plan mainly because it was linked to the nitrogen-using wheat crop. A manure constraint was also included for the farm based upon the size of the dairy herd. The shadow prices are presented in Table IX.

CONCLUSIONS

One objective of this paper was to demonstrate an integrated technical and economic approach to the study of alternative cropping systems. Such an approach and, in particular, this area of study requires interdisciplinary action.

Table VIII. Summary of LP runs: Farm B.

	Soybean Price		
Corn price = $2.36/bu	($5.17/bu)	($5.71/bu)	($6.56/bu)
Nitrogen price = 17¢/lb:			
Crop mix (ac)			
Conventional corn	600	518	300
Soybeans	–	82	300
Net income ($/yr) (optimum plan)	44,565	45,935	53,375
Nitrogen price = 34¢/lb:			
Crop mix (ac)			
Conventional corn	307	300	300
Soybeans	167	300	300
Corn with vetch	126	–	–
Net income ($/yr) (optimum plan)	28,835	36,210	48,450
Nitrogen price = 51¢/lb:			
Crop mix (ac)			
Corn with vetch	300	300	300
Soybeans	300	300	300
Net income ($/yr) (optimum plan)	22,427	30,041	42,026

Table IX. Summary of LP runs: Farm C with dairy operation.

	Nitrogen Price (¢/lb)		
Crop Acreages	17	34	51
Conventional corn	52	0	0
Soybeans	0	200	200
Wheat	110	67	67
Alfalfa	330	200	200
Corn following alfalfa	0	0	0
Corn following alfalfa with manure	110	67	67
Corn with manure	108	133	133
Net income ($/yr)	88,836	86,972	85,889
Shadow price for manure ($/$10^3$ gal)	4.75	5.26	5.26

The use of animal waste as a complete fertilizer is profitable under current prices. However, this depends to a large extent upon timeliness and machinery constraints. The use of vetch as a source of nitrogen for corn is not profitable under current prices. However, the economic analysis did not value the possible benefits due to reduced soil erosion.

The key resource trade-off discovered in this analysis involved energy and labor. The adoption of less energy-intensive cropping technologies enables a lower energy input per unit of land. However, it also results in a higher ratio of labor to energy. Thus, a key consideration in evaluating the possibilities for the adoption of energy-saving cropping technologies would be the supply of labor and the competing demands for that labor within the agricultural sector.

ACKNOWLEDGMENTS

This paper was developed from research supported by the National Science Foundation Grant number SIA 75-18726, "Analysis of Energy Efficiency in Various Methods of Production of Major Crops in the Corn Belt." Much of this paper is based upon the work of V. B. Mayrose, B. A. McCarl, C. H. Noller, J. C. Nye, R. C. Pickett and A. L. Sutton.

REFERENCES

1. Heichel, G. H. "Agricultural Production and Energy Resources," *Science* 64:34 (1975).
2. Griliches, Z. "The Demand for Fertilizers: An Economic Interpretation of a Technical Change," *J. Farm Econ.* (1958).
3. Committee on Agricultural Production Efficiency. "Energy and Agricultural Productivity," in *Agricultural Production Efficiency*, National Academy of Sciences (1975), p. 111.
4. Booz-Allen and Hamilton. "Energy Use in the Food System," Federal Energy Administration, Food Industry Advisory Committee (May 1976).
5. Doering, O. C., *et al.* "Energy and Resource Alternatives for Different Cropping Systems," NSF Project Grant #SIA 75-18726, Project Report in Progress, Purdue University (June 1, 1977).
6. Muller, R. E. "Status of the Agricultural Energy Input-Output Simulator," Agricultural Engineering Department, Purdue University (August 1976).
7. Dale, R. F. "An Energy-Crop Growth Variable for Identifying Weather Effects upon Corn Growth and Yield," Preprint from the 13th Agriculture and Forest Meteorology Conference, Purdue University, West Lafayette, Indiana, April 1977, pp. 85-86.
8. Thompson, L. M. "Weather and Technology in the Production of Corn and Soybeans," CAED Report 17, The Center for Agricultural and Economic Development, Iowa State University Press, Ames, Iowa (1963).

9. Brink, L., B. A. McCarl and D. H. Doster. *Methods and Procedures in the Purdue Crop Budget Model B-9: An Administrator's Guide*, Station Bulletin No. 121, Department of Agricultural Economics, Agricultural Experiment Station, Purdue University, West Lafayette, Indiana (March 1976).
10. Mayrose, V. B., *et al.* "Computer Model for Analysis of Fossil-Fuel Energy Inputs in Livestock Production Systems," Department of Animal Sciences, Purdue University (1976).
11. Livestock Wastes Subcommittee of the Midwest Plan Service, G. L. Pratt, Chairman. "Livestock Waste Facilities Handbook," Midwest Plan Service, Iowa State University, Ames, Iowa (1976).
12. de Janvry, A. "A Socioeconomic Model of Induced Innovations for Argentine Agricultural Development," *Quart. J. Econ.* 87:410-35 (1973).
13. Koening, H. E., *et al.* "Energy and Agriculture," Paper presented to 1976 Summer Workshop on Energy Extension Service, University of California, Berkeley, California (1976).

23

A LINEAR PROGRAMMING MODEL FOR DAIRY FARM NUTRIENT MANAGEMENT

D. A. Haith
Departments of Agricultural Engineering and
Environmental Engineering
Cornell University
Ithaca, New York

D. W. Atkinson
U.S. Department of Defense
Washington, D.C.

INTRODUCTION

The management of nutrients is the central activity of a dairy farm. The feeding of animals, production of milk and growing of crops are significant elements of nutrient management. In addition, efficient use of soil fertility, commercial fertilizers and animal manures are aspects of the farm's overall nutrient budgets. The environmental impacts of dairy farming are to a large extent determined by the losses of nutrients, particularly nitrogen and phosphorus, from the farm to water bodies. The farmer's income and his contribution to water pollution are thus greatly dependent on nutrient management decisions. It is difficult to make or evaluate such decisions independently of one another since they are components of the farm's total nutrient budgets. For example, the choice of a particular dairy cow ration will influence the selection of crops which in turn affects fertilization and manure disposal schedules and hence nitrogen and phosphorus losses from the farm. In essence, the dairy farm is a complex system of interacting components or processes, and decisions related to feed purchases, crop selection, herd size, fertilization and manure handling must be evaluated for their effects on this total system.

Given a continuing concern for both food production and water quality, an analytical framework for determining the effects of dairy farm management decisions on farm income and water quality is desirable. Linear programming models have long been used in financial planning analyses of farms.[1-3] These models and other nonlinear (simulation) models[4] have typically provided a good deal of financial budgeting detail but have omitted many of the factors which may result in water pollution from the farm, including manure disposal, erosion, runoff and unused plant nutrients. Linear programming has been used to determine optimal manure disposal programs,[5,6] but the effects of disposal on water pollutant losses from the farm and on the overall farm operation were not considered. The relationships between income and nutrient and sediment losses in agricultural watersheds have been studied using linear programming,[7,8] but these watershed models provide limited information for evaluation of management options on the smaller scale of the individual farm.

Coote[9] developed a comprehensive linear programming model of a dairy farm which incorporated the detailed financial budgeting of the earlier linear programming formulations[1] and added mathematical relationships for the prediction of erosion and water-borne losses of nitrogen and phosphorus from the farm. Coote's model provided a sound basis for the investigation of the impacts of dairy farm management on both income and water quality. The pollutant loss predictions were verified using field data,[9-11] and the model was used to determine the effects of possible waste management regulations on farm income.[9,10] The model has many other potential uses, including evaluations of the effects of changes in fertilizer, milk and feed prices on income and pollutant losses. However, the comprehensive nature of the model imposes some inherent limitations on its applicability. It is a large and complicated model with substantial data requirements and can be applied only after a large investment of time and money by potential users.

The purpose of the present study was to develop a simpler and more accessible version of Coote's model which could still be used to investigate the relationships between management decisions, income and pollutant losses for a dairy farm. This new model, which is called HDF1, was developed by focusing on the major nutrient cycles of the dairy farm and on decisions which have the most significant effect on income and pollutant losses. The result, which is presented in the next section, is a compact linear programming model which is an order of magnitude smaller than Coote's model with correspondingly more modest data requirements. The key test for the model is its ability to approximate the results of the larger model, particularly with regard to pollutant losses, since that portion of the earlier model has been tested with field data. A comparison of results from the two models is presented herein.

MODEL DEVELOPMENT

Decision Variables

The basic decisions or variables included in HDF1 are cropland areas, manure and nitrogen fertilizer application rates and the size of milking herd:

X_{ij} = Area of soil i planted to crop (or rotation) j (ha)
M_{ij} = Manure spread on soil-crop X_{ij} (ton)
N_{ij} = Fertilizer nitrogen applied to soil-crop X_{ij} (kg)
H = Number of cows in the dairy herd

There are I soils and J crops, with the Jth crop, X_{iJ}, always being pasture. Since only certain soil-crop combinations will be feasible or reasonable on any farm, we have

$$(1 - \alpha_{ij})\, X_{ij} = 0 \qquad \forall_{i,j} \tag{1}$$

where α_{ij} = 1 if the soil-crop combination (i,j) is permitted and zero otherwise. Also, cropland on each soil must be limited to the tillable area (ha), L_i, of that soil

$$\sum_{j=1}^{J} X_{ij} \leqslant L_i \qquad \forall_i \tag{2}$$

Manure spreading may not be allowed in all cases, and hence

$$(1 - \beta_{ij})\, M_{ij} = 0 \qquad \forall_{i,j} \tag{3}$$

where β_{ij} = 1 if manure can be spread on land X_{ij}, and zero otherwise. A second constraint on manure spreading is required to assure that M_{ij} = 0 whenever X_{ij} = 0:

$$-1000X_{ij} + M_{ij} \leqslant 0 \qquad \forall_{i,j} \tag{4}$$

The "1000" on the left of Constraint 4 is an arbitrarily large number greater than the maximum possible manure-spreading rate.

A manure mass balance is given by

$$\sum_{i=1}^{I} \sum_{j=1}^{J} M_{ij} - mH = 0 \tag{5}$$

where m is the amount of manure to be disposed of (including that of animals such as replacement heifers which are not part of the milking herd) divided by the herd size. The herd size may be constrained to some maximum number, Hmax, based on the available housing facilities or other considerations:

$$H \leqslant Hmax \tag{6}$$

Animal Feeding

Although dairy cows have need for many different nutrients, it is assumed that the on-farm crops are grown principally to help supply the animals' net energy and digestible protein requirements

E = Net energy requirement to be satisfied from on-farm crops (mcal per cow)
D = Digestible protein requirement to be satisfied from on-farm crops (kg per cow)

These definitions imply *a priori* decisions on feed purchases; *i.e.*, the linear programming model will not determine optimal feed purchases. The energy and protein requirements may include the needs of replacement heifers and other stock as desired. For example, if the ratio of replacement heifers to milking cows is 1:4, the nutrient requirements E and D are those of a cow plus one quarter of the heifer requirement. Stated in another way, E(D) is the total net energy (digestible protein) required from crops by all animals on the dairy farm divided by the number of cows in the milking herd.

Crop yields and nutrient contents are assumed known and specified by

Y_{ij} = Yield of crop (or rotation) j on soil i (units/ha)
e_{ij} = Net energy content of Y_{ij} (mcal/unit)
d_{ij} = Digestible protein content of Y_{ij} (kg/unit)

Net energy and protein balances for the farm are thus provided by

$$\sum_{i=1}^{I} \sum_{j=1}^{I} e_{ij} Y_{ij} X_{ij} - EH \geqslant 0 \tag{7}$$

and

$$\sum_{i=1}^{I} \sum_{j=1}^{I} d_{ij} Y_{ij} X_{ij} - DH \geqslant 0 \tag{8}$$

In addition, it is realistic to limit the amount of nutrients which could be supplied by pasture because of seasonal availability of pasture and its proximity to the barns:

$$\sum_{i=1}^{I} e_{iJ} Y_{iJ} X_{iJ} - \frac{ep}{100} EH \leqslant 0 \tag{9}$$

where ep is the maximum percentage of net energy which may be supplied by pasture.

Crop Nitrogen Requirements

A crop nutrient balance is constructed for nitrogen only. Potassium is not considered a potential water pollutant, and the total amount of phosphorus in the soil is generally not greatly affected by fertilizer or manure applications. Crop needs for potassium and phosphorus are not ignored, however, since the costs of these fertilizers are included in a later segment of the model.

The nitrogen requirement of crop j on soil i is specified by NR_{ij} (kg/ha). The nitrogen requirement will vary with soil fertility, and hence NR_{ij} is the requirement for the crop over and above the nitrogen provided by mineralization of soil organic matter. The nitrogen applied to the crop-soil area (i,j) is given by

$$\left(1 - \frac{n\ell_{ij}}{100}\right) nm\, M_{ij} + \left(1 - \frac{n\ell_{ij}}{100}\right) N_{ij}$$

where nm is the nitrogen content of the manure and $n\ell_{ij}$ is the percent of the applied nitrogen which is lost due to ammonia volatilization and denitrification. In order to supply the crop nitrogen requirements we must have

$$-NR_{ij} X_{ij} + \left(1 - \frac{n\ell_{ij}}{100}\right) nm\, M_{ij} + \left(1 - \frac{n\ell_{ij}}{100}\right) N_{ij} \geqslant 0 \quad \forall_{i,j} \tag{10}$$

Constraint 10 is obviously an oversimplification of the soil nitrogen balance. It is assumed that all manure nitrogen is available for crop needs and that the same percentages of manure and fertilizer nitrogen are volatilized.

Pollutant Losses

The following environmental parameters are estimated by the model for each cropped area (including pasture):

1. potential nitrogen loss: the difference between applied nitrogen and that used by the crops;
2. eroded soil;
3. particulate phosphorus in the eroded soil;
4. dissolved nitrogen in runoff waters; and
5. dissolved phosphorus in runoff waters.

Each of the above can be restricted to a maximum average annual value per ha. For example, the potential nitrogen loss constraint is

$$-\sum_{i=1}^{I}\sum_{j=1}^{J} NR_{ij} X_{ij} + \sum_{i=1}^{I}\sum_{j=1}^{J} nm\, M_{ij} + \sum_{i=1}^{I}\sum_{j=1}^{J} N_{ij} \leqslant L_N \sum_{i=1}^{I} L_i \qquad (11)$$

where L_N is the maximum allowable potential nitrogen loss (kg/ha). It should be noted that volatilization losses are included in the above and that it is assumed that any tillable land self idle will not produce any nitrogen losses (*i.e.*, L_N is multiplied by $\sum_{i=1}^{I} L_i$ rather than $\sum_{i=1}^{I}\sum_{j=1}^{J} X_{ij}$).

The erosion, or soil loss constraint is

$$\sum_{i=1}^{I}\sum_{i=j}^{J} s_{ij} X_{ij} \leqslant L_S \sum_{i=1}^{I} L_i \qquad (12)$$

where s_{ij} is the erosion (ton/ha) from land area X_{ij} as determined by the universal soil loss Equation 14,[12] and L_S is the maximum allowable per ha soil loss.

Particulate phosphorus loss is restricted by

$$\sum_{i=1}^{I}\sum_{j=1}^{J} p_i s_{ij} X_{ij} \leqslant L_P \sum_{i=1}^{I} L_i \qquad (13)$$

where p_i is the phosphorus content of soil i (kg/ton). This can be multiplied by an enrichment ratio, if data are available, to account for the higher content of phosphorus in eroded than *in situ* soil. The maximum allowable particulate phosphorus loss is L_P (kg/ha).

Runoff losses of soluble nitrogen and phosphorus are modeled as the product of estimated runoff and nutrient concentrations:

$$0.1 \sum_{i=1}^{I}\sum_{i=j}^{J} (rs_{ij}\, ns_j + rw_{ij}\, nw_{ij}) X_{ij} \leqslant L_{RN} \sum_{i=1}^{I} L_i \qquad (14)$$

$$0.1 \sum_{i=1}^{I}\sum_{i=j}^{J} (rs_{ij}\, ps_j + rw_{ij}\, pw_j) X_{ij} \leqslant L_{RP} \sum_{i=1}^{I} L_i \qquad (15)$$

where

rs_{ij}, rw_{ij} = summer and winter runoff, respectively, from land area X_{ij} (cm)

ns_j, nw_j = soluble nitrogen concentration in summer and winter runoff, respectively, from land planted to crop j (mg/l)

ps_j, pw_j = soluble phosphorus concentration in summer and winter runoff, respectively, from land planted to crop j (mg/l)

L_{RN}, L_{RP} = maximum allowable losses of soluble nitrogen and phosphorus, respectively, in runoff (kg/ha)

The constant (0.1) is a conversion factor. Summer runoff is estimated using the U.S. Soil Conservation Service (SCS) curve number runoff equation,[13] and winter runoff can be calculated using a degree-day snowmelt equation in conjunction with the SCS equation.[14] Phosphorus and nitrogen concentrations are average values based on field studies for the relevant crops.[15]

The above approach for estimating soluble nutrient losses is not based on mass balance models of the soil nutrient budgets as in Coote's original model, nor does it in any sense match the kinetic modeling of nutrient transformations found in event-based simulations.[16] In this sense, the approach is less than rational when compared with more elaborate models. However, as noted elsewhere[15] the predictions given in Constraints 14 and 15 have operational advantages:

1. They are based on readily available data.
2. They are sensitive to variations in soils, crops and climate.
3. Runoff estimates are event-based; *i.e.,* runoff is determined for each storm and then summed for winter and summer.
4. Runoff predictions are sensitive to management practices such as contouring, return of crop residues, strip cropping, etc., and hence the effects on these practices on pollutant losses can be estimated.

Farm Income

Assuming that the farmer's objective is maximization of net income, Z ($/yr), the objective function for the linear programming model is

$$Z = rH - \sum_{i=1}^{I} \sum_{j=1}^{J} c_{ij} X_{ij} - c_m \sum_{i=1}^{I} \sum_{j=1}^{J} M_{ij} - c_n \sum_{i=1}^{I} \sum_{j=1}^{J} N_{ij} \quad (16)$$

where

r = net return per dairy cow exclusive of costs of farm crops, manure disposal and nitrogen fertilizer ($)

c_{ij} = cost of cropping soil i with crop j exclusive of nitrogen fertilizer and manure disposal costs ($/ha)

c_m = cost of manure disposal ($/ton)

c_n = purchase cost of nitrogen fertilizer ($/kg of nitrogen)

Model Summary

The complete model is

$$\text{Max } Z$$

subject to Constraints 1 through 15 and the usual nonnegativity restrictions on decision variables (H, X_{ij}, M_{ij} and N_{ij}). The model is solved on an annual basis, using either average annual weather, yields and prices data or data selected for a particular year of record. General data sources for the model are soil surveys, farm cost accounts, daily weather data and tabulated parameters for the universal soil loss equation[12] and the SCS runoff equation.[13]

MODEL EVALUATION

The linear programming model, HDF1, satisfies the objective of simplicity. Data requirements are modest and the computer costs of runs are typically small ($1 to $5). The value of the model ultimately depends on the accuracy of its predictions, however, and although it might be desirable to sacrifice some accuracy for convenience and simplicity, there are limits to such tradeoffs. The issue explored in this section is whether or not this simple model produces results comparable to those obtained by a much more extensive and supposedly more realistic model. A comparison is made between the results obtained from HDF1 and those obtained from the larger Coote model. It is assumed that the accuracy of Coote's model is not seriously in question, since it has been validated with field data.[10,11]

The farm which was studied was the "typical" dairy farm in West Jefferson County, New York, which was modeled earlier by Coote[9] and Coote *et al.*[10,11] This is a farm with relatively poor soils, flat slopes and 124 ha of tillable land. The results of applying Coote's model to the farm are provided in Coote *et al.*[11] and input data are given in Coote.[9] Data for the new model (HDF1) are given below.

Soil and Crop Data

The farm contains the four basic soil associations listed in Table I. Three rotations are recommended for these soils: corn-oats-alfalfa-alfalfa (COAA), corn-oats-grass-grass-grass (COGGG), and continuous pasture. The corn is

Table I. Soil associations.[9]

i	Soil Association	Tillable Area (ha) L_i	Phosphorus Content (kg/ton) p_i
1.	Rockland-Panton (R/P)	26	0.90
2.	Livingston-Panton (L/P)	5	0.90
3.	Acid Sands (A/S)	9	0.12
4.	Panton-Vergennes (P/V)	84	0.94

grown for silage and the alfalfa for hay. Although Coote permitted the grass to be either hay or grass silage, it is assumed that all grass is silage in the present case. The possible soil-crop combinations are listed in Table II.

Table II. Possible soil-crop combinations, α_{ij} (α_{ij} = 1 if combination is possible, α_{ij} = 0, otherwise).

	Crop j =		
Soil	COAA	COGGG	Pasture
1. R/P	0	0	1
2. L/P	1	1	0
3. A/S	0	1	1
4. P/V	1	1	0

It is assumed that manure may be spread on both rotations but not pasture. This differs from Coote's model which dealt with each crop separately and did not allow manure spreading on alfalfa. The entire rotation is treated as a single crop in HDF1 and hence manure spreading cannot be limited to certain years of a rotation.

Nutrients produced from the crop rotations are given in Table III. Since the Coote model provided for total digestible nutrient (TDN) rather than

Table III. Nutrients produced from cropping options (kg/ha).

	$e_{ij}\ Y_{ij}/d_{ij}\ Y_{ij}$ [a]		
Soil	COAA	COGGG	Pasture
1. R/P	–	–	1458/270
2. L/P	2532/356	2111/204	–
3. A/S	–	2111/204	1458/270
4. P/V	3386/491	2853/280	–

[a]Averaged over the rotation. Net energy, e_{ij}, measured as total digestible nutrients.[9]

net energy requirements for the dairy herd, the net energy content of crops (e_{ij}) is measured by TDN in this application of HDF1. A pasture limitation of ep = 17% is based on the maximum allowable pasture area included in Coote's model. Crop nitrogen requirements, NR_{ij}, are given in Table IV.

Table IV. Crop nitrogen requirement, NR_{ij} (kg/ha) averaged over rotation.[9]

Soil	Crop		
	COAA	COGGG	Pasture
1. R/P	–	–	0
2. L/P	28	33	–
3. A/S	–	65	22
4. P/V	56	65	–

Animal Nutrition and Manure

Nutrient requirements and manure production are given in Table V. The TDN and digestible protein values are based on 60% of the forage requirement for cows and replacement heifers (one heifer for every four cows). All grain was purchased in Coote's West Jefferson model, and up to 50% of the required forage was also allowed to be purchased. The model then solved for optimal feed purchases. This is not possible in the current model, and it was arbitrarily assumed that 40% of the required forage would be purchased. The manure production rate includes manure from heifers during the period of time in which they are housed (9 tons per heifer @ 1/4 heifer per cow).

Table V. Dairy herd nutrition and manure production.[9]

Net Energy Requirement, E (kg of TDN per cow)	2100
Digestible Protein Requirement, D (kg per cow)	313
Manure Production, m (ton per cow)	19
Nitrogen Content of Manure, nm (kg per ton)	5

Pollutant Loss Parameters

Most of the crop nitrogen is supplied by manure on the West Jefferson farm, and the nitrogen volatilization rate, $n\ell_{ij}$, was set at 35% for all i and j, corresponding to the ammonia content of the manure. Soil losses were estimated from the universal soil loss equation:

$$s_{ij} = R\,K_i\,LS_i\,C_{ij}\,P_{ij} \tag{17}$$

where as usual R, K_i, LS_i, C_{ij} and P_{ij} are rainfall erosivity, soil erodability, topographic factor, cover/management factor and practice factor, respectively, Soil slope and erodability (K_i) were obtained from Coote[9] and remaining parameters were obtained from tabulated values.[12] An average slope length of 91 m (300 ft) was assumed, the cover/management factor was averaged over each rotation and cross slope cropping also assumed. The resulting soil losses are given in Table VI.

Table VI. Average annual soil losses, s_{ij} (ton/ha).

	Crop		
Soil	COAA	COGGG	Pasture
1. R/P	–	–	0.4
2. L/P	1.4	1.2	–
3. A/S	–	1.1	0.2
4. P/V	2.9	2.4	–

Runoff estimates were obtained from Coote's precipitation, antecedent moisture and curve number data.[9] Winter runoff was estimated directly using the curve number equation; *i.e.*, all snow was treated as rain. "Winter" was considered to be December-April. The runoff estimates are given in Table VII.

Table VII. Summer and winter runoff (cm) (averaged over rotation).

	rs_{ij}/rw_{ij}		
Soil	COAA	COGGG	Pasture
1. R/P	–	–	3.8/7.8
2. L/P	5.7/10.09	5.0/8.2	–
3. A/S	–	0.02/0.37	0.00/0.08
4. P/V	5.7/10.0	5.0/8.2	–

Soluble nitrogen and phosphorus concentrations were based on midrange values reported in Haith and Dougherty[15] and are listed in Table VIII for individual crops. Concentrations for the rotations are averages weighted by the runoff curve numbers of each crop (Table IX).

Table VIII. Soluble nutrient concentrations in cropland runoff.[15]

	Soluble Nitrogen Concentrations (mg/l)		Soluble Phosphorus Concentrations (mg/l)	
	Summer	Winter[a]	Summer	Winter[b]
Corn	4.4	8.9	0.90	0.05
Oats	5.3	8.9	0.24	0.05
Alfalfa, Grass, Pasture[c]	1.0	1.0	0.22	0.05

[a]Fallow conditions for corn and oats.
[b]Fallow conditions for all crops.
[c]Pasture concentrations assumed to be same as hay.

Table IX. Nutrient concentrations used in HDF1.

	Soluble Nitrogen Concentrations (mg/l)		Soluble Phosphorus Concentrations (mg/l)	
	Summer	Winter	Summer	Winter
Crop	ns_j	nw_j	ps_j	pw_j
1. COAA	3.0	5.2	0.41	0.05
2. COGGG	2.6	4.5	0.37	0.05
3. Pasture	1.0	1.0	0.22	0.05

Economic Data

All income and cost data for the dairy farm is summarized in the four types of economic parameters in Equation 16. Economic budgeting in most farm linear programming models and Coote's in particular is considerably more detailed, and hence Coote's economic data could not be used directly. The net income predicted in the earlier study of the West Jefferson farm was exclusive of the value of labor by the farm family (of which there were 3305 hr available per year) and the costs of interests and taxes on the farmland. Income was based on 1970 prices. These conventions were followed in this application of HDF1.

The net return per cow, r, was estimated by the following steps.

1. Annual return per cow excluding labor, forage and land costs and including heifer raising costs (1/4 heifer per cow) was obtained from 1973 Farm Cost Accounts.[17]
2. Based on a total labor requirement of 55 hr/yr per milking cow (including 1/4 heifer), a purchased labor cost ($3.60/hr) was estimated for herd sizes requiring more than 3305 hr of labor.
3. Return and costs were adjusted back to 1970 using milk price records and a "prices paid by farmers" index.[17]
4. Costs of 40% forage purchase (hay at $36/ton[9]) and the labor costs were subtracted from return per cow to give the net returns per cow given in Table X.

Table X. Net returns per dairy cow.

Herd Size	Return per Cow, r ($/yr)
50	298
70	276
90	247
110	228
130	215
150	206

Cropping costs, c_{ij}, were more straight forward, and are the 1973 Cost Account values less land and nitrogen fertilizer costs averaged over the rotation, and adjusted to 1970 prices. Land improvement costs (primarily liming) from Coote[9] were added and the results given in Table XI. Manure spreading costs were taken as the average cost reported in the earlier modeling results for the West Jefferson farm, c_m = $1.36/ton. Nitrogen fertilizer costs, c_n, were $0.22/kg in 1970.[9]

Table XI. Cropping costs, c_{ij} ($/ha).

Soil	Crop		
	COAA	COGGG	Pasture
1. R/P	–	–	10*
2. L/P	175	186	–
3. A/S	–	191	41
4. P/V	180	191	–

Model Comparisons

The previous runs with the large model did not restrict pollutant losses so the loss limits L_N, L_S, L_P, L_{RN}, L_{PN} were all set at large values which would not constrain the solution. The solutions to HDF1 are given in Table XII. The optimal herd size was always at its maximum value. As is seen

Table XII. Model solution for various herd sizes (only nonzero decision variables given).

	Herd Size, H = Hmax					
	50	70	90	110	130	150
X_{13} (R/P, Pasture) (ha)	12.1	16.9	21.7	26.0	26.0	26.0
X_{33} (A/S, Pasture) (ha)	–	–	–	0.6	5.5	9.0
N_{33}/X_{33} (N Fertilizer Rate on X_{33}) (kg/ha)	–	–	–	22.0	22.0	22.0
X_{41} (P/V, COAA) (ha)	25.7	36.0	46.3	56.6	66.9	77.8
M_{41}/X_{41} (Manure on X_{41}) (ton/ha)	36.9	36.9	36.9	36.9	36.9	36.6

from Table XII, animal forage requirements are always met by a combination of pasture and the COAA rotation. Table XIII lists a comparison of results obtained by HDF1 and the more complicated model. Income, cropped areas and pollutant losses are in relatively close agreement at most herd sizes. The divergence in income at herd sizes greater than 50 cows may be due to the more realistic matching of animal housing facilities to herd sizes which is included in Coote's model. The most significant difference in the two results is in the predicted losses of soluble nitrogen in runoff, with HDF1 apparently consistently overestimating losses. The reasons for the differences are not obvious, although in both models runoff losses are a small fraction of the potential nitrogen losses. The size of the latter is cause for some concern. Considering the potential nitrogen loss predicted by HDF1 for a 150-cow herd, 9893 kg, if the 35% ammonia volatilization loss and 565 kg runoff loss are subtracted, approximately 5800 kg of nitrogen is unaccounted for on the COAA rotation (roughly 75 kg/ha). The ultimate disposition of this nitrogen is not predicted by either model. Presumably, some portions are leached into the ground water and/or denitrified and the remainder is in the form of organic nitrogen in the manure which will replace the soil organic nitrogen which has been mineralized.

In general, the new model gives results which are comparable to the earlier Coote model. Given the order of magnitude reduction in size and data needs achieved by HDF1, the trade-off of accuracy for size seems reasonable.

Table XIII. Comparison of model results–complex (Coote[9] and Coote *et al.*[10,11]) vs simple (HDF1) model.

	Herd Size, H = Hmax											
	50		70		90		110		130		150	
	Coote	HDF1	Coote	HDF1	Coote	HDF1	Coote	HDF1	Coote	HDF1	Coote	HDF1
Net Income ($/yr)	8644	8854	7984	10856	8025	11348	8540	11759	8887	12034	9231	12346
Land in COAA (ha)[a]	27	26	38	36	49	46	60	57	71	67	81	78
Potential Nitrogen Loss (kg)	4315	3309	5577	4632	6831	5956	8100	7279	9375	8602	10656	9893
Soil Loss (tons)	54	79	87	111	122	143	150	174	180	206	209	237
Particulate Phosphorus Loss (kg)	72	75	116	104	163	134	200	163	240	192	278	222
Soluble Nitrogen Losses in Runoff (kg)	134	191	166	267	198	343	227	418	258	490	289	565
Soluble Phosphorus[b] Losses in Runoff (kg)	–	9	–	12	–	16	–	19	–	22	–	25

[a]Rotation COGGG never in optimal solution for either model.
[b]Not predicted by Coote model.

APPLICATIONS

The efficient dairy farm manager recognizes that he cannot change his crop rotation or select a new manure handling system without evaluating the effects of such decisions on his overall farm operation. This partially explains why the results of agricultural research often move into practice very slowly. A new crop variety may give better yields, but the farmer must determine the effect of the new crop on his management system in order to see whether or not a significant improvement in farm income will result from the new crop. The development of such knowledge or experience often requires a lengthy period of testing and demonstration. There would be obvious advantages in being able to predict at least partially the effects of a new crop variety on the farming system prior to on-farm testing and demonstration. For example, the new crop may have different protein levels and require different fertilization patterns than the older varieties. How will the combined effects of yields, protein content and fertilization change the farm management decisions? Analogous concerns would apply to new manure management practices, improved animal feeding programs and introduction of different programs of dairy herd management.

To at least a limited extent, mathematical models of the dairy farm can provide information on the effects of implementing new practices on farm income. Such models will be most useful if they are readily understandable to the engineer, agronomist, animal scientist or plant scientist. Stated another way, if the models are so complicated and have such extensive data requirements that they can be used only after a major investment of time and money by systems analysts and computer specialists, they are not likely to be valuable to many researchers. The linear programming model presented in this chapter is a relatively straightforward formulation, however, and there appears to be little apparent reason why the model could not be incorporated into dairy research programs.

The linear programming model HDF1 provides a reasonably detailed accounting of soil erosion and water-borne losses of nitrogen and phosphorus from the dairy farm. The model can hence be used to estimate the environmental impacts of changes in farm management. Conversely, it can also determine the effects of certain environmental management regulations on the dairy farm, including manure disposal restrictions, mandatory implementation of soil and water conservation practices and fertilizer regulations. The effects of a tax on fertilizers and limits on nitrogen, phosphorus and soil losses may similarly be investigated. For example, it would be possible to determine effects of a 50% reduction of nutrient losses on the incomes of dairy farms of varying characteristics. The linear programming model can thus be an efficient tool for studying policy issues related to environmental quality and farm income.

SUMMARY

A relatively simple dairy farm linear programming model has been developed and tested. The model focuses on the nutrient budgets of the dairy farm and determines an optimal set of cropping, manure spreading, nitrogen application and herd size decisions subject to constraints on farm losses of nitrogen, phosphorus and eroded soil. The model is a more modest version of an earlier, detailed farm management model which was tested using field data. The new model, HDF1, has been evaluated by comparing its results with those obtained by the earlier model. Although HDF1 is much smaller and has fewer data requirements than its predecessor, the results obtained from both models are comparable.

The model HDF1 can be used to determine the effects of dairy farm management decisions on farm income and on water pollutant losses from the farm. It is also a suitable tool for the evaluation of environmental policy decisions on farm management. Given its simplicity, the model is not recommended for detailed farm budgeting or planning studies. The cost and return estimating procedures of the model do not preserve the sophistication (and hence accuracy) of earlier linear programming models used for budgeting purposes.[1] However, the essential nutrient interactions of the dairy farm are included, and the model does appear to be a reasonable mechanism for the screening of research and policy options to determine their likely impact on both the dairy farm and the environment.

REFERENCES

1. Barker, R. "Use of Linear Programming in Making Farm Management Decisions," Bull. 993, Cornell Univ., Agr. Exp. Sta., Ithaca, New York (1964).
2. Heady, E-O. and W. Chandler. *Linear Programming Methods* (Ames, Iowa: Iowa State University Press, 1958).
3. McAlexander, R. H. and R. F. Hutton. "Linear Programming Techniques Applied to Agricultural Problems," Penn. Agr. Exp. Sta., Agr. Econ. & Rural Soc. Bull. No. 18. Penn. State Univ., State College, Pennsylvania (1959).
4. Tate, D. G. "Financial Decision-Making (for) Nova Scotia Dairy Farms," *Canadian J. Agric. Econ.* 20(1):66-79 (1972).
5. Dodd, V. A., D. F. Lyons and P. D. Herlihy. "A System of Optimizing the Use of Animal Manures on a Grassland Farm," *J. Agric. Eng. Res.* 20:391-403 (1975).
6. McKenna, M. R. and J. H. Clark. "Economics of Storing, Handling and Spreading of Liquid Hog Manure for Confined Feeder Hog Enterprises," *Proceedings 2nd Cornell Waste Management Conf.* (1970), pp. 98-110.

7. Casler, G. L. and J. J. Jacobs. "Economic Analysis of Reducing Phosphorus Losses from Agricultural Production," in *Nitrogen and Phosphorus Food Production Waste and the Environment*, K. S. Porter, Ed. (Ann Arbor, Michigan: Ann Arbor Science Publishers, Inc., 1975).
8. Onishi, H. and E. R. Swanson. "Effect of Nitrate and Sediment Constraints on Economically Optimal Crop Production," *J. Environ. Qual.* 3(3):234-238 (1974).
9. Coote, D. R. "Animal Waste Disposal and Its Impact on Dairy Farms in Two Regions Dominated by Different Soils," Unpublished Ph.D. Thesis, Cornell University (University Microfilms, Ann Arbor, Michigan) (1973).
10. Coote, D. R., D. A. Haith and P. J. Zwerman. "Environmental and Economic Impact of Nutrient Management on the New York Dairy Farm," *Search* 5(5), Cornell University, Ithaca, New York (1975).
11. Coote, D. R., D. A. Haith and P. J. Zwerman. "Modeling the Environmental and Economic Effects of Dairy Waste Management," *Trans. ASAE* 19(2):326-331 (1976).
12. Stewart, B. A., D. A. Woolhiser, W. H. Wischmeier, J. H. Caro and M. H. Frere. "Control of Water Pollution from Cropland, Vol. I," U.S. EPA Report EPA-600/2-75-026a (1975).
13. Ogrosky, H. O. and V. Mockus. "Hydrology of Agricultural Lands," in *Handbook of Applied Hydrology*, V. T. Chow, Ed. (New York: McGraw-Hill, 1964).
14. Stewart, B. A., D. A. Woolhiser, W. H. Wischmeier, J. H. Caro and M. H. Frere. "Control of Water Pollution from Cropland, Vol. II," U.S. EPA Report EPA-600/2-75-026b (1976).
15. Haith, D. A. and J. V. Dougherty. "Nonpoint Source Pollution from Agricultural Runoff," *J. Environ. Eng. Div., ASCE* 102(EE5):1055-1069 (1976).
16. Donigian, A. S., Jr. and N. H. Crawford. "Modeling Pesticides and Nutrients on Agricultural Lands," U.S. EPA Report EPA-600/2-76-043 (1976).
17. Snyder, D. P. "Farm Cost Accounts," Report A.E. Res. 74-13, Dept. of Agr. Econ., Cornell Univ., Ithaca, New York (1974).

APPENDIX–NOTATION

α_{ij} = 1 if soil-crop combination (i,j) is permitted, and 0 otherwise
β_{ij} = 1 if manure can be spread on soil-crop combination (i,j) and 0 otherwise
C_{ij} = Crop/management factor for crop j, soil i
c_{ij} = Cost of cropping soil i with crop j exclusive of nitrogen fertilizer and manure disposal costs ($/ha)
c_m = Cost of manure disposal ($/ton)
c_n = Purchase cost of nitrogen fertilizer ($/kg)
D = Digestible protein requirement to be satisfied from on-farm crops (kg/cow)
d_{ij} = Digestible protein content of crop j grown on soil i (kg/unit of yield)
E = Net energy requirement to be satisfied from on-farm crops (mcal/cow)
e_{ij} = Net energy content of crop j grown on soil i (kg/unit of yield)
ep = Maximum percentage of net energy which may be supplied by pasture

H = Number of cows in the dairy herd
$Hmax$ = Maximum herd size
I = Number of soils
J = Number of crops ($j = J$ corresponds to pasture)
K_i = Erodability of soil i
L_i = Tillable area of soil i (ha)
L_N = Maximum allowable potential nitrogen loss (kg/ha)
L_P = Maximum allowable particulate phosphorus loss (kg/ha)
L_{RN} = Maximum allowable loss of soluble nitrogen in runoff (kg/ha)
L_{RP} = Maximum allowable loss of soluble phosphorus in runoff (kg/ha)
L_S = Maximum allowable erosion (ton/ha)
LS_i = Topographic factor, soil i
M_{ij} = Manure spread on soil-crop X_{ij} (ton)
m = Manure production (ton/cow)
N_{ij} = Fertilizer nitrogen applied to soil-crop X_{ij} (kg)
NR_{ij} = Nitrogen requirement for crop j on soil i in excess of nitrogen supplied by mineralization of soil organic matter (kg/ha)
$n\ell_{ij}$ = Percent of applied nitrogen volatilized
nm = Nitrogen content of manure (kg/ton)
ns_j = Soluble nitrogen concentration in summer runoff from crop j (mg/l)
nw_j = Soluble nitrogen concentration in winter runoff from crop j (mg/l)
P_{ij} = Erosion control practice factor for crop j, soil i
p_i = Phosphorus content of soil i (kg/ton)
ps_j = Soluble phosphorus concentration in summer runoff from crop j (mg/l)
pw_j = Soluble phosphorus concentration in winter runoff from crop j (mg/l)
R = Rainfall erosivity
r = Net return per dairy cow exclusive of costs of farm crops, manure disposal and nitrogen fertilizer ($)
rs_{ij} = Summer runoff from soil-crop X_{ij} (cm)
rw_{ij} = Winter runoff from soil-crop X_{ij} (cm)
s_{ij} = Erosion from soil-crop X_{ij} (ton/ha)
X_{ij} = Area of soil i planted to crop (or rotation) j (ha)
Y_{ij} = Yield of crop j on soil i (units/ha)
Z = Net income of the dairy farm ($/yr)

SECTION V

METHANE GENERATION

24

THE ROLE OF AN ANAEROBIC DIGESTER ON A TYPICAL CENTRAL IOWA FARM

R. J. Smith
Agricultural Engineering Department
Iowa State University
Ames, Iowa

R. L. Fehr
Agricultural Engineering Department
University of Kentucky
Lexington, Kentucky

J. A. Miranowski and E. R. Pidgeon
Agricultural Economics Department
Iowa State University
Ames, Iowa

INTRODUCTION

In the Midwest, livestock and grain production are commonly conducted as joint enterprises. Enhanced awareness of the detrimental effects of uncontrolled manure disposition, coupled with one wet harvesting season when LP gas was in short supply, has focused attention on anaerobic digestion of manure. There are some who hope that controlled anaerobic treatment of manure will solve the joint problems of agricultural pollution and dependence on imported energy. Unfortunately, what is scientifically possible may not prove economically viable. Moreover, there are many pitfalls on the path between the scientist's laboratory and the engineer's full-scale plant. One mandate of the College of Agriculture at Iowa State University is to develop and propagate knowledge that will improve farm productivity and profitability in the state. The study described in this chapter was an attempt to assess not only the engineering aspects of farm-scale anaerobic digestion, but also the economic impact.

Agricultural practices vary widely over the U.S., and this study is limited to conditions existing in central Iowa. Some of the engineering design information has wider application, but the results of the economic modeling should not be taken too far out of context. This study is hypothetical, not experimental; we shall, however, draw on experimental data from a concurrent study with a 100-gal pilot unit being used on beef manure.

OBJECTIVES

1. To establish the characteristics of a typical central Iowa farm.
2. To analyze fossil-fuel and electrical-power use on the farm for each unit process.
3. To design unit processes for handling all aspects of livestock production, residue processing and produced fuel use when anaerobic digestion for methane production was incorporated.
4. To assemble cost data for use in economic-optimization models.
5. To apply linear-programming techniques to the farm operation to predict profitability and to show what flexibility of response the producer will have to fluctuating markets for his grain and animal products.
6. To produce preliminary plans for implementing anaerobic digestion on a typical 320-ac farm.

A TYPICAL CENTRAL IOWA FARM

The characteristics of the representative central Iowa farm were established by using data from the 1974 Iowa Farm Business Summary for Central Iowa.[1] This particular source uses farm-record data from the Iowa Farm Business Associations. Participants in the Association represent above-average farm management, management capabilities that will be required when adopting exotic techniques such as on-farm methane production or excreta-silage feeding.

Typical cropping activities include corn, soybeans, oats and alfalfa hay. The average farm in this region has 312 ac of land. Approximately 90% of the land on an average farm is used for crop production. The soil association is Clarion-Nicollet-Webster and is relatively uniform, except for river-bottom areas.

In 1974, close to half of the farm income emanated from livestock sales. The major livestock activities were beef and hogs, with limited numbers of dairy, poultry and sheep. The average-sized farm in central Iowa farrowed 31 litters of pigs and marketed 73 beef cattle, but 1974 was not necessarily a typical livestock year.

The Iowa Farm Business Association did not gather data on farm equipment and livestock-production facilities. On the basis of our knowledge of the area, plus discussions with extension specialists, we decided that four-row planting and harvesting equipment was predominant. Also, adequate storage capacity was assumed available for crop-production activities. There is a steady trend towards keeping swine in a controlled environment, but most beef production remains in open lots. Because this study looks to the future, we felt that we were entitled to postulate that a 50-sow farrow-to-finish unit using total confinement, and a flush-flume building of 300-head capacity for beef could be expected on a 320-ac farm if meat prices were favorable.

The preceding assumptions relating to livestock-production facilities are not necessarily typical for the average farm size chosen; almost by definition, however, innovators are atypical. Although confinement housing is more likely to be found in larger operations, the unit costs used in the economic model were representative of the size of units postulated. Unit costs were constant in the model because the range of sizes was limited.

MANURE MANAGEMENT IN IOWA

Iowa is subjected to extremes of temperature. Lows may exceed -20°F and highs 100°F. Confinement production of beef and swine can be partly justified by improved animal production; another, perhaps more significant, benefit is an improved working environment for the operator. Consequently, we have limited our study to anaerobic digestion of manure from housed animals. We feel strongly that anaerobic digestion is a sufficiently arcane process already, without complicating it further by attempting to digest the heterogeneous material obtainable from open-lot scrapings. It is recognized that the prerequisite of housing may limit adoption of digestion simply because there is a massive investment increase relative to open-lot animal production. Even with a commitment to housing, anaerobic digestion has to compete with raw-manure storage schemes (above or below ground) or anaerobic lagoons. Currently, beef buildings of small size (*e.g.*, less than 250 head) and most housed-swine operations use deep-pit storage and landspreading with a tank wagon. The labor required for this system is high; consequently, larger operations are tending to install hydraulically cleaned buildings, with anaerobic lagoons as a source of renovated flush water. The digester is more complex than a pit and spreader, and its economic benefits will have to be very obvious before it will be adopted by smaller operators. The anaerobic lagoon does not have a good reputation in Iowa; consequently, it is the operators of large swine and beef facilities who are most interested in an alternative treatment system that is compatible with flushing yet does not have the odor associated with lagoons.

METHANE AS A FUEL

Digester gas is a mixture of methane and carbon dioxide. Technically, though perhaps not economically, the carbon dioxide can be removed by scrubbing, leaving essentially pure methane. We first addressed ourselves to possible methane uses on the farm. Methane can be stored under pressure in a mobile tank and used for vehicles. Further inspection of this approach, however, shows that it is not realistic for current mobile equipment. Few engines on farm equipment are now less than 75 hp, and 50-gal fuel tanks are quite common. The critical temperature of methane is -116°F; hence, liquefaction by compression at ambient temperatures is impossible. We were unable to find a process that would chemically change methane to a liquid fuel and be practical on the farm. We examined solution of methane in various hydrocarbons,[2-4] but the published data showed that, at a given pressure, a given mass of CH_4 took more total volume with its solvent than in isolation. We attempted to estimate the fraction of the gas energy required to compress methane to various pressures if the digester gas powered an engine-generator and if this, in turn, powered an electrical compressor. Table I shows that the penalty is quite severe.

Table I. Percentage of the calorific value of methane used in storing the gas under pressure by using a gas-powered engine, a generator, an electric motor and a compressor.

		Energy	
Final Pressure (lb/in.^{2} abs)	Storage Volume (ft^3/lb at 68°F)	Ideal Compression (Btu/lb)	Lost Overall in Practical System (%)
115	3.7	170	6.5
215	1.7	200	7.6
315	1.1	240	9.1
415	0.8	269	10.2
515	0.65	294	11.2

We consider that a pressure of 515 lb/in.^{2} abs would be quite hazardous on the farm, and even at this pressure the volume would only be reduced to 1/34 of that at atmospheric pressure. A pressure tank would contain about 4,600 Btu/gal, which is a poor energy density compared with diesel fuel at 120,000 Btu/gal.

Thus, we assume that the gas from the digester is best used as generated. Removal of hydrogen sulfide is simple, after which the gas would fuel an internal-combustion engine coupled to a generator. Neyeloff and Gunkel[5] ran a series of tests using a spark-ignition engine on various mixtures of CH_4 and CO_2. Although a 67/33 mixture caused a 40% loss of output in a conventional engine having a compression ratio of 7.5:1, digester gas could be used in engines having up to a 13:1 compression ratio. At 15:1, power output was improved so that only a 20% penalty was observed. That digester gas can be used in high-compression-ratio engines is amply validated by the many converted diesel engines found in municipal wastewater-treatment plants.

In addition to mechanical energy, a very useful quantity of thermal energy can be recovered from the engine. Thermal energy can be recovered from the cylinder/head cooling jacket, the exhaust and the lubricating oil. ASHRAE[6] discussed heat recovery from internal combustion engines. If the coolant were held in the range of 180 to 200°F, ASHRAE indicated that about 30% of the fuel-energy input appeared in the cylinder/head coolant and that about 18% might be recovered from the exhaust. Because Neyeloff and Gunkel[5] determined that only 20% of the fuel energy could be expected as shaft energy, we consider that more than 48%, perhaps 55%, of the fuel energy should be available as usable heat in the engine-cooling water.

The hot coolant from the engine would be used to heat a water-storage tank, and the water from this tank would be available to heat the digester and the swine-confinement buildings. The electrical output would be fed into the farm electrical system. Electrical controls are available, but safety aspects and problems with rate structures would require attention if on-farm, methane-powered generators became commonplace.

THE DIGESTER

General

The literature on anaerobic digestion of human waste is vast. Although the biology and biochemistry of digestion are well known, digesters are still regarded as the most temperamental devices in a waste-treatment plant. Biological fundamentals gleaned from municipal practice undoubtedly apply to livestock wastes, but some of the physical design criteria may not. The number of full-sized anaerobic digesters built to handle livestock manure, at least those which have been reported in the scientific literature, is very small. It is our contention that digestion of animal manures presents some problems and anomalies not encountered when digesting human wastes.

Mixing

Manure from ruminants contains long hair, whole corn kernels and large (1 to 2 in.) plant fragments. Work by Hein *et al.*[7] showed that gas-lift mixing was unsatisfactory for beef wastes. Yet, a recent communication from Missouri[8] indicates that the full-scale unit there, working on swine manure, is successfully using intermittent gas mixing. We believe that each species may require a different amount of mixing energy and that mixing energy may be a much larger fraction of the energy available from the gas than is true for municipal waste.

There are few references to mixing energy for digesters in the literature. There seem to be two extreme criteria for mixing; one is prevention of gross sludge or scum formation, and the other is continual exposure of bacteria to new substrate and removal of metabolic by-products. In practice, working municipal digesters usually aim for the first criterion, but microbiologists aim for the second. The range of values that can be found is large. Morgan[9] reported that a 150-ft^3 unit handling municipal sludge worked very well when gas was recirculated at 7.5 to 15 ft^3/min/1000 ft^3. Hein *et al.*[7] found that 15 ft^3/min/1000 ft^3 was essential for beef manure, that 22.5 ft^3/min/1000 ft^3 was better, and that 5 ft^3/min/1000 ft^3 led to stratification and failure. Converse *et al.*[10] successfully used 3-min bursts of 150 ft^3/min/1000 ft^3 every 30 min when digesting dairy manure. Pfeffer[11] has examined the digestion of ground municipal solid waste and indicated that 0.19 hp/1000 ft^3 would suffice for mixing a 4% slurry. If we take 7.5 ft^3/min/1000 ft^3 and assume that the gas compressor works against a total head of 30 ft of H_2O and has a thermodynamic efficiency of 70%, then the mixing power would be 0.46 hp/1000 ft^3.

An alternative to gas-lift mixing is use of a propeller or turbine. It is perhaps significant that two of the more successful livestock-waste digestion studies reported recently have chosen this method. Lapp *et al.*[12] employed mixing paddles in their 80-ft^3 digester, which was handling swine wastes. Unfortunately, no energy-input information was given. Data from Robertson *et al.*[13] showed that 4.5 hp/1000 ft^3 was required to mix the 480-ft^3 digester used in their swine-waste study.

A further factor is size; Malina and Miholits[14] present a table of gas-recirculation data for full-scale municipal units. A 20-ft diam digester required 8.8 ft^3/min/1000 ft^3, but a 110-ft-diam unit required only 1 ft^3/min/1000 ft^3; the corresponding power figures were 0.73 hp/1000 ft^3 and 0.07 hp/1000 ft^3. It seems reasonable that size should be a factor. A given particle has a unique settling velocity. As the digester size increases, the size of eddy turbulence increases, and less power per unit-volume is required to achieve an eddy velocity equal to the settling velocity of the particle.

We conclude that livestock manure from ruminants, if digested without any size reduction, will require more power per unit-volume than will swine manure, and swine manure will require more power than municipal sludge. There will probably be a reduction in power requirement as digester size increases, but present information is totally inadequate to predict what this relation may be. We believe that the relatively small unit proposed for this farm should be provided with turbine mixing and that this should be overpowered but run intermittently. Because the criterion for power is so vague, "overpowered" has doubtful meaning, but some estimate of power for turbine mixing can be obtained from Zweitering.[15] A numerical value is given in the section on sizing the digester.

Gas Production

The biochemistry of methane production has been well handled in other references;[16-19] however, the potential energy output is not often derived explicitly. Conventionally, digester loading has been expressed in units of mass of volatile solids per unit volume of digester per unit time. Experience with municipal digesters that are mixed and held at 95°F shows that loading rates from 0.1 to 0.4 lb VS/ft^3/day have been used successfully.[20] Loading rates for livestock wastes have been reviewed by Miner and Smith[21] and by Shadduck and Moore;[22] the range of values was similar to municipal rates. Our current work with beef manure indicates that 0.4 lb VS/ft^3/day seems possible but is unstable; we can achieve longer experimental periods with loading rates of 0.3 lb VS/ft^3/day. Using volatile solids as a loading criterion is perhaps an unfortunate choice. Because the process is totally anaerobic, the COD of the influent must equal the COD of the effluent and the methane. Information on COD reduction in a livestock-waste digester is scarce. The review by Miner and Smith[21] showed values from 35 to 70%, depending on species. It seems probable that manure from animals that are not fed roughage will contain a greater fraction of digestible matter. Our pilot-plant studies over 3 years with beef manure indicate that, at detention times from 8 to 12 days, only 45% of the COD is converted to methane. The stoichiometric relations for oxidation of methane yield the relation that 1 lb of COD is equivalent to 6 ft^3 of CH_4 at 68°F, or 6000 Btu when expressed as energy. Consequently, all this information can be put together to show that a typical livestock-waste digester fed primarily beef manure will produce

$$0.3 \times 1000 \times 1.1 \times 0.45 \times 6000 = 891{,}000 \text{ Btu/day/1000 ft}^3$$
$$= 14.6 \text{ hp/1000 ft}^3$$

(the 1.1 factor converts VS to COD)

where the Btu/day to hp conversion is purely numerical and does not account for the thermodynamic or mechanical efficiency of any conversion equipment. Figure 1 shows the energy partition that we consider reasonable for an anaerobic digester fed predominantly ruminant wastes. An overall conversion efficiency of 16% between the energy in the gas and the electrical energy, which could be delivered at the terminals of an engine/generator, is postulated. We can now estimate the amount of mixing energy that would be available. If we accept that up to 50% of the generator output could be diverted to the mixer motor (efficiency 90%), then the mixer should not consume more than 1.1 hp/1000 ft^3. This value is less than any of the published

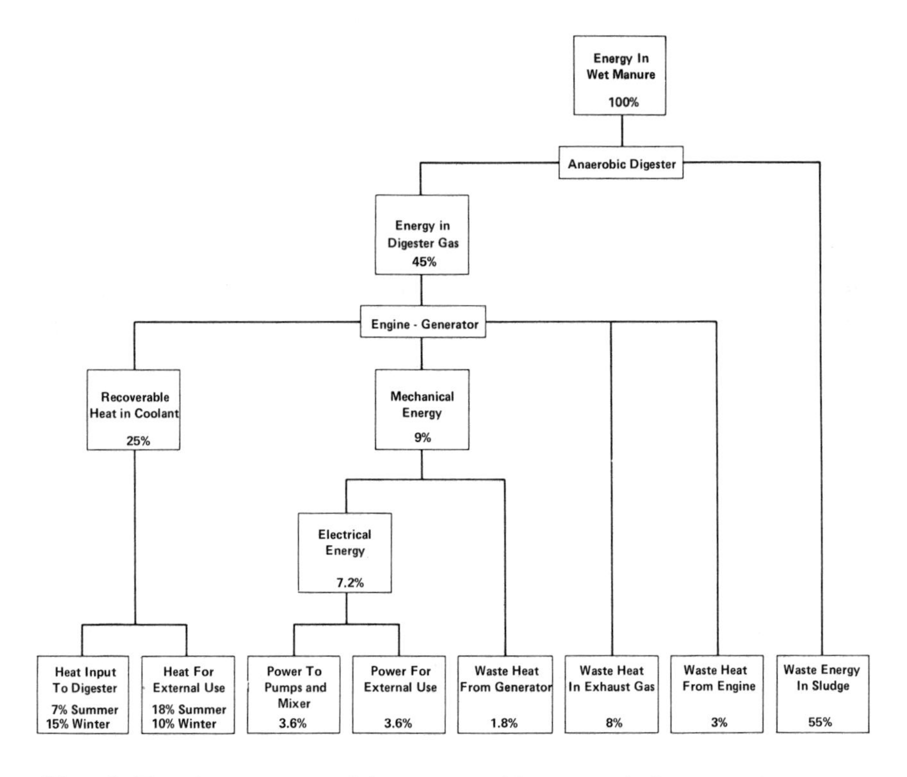

Figure 1. Tentative assessment of the energy partition expected when recovering energy from livestock wastes using anaerobic digestion, based on the energy contained in the manure dry matter.

figures of which we are aware for mixing pilot-scale digesters fed livestock manure. We believe that mixing is necessary for a livestock digester, but our calculations show clearly that the amount of energy available for this purpose is not large; consequently, we feel that mixer design is of paramount importance.

Digester Size for a 320-Acre Farm

The farm would have a cow-calf herd, managed in an open lot and on pasture; no manure from this herd would be available for the digester. Current economics preclude cow-calf production in housed confinement. The calves from this herd would be fattened in a flush-flume beef building of 300-head capacity. Other cattle would be brought in to keep the building full at all times. The 50-sow farrow-to-finish operation would practice year-round farrowing. Table II shows the expected manure production available to the digester. These values are taken from Miner and Smith.[21] Although

Table II. Average daily manure production of the housed livestock.

Month	Cattle (lb/day)	Swine (lb/day)	Total (lb/day)
January	13800	2700	16500
February	13800	2700	16500
March	13100	2700	15800
April	11100	2700	13800
May	12700	2700	15400
June	13700	2700	16400
July	14700	2700	17400
August	15800	2700	18500
September	12600	2700	15300
October	14100	2700	16800
November	13800	2700	16500
December	12500	2700	15200
Average	13475	2700	16175

the idea is not universally accepted, we believe that ammonia toxicity will limit the solids concentration that may be used. Our own experience has been that digestion deteriorates rapidly in NH_3-N $>$ 1500 mg/l; consequently, we propose diluting the manure with an equal volume of fresh water to give an influent slurry solids content of 5.6%. If the detention time is 10 days, then the data presented in Table II require the liquid volume of the digester to be 5200 ft^3. The resulting loading rate is 0.29 lb VS/day/ft^3. The digester will be 18 ft in diameter and will have a liquid depth of 20.5 ft. We used the formulae developed by Zweitering[15] to estimate the power required by a turbine mixer if the largest particle were 0.08 in. and had a specific gravity of 1.5. It is recognized that this particle size is less than that of whole corn kernels and silage fragments. Zweitering's formulae predict that the power required will be 2.2 hp, or 0.42 hp/1000 ft^3. We propose that 5 or 7.5 hp be installed and run intermittently.

Energy Budget

The digester is insulated with 12 in. of polyurethane foam. We have estimated the heat required to maintain the digester at 95°F and the heat required to raise the influent to 95°F. Table III shows that the heat required for slurry heating far exceeds the heat for maintenance; consequently, the 12 in. of foam could probably be reduced to 6 in.

The monthly amounts of thermal and electrical energy that we calculate would be available from the system are shown in Table IV. We estimate that a 25-hp engine could use all the gas produced at the heaviest manure-loading

Table III. Monthly heat supply required to maintain a digester at 95°F, based on inflow rate of 520 ft^3/day.

	Temperature		Heat		
Month	Outside Air (°F)	Influent[a] (°F)	To Influent (10^6 Btu)	Through Boundaries (10^6 Btu)	Total Required (10^6 Btu)
January	19	40	53.3	2.91	56.2
February	24	40	48.2	2.46	50.7
March	34	40	53.3	2.34	55.6
April	49	44	47.8	1.71	49.5
May	61	56	37.8	1.3	39.1
June	70	65	28.1	0.93	29
July	74	69	25.2	0.81	26
August	72	67	27.1	0.88	28
September	63	58	34.7	1.19	35.9
October	54	49	44.6	1.57	46.2
November	37	40	51.6	2.15	53.8
December	24	40	53.3	2.72	56

[a]The temperatures are based on local experience, 40°F at air temperatures less than 45°F and air temperature minus 5°F at higher air temperatures.

Table IV. Net energy yields on a monthly basis available from an internal-combustion engine-generator powered by digester gas.

	Thermal			Electrical		
Month	Gross (10^6 Btu)	Internal (10^6 Btu)	Net (10^6 Btu)	Gross (kWh)	Internal (kWh)	Net (kWh)
January	81.6	56.2	25.4	6890	4050	2840
February	73.7	50.7	23	6220	3650	2570
March	78	55.6	22.4	6580	4050	2530
April	65.6	49.5	16.1	5540	3920	1620
May	76	39.1	36.9	6410	4050	3360
June	78.5	29	49.5	6620	3920	2700
July	86.2	26	60.2	7270	4050	3220
August	91.8	28	63.8	7750	4050	3700
September	73	35.9	37.1	6160	3920	2240
October	83	46.2	36.8	7010	4050	2960
November	79	53.8	25.2	6660	3920	2740
December	75	56	19	6320	4050	2270

rate. The results in Table IV assume that the generator output always can be absorbed by the farm and that fluctuations in gas supply also can be managed by the engine-generator system and its controls.

ENERGY USE ON THE FARM

General

There are three forms in which energy, other than feed, is used on the farm: (a) fuel for mobile equipment, (b) thermal energy for temperature control and (c) electrical energy for motors. We have expressed energy use in these categories because they are largely independent of each other, and we have used units of gallons of diesel fuel, 10^6 Btu and kWh for the three forms. No attempt was made to assess device-use efficiency or to express electrical energy in the form of primary-fuel use. Inasmuch as we are concerned only with on-farm energy substitution, we have not attempted to look at off-farm energy used in manufacturing equipment. As explained earlier, we see the digester producing thermal energy in the form of hot water at about 160°F and electrical energy. Because, in our opinion, methane is not suitable for mobile use, we have not presented information on energy for crop production. Diesel appears in these energy calculations where the power is provided by a tractor PTO. Two examples are loading a tower silo and powering an irrigation pump for a big gun.

Corn Drying

There are two main types of corn drying; either hot air is used for a short time or warm air is used for several days. We have not analyzed high-temperature drying because it does not lend itself to the forms of energy substitution available from the digester. We are assuming that a future mixed farm will wish to minimize energy consumption; hence, we postulate maximum use of high-moisture corn storage and use of low-temperature drying. Although low-temperature drying has tended to use electric heating, we do not feel that this is necessary because any form of thermal energy could be used if the equipment were available. We do recognize that low-temperature drying cannot be used every year. Local opinion[21] indicates that corn cannot be more than about 22% moisture content (m.c.) for low-temperature drying. The farm will dry about 12,000 bu/yr. Our estimate of the energy requirements are presented in Table V.

Feed Processing

High-moisture corn will be used for finishing both the beef and swine, but the feeder pigs and breeding stock will be fed 14 to 15% m.c. dried corn. This choice was made because high-moisture corn spoils readily if left in feeders too long. Daily transport of feed is realistic for the finishing animals, but not for the other parts of the herd. We tried to assess conveying and

Table V. Electrical and thermal energy required for low-temperature corn drying (22 to 14% moisture content) of 12,000 bu.[a]

Month	Number of Drying Days	Fan Power (hp)	Electrical Consumption (kWh)	Thermal Consumption (10^6 Btu)
October	12	2 x 5	2880	45.2
November	13	2 x 5	3120	48.9

[a]Total energy input of 0.35 kWh/bu for each point of moisture removed. 1 hp requires kW input. Thermal component = Total - Electrical (1 kWh = 3410 Btu).

grinding energies involved in feed processing. The results for the swine herd are included in Table VI and, for the beef herd, in Table VII.

Table VI. Total monthly energy demand for feeding a 50-sow herd to market at 220 lb using 25%-moisture corn for the finishing animals and 15% for the remainder. Results exclude drying.

Month	Diesel (gal)	Electricity (kWh)
January	-	94.9
February	-	85.8
March	-	94.9
April	-	91.8
May	-	94.9
June	-	91.8
July	-	94.9
August	-	94.9
September	-	91.8
October	-	124.1
November	-	91.8
December	-	94.9

Environmental Control for Pork Production

It is generally accepted that the performance of a pig is optimum over a rather narrow range of temperature.[24] At present, swine buildings are heated and ventilated, but air conditioning during the summer is not justified economically. Breeding stock benefit from air conditioning, and it is becoming more common to use zone air conditioning for sows during farrowing. Predicting the energy that will be used by a housed swine herd is currently an active research topic.[25-28] Energy-use predictions are complicated by divergence between the psychrometic approach to ventilation (that is, ventilation

Table VII. Total monthly energy demand for feeding beef animals[a] to market at 1100 lb, using 25%-moisture corn and 65%-moisture corn silage.

Month	Corn[b] (lb)	Stover[b] Silage (lb)	Silo Loading (gal)	Silo Unloading and Transport (kWh)
January	126900	142800	-	240
February	120100	145000	-	241
March	135600	168400	-	279
April	131200	163000	-	270
May	135600	168400	-	279
June	131200	163000	-	270
July	135600	168400	-	279
August	135600	168400	-	279
September	131200	163000	-	270
October	136700	170200	171	281
November	127300	150600	170	251
December	126900	142800	-	240

[a]Annual production: 3 x 90 head from 450 to 1100 lb + 90 head from 650 to 1100 lb + 90 head from 800 to 1100 lb.

[b]Wet weight.

is used to control relative humidity) and the practical approach in which thermostats control ventilation. To simplify the calculations, we assumed that improved control availability will soon allow fan control from humidistats. Thus, the results presented in Table VIII are based on moisture control during cold weather. In hot weather, accepted maximum ventilation rates were used.[29] The efficiency of fans varies quite widely; a review of several manufacturers' data suggest that 7500 ft^3/min/hp is a reasonable figure. Each building was well insulated (R = 15 h ft^2 °F/Btu in the walls and 25 h ft^2 °F/Btu in the ceiling). Limited air conditioning for sows during farrowing was specified. The values in Table VIII probably are a little optimistic at present because many producers will overventilate buildings during cold weather.

Farmhouse

We have made no attempt to obtain energy-use data for real farmhouses because there are too many older Iowa farmhouses built before modern insulation materials were available. Today's farmer lives in a house of substantially the same design as those found in the suburbs. Hittman Associates, Inc.[30] performed a very detailed study of energy use in a typical suburban house. We took values from this study and made some adjustment for climatic conditions and greater clothes-washer use. The total energy demand of a new Iowa farmhouse is shown in Table IX.

Table VIII. Monthly electrical and heating loads for a life-cycle swine unit producing about 750 finished pigs/yr at 220 lb.

Month	Outside Temperature (°F)	Electrical Energy (kWh)	Thermal Energy (10^6 Btu)
January	19	267	26.8
February	24	268	21.5
March	34	428	21.9
April	49	1404	8.3
May	61	2919	3.7
June	70	3381	0
July	74	3940	0
August	72	3493	0
September	63	2831	3.7
October	54	1506	6.9
November	37	491	21.8
December	24	299	24.2

Table IX. Total monthly energy demand of a modern Iowa farmhouse.[a]

Month	Thermal (10^6 Btu)	Electrical (kWh)
January	18	717
February	13.1	717
March	13.4	717
April	7.1	717
May	5.9	885
June	5.7	1287
July	2.6	1546
August	2.6	1373
September	3.2	1238
October	6.6	717
November	12.6	717
December	17.4	717

[a]This figure refers to output performance and does not account for fuel use inefficiencies.

Land Application of Liquid Residues

Although recycling manure or digested manure directly through animals by making a silage with crop residues has received considerable attention recently, it seems unlikely that any manure-handling scheme will have absolutely no effluent. We took a conservative approach and assumed that all the digester effluent must be returned to the land. Because the manure has been diluted and digested, it can be applied to the land by using a travelling big gun. In our energy calculations, we have postulated that the digested effluent would be held in an open-earth pond. Under such storage conditions, about 75% of the nitrogen would be lost through volatilization.[31] In Iowa, livestock producers may only apply 250 lb N/ac annually on a regular basis. Such an application rate makes poor use of P and K but is typical practice in Iowa. Making all these assumptions showed that 35 ac were required and that 1.5 in. would be applied annually. Using manufacturers' information for rain guns indicated that 219 gal of fuel would be required (PTO irrigation pump). Of this amount, 2/5 was budgeted to April and 3/5 to October.

THE DIGESTER SYSTEM FOR A 320-ACRE CENTRAL-IOWA FARM

Digester Design

If, as we firmly believe, mixing is necessary for reliable operation, then the shape of the digester becomes most important. This is not the forum for examining complete-mix vs plug-flow concepts; suffice it to say that our experience with deep pits and plugged sewer lines makes us very suspicious of any design that claims beef manure will flow down a tube by gravity. Municipal designs generally are in the form of a cylinder, with the axis vertical. This seems a very logical design because circular walls are inherently stronger than plane ones. Although slip-formed concrete silos are now available for on-farm construction, we decided to look at a grain bin as a liquid container. We did not choose a concrete silo because the investigators at the University of Missouri had successfully constructed such a unit,[32] and we did not wish to duplicate this effort. A well-known manufacturer makes a glass-lined manure-storage tank, but we did not wish to become committed to one manufacturer's product in our preliminary planning. Another reason for not wishing to pick a proprietary tank was that we wished to be free to select the height-to-diameter ratio that would best suit us.

If the hoop stress on a cylindrical tank holding a given volume of liquid is examined, it is found that the larger the ratio of diameter to height, the smaller is the stress. Unfortunately, such geometry runs counter to mixing requirements. Zwietering[15] indicated that the power required to suspend a given particle increases with diameter but that height has little effect. We derived a simple relation between hoop stress and diameter-to-height ratio and concluded that an acceptable compromise ratio was 0.88. Because the digester was to hold 5200 ft^3, the diameter required is 18 ft, and the depth is 20.5 ft. We have examined the stresses on the bolts holding the steel sheets together, and a viable design is possible with 10-gauge sheet for the bottom 5 ft.

Municipal digesters use floating covers because the variable-volume gas space allows gas storage and lessens the possibility of negative pressures being developed within the digester. Although the safety features of a floating cover are fully recognized, we decided that a fixed cover would cost less. Short-term gas storage in the form of a rubber bag could be arranged externally and would provide protection against negative pressures. There obviously are many ways of making a cover; we have chosen to use a stressed-skin wooden panel. The space between the stringers would be filled with insulation. After construction of the cover onsite, the bottom face would be fiberglassed to reduce gas permeability. The cover would be held to the steel walls by an angle iron rolled to fit the bin. We hope that gas tightness of the whole structure can be achieved by using caulking compounds in all seams and an internal coating of bituminous epoxy after assembly.

Although municipal digesters have moved away from internal heating coils, we think that internal coils deserve reexamination in light of our experiences with plastic pipe and livestock manures. Plastic pipe resists biological and chemical attack more readily than does steel pipe, so we believe that an internal, pipe-coil heat exchanger warrants further investigation. Work by Abdalla[33] showed that the heat transfer coefficients for plastic pipes surrounded by water are not greatly inferior to those for steel. We conclude that the thermal resistance is largely a function of the two film coefficients. Figures 2 and 3 show a cross section of the digester and certain fastening details. Table X shows the estimated materials cost.

The digester would be loaded by a pipe entering through the floor and would discharge through the vented siphon shown in Figure 4. The drain pipe at the center of the floor (Figure 2) would be used for intermittent solids removal as necessary. We are not completely sure if the scum breaker is entirely necessary. Its omission would allow us to use a shaft seal, consisting of a sheath dipping into the liquid, instead of a graphite packing as shown in Figure 2.

Manure Handling

Iowa State University agricultural engineers are rather firmly committed to hydraulic manure handling, and our design reflects this. If dilution water is needed, such handling fits well into the design. The flow diagram in Figure 5 shows the components. The settling/holding tank would hold 12 hr of manure production in addition to an equal amount of dilution water. The supernatant from this tank would be used to flush the manure from the beef and swine buildings. Pump B would deliver the contents of the settling/holding tank to the digester once every 12 hr. We find that 12 hr is the minimum interval that we can use and still have an adequate volume of supernatant for flushing all the buildings. We have shown a covered earth

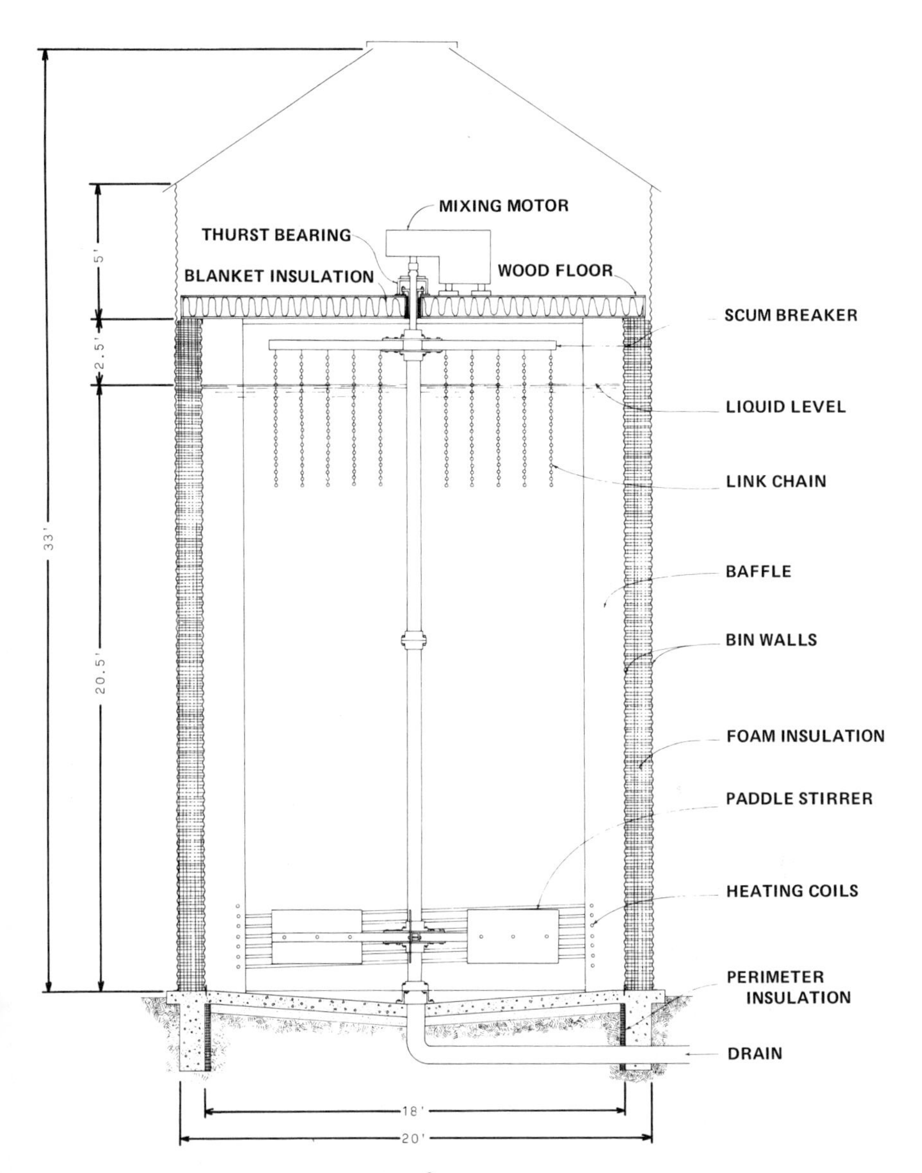

Figure 2. Cross section of a 5200-ft^3 pm-farm digester constructed from corrugated steel sheets (grain-bin material).

storage pit for the supernatant. The cover would be a plastic membrane. Although such a cover is desirable to contain any residual odor, to trap any extra gas produced and to control ammonia volatilization, its economic viability is very doubtful at present.

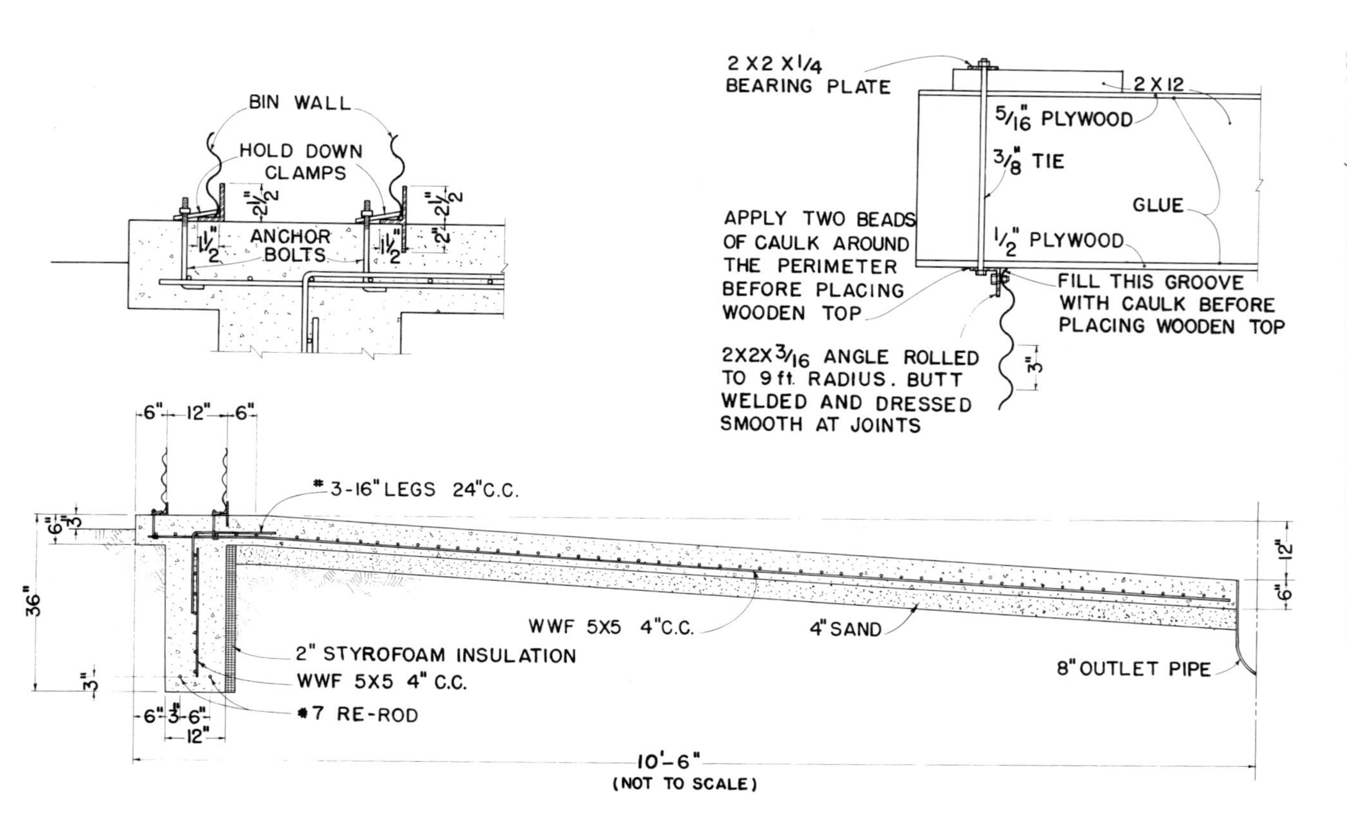

Figure 3. Fastening details for an on-farm anaerobic digester constructed from corrugated-steel sheets.

Table X. Estimated cost of the component parts of a 5200-ft^3 digester using two concentric steel shells and insulated with polyurethane foam.

Item	Cost ($)
Inner Bin Wall	3060
Wood Top	745
Interior Coating	340
Heating Coil	410
Baffle	130
Plumbing	1000
Paddle Stirrer	750
Stirrer Motor and Gear Reducer	850
Outer Bin Wall	2065
Roof	530
Concrete	950
Insulation	1005
Total	11,835

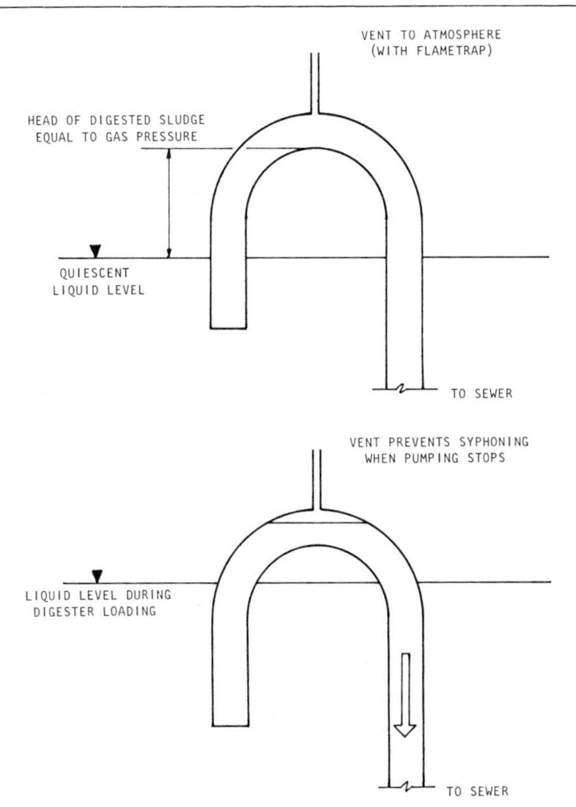

Figure 4. Outlet device allowing automatic discharge of digester contents during loading period.

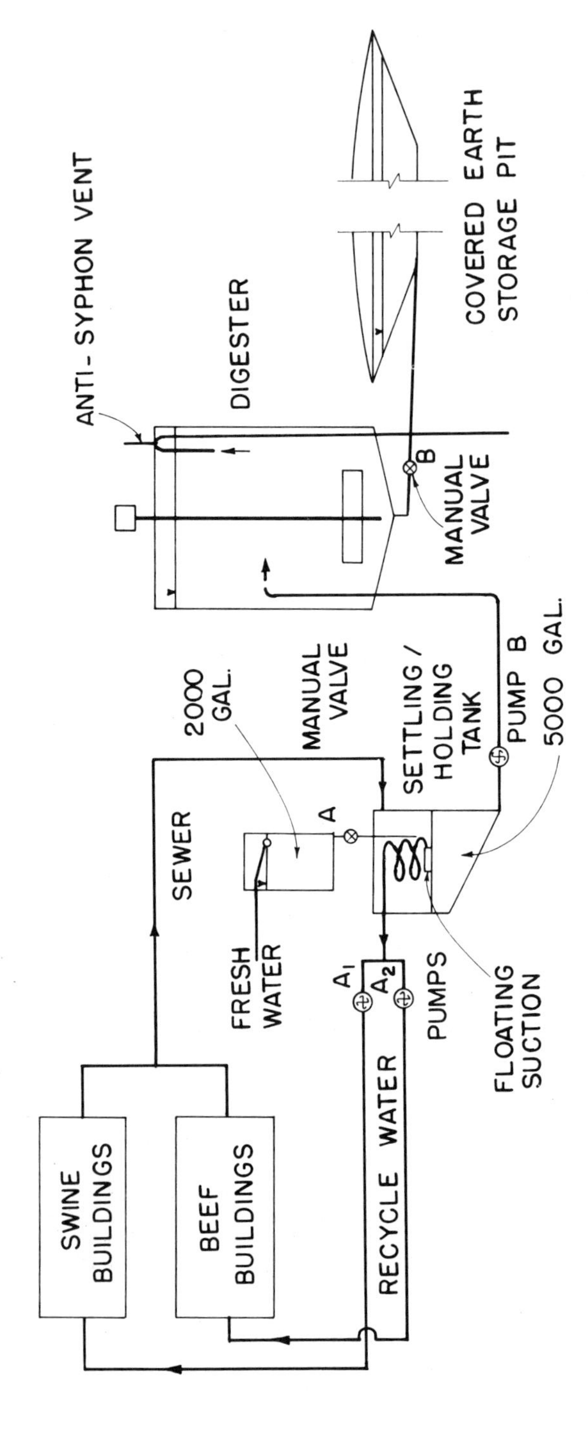

Figure 5. Flow diagram for manure processing on a 320-acre Iowa farm that will use anaerobic digestion for energy recovery and pollution control.

ECONOMIC ANALYSIS

Introduction

The economic analysis was not concerned solely with the role of the digester. We also analyzed the use of livestock wastes and crop residues as feed resources for beef cattle. Energy recovery as methane and energy recovery through refeeding are seen as complementary strategies, both possibly leading to reduced energy use for the same food production.

To execute the economic feasibility analysis, a linear-programming model was used to select the optimum plan for our representative energy-self-reliant farm. Given a set of alternative activities with varying levels of operation, a specific numerical objective function and limited resource inputs (*e.g.,* land, buildings, labor and capital), this computational technique optimizes the activity mix. Although a number of different objective functions may be considered, this chapter will only discuss maximizing profits for the farm firm, subject to resource and energy restraints.

Another feature of linear programming that is used in this analysis is parametric variation of energy prices (including nitrogen fertilizer price), while holding the remaining assumptions unchanged. As energy prices rise, the new energy-saving technologies may enter the optimal activity mix. These alternative solutions allow us to ascertain the impact of present and future energy scenarios on the feasibility of methane digestion and excreta feeding.

There are a number of basic assumptions associated with linear programming:

1. The objective function must be linear.
2. Resources and activities must be divisible and finite in number.
3. Linear relationships exist between activities and resources; a proportionality exists between activity levels and resource use.
4. The decision variables must be nonnegative.
5. The model is deterministic in the sense that all values are assumed known with certainty.

Resource, Price and Cost Characteristics

The characteristics chosen for a typical central Iowa farm (Table XI) were average for participating farms in the Iowa Farm Business Association Central Region.[1] The initial model solutions are based upon the input costs and output prices reported in Table XII. These prices and input costs were average for central Iowa during the first half of 1976. The estimates come from extension publications and personnel, farm supply firms, and market-reporting services. The initial prices employed should be viewed critically because optimal programming solutions may be sensitive to these assumptions.

Table XI. Characteristics of a representative 320-acre central Iowa farm.

250 ac	Nicollet-Webster soil
40 ac	Clarion soil
30 ac	Farmstead
25 sows	Farrow-to-finish confinement system
300 steers	Beef-confinement feeding system
Farmer plus full-time hired man	Labor
Machinery	Four-row planting and harvesting

Table XII. Output prices and input costs assumed for initial solution of linear programming model.

Item	Value ($/unit)
Output Prices	
Corn–Fall (24% moisture)	2.04/bu
Corn–Spring (12% moisture)	2.34/bu
Soybeans–Fall	4.40/bu
Soybeans–Spring	4.60/bu
Oats–Fall	1.43/bu
Alfalfa Hay	40/ton
Feeder Pigs	46/pig
Market Hogs	48/cwt
Market Steers	45/cwt
Input Costs	
Anhydrous Ammonia	0.122/lb
Diesel Fuel	0.375/gal
Electricity	0.04/kWh
LP Gas	9.33/gal
Feeder Pigs	48/pig
Steer Calves	44/cwt
Yearling Steers	43/cwt
Alfalfa Hay	45/ton
Corn–Spring	2.59/bu
Oats–Spring	1.45/bu

Empirical Solutions at Four Levels of Energy Price

Multiplicative price increases were applied to diesel fuel, LP gas and electricity, but the cost of anhydrous ammonia was increased by only 25% of the direct energy-price increase. The profit-maximizing solutions at the various energy prices are presented in Table XIII. Even though energy-price

Table XIII. Optimal solutions associated with alternative energy-price levels, assuming no land rental.

Item	Level of Activity			
Energy Price Multiplier	0	2	5	10
Corn-Soybeans Low Fertilization (ac)	–	–	–	67
Continuous Corn High Fertilization (ac)	250	250	–	–
Corn-Corn-Soybeans High Fertilization (ac)	–	–	250	183
Corn-Oats-Meadow-Meadow High Fertilization (ac)	–	–	14	40
Corn-Corn-Oats-Meadow High Fertilization (ac)	40	40	26	–
Hog Farrowings (litters)	100	100	100	100
Finishing Market Hogs (head)	700	700	700	700
Feeder Calves Fed Excreta Period 11 (head)	7	7	9	27
Feeder Calves Fed Excreta Period 1 (head)	–	–	42	154
Feeder Calves Fed Corn Grain Exclusively (head)	156	156	249	120
Feeder Yearlings Fed Excreta Period 11 (head)	137	137	–	–
Cow/Calf Fed Excreta (units)	50	50	94	–
Methane Digester (operation level)	–	–	–	68%
Methane Digester (income penalty $)	5,549	4,495	1,211	–
Net Returns ($)	57,512	49,954	31,014	10,059
10^6 Btu	2,228	2,228	1,500	1,230

rises will effect other price adjustments in the economic system, nonenergy prices were assumed constant for simplicity in this analysis. In this particular set of solutions, the farmer's objective is assumed to be profit maximization. The model will select methane digestion and excreta-silage feeding if these activities add to farm profitability through lower costs or greater returns.

Alternatively, these activities may be forced into the solution to determine the income penalty associated with their inclusion.

Given current energy prices, the methane digester does not enter the optimal activity mix. If the methane digester is forced into the solution operating at capacity, it will reduce net returns by $5549. Thus, the current energy-price situation does not encourage the adoption of this technology. Rising energy prices are expected to alter the optimum mix of activities and to improve the profitability of energy-saving technologies. The solutions included in Table XIII are for energy prices 2, 5 and 10 times current prices. Interestingly, a doubling of energy prices has no impact on the optimal mix of activities, although net returns do drop. Such insensitivity is easily explained by the relatively small proportion of total production costs accounted for by energy outlays.[34]

Extreme changes in energy prices lead us to the conclusion that the assumption of fixed prices for supplies and resources other than energy and nitrogen may be less than realistic. At a tenfold energy price increase, the output of the farm will drop; if such a drop in production were common to a large number of producers, then prices for the farm produce would rise.

The methane digester does not enter the activity mix of the optimal solution until a tenfold price rise occurs. Even at this price level, the digester operates at only 68% of capacity. The economic analysis assumes that the digester is divisible and that the operational level or scale is a percentage of the 5200-ft^3 unit described previously. This assumption is equivalent to assuming a constant average-cost function with no economies-of-size involved. Presently, insufficient data are available to determine if economies-of-size do exist for the proposed on-farm design. Constant average costs do not seem unrealistic within the restricted range of sizes considered here. The economic feasibility of on-farm methane digestion, employing the given technological assumptions, is questionable unless highly significant energy-cost increases are anticipated. Improvements in digester technology, unless dramatic, do not seem likely to affect the role of a digester in the short run. In the long run, however, severe energy-price rises combined with other factors might make adoption of digestion technology economically viable.

CONCLUSIONS

Energy Substitution

Table XIV summarized our estimates of energy use in the three categories: diesel fuel, heat and electricity. The diesel fuel is shown for information only because, as discussed earlier, we do not see methane being used as a

Table XIV. Summary of the monthly energy needs for a 320-acre, central Iowa farm, and the amount of energy that could be supplied by an engine generator fueled by digester gas.

	Input															Output	
				Feed Processing[c]				Swine Environmental Control								Digester Net Production	
	Corn[a]	Drying[b]		Swine		Beef				Farmhouse		Effluent[d]	Total				
Month	(gal[e])	(kWh[f])	(10^6 Btu[g])	(gal)	(kWh)	(gal)	(kWh)	(kWh)	(10^6 Btu)	(kWh)	(10^6 Btu)	(gal)	(gal)	(kWh)	(10^6 Btu)	(kWh)	(10^6 Btu)
January	–	–	–	–	94.9	–	240	267	26.8	717	18	–	–	1319	44.8	2840	25.4
February	–	–	–	–	85.8	–	241	268	21.5	717	13.1	–	–	1312	34.6	2570	23
March	–	–	–	–	94.9	–	279	428	21.9	717	13.4	–	–	1519	35.3	2530	22.4
April	–	–	–	–	91.8	–	270	1404	8.3	717	7.1	88	88	2483	15.4	1620	16.1
May	202	–	–	–	94.9	–	279	2919	3.7	885	5.9	–	202	4178	9.6	3360	36.9
June	–	–	–	–	91.8	–	270	3381	–	1287	5.7	–	–	5030	5.7	2700	49.5
July	–	–	–	–	94.9	–	279	3940	–	1546	2.6	–	–	5860	2.6	3220	60.2
August	–	–	–	–	94.9	–	279	3493	–	1373	2.6	–	–	5240	2.6	3700	63.8
September	–	–	–	–	91.8	–	270	2831	3.7	1238	3.2	–	–	4431	6.9	2240	37.1
October	910	2880	45.2	35	124.1	171	281	1506	6.9	717	6.6	131	1247	5508	58.7	2960	36.8
November	910	3120	48.9	–	91.8	170	251	491	21.8	717	12.6	–	1080	4671	83.3	2740	25.2
December	–	–	–	–	94.9	–	240	299	24.2	717	17.4	–	–	1351	41.6	2270	19
Total	2022	6000	94.1	35	1146.5	341	3179	21227	138.8	11348	108.2	219	2617	42900	341.1	32750	415.4

[a]Fuel for field work, harvesting, and transport only, 260 ac.
[b]Low temperature corn drying for 1200 bu, 22% to 14%.
[c]Diesel fuel is used to load tower silos with silage and high moisture corn.
[d]Big gun irrigation of liquid effluent from waste-handling system.
[e]Gallons of diesel fuel for an internal combustion engine.
[f]Electrical energy at the terminals of an electric motor or generator.
[g]Thermal energy input does not include an allowance for device efficiency.

mobile fuel in the short run. Figure 6 shows clearly that the demand for heat energy is about 6 months out of phase with the heat available from the engine-generator.

Much of the energy required by fixed equipment on the farmstead can be satisfied by the output of the engine-generator. Although this could supply 76% of the farmstead electrical demand on an annual basis, excess energy is available in the winter and too little in the summer. The winter excess could easily be absorbed by providing dual heating sources so that electrical heat could supplement the hot water available from the engine-cooling system.

The electrical-energy output of the digester could also be increased by supplementing the manure with a slurry made from finely ground corn stover and water. Although there is no conceptual reason why this approach should not be successful, we could not find enough information to allow us to estimate the fineness of grind required or the energy for such grinding. With present technology, it would seem that more electrical energy could be recovered from corn stover by direct burning in a large power plant than from anaerobic digestion.[33] The digestion alternative should not be ignored, but field-scale tests of stover grinding and digestion are essential before an intelligent assessment of this energy-recovery technology can be made.

The amount of hot water available from the digester-engine system could be increased quite dramatically. It was shown in Table III that the amount of heat for heating the cold influent slurry far exceeds the heat for maintaining the unit at 95°F. A heat exchanger is needed that would allow transfer of heat from the effluent to the cold influent. Unfortunately, we are not aware of any low-cost heat exchangers that could handle such heterogeneous materials as sludge and digested effluent without major fouling.

Other components of on-farm digestion, alluded to only superficially in this report, are the control systems required to ensure automatic operation. Designing systems that switch between three sources of heat—that is, between oil (or gas), hot water and electricity—obviously is possible, but their cost was not included in the economic model because the design of such controls will be a major project. Another area requiring special attention will be control of the engine-generator as the gas supply varies and as the farm electrical load varies.

Economic Analysis

The linear programming solutions for the representative central Iowa farm indicated insensitivity to moderate increases in energy prices. Because direct energy costs account for less than 10% of total agricultrual production costs, these results were not unexpected. Yet, this finding is particularly significant in light of recent concern about the impacts of rising energy prices on agricultural output. At least for our particular situation, the concern may be

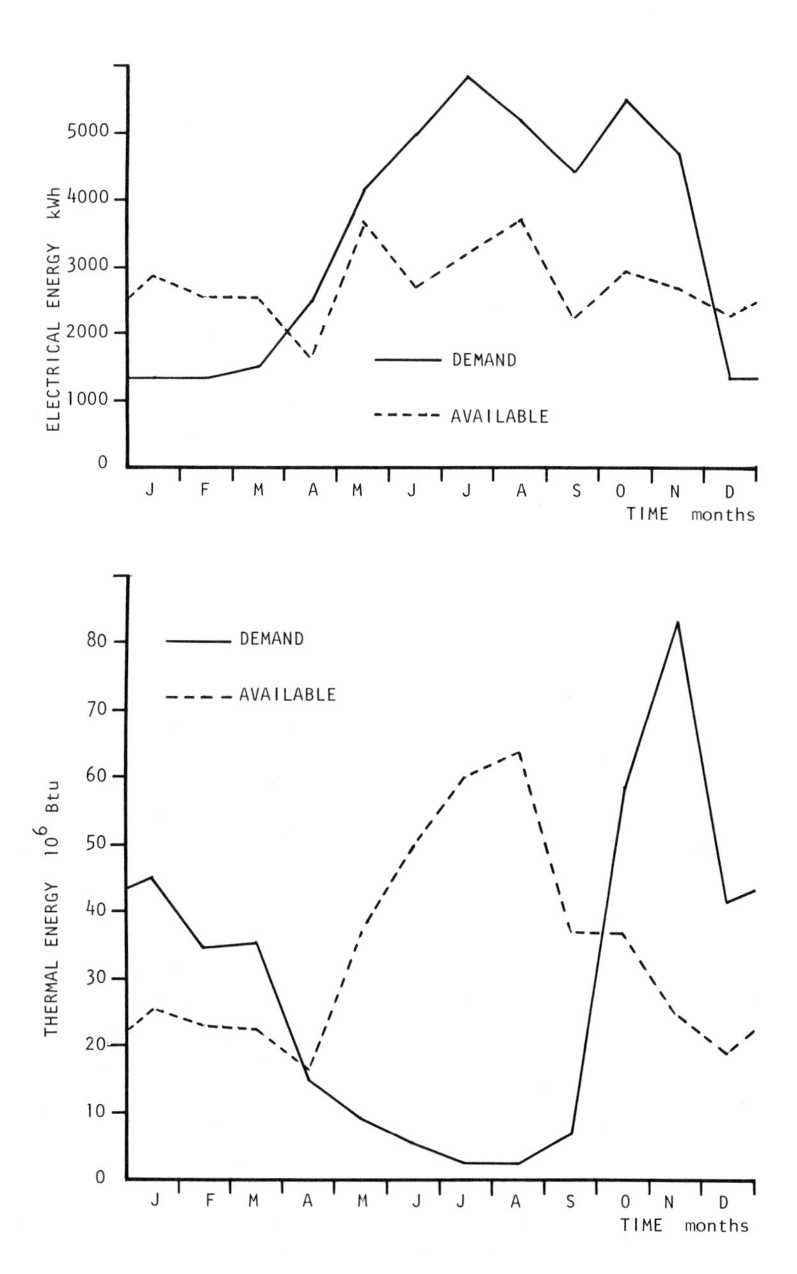

Figure 6. Electrical and thermal needs of a farmstead as they vary over an annual period contrasted with the energy available from an engine-generator fueled by digester gas.

unwarranted. When energy costs increase quite drastically, more significant changes in the activity mix do occur, but the accuracy of these results is questionable because only energy-price impacts on a representative farm were considered.

The economic feasibility of an on-farm methane-digester system, premised upon the assumptions of this analysis, is questionable within a responsible range of energy-price rises. Other economic evaluations have obtained similar results for different farm activities.[36] Yet it is important to observe that the digester system may become feasible under specific circumstances. Farmers may have multidimensional utility functions in which they include factors other than profit maximization. They may place a significant weight on having alternative energy supplies during periods of rationing or shortages. Likewise, they may believe the technology will become more profitable as their knowledge of the technology evolves over time. Discounting backwards these anticipated benefits, the farmer may make the decision to adopt the digestion system.[37]

Finally, one additional caveat is essential. These results apply to a representative central Iowa farm. This analysis has presumed certain resource restraints and input and output prices that influenced the solutions generated. The results may be generalized to other farms with similar price and resource environments. Yet these findings will not necessarily hold for the region, state or other division. The analysis is microeconomic in scope and utilizes a partial-analysis approach. It postulates isolation from the supply and market adjustments that would be forthcoming in a general equilibrium framework if the digester and excreta-feeding activities were adopted by a significant share of farmers, or if energy-price increases and restrictions were economy-wide.

General Comments

In spite of the very subdued enthusiasm shown for a methane digester in this report, the unit may still have a place on the farm. It is most unlikely that it can be justified by its energy-recovery potential alone. Perhaps other selling points will be that the odor from a digester, operating well, is very slight; the stabilization process itself can certainly be made energy self-sufficient; and digestion systems seem compatible with other labor-saving improvements in confined-livestock production (*e.g.*, hydraulic manure transport and land application by irrigation). Another attractive feature is that very little nitrogen is lost during stabilization. Returning the sludge to the land may not take the most advantage of the nutrient-retention capabilities of digestion; refeeding to ruminants may prove a better alternative.

Finally, a word of caution is desirable. What is scientifically achievable may not be entirely desirable on the farm. The modern farmer is very far removed from the peasant, but he still is not a superman. Field cultivations and harvesting are being performed by increasingly more complex machines that are difficult, if not impossible, to service on the farm. Keeping livestock in confinement has not been fully accepted because a very high level of management is required to keep the units performing at the level that maximizes profit. Adding another complex device—perhaps not fully understood by engineers, let alone farmers—will further burden the farmer with maintenance and management requirements that are far removed from his primary interests: raising crops and animals. Farming in the U.S. has been successful, due in part to one family's ability to manage a very productive enterprise. If the equipment required to improve the productivity of a farm becomes too complex, the farmer will not adopt it. The long-term consequences of surrounding a farmer with more and more complex equipment need thorough examination.

ACKNOWLEDGMENTS

This is Journal Paper No. J-8819 of the Iowa Agriculture and Home Economics Experiment Station, Ames, Iowa. This work was supported by Grant No. 12-14-1001-597 from the Agricultural Research Service of the U.S. Department of Agriculture; it was also part of the Iowa 2126 project contribution to NC-93 research.

REFERENCES

1. Iowa State University. "1974 Farm Business Summary for Central Iowa," Iowa Cooperative Extension Service FM 1704 (1975).
2. Kohn, J. P. and W. F. Bradish. "Multiphase and Volumetric Equilibria of the Methane-*n*-Octane System at Temperatures between -110°C and 150°C," *J. Chem. Eng. Data* 9:5-8 (1964).
3. Puri, S. and J. P. Kohn. "Solid-Liquid-Vapor Equilibrium in the Methane-*n*-Eicosane and Ethane-*n*-Eicosane Binary Systems," *J. Chem. Eng. Data* 15:372-374 (1970).
4. Shim, J. and J. P. Kohn. "Multiphase and Volumetric Equilibria of Methane-*n*-Hexane Binary System at Temperatures Between -110°C and 150°C," *J. Chem. Eng. Data* 7:3-8 (1962).
5. Neyeloff, S. and W. W. Gunkel. "Methane-Carbon Dioxide Mixtures in an Internal Combustion Engine," in *Energy, Agriculture and Waste Management,* Proc. Cornell Agricultural Waste Management Conference (1975), pp. 397-408.
6. American Society of Heating, Refrigerating, and Air-Conditioning Engineers. *Systems* (New York: ASHRAE, 1970).
7. Hein, M. E., R. J. Smith and R. L. Vetter. "Anaerobic Digestion of Beef Manure and Corn Stover," ASAE 75-4542, American Society of Agricultural Engineers, St. Joseph, Michigan (1975).
8. Fischer, J. R., Agricultural Research Service, U.S. Department of Agriculture, University of Missouri, Columbia, Missouri, personal communication.

9. Morgan, P.F. "Studies of Accelerated Digestion of Sewage Sludge," *Sew. Ind. Wastes* 26:462-476 (1954).
10. Converse, J. C., J. G. Zeikus, R. E. Graves and G. W. Evans. "Dairy Manure Degradation under Mesophilic and Thermophilic Temperatures," ASAE 75-4540, American Society of Agricultural Engineers, St. Joseph, Michigan (1975).
11. Pfeffer, J. T. "Anaerobic Processing of Organic Refuse," in *Proc., Bioconversion Energy Research Conference,* National Science Foundation G139215 (1973), pp. 31-39.
12. Lapp, H. M., D. D. Schulte, E. J. Kroeker, A. B. Sparling and B. H. Topnik. "Start-Up of Pilot-Scale Swine Manure Digesters for Methane Production," in *Managing Livestock Wastes,* Proc., International Symposium, American Society of Agricultural Engineers PROC-275 (1975), pp. 234-237, 243.
13. Robertson, A. M., G. A. Burnett, P. N. Hobson, S. Bousfield and R. Summers. "Bioengineering aspects of Anaerobic Digestion of Piggery Wastes," in *Managing Livestock Wastes,* Proc., International Symposium, American Society of Agricultural Engineers PROC-275 (1975), pp. 544-548.
14. Malina, J. F., Jr. and E. M. Miholits. "New Developments in the Anaerobic Digestion of Sludges," in *Advances in Water Quality Improvement,* E. F. Gloyna and W. W. Eckenfelder, Jr., Eds. (Austin, Texas: University of Texas Press, 1968), pp. 355-379.
15. Zwietering, T. N. "Suspension of Solid Particles in Liquid by Agitators," *Chem. Eng. Sci.* 8:244 (1958).
16. McCarty, P. L. "Anaerobic Waste Treatment Fundamentals. Part I. Chemistry and Microbiology," *Pub. Works* 95(9):107-112 (1964).
17. McCarty, P. L. "Anaerobic Waste Treatment Fundamentals. Part 2. Environmental Requirements and Control," *Pub. Works* 95(10):123-126 (1964).
18. McCarty, P. L. "Anaerobic Waste Treatment Fundamentals. Part 3. Toxic Materials and Their Control," *Pub. Works* 95(11):91-94 (1964).
19. Hobson, P. N., S. Bousfield and R. Summers. "Anaerobic Digestion of Organic Matter," *Crit. Rev. Environ. Control* 4:131-191 (1974).
20. Metcalf and Eddy, Inc. *Wastewater Engineering* (New York: McGraw-Hill Book Company, 1972).
21. "Livestock Waste Management with Pollution Control," J. R. Miner and J. R. Smith, Ed., North Central Regional Research Publication 222, Midwest Plan Service Handbook MWPS-19, Midwest Plan Service, Iowa State University, Ames, Iowa (1975).
22. Shadduck, G. and J. A. Moore. "The Anaerobic Digestion of Livestock Wastes to Produce Methane. 1946 - June, 1975," A Bibliography with Abstracts, Agricultural Engineering Department, University of Minnesota, St. Paul, Minnesota (1975).
23. Kline, G. L., Agricultural Research Service, U.S. Department of Agriculture, Iowa State University, Ames, Iowa, personal communication.
24. Hazen, T. E. and D. W. Mangold. "Functional and Basic Requirements of Swine Housing," *Agric. Eng.* 41:585-590 (1960).
25. Carpenter, G. A. and J. M. Randall. "The Interpretation of Daily-Temperature Records to Optimize the Insulation of Intensive Livestock Buildings," *Agric. Meteorol.* 15:245-255 (1975).

26. Schuler, R. T. and D. Broten. "Computer Simulation of Animal Ventilation," ASAE 76-4023, American Society of Agricultural Engineers, St. Joseph, Michigan (1976).
27. Spillman, C. K. and J. P. Murphy. "Effect of Farrowing House Temperature on Energy Use and Litter Size," ASAE 76-4044, American Society of Agricultural Engineers, St. Joseph, Michigan (1976).
28. Stevens, G. R., J. A. DeShazer, T. L. Thompson and N. C. Teeter. "Environmental Control for Swine Housing Based on Energy Conservation and Animal Performance," ASAE MC-76-703, American Society of Agricultural Engineers, St. Joseph, Michigan (1976).
29. Midwest Plan Service. "Structures and Environment Handbook," Midwest Plan Service, Iowa State University, Ames, Iowa (1975).
30. Hittman Associates, Inc. "Residential Energy Consumption Single Family Housing," Report HUD-HAI-2, U.S. Department of Housing and Urban Development (1973).
31. Vanderholm, D. H. "Nutrient Losses from Livestock Waste During Storage, Treatment and Handling," in *Managing Livestock Wastes,* Proc., International Symposium, American Society of Agricultural Engineers PROC-275 (1975), pp. 282-285.
32. Sievers, D. M., J. R. Fischer, N. F. Meador, C. D. Fulhage and M. D. Shanklin. "Engineering and Economic Aspects of Farm Digesters," ASAE 75-4541, American Society of Agricultural Engineers, St. Joseph, Michigan (1975).
33. Abdalla, A. H. M. "Thermal Conductivity of Polyethylene Pipes Embedded in Concrete," Ph.D. Thesis, Iowa State University, Mic. 73-9421, University Microfilms, Ann Arbor, Michigan (1972).
34. United States Senate. "The United States Food and Fiber Sector: Energy Use and Outlook," prepared by the Economic Research Service of the United States Department of Agriculture for the Subcommittee on Agricultural Credit and Rural Electrification of the Committee on Agriculture and Forestry, U. S. Government Printing Office, Washington, D.C. (1974).
35. Buchele, W. F., Agricultural Engineering Department, Iowa State University, Ames, Iowa, unpublished communication.
36. Slane, T. C. "An Economic Analysis of Methane Generation on Commercial Poultry Farms," M.S. Thesis, University of Massachusetts, Amherst, Massachusetts (1974).
37. Willis, C. E. and R. L. Christensen. "Measurement of External Learning Benefits from Methane Digestion on Commercial Poultry Farms," in *Proceedings, Conference on Energy and Agriculture, St. Louis, Missouri 1976* (in press).

25

EXPERIENCES FROM OPERATING FULL-SIZE ANAEROBIC DIGESTER

S. Persson
H. D. Bartlett
Agricultural Engineering Department

R. W. Regan
Civil Engineering Department

A. E. Branding
Dairy Science Department
The Pennsylvania State University
University Park, Pennsylvania

INTRODUCTION

The Penn State biogas generator was designed to process the gutter discharge from 50 lactating Holstein dairy cows. Components of this system include a heated anaerobic digester with floating dome and recirculated biogas agitation. Slurry is fed in at the bottom and discharges through a standpipe overflow tube (Figure 1). The manure is watered at the gutter discharge to make a slurry of approximately 10% solids (dry basis) to allow hydraulic transport through the digester. The biogas generation process requires equipment components to contain and handle the flow of liquid, gaseous and heat media.

BIOGAS GENERATOR DESIGN

Structural and equipment elements were selected from prefabricated units and, wherever possible, from commercially available farm equipment components.

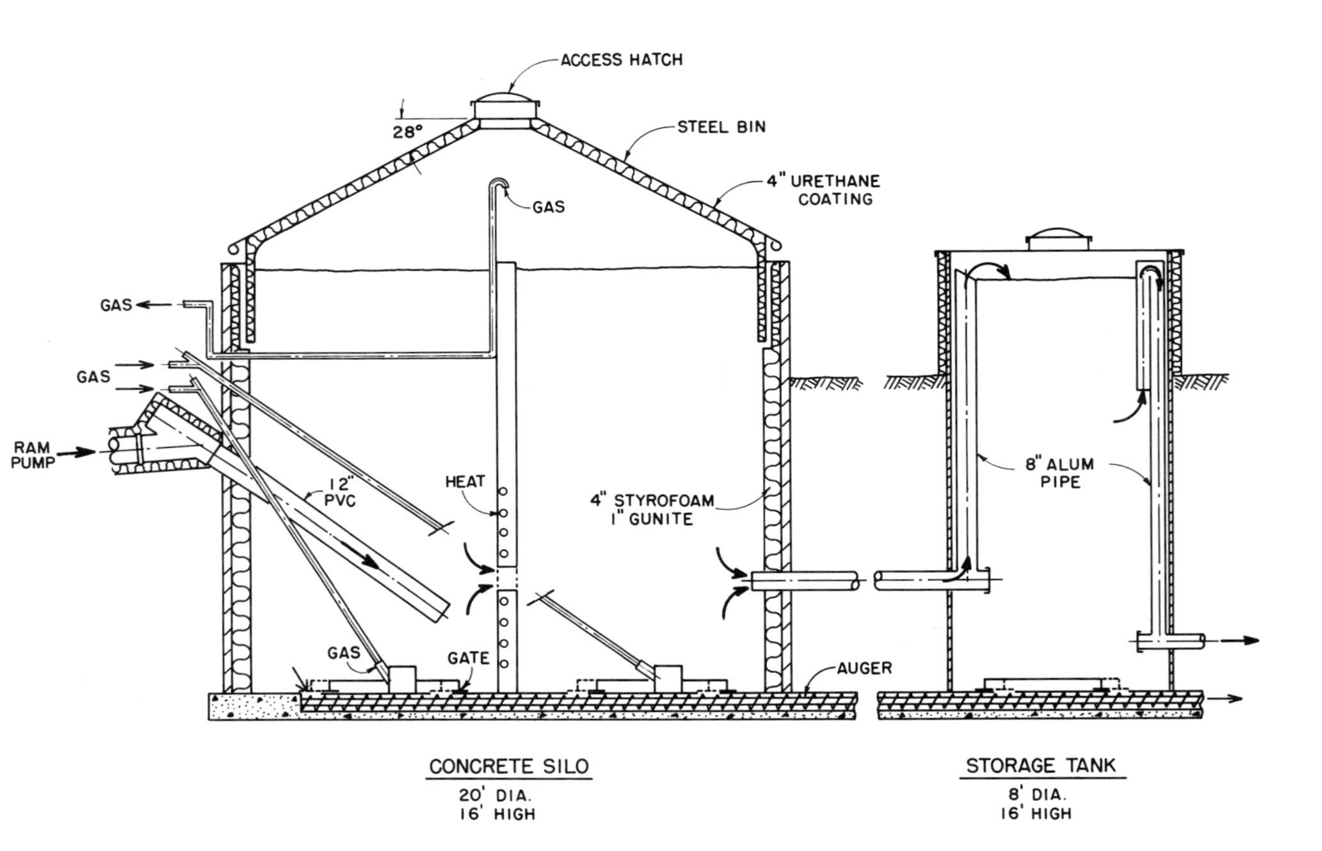

Figure 1. Biogas generator (Agricultural Engineering Department, The Pennsylvania State University).

Digester Construction

The structural shell of the digester is made of prefabricated concrete panels with steel hoops for a 20-ft-diam by 16-ft-deep silo. It is insulated on the interior with 4 in. of styrofoam applied as a double layer of 2-in.-thick styrofoam sheets. The styrofoam is plastered with gunite (about 1 in. thick). Heat pipes were cast in concrete panels which were installed in the divider wall to supply the heat at the center of the digester. The gas collection dome is a 40-in. wall section and roof assembly of an 18-ft-diam prefabricated steel grain bin. The dome is supported on the top of the divider wall but floats on the gas that is produced. Dome insulation is 3 in. of urethane foam sprayed onto the inside of the roof and walls. The urethane also provides the gas seal for the seams in the metal roof and wall sections.

Slurry System

The gutter cleaner discharge, consisting of animal excrement, spent sawdust bedding and rejected feed (hay and silage), is delivered to a ram manure pump hopper where fresh water is added to make a slurry to allow hydraulic flow through the digester. The winter 1975 digester supply system used an electrically driven centrifugal manure pump which recirculated water as the manure was added to form a slurry which was pumped into an elevated tank and discharged by gravity to the bottom of the digester. This digester feed system required dilution to about 8% solids for pumping.

It was replaced in 1976 by a ram pump located under the gutter cleaner discharge, which forces the slurry into the digester against an 8-ft head of slurry. Reciprocating action of the plunger through the manure-water mixture makes a well-blended mixture entering the digester. The ram discharge line delivers the slurry in close proximity to the heated wall panels so that the slurry is heated as quickly as possible. Dilution to attain a 12% solids content slurry has proved satisfactory for hydraulic flow through the digester. Thus, the detention time for this size digester was approximately 15 days for the winter 1976 studies and 25 days for the winter 1977 tests.

Continuous agitation by recirculating the biogas produces a well-blended effluent which overflows readily out of the digester standpipe into the storage tank. Also, it flows readily through the bottom opening of the storage tank after standing for one week.

Sludge removal ports are provided in the bottom of both sides of the digester and in the storage tank. The sludge discharge ports are opened and closed by sliding valves which are operated by a hydraulic cylinder controlled by hydraulic valves at a remote location. The openings allow the sludge to flow into a circular channel in which a hydraulic motor-powered auger conveyor is installed. Thus, the mineral components which settle readily are removed periodically to prevent their build-up, which reduced the digester capacity.

Gas System

The gas system consists of an electrically driven milker vacuum pump (DeLaval No. 74) which draws the biogas from the top of the gas collection dome and compresses it sufficiently to overcome the fluid pressure head of the slurry and force it through the slurry for agitation.

Frequent agitation is necessary to maximize heat transfer for attaining uniform temperature and microbial population, to minimize settling and to prevent crust formation which inhibits gas release from the media.

Three gas diffusers are positioned at the bottom of the digester–two in the input or primary section and one in the secondary section. Distribution of the recirculated gas to the respective diffusers can be adjusted by throttling valves in the manifold outlet lines to each diffuser. Gas pressure at the pump outlet is set at 35 mm Hg, which provides active bubbling through the slurry surface and maintains relatively uniform slurry temperatures. The pressure line is tapped for a boiler gas supply line which is equipped with a pressure regulator valve.

Heating System

A standard gas-fired hot water boiler (American Standard, GPM-3, Series 1B-J1, 60,000 Btu/hr) has supplied sufficient heat to maintain proper digester conditions when input was almost at freezing. The heat exchange surface is two loops, approximately 90 ft each, of 3/4-in. steel pipe cast in the concrete divider wall panels. A mixing valve (Powers, No. 11 Regulator) is installed to blend the boiler outlet water with the cooled return water to maintain constant preset temperature delivery to the heat exchanger at approximately 140°F maximum. Higher temperatures were avoided to obviate exchanger surface temperatures which might inhibit growth of methane-forming organisms or cause the slurry to cake on the heat exchange surface.

The boiler is fired with LP gas for start-up and switched to biogas when sufficient quantity is produced to operate the boiler for several hours at a time. Thereafter, the boiler is operated on biogas entirely except for the pilot lighter. LP gas is used continually to supply the pilot.

Hydraulic System

The hydraulic cylinders to operate the gates in the bottom of the digester and the hydraulic motors which power the sludge removal auger are operated by an electrically driven hydraulic pump. A tractor hydraulic system may be substituted but may be somewhat inconvenient to schedule and/or move to the site.

The ram pump comes equipped with a hydraulic power package which is driven by an electric motor.

OPERATIONAL EXPERIENCES

The present system design requires considerable manual supervision of the slurry feeding and temperature regulation. Attention to temperature regulation can be minimized by the use of more automatic control devices. However, reduction in attention to slurry feeding requires barn management practices that allow a minimum of bedding and rejected feed to enter the system; it also requires elimination of extraneous items which are often deposited in the gutter.

This system has been plagued with excessive amounts of waste feed and sawdust with sizable wood chips which have interfered with manure pump operation. Chips jam between the centrifugal pump impeller and its housing, causing stoppage. They also interfere with operation of the valves in the ram pump, preventing them from closing completely, which results in backflow of slurry from the digester when the plunger is in the retracted position.

The slurry feed system needs to be housed in a freeze-proof shelter. It was found necessary to install heating cable and insulation around the ram pump switching valve to insure its operation at temperatures much below freezing. No problems were encountered with the plunger freezing to the housing when the pump hopper was emptied after each cleaning. Slight seepage around the rear of the ram has been experienced and must be managed.

Experience with the initial digester design demonstrated the need to avoid 90-degree turns in the slurry feed and outflow lines as well as the gas agitation lines to allow mechanical cleaning in the event of stoppage. Modification to the original design incorporated external access to these lines by installing "Y" or "T" fittings before lines enter the digester. Whenever fluid pressure is applied to ruminant animal manure slurry in pipes, it tends to compact and cause stoppage. Barn snow (calcite grit), which is used for sanitation and to overcome slippery floor conditions, aggravates stoppage problems at the bottom of the digester. Reject feed tends to float and accumulates into a thick mat on the digester surface.

A principal limitation in the system has been gas leakage from the collection dome. This design relies on the urethane insulation for the gas seal. However, this has not been entirely satisfactory; apparently the lack of rigidity in the steel roofing, or the differential thermal expansion of the urethane and steel, causes cracks which allow gas leakage. A vacuum relief valve should be installed in the system to prevent danger of collapsing the dome in case of pulling negative pressure in the dome.

Severe corrosion problems have been encountered, especially with the biogas lines and related equipment. Whenever possible, plastic pipe and components should be used. However, caution must be used where elevated temperature may occur, such as in connections to the compressor. Unprotected steel will corrode rapidly, which could cause gas leaks or early structural failure of components. Gas meters have not operated dependably under continuous use.

Undetected gas leaks are potential explosion, fire or asphyxiation dangers. Safety features such as flame traps in the gas system, pressure relief valves in the hydraulic system and thermal overload protection devices in the heating system are essential. Large, vividly colored "Danger" signs are posted around the biogas production system. All workers should be instructed to have an awareness of the dangers around the system.

SUMMARY OF OPERATION

The biogas generation system was filled with municipal sewage sludge for seeding in late December, and increasing weekly increments of manure were added until the total manure production from the barn was supplied to the digester by the end of January 1977. By that time, biogas production was sufficient to operate the boiler continuously. Thereafter, gas in excess of the heating system needs was produced. However, gas leaks which developed in the dome have prevented accurate determination of the total gas production.

Following are approximate values for the slurry input and gas output of the system as operated from February 1 to April 15, 1977.

Slurry System: (twice daily feeding by ram pump, 5.75 gal/stroke at 5 strokes/min)

Total input = 1150 gal/day (includes 540 gal fresh water)
= 155 ft^3/day = 32.2 m^3/day

Analysis	Input	Effluent
Solids (%)	12	6
COD (mg/l)	80,000	60,000
Total N (% d.b.)	2.4	4.6
pH	7.5	7.8

Heat System: (LP-biogas hot water boiler, 60,000 Btu/hr)

Boiler input = 50 ft^3/hr to 90 ft^3/hr, depending on weather conditions
Gas agitation = 2 hp electric motor (continuous operation)
Water circulation = 1/3 hp electric motor (continuous operation)

Biogas Production: Values based on boiler use by calibration (burner manifold pressure vs metered flow) plus draw off through gas meter.

Total = 2200 ft^3/day (62 m^3/day)

Methane = 62%

ACKNOWLEDGMENTS

This project was funded in part with funds received from The Pennsylvania Fair Fund administered by The Pennsylvania Department of Agriculture.

26

METHANE PRODUCTION DURING TREATMENT OF FOOD PLANT WASTES BY ANAEROBIC DIGESTION

L. van den Berg and C. P. Lentz
Division of Biological Sciences
National Research Council of Canada
Ottawa, Ontario, Canada

INTRODUCTION

Energy considerations and environmental concerns have combined in recent years to increase interest in anaerobic digestion of wastes resulting from agricultural production and food processing. Many of these wastes, such as feedlot manure and fruit peeling wastes, are relatively concentrated in organic substances, making them difficult to dispose of by aerobic treatment (very high oxygen demand) or by land application (odors). Anaerobic digestion, by being capable of converting organic substances to methane and in the process leaving a residue suitable for land application or further processing, would be an ideal method of treatment. While anaerobic digestion has been used for these purposes in municipal sewage treatment systems for many years, its use in North American agriculture and industry has been limited.

Several factors appear to have contributed to the limited use of anaerobic digestion: relatively slow rates of conversion of organic substances to methane, necessitating large digesters; propensity of the digestion to fail—for example, because of the presence of heavy metals or of overloading; the need for long retention times because many of the methanogenic bacteria are growing very slowly; and the need to maintain the digester contents at the mesophilic temperature optimum, *i.e.,* close to 35°C. In practice, for example, many municipal digesters operate at a volatile solids loading rate of less than 1.5 kg/m^3-day at a hydraulic retention time of over 30 days and an efficiency of conversion of 40% or less. This means a methane production rate of less than 0.5 m^3/m^3-day. Reported rates of methane production from animal manures are similar (0.4 to 0.7 m^3/m^3 day).[1-3] Under these conditions the capital investment per unit volatile solids loading rate is high and the energy in the methane produced may not be sufficient to maintain digester temperature and adequate mixing.

Some studies have been made to overcome these limitations. More efficient mixing of digester contents, particularly when combined with a return of the suspended solids to the digester from a settling unit (as in the anaerobic contact process), reduces the propensity to failure and allows higher loadings, especially when the waste material is relatively dilute.[4] Higher loadings are also possible when digesters are operated at the thermophilic optimum of 50 to 55°C, although, of course, substantially more energy is required to maintain this temperature.[5,6] In municipal digestion systems, heating sewage sludge prior to digestion has also been found to increase its digestibility and the production of methane but, of course, requires a higher energy input.[7] These findings do not appear to have found widespread application as yet.

Results obtained previously in this laboratory with food processing wastes using the anaerobic contact process at 35°C have shown that rates of methane production several times those common in many sewage treatment plants are feasible.[8,9] The cost of methane production from biomass is closely related to the rate of methane production per unit digester volume;[1] hence, high rates of methane production as demonstrated in our earlier work would greatly increase commercial interest. The earlier work was limited to two food processing plant wastes, and limited information was obtained on the effect of factors involved in maintaining these high rates of methane production over long periods of time.

The main purpose of the present study was to determine the long-term maximum performance level of the anaerobic contact process when used with widely differing food processing plant wastes at the mesophilic optimum, 35°C. Performance was judged by rate of methane production per unit digester volume, as well as by the usual waste-treatment parameters of volatile solids and chemical oxygen demand removal, volatile acid content and suspended solids content in the effluent. Factors affecting performance included in the study were nutrient supply, hydraulic retention times, overloading with volatile solids and length of interruption of waste addition. Food processing plant wastes included were bean blanching waste, pear peeling waste (mechanical), rum stillage waste and potato peeling waste ("Dry-caustic" peeling). For comparison with the last-named waste, which contains high concentrations of sodium hydroxide, a waste made up of ground potatoes in water also was used. An additional objective of the present study was to determine the rate at which a digester could be adapted from one waste to another. Results with rum stillage waste were published previously,[10] as well as some of the results obtained with pear peeling waste.[8,9]

EXPERIMENTAL

Fermenter Design and Operation

Details of design and operation of the fermenter units used have been described previously.[9] Briefly, each fermenter unit (Figure 1) was composed of a 30-liter cylindrical fermenter, a 6-liter settling flask (modified Erlenmeyer), a refrigerated feed tank and suitable peristaltic pumps and variable-speed stirrers. The fermenter was provided with moderate agitation to ensure adequate mixing without dispersing flocculated microorganisms and thereby reducing their settleability. Slow stirring at the bottom of the settling flask prevented chaneling in the settled material in the vicinity of the return tube and reduced the tendency for gas bubbles to become entrapped. Waste of the required strength was pumped into the fermenter continuously at the desired rate and effluent overflowed from the settling flask at the same rate. Rates of pumping of fermenter liquid into the settling flask and of settled sludge back into the fermenter (recirculation) were adjusted by trial for each set of conditions to provide optimum settling of the floc. Excess sludge (undigested waste particles and excess microorganisms) was removed when necessary to maintain the suspended solids content in the fermenter below 3%. Settling often became a problem at higher solids contents.

Fermenters were originally filled with liquid obtained from an active sewage sludge digester and the selected waste added first at a relatively low rate (such as one kg volatile solids per m^3 per day). As volatile acid analysis indicated that the fermenter was adapting, the loading rate was increased. A similar procedure was followed when fermenters were changed over from one waste to another and when fermenters were fed again after a period of interruption of waste addition.

Waste Characterization and Preparation

Wastes were obtained from commercial processing plants in as concentrated a form as possible and stored frozen until used. Bean blanching waste (produced during the processing of white beans) was used as received. Pear peeling waste (produced during mechanical pear peeling) was homogenized in a Waring blender, screened to remove fibers that might plug the tubes in the peristaltic pumps and diluted as required. Potato peeling waste (from a

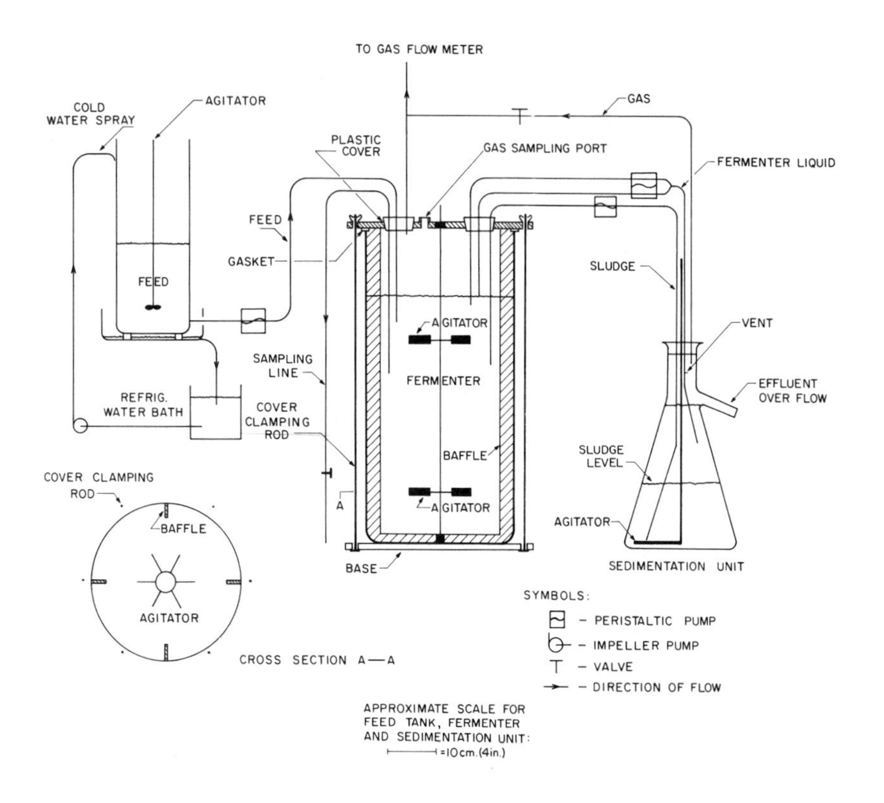

Figure 1. Schematic outline of fermenter set-up.

Notes: (1) Complete set-up, except for refrigerated water bath, was kept in a temperature-controlled room (35°C, 95°F).

(2) Fermenter consisted of 12-in. x 24-in. o.d. glass jar, with ground glass edge and 0.5-in. plastic cover. Baffles were 1.25 in. wide. Agitators were 4.75-in. diam with each blade measuring 0.5 in. x 1.75 in. and were mounted on an 0.5-in.-diam rod turning in Teflon* bearings both in the cover and at the bottom of the fermenter. Stirrer speed was variable from 50 to 800 rpm.

(3) Sedimentation unit consisted of a 6-liter Erlenmeyer flask. Fermenter liquid inlet tube was vented to avoid disturbing of effluent layer by gas bubbles. Agitation was provided by a single-bladed rubber impeller (4 in. long, tapering from 0.5 in. wide at shaft to about 0.25 in. wide at the end), rotating at 30 to 45 rpm depending on suspended solids content.

(4) The level in the fermenter was controlled within narrow limits by the position of the inlet tubes to the pump transferring fermenter liquid to the sedimentation unit. The level in the sedimentation unit was controlled by the overflow outlet.

*Registered trademark of E. I. duPont de Nemours & Company, Inc., Wilmington, Delaware.

"Dry-caustic" peeling process) was diluted to bring the sodium content down from about 6000 to 8000 mg/l to 3000 mg/l. For comparison, a low-sodium potato waste of about the same organic strength was prepared from comminuted potatoes. Rum stillage waste from a blackstrap molasses spirit processing distillery was used without dilution. Where required, nitrogen was added as ammonium bicarbonate, phosphate as sodium and potassium phosphate, sulfate as potassium sulfate, and complex nutrients in the form of yeast extract. Typical compositions of wastes as generally used are presented in Table I.

Table I. Composition of typical wastes (contents in mg/l).[a]

	Bean Blanching Waste	Pear Peeling Waste[b]	Potato Waste		Rum Stillage Waste[e]
			Dry-Caustic Peeling[c]	Whole Potatoes[d]	
Total Solids	21,000	42,000	42,000	60,000	58,000
Total Volatile Solids	18,000	40,000	36,000	57,000	43,000
Suspended Solids	6,000	18,000	7,000	43,000	3,900
Volatile Suspended Solids	5,000	17,000	7,000	43,000	3,600
Chemical Oxygen Demand	20,000	50,400	38,400	70,000	53,000
Kjeldahl Nitrogen	800	900	1,200	1,000	1,050
Phosphate (as P)	200	180	120	200	210
Sulfate (as SO_4)	–	270	–	–	1,300
Sodium	10	150	3,000	<10	680
Potassium	1,400	440	900	1,450	4,800
Calcium	–	–	–	–	1,400
Magnesium	–	–	–	–	680
Alkalinity, as $CaCO_3$	1,900	3,300	7,700	–	9.500
COD:N:P	100:4:1	280:5:1	320:10:1	350:5:1	250:5:1

[a]Addition of 1 g yeast extract, when used, added approximately 850 mg volatile solids, 70 mg Kjeldahl nitrogen and less than 30 mg potassium, 20 mg phosphorus and 2 mg sodium.

[b]Diluted from concentrated pear waste, with 4.8 g NH_4HCO_3, 0.5 g K_2SO_4, 0.4 g KH_2PO_4 and 0.42 g Na_2HPO_4 added per liter after dilution. Other dilutions made as required, but same amounts of salts added per liter after dilution unless otherwise noted.

[c]Diluted from concentrated potato waste, with 3 g NH_4HCO_3 added per liter after dilution. Other dilutions made as required but same amount of salt added per liter after dilution.

[d]Finely ground whole potatoes in water. Waste was heated before use sufficiently to prevent starch from settling.

[e]Waste as received but with 1.95 g NH_4HCO_3 and 0.24 g Na_2HPO_4 added per liter.

Determination of Maximum Rates of Methane Production

In determining maximum rates of methane production the volatile solids loading rate (VSLR) was increased in relatively small steps (20 to 25%) over a period of several weeks until the fermenter was unable to cope with the load. Fermenters were operated for at least one week at each loading rate at reasonably constant volatile acids content to ensure that steady-state conditions had been established. When the maximum VSLR had been exceeded, steady-state conditions could not be obtained and the volatile acid content continued to increase to levels above 1500 mg/l even with frequent interruptions of waste addition (one day maximum at a time) to allow the methane-forming bacteria to adjust. The rate of methane production at the highest rate of feed addition possible with a reasonably constant volatile acid content prior to overloading represents the maximum rate of methane production quoted in this chapter.

The effects of the following factors on maximum rate of methane production were studied: nutrient content of waste, hydraulic retention time, overloading with volatile solids and length of interruption of waste addition. Nutrients studied were ammonia nitrogen and phosphate (COD:N:P ratio varied from 100:5:1 to 2000:10:1), sulfide or sulfate (up to 150 mg S/l) and yeast extract (up to 4 g/l). The latter was used because the variety and complexity of bacterial nutrients in it made it likely that any limitation on rate of methane production would be caused by factors other than nutrient supply. Hydraulic retention times were varied (mainly with pear waste, and to a lesser extent with potato waste) by diluting wastes while maintaining added nutrient concentration. During volatile solids overloading tests, the maximum rate of methane production prior to overloading was compared with the maximum rate after overloading. To determine the effect of interruption of waste feed on fermenter performance (with pear waste only), maximum rates of methane production were determined following one to seven months of interruption. During this period the fermenters were agitated continuously.

Methods of Analysis

Methods of analysis and frequency of measurements were essentially as described previously:[9,10] pH measurements (indicating fermenter condition) were made daily, volatile acids content was determined by gas chromatography one to four times per week depending on fermenter conditions, while solids, COD, rate of gas production, gas composition and alkalinity of fermenter liquid and effluent, used mostly as indicators of fermenter performance, were determined once every one or two weeks during steady-state operation. Kjeldahl nitrogen and phosphate and sulfide contents were determined as required in studies on nutrient addition.

RESULTS

Maximum rates of methane production depended on type of waste and nutrient added (Table II). All wastes could be digested with only inorganic salts added, but the maximum rates of methane production were relatively low (less than 1 m^3/m^3-day), except for bean blanching waste, which appeared to be nutritionally adequate. With bean blanching waste, rates of methane production over 2 m^3/m^3-day were possible at a 2-day hydraulic retention time for 1 to 2 weeks, but decreased settleability of the microorganisms resulted in washout. Settleability was not a major factor in limiting the rate of methane formation from pear and potato peeling wastes and rum stillage waste without yeast extract, because the limit was reached at solids retention times of 27 to 90 days. It is of interest that pear waste was deficient in sulfur and was virtually indigestible without addition of sulfate or sulfide. Addition of 0.5 g K_2SO_4/l was adequate, yielding 0.5 to 3 mg soluble sulfide in the fermenter liquid.

Addition of complex bacterial nutrients in the form of yeast extract increased the maximum rate of methane production by a factor of 2 to 3.5, depending on type of waste (Table II). Addition of yeast extract also enabled digestion of more dilute wastes (shorter hydraulic retention times) and shorter solids retention times, presumably because of a faster growth rate of methanogenic microorganisms. The maximum rates of methane production did not appear to be limited by nutrient supply, since higher rates could not be obtained by greater addition of yeast extract. For potato waste a major factor limiting the rate of methane formation was the lack of settling of the biomass, reducing the solids retention time essentially to the hydraulic retention time. Hence, loading rates were limited to hydraulic retention times over about 10 days. For pear waste, hydraulic retention times as short as 2 days could be used, but limited digestibility of the cellulose-containing stone cell particles, causing a high suspended solids content in the fermenter, was a limiting factor in overall performance.

Studies on the need for nitrogen and phosphate for maximum methane production indicated that a COD:N:P ratio in the feed of about 300:5:1 was generally adequate. With potato peeling waste containing a high sodium content, a higher nitrogen content was beneficial (COD:N:P ratio of 300:10:1), presumably because it antagonizes sodium. With a COD:N:P ratio of 300:5:1, the contents of soluble Kjeldahl nitrogen and phosphate-phosphorus in the fermenter liquid were generally over 200 and 50 mg/l, respectively. With pear peeling waste, soluble Kjeldahl nitrogen and phosphate phosphorus contents in the fermenter liquid over 100 and 20 mg/l, respectively, were found not to limit maximum rates of methane production.

Table II. Maximum rates of methane production and associated volatile solids loading rates and hydraulic and solids retention times from vegetable processing wastes without and with added yeast extract.

	Waste[a]				
			Potato		
Characteristic	Bean Blanching	Pear Peeling	Dry-Caustic Peeling	From Whole Potatoes	Rum Stillage
Waste without added yeast extract					
Methane production rate (m^3/m^3-day)	1.9	0.8	0.6	0.5	0.7
Volatile solids loading rate (kg/m^3-day)	6.6	3.0	2.4	1.6	2.4
Hydraulic retention time (days)	3.0	10	20	15	18
Solids retention time (days)	13	80	27	40	$\geqslant$90
Waste with added yeast extract[b]					
Methane production rate (m^3/m^3-day)	–	1.4	1.1	1.3	2.4
Volatile solids loading rate (kg/m^3-day)	–	5.5	4.3	4.5	8
Hydraulic retention time (days)	–	2	9	13	5
Solids retention time (days)	–	$\geqslant$20	11	15	40

[a]Total volatile solids content of waste used (mg/l) can be calculated by calculating hydraulic residence time x volatile solids loading rate x 1000.
[b]1.5 g yeast extract per liter was added to pear peeling and rum stillage waste, 4 g/l to potato waste.

Comparison of fermenter performance characteristics for different wastes at the same VSLR (Table III) shows that differences in rate of methane production were small (1.0 to 1.2 m^3/m^3-day) in spite of large differences in fermenter and effluent characteristics. Fermentation of bean blanching, pear peeling and rum stillage wastes resulted in reasonably good sludge settleability, low effluent solids content, high fermenter suspended solids content and long solids retention times. With potato waste, little settling took place, effluent quality was poor and solids retention times short. Poor settling was generally not caused by rising gas bubbles but appeared related to floc characteristics.

Table III. Comparison of performance characteristics for different wastes during steady-state operation at a volatile solids loading rate of 3.5-4.5 kg/m^3 day (0.22-0.28 lb/ft^3 day).

Performance Characteristic	Waste Digested				
			Potato[b]		
	Bean Blanching	Pear[a] Peeling	Dry-Caustic Peeling	From Whole Potatoes	Rum Stillage[a]
Hydraulic Retention Time (days)	4.2	5	9	13	10
Feed					
Volatile Solids Content (ppm)	18,000	20,000	39,000	60,000	44,300
Chemical Oxygen Demand (ppm)	20,000	25,000	42,000	75,000	54,600
Fermenter					
Volatile Suspended Solids Content (ppm)	11,000	30,000	5,500	8,000	14,000
Volatile Acids Content (ppm)	100-500	0-100	300-500	100-300	200-400
Effluent					
Volatile Solids Content (ppm)	4,500	1,500	13,700	13,000	9,000
Volatile Suspended Solids Content (ppm)	2,000	1,200	4,700	7,000	1,500
Chemical Oxygen Demand (ppm)	2,400	1,800	16,000	16,000	11,000
Solids Retention Time[c] (days)	23	>100	11	15	80
Efficiency[d]					
Chemical Oxygen Demand Removal (%)	88	91	62	79	80
Volatile Solids Removal (%)	75	93[e]	65	78	80
Methane Production[f] (m^3/m^3 day)	1.2	1.0	1.0	1.2	1.2

[a]Containing 1.5 g yeast extract per liter.
[b]Containing 4 g yeast extract per liter.
[c]Hydraulic retention time x (fermenter volatile suspended solids content)/(effluent volatile suspended solids content).
[d](Content in feed - content in effluent)/(content in feed) x 100.
[e]Solids accumulating in the digester (not digested) amounted to about 30% of feed volatile solids, hence 93-30 = 63% of the volatile solids were destroyed.
[f]Methane production was 0.4 m^3 methane (1 atm., 35°C)/kg volatile solids destroyed (± 10%).

Starting a fermenter with sewage sludge as inoculum, or changing a fermenter from one waste to another, could readily be done with nutritionally balanced wastes. Figure 2 presents typical start-up results with pear waste containing yeast extract. Maximum possible stepwise increases in loading rate during acclimation were 20 to 40% per week. As shown in Figure 2, volatile acids content (mainly acetic acid) increased immediately after a stepwise increase in feed rate, but apparently methanogenic bacteria subsequently increased in number and activity and reduced volatile acid contents to low levels.

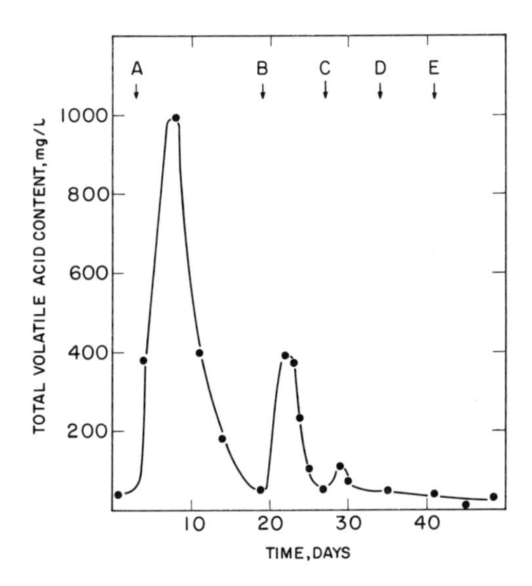

Figure 2. Typical changes in volatile acids content of fermenter liquid during start-up from sewage digester sludge. At A: pear waste feed added (at a hydraulic retention time of 5 days): 1.0 kg volatile solids/m^3 day. B: increased to 1.5 kg/m^3-day; C: increased to 2.1 kg/m^3-day; D: increased to 3.0 kg/m^3-day; E: increased to 4.0 kg/m^3-day.

Increasing the feed rate above the maximum rates presented in Table II led to permanent changes in fermenter performances. Instead of an increase and subsequent decrease in volatile acids following the increased rate of feed addition (such as in Figure 2), volatile acids continued to increase and failure occurred if the rate of feed addition was not sharply reduced. Subsequently, the fermenter could be operated at only about half the earlier maximum VSLR. Only reinoculation with suspended solids from another active fermenter or from an active sewage sludge digester restored the maximum VSLR to its earlier level.

Other factors similarly reduced the maximum VSLRs presented in Table II. Feeding waste without nutrients prior to feeding a nutritionally adequate waste, feed interruption for more than 3 months (up to 3 months had little effect), use of very short hydraulic residence times (less than 2 days) and of short solids retention times (less than 10 days) reduced fermenter performance permanently when previously optimum conditions were restored. Nutrient addition, such as extra yeast extract, was not beneficial in these circumstances, but reinoculation was necessary to restore the high maximum VSLRs presented in Table II.

The need for reinoculation, which suggests some form of microbiological instability, demonstrated itself also when fermenters were operated under constant conditions of feed composition and hydraulic retention time (5 days or longer) at a reasonably high, but constant, VSLR (4 kg/m^3-day or higher) over long periods of time (up to 12 months). From time to time (3- to 12-month interval), the volatile acid content would increase for no apparent reason and the loading rate would have to be reduced to below 2.5 kg/m^3-day to avoid complete failure. Reinoculation would be necessary to obtain the earlier VSLR.

DISCUSSION AND CONCLUSIONS

Results show that the anaerobic contact process at 35°C is capable of producing over 2.0 m^3 methane per m^3 fermenter volume per day from food processing plant wastes, with VSLRs of up to 8 kg/m^3-day and efficiencies in converting organic solids to methane up to 80%. These rates of methane production are several times the rates commonly achieved in practice, and results do not rule out the possibility of higher rates yet. Such high rates, if achieved in practice, would drastically alter present assumptions in cost calculations of methane production from biomass[1] and in all likelihood make methane produced by anaerobic digestion competitive with methane presently obtained from dwindling fossil-fuel sources.

Whether or not the high rates of methane production obtained in this study can be achieved in practice will depend largely on the solution of four major problems:

1. At continuous high rates of methane production the fermentation is unstable, requiring frequent reinoculation from other active digesters. Results of tests have indicated that the methanogenic organisms are affected. The causes of this microbiological instability are as yet unknown. The methanogenic organisms involved are mainly those which convert acetic acid to methane and which produce about two-thirds of the methane. Few of these organisms have been isolated in pure culture, and little is known about their characteristics and requirements.

2. Many wastes require addition of unknown nutrients for high rates of methane production. Results of limited tests have indicated that these nutrients, which are obviously present in yeast extract, cannot be provided by adding proteins or amino acids (the potato waste in this study contained protein), common vitamins or salt. Since the unknown nutrients have such a dramatic effect on methane production in food processing waste fermentations, it is likely that they could also be used to improve other methanogenic fermentations, as well as to accelerate acclimation changes, provided that their composition was known.
3. The bacteria often flocculate and settle poorly. Results obtained indicate that poor settling may occur even when gas formation is eliminated as a factor. Since retention of bacteria in the anaerobic contact process is essential for optimum performance, information on factors affecting flocculation and settling of the anaerobic bacteria involved in the fermentation is needed.
4. In cellulose-containing wastes, the rate of cellulose hydrolysis often limits the rate of methane production. It is known that the presence of lignin reduces the rate of cellulose breakdown and that rate of agitation affects cellulose breakdown, but relatively little is known about the extent to which other factors affect cellulose hydrolysis under anaerobic conditions.[11] Conversion of over 80% of the cellulose present in some waste has been shown to be feasible.[11]

Studies of a basic nature are presently underway in this laboratory to elucidate these problem areas. The potential benefits of anaerobic digestion for both energy production and waste treatment indicate the need for solutions to the problems, while results obtained in the present study indicate that solutions are possible.

ACKNOWLEDGMENT

The authors thank The Canadian Canners, Ltd., for providing the wastes, and Miss K. A. Lamb and Mr. E. A. Rooke for technical assistance.

REFERENCES

1. Ifeadi, C. N. and J. B. Brown, Jr. "Technologies Suitable for Recovery of Energy from Livestock Manure," Proc. Cornell Agric. Waste Management Conf. (1975), pp. 373-396.
2. Fisher, J. R., D. M. Sievers and C. D. Fulhage. "Anaerobic Digestion of Swine Wastes," Proc. Cornell Agric. Waste Management Conf. (1975), pp. 307-316.
3. Kroeker, E. J., H. M. Lapp, D. D. Schulte and A. B. Sparling. "Cold Weather Energy Recovery from Anaerobic Digestion of Swine Manure," Proc. Cornell Agric. Waste Management Conf. (1975), pp. 337-352.
4. McCarty, P. L. "Anaerobic Waste Treatment Fundamentals, Parts 1, 2 and 3," *Public Works* 95(9):107-112, (10):123-126, (11):91-99 (1964).

5. Garber, W. F., G. T. Ohara, J. E. Colbaugh and S. K. Raksit. "Thermophilic Digestion at the Hyperion Treatment Plant," *J. Water Poll. Control Fed.* 47:950-961 (1975).
6. Smart, J. and B. I. Boyko. "Full-Scale Studies on the Thermophilic Anaerobic Digestion Process," report on project 73-1-29, published by Ontario Ministry of the Environment, Pollution Control Branch, 135 St. Clair Ave. West, Toronto, Ontario, Canada, M4V 1P5 (1973).
7. Haug, R. T., D. C. Stuckey, J. M. Gossett and P. L. McCarty. "The Effect of Thermal Pretreatment on Digestibility and Dewaterability of Organic Sludges," paper presented at the 49th Annual Conference, Water Pollution Control Federation, Minneapolis, Minnesota (1976).
8. van den Berg, L. and C. P. Lentz. "Anaerobic Digestion of Pear Waste: Laboratory Equipment Design and Preliminary Results," *Can. Inst. Food Technol. J.* 4:159-165 (1971).
9. van den Berg, L. and C. P. Lentz. "Anaerobic Digestion of Pear Waste: Factors Affecting Performance," Proc. 27th Purdue Industrial Waste Conf. (1972), pp. 313-323.
10. Roth, L. A. and C. P. Lentz. "Anaerobic Digestion of Rum Stillage," *Can. Inst. Food Technol. J.* 10:105-108 (1977).
11. Khan, A. W. "Anaerobic Degradation of Cellulose by Mixed Cultures," *Can. J. Microbiol.* (in press, 1977).

27

ANAEROBIC FERMENTATION OF ANIMAL WASTES: DESIGN AND OPERATIONAL CRITERIA

G. R. Morris
Environmental Sciences Division, Envirex, Inc.
Milwaukee, Wisconsin

W. J. Jewell
Department of Agricultural Engineering
Cornell University
Ithaca, New York

R. C. Loehr
Environmental Studies Program
Cornell University
Ithaca, New York

INTRODUCTION

To date (1977), despite widespread application of the anaerobic fermentation process in municipal wastewater treatment, limited information has been made available concerning the process fundamentals of anaerobic systems in the treatment of animal wastes. Without such information, a quantitative basis of design, *i.e.,* a mathematical model, cannot be developed which will be universally applicable to the prediction of the overall process behavior for different animal wastes.

A limited number of studies have been performed during the past several years to determine the effects of the operational variables upon the anaerobic treatment of animal wastes. The information obtained from these investigations has been invaluable to the basic understanding of the factors involved. However, little or no attempt has yet been made to bridge the gap between these basic studies and the development of mathematical relationships which will permit the use of these results to predict design and operational criteria for anaerobic systems treating animal wastes.

If current anaerobic fermentation systems are to be designed and operated to meet the increasing demands placed upon them, it will be necessary to fully develop and understand the process fundamentals in order that mathematical relationships can be established. To be more precise, it is the effect of such parameters as retention time and substrate concentration on the performance of the treatment process that needs to be determined. Thus, the comprehensive laboratory and theoretical investigations which follow were designed to achieve the following objectives:

1. to evaluate the effects of process design or operational variables on the anaerobic fermentation process;
2. to determine reaction relationships in order to predict the effluent quality from an anaerobic system treating animal wastes; and
3. to develop mathematical expressions which will permit the use of the reaction relationships for the design and operation of effective anaerobic waste treatment systems.

Since dairy cow manure constitutes a major portion of farm animal wastes and is a fairly good illustration of the complexity involved in animal wastes, it was decided that dairy manure would be an appropriate substrate.

BACKGROUND

A review of the anaerobic fermentation process requires an investigation of the environmental and operational factors that are involved. Factors influencing anaerobic fermentation are listed in Table I. Several comprehensive literature reviews have summarized the present knowledge relating to the fundamentals of anaerobic digestion.[1-3] Some of the pertinent fundamentals will be briefly reviewed to provide some background information and guidelines in order to develop operational and general design criteria for this study.

Table I. Environmental and operational factors affecting anaerobic fermentation.

Environmental Factors	Operational Factors
pH	Composition of Organic Substrate
Alkalinity	Retention Time
Volatile Acid Concentration	Concentration of Substrate
Temperature	Organic Loading Rate
Nutrient Availability	Degree of Mixing
Toxic Materials	Heating and Heat Balance

Retention Time–Organic Loading Rate–Solids Concentration

The relationship between volumetric organic loading, retention time and solids concentration has been developed by Sawyer[4] for municipal sludges. Figure 1 illustrates organic loading as a function of the retention time and the total solids concentration in the influent waste.

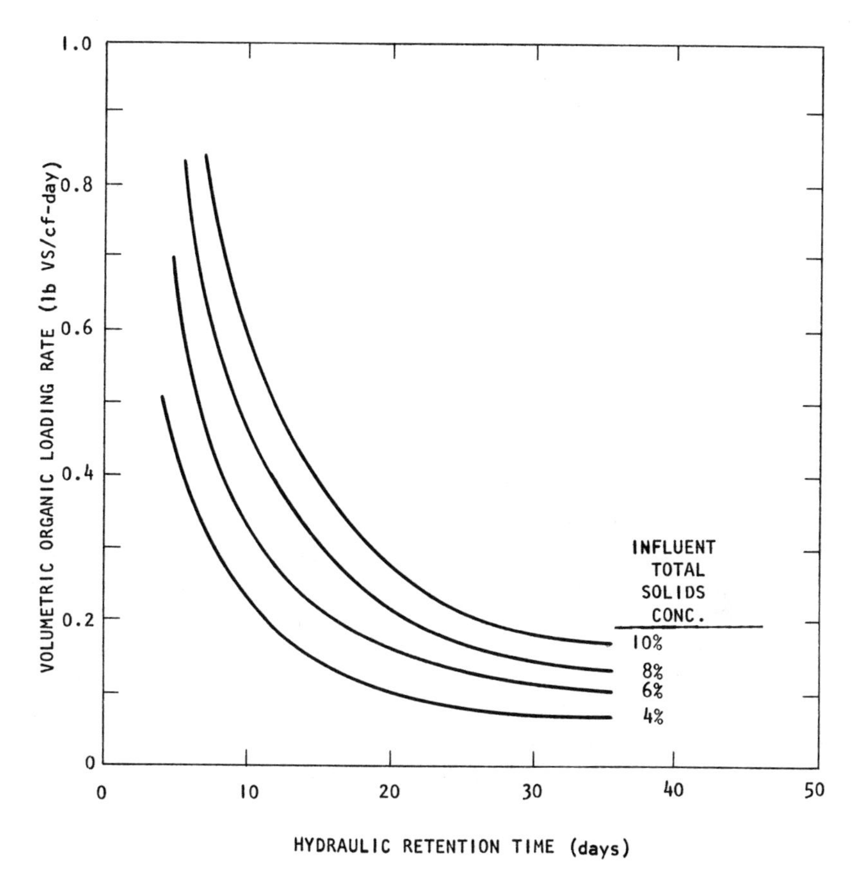

Figure 1. Relationship between volumetric organic loading rate, influent solids concentration and hydraulic retention time (adapted from Sawyer[4] where TVS = 88% TS).

The rate at which organic waste is supplied to the digestion tank is referred to as the volumetric organic loading rate (VOLR). It is commonly expressed in terms of pounds of volatile solids added per cubic foot of digester capacity per day (lb $VS/f^3/day$). As shown in Figure 1, different loading rates can be obtained either by changing the rate of flow through the digester or by altering the influent solids concentration of the organic waste. In practice, the most common method of changing the loading rate has been to change the flow rate, which would affect the hydraulic retention time (HRT) of the digester.

The concept of hydraulic retention time has been a major parameter in the design of all waste treatment processes.[5] Theoretically, the HRT represents the time the waste stays in the unit and can be determined by dividing the volume of the unit by the quantity of waste flowing through the unit per unit time. The HRT is synonymous with the solids retention time (SRT) for a nonrecycle system. The SRT is one of the most important operational factors controlling the design of an anaerobic fermentation system. It represents the average retention time of microbes in the system and can be determined by dividing the pounds of VS in the system by the pounds of VS leaving the system.[6] Factors influencing the SRT are the temperature at which the process occurs, the VS loading to the digester, the volatile percentage of the total solids (TS), the TS concentration in the raw waste and the degree of stabilization desired. Reported minimum SRT values for municipal sludge digestion, where the methane bacteria will not wash out, are in the range of 5 days at temperatures near 35°C.[7] Usually, SRTs of 10 to 30 days are employed to provide a safety factor for anaerobic treatment of municipal sludges.

In general, a high influent solids concentration is considered desirable in digester operations because:[8] (a) it conserves heat due to the lesser amount of water present; (b) it promotes microbial efficiency; and (c) it increases the retention periods. However, there are restrictions on the degree of concentration due to biological and physical limitations, such as maintenance of adequate mixing in the digester and the difficulty of pumping the waste. The equilibrium concentration of the solids in the digester is dependent on the VS of the raw waste and the fraction of solids destroyed. The optimum total solids concentration in the digester should probably not exceed 6 to 8%.[4] Therefore, stabilization of the influent solids should approximate this.

EXPERIMENTAL PROGRAM

The experimental program was set up to examine the following: (a) the biodegradability of dairy cow manure and (b) the influence of retention time, volumetric organic loading rate and influent solids concentration on the anaerobic fermentation process. The experimental program that was employed in this study is summarized in Table II.

To accomplish these objectives, two types of bench-scale, manually mixed anaerobic fermentation systems (Figure 2) were constructed and operated under controlled laboratory conditions. Both systems were operated at a

Table II. Program of experimentation.

Semicontinuous Digestion Systems

Seed: Ithaca Sewage Treatment Plant
Feeding Schedule: Once Daily
Mixing: Manually–Three Times Daily
Volume: 15 liters
Temperature: 32.5°C
Operational Variables:

Volumetric Organic Loading Rate[a] (lb VS/f^3/day)

HRT (days)	Influent Total Solids Concentration (% TS)			
	4	6	8	10
30	0.072	0.109	0.145	0.182
20	0.109	0.164	0.218	0.272
10	0.218	0.327	0.436	0.544
5	0.436	0.653	0.872	1.089
2.5	-	-	1.743	-

[a]Based on TVS = 87.2% TS

Time to Steady State: 2.5 to 3 HRTs
Time at Steady State: 1 to 2 HRTs

Batch Digestion Systems

Seed: Effluent from 20-day HRT Semicontinuous Digesters
Mixing: Manually–Once Daily
Volume: 1.5 liters–Initially
Temperature: 32.5°C
Operational Variables:
HRT: 40, 50, 60, 80, 100 days
S_o: 4, 6, 8, 10% TS

temperature of 32.5°C. Dairy cow manure from Cornell University's Animal Science Teaching and Research Center (ASTARC) at Harford, New York, was used as the substrate material throughout this study. Operational variables investigated were HRT and influent solids concentration. Following a period of operation sufficient to achieve steady-state conditions, process performance was evaluated in terms of methane production and degree of degradation. Steady state in this context was identified by the following conditions: uniform reductions of COD and VS; uniform production of methane gas; and the surpassing in time of 2.5 to 3 HRTs.

When steady-state operation had been achieved, the following analyses were performed on the digested samples: TS, VS, COD, pH, alkalinity, TKN

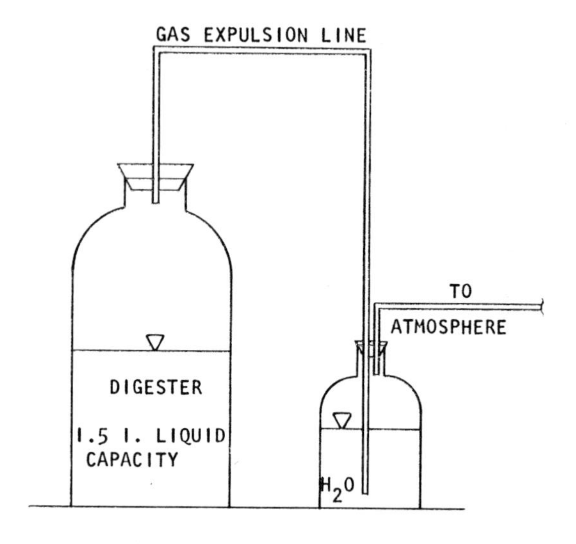

a. Batch anaerobic digestion system.

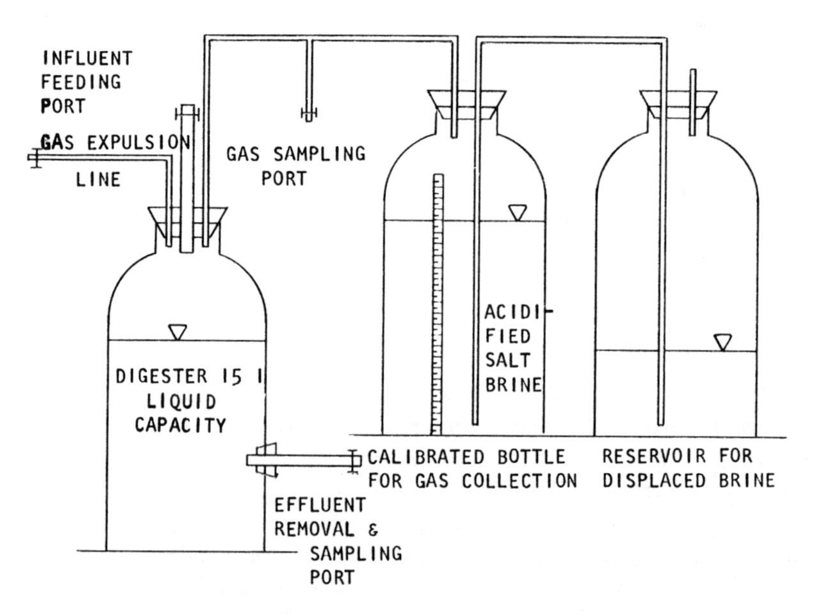

b. Semicontinuous anaerobic digestion system.

Figure 2. Schematic diagram of anaerobic digestion systems.

and NH_3-N. Analyses of the composition of the gas produced were also performed. A majority of the analytical tests used are analyses commonly performed in environmental research laboratories and are described in detail in *Standard Methods for the Examination of Water and Wastewater*, thirteenth edition.[9] Prakasam *et al.*[10] later adapted these analyses for agricultural wastes.

EXPERIMENTAL RESULTS

The experimental results were evaluated in terms of (a) characteristics of dairy cow manure; (b) biodegradability of dairy cow manure; and (c) digester performance at steady-state conditions.

Characteristics of Dairy Cow Manure

Fresh dairy manure is somewhat variable in character as illustrated by the range of results presented in Table III. A comparison of parameter values determined in this study with respect to reported values[11-15] indicated that there is considerable similarity among these values. Therefore, the dairy manure used as the substrate in this study can be considered representative of dairy manure as reported in the literature.

Table III. Summary of raw dairy manure characteristics.[a]

Parameter	Average	Range
TS_{raw} (%)	13.3	12.0 - 15.1
VS (% TS)	87.2	83.8 - 89.1
TCOD (g/g TS)	1.10	0.920 - 1.26
SCOD (mg/g TS)	264.	180. - 406.
pH	7.3	7.1 - 7.7
Alkalinity (mg/g TS)	110.	78.8 - 148.
TKN (mg/g TS)	48.7	39.5 - 57.6
NH_3-N (mg/g TS)	18.4	13.5 - 28.0

[a]This summary represents the average values from analyses conducted on 17 samples.

Biodegradability of Dairy Cow Manure

Reference to the biodegradable and refractory fractions of the organic substrate will be made throughout this study. The refractory fraction (R) is defined as that portion of an initial quantity of organic material which is resistant to biological degradation and remains undergraded at the time when the rate of degradation has decreased to a level so as to be insignificant from an engineering standpoint.[16]

The refractory fractions used in the measurements of the biodegradability of dairy manure were determined from the portions of the initial VS that remained as HRT approached infinity. The R of the influent VS concentration was determined graphically for the four substrate concentrations investigated. An example is shown in Figure 3. The R averaged 0.575.

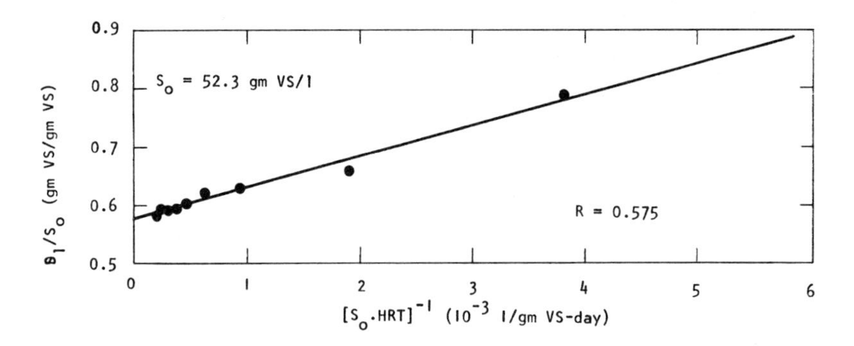

Figure 3. Graphical analysis of VS data to determine R.

In other words, the biodegradability of dairy cow manure was 42.5% of the influent VS concentration. This relatively low biodegradability fraction (1-R) indicates the resistance of the organic fraction of manure to biological degradation. The principal explanation for this resistance can be attributed to the high percentage of indigestible plant cell wall constituents present in the manure. Organic material that is contained in the plant cell walls include intricate poly-saccharide-lignin complexes which are characteristically more difficult and slower to degrade than many other organics.[17]

Digester Performance at Steady-State Conditions

In order to obtain the stated objectives of this investigation, each anaerobic digester was operated with a different combination of substrate concentration and HRT. A total of seventeen different combinations of substrate concentration and HRT were investigated. The following results are selected average results to illustrate representative performance characteristics.

Effluent Quality Characteristics

Figure 4 illustrates the effect of HRT on the extent of biodegradation of the total and biodegradable portions of VS. As noted, VS effluent concentrations are presented with respect to their corresponding influent substrate

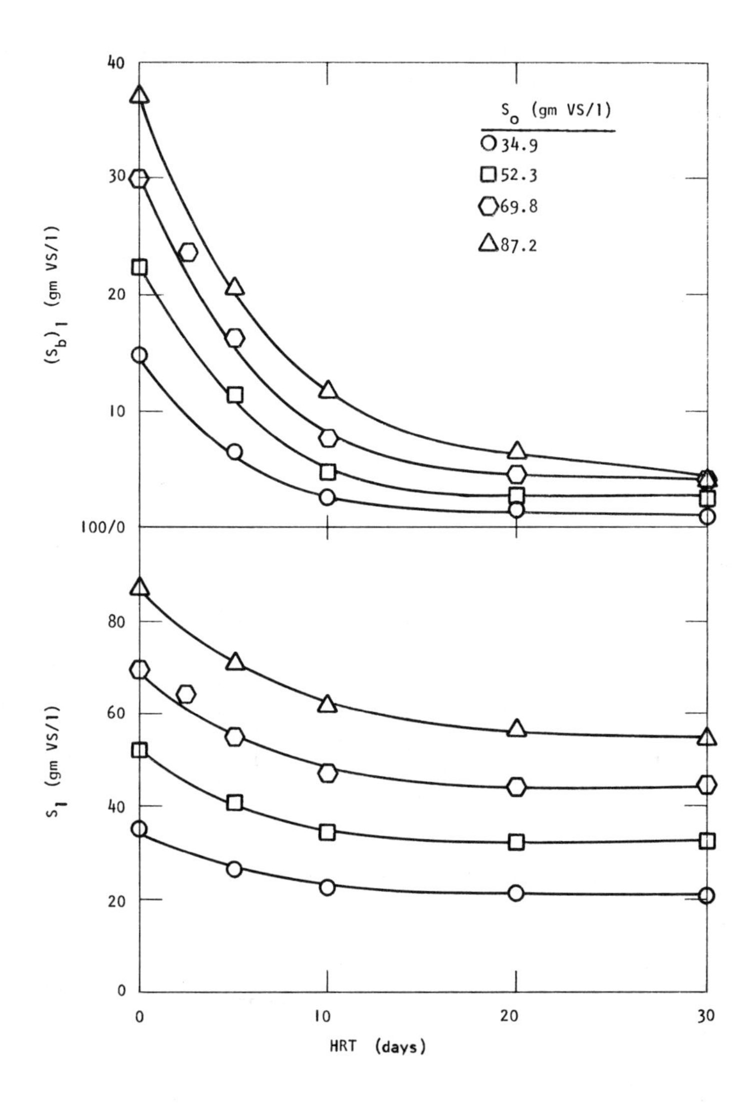

Figure 4. Effect of HRT on the biodegradation of BVS and TVS.

concentrations. An important observation that should be made from the data presented in Figure 4 is that there was a continual decrease in the effluent VS concentration as HRT was increased. It should also be noted that the rates of reduction of the VS in the manure began to slow down at about 7.5 days HRT and beyond 12.5 days HRT, the rates of reduction were insignificant when compared with their initial rates. The HRT at which this transition in the effluent concentration curve occurred was observed to be approximately 10 days.

Methane Production Characteristics

Both total gas and methane gas daily production rates are shown in Figure 5a. Here, production is expressed as liters of respective gas produced

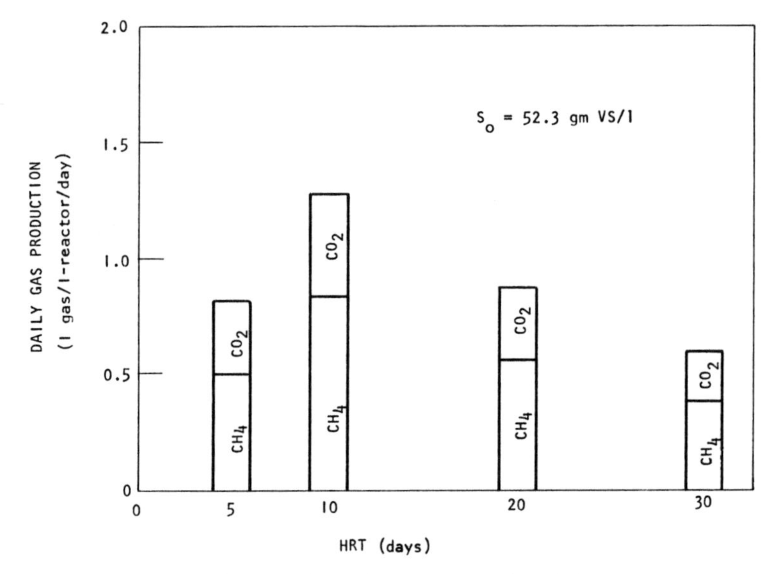

a. Effect of HRT on daily gas production rates.

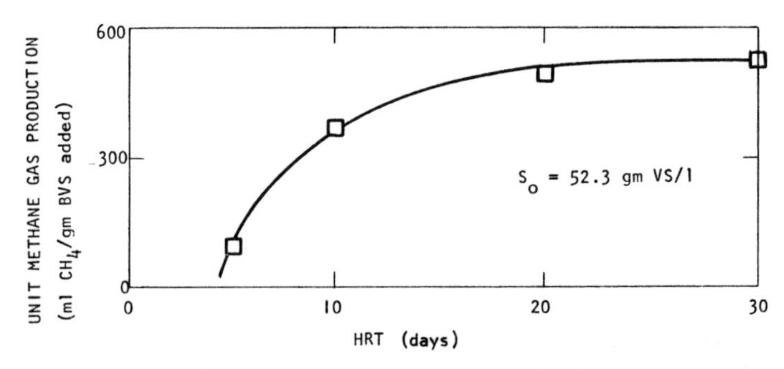

b. Effect of HRT on unit methane production rates.

Figure 5. Effect of HRT on methane production. **A.** Effect of HRT on daily gas production rates; **B.** Effect of HRT on unit methane production rates.

per liter of reactor per day. There was very little variation in the methane content of the fermentation gas, although changes in total methane production did occur. The average composition of the fermentation gas produced was approximately 63% methane. Methane production can also be expressed in terms of milliliters of methane evolved per gram of BVS (biodegradable volatile solids) added. Figure 5b illustrates the effect of HRT on this rate of methane production.

Treatment Efficiency

The development of removal efficiency as a function of HRT is shown in Figure 6. This figure illustrates that at each S_0 investigated, the percent reduction in VS increased as the HRT increased. This effect becomes less apparent as the HRT is lengthened beyond 10 days. It should be noted that this observation is closely associated with the relationships presented earlier for effluent VS concentration (Figure 4) and volume of methane produced per gram of BVS added (Figure 5b), as all three are interrelated.

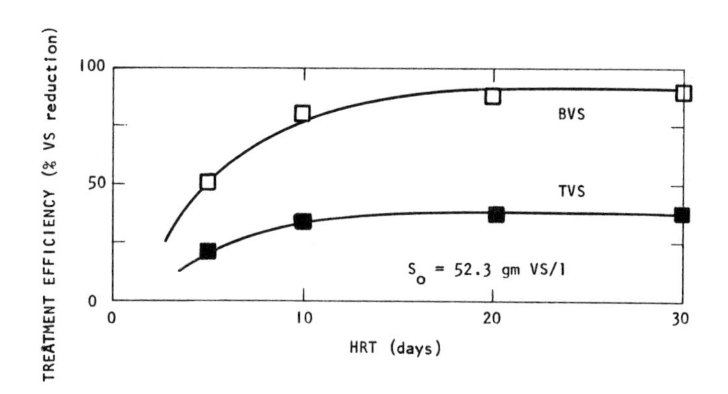

Figure 6. Effect of HRT on percent reduction of VS.

MATHEMATICAL DESIGN APPROACH FOR ANIMAL WASTE TREATMENT

Development of a Mathematical Relationship

Initial considerations in the development of this study called for a better understanding of how the process design variables affect the anaerobic fermentation process. Thus, an important observation made in this investigation was that the decomposition rate of VS could be expressed as a function of the process design variables. Analysis of the data for the relationship between HRT, influent VS concentration (S_o) and VOLR with respect to the process performance characteristics consistently displayed predictable patterns.

The relationship between the process design variables and the extent of biodegradation of VS is presented in Figure 7, showing that $(S_b)_1$ is

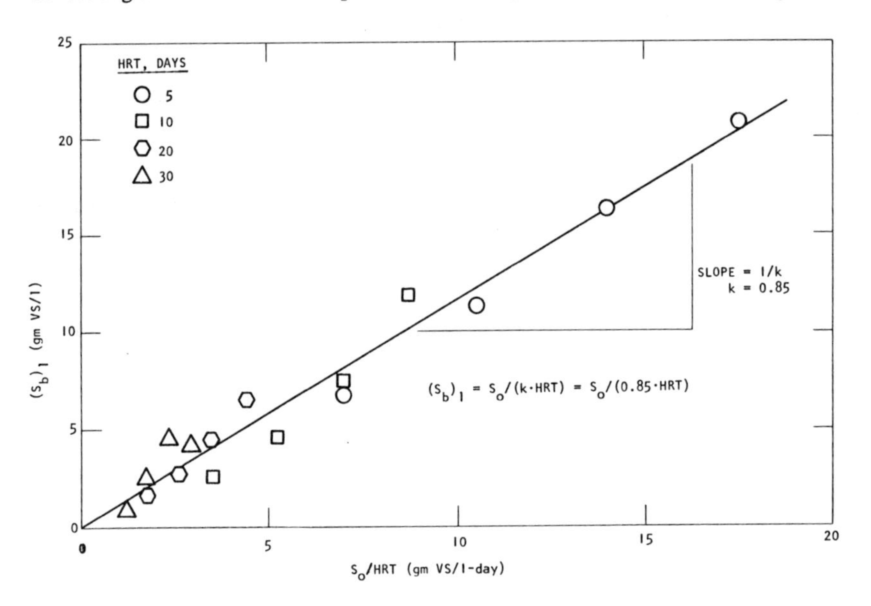

Figure 7. Effect of S_o and HRT upon the biodegradable effluent VS concentration.

dependent upon both the influent VS concentration and the HRT of the system. Thus, $(S_b)_1$ is observed as a linear function of S_o/HRT or VOLR. This relationship can be expressed mathematically as:

$$(S_b)_1 = S_o/(k \cdot HRT)$$

or:

$$S_1 = [S_o/(k \cdot HRT] + R\ S_o$$

The coefficient k can be defined as the rate of substrate removal expressed as days^{-1}. This coefficient represents the inverse of the slope of the line.

Figure 8 presents an excellent illustration of how these process design variables affect the effluent TVS concentration. The array of straight lines in this figure supports the two previously discussed relationships showing that $(S_b)_1$ is dependent upon both S_o and HRT. First, the dotted lines representing a constant S_o show that a decrease in HRT results in a linear increase in S_1 since $(S_b)_1$ is inversely proportional to HRT, and the refractory portion

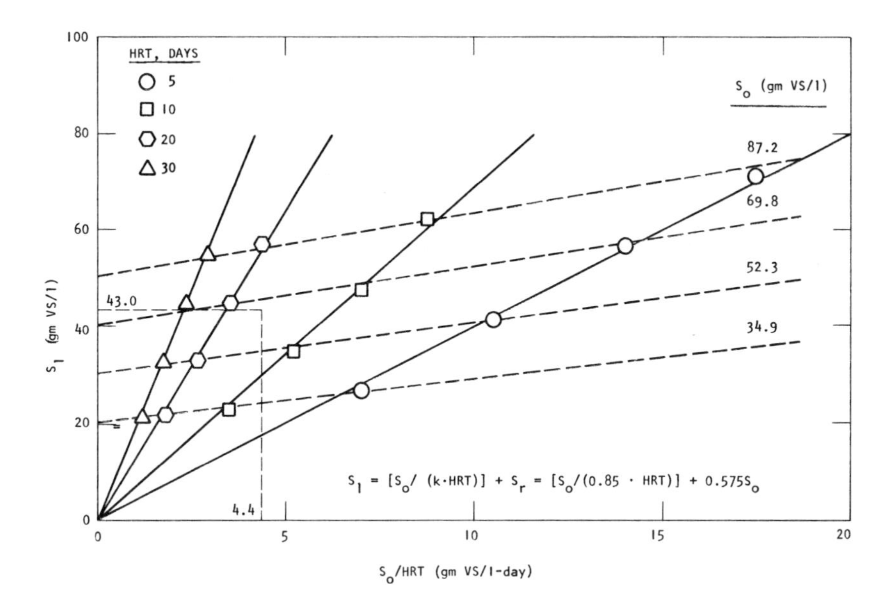

Figure 8. Effect of S_o and HRT upon the total effluent VS concentration.

(S_r) is unaffected by the treatment process. This explains why these lines are parallel and approach their respective refractory VS concentrations at the ordinate intercept when HRT approached infinity. The constant HRT values represented by solid straight lines show that an increase in S_o results in a proportional increase in S_1. This increase in S_1 is due to two factors: (a) a linear increase in biodegradable effluent VS concentration since $(S_b)_1$ is linearly related to S_o at constant HRTs; and (b) an increase in the influent refractory VS concentration. The increase in S_1 was proportional to the increase in S_o because both $(S_b)_1$ and S_r are proportional to S_o. Since S_o was increased by a constant factor, this explains why the constant S_o lines and the constant HRT points are equidistant.

It may be seen from data shown earlier in Figure 4 that the HRT of the system is the most critical parameter in the evaluation of digester performance and the quality of the end product. It should be mentioned at this time that the VOLR cannot be considered a critical design parameter because of two factors. First, since VOLR is a variable function of S_o and HRT (VOLR = S_o/HRT), it is incapable of quantitatively describing the operational variables of the treatment process. For example, if the substrate concentration and HRT are increased by the same factor, the loading rate would remain constant; however, the HRT would be greatly altered, thereby affecting the design and operation of the treatment process. Secondly, in most cases, the substrate concentration is preestablished, thereby requiring HRT to become the major design and operational variable. The importance of HRT in the design of anaerobic fermentation systems was first hypothesized by Rankin[18] in his study of conventional digestion systems and is now substantially supported by many investigators.[7,19-22]

It is necessary, from an economic viewpoint, to achieve maximum volumetric utilization of the digester capacity compatible with the selected design regime.[20] Optimum utilization of the volumetric capacity of a conventional digestion system can be achieved by increasing the influent solids concentration. Thus, the desirability and importance of maximizing the concentration of the substrate correlates with maximizing the volumetric waste utilization capacity of a system. Although a high influent solids concentration is considered desirable, there are restrictions on the degree of substrate concentration due to biological and physical limitations. In the treatment of dairy manure, the data had indicated that effective waste stabilization can be accomplished at influent solids concentrations as high as 87.2 g VS/liter or 10% total solids (TS). However, a 10% TS concentration might be considered the practical upper limit due to maintenance of adequate mixing in the digester and the difficulty of pumping the fibrous solids contained in the manure.

Mathematical Design Approach

There has been an increasing need in anaerobic treatment technology for the adaptation of design models to the anaerobic fermentation of animal wastes. The relationships necessary to bridge this gap have been investigated, developed and assembled in this study. The engineering significance of this investigation lies in using these relationships in the development of design and operational criteria for anaerobic waste treatment systems.

Digester design and operational criteria can be determined and established through the application of the mathematical relationship developed in this study. The design approach is comprised of the expressions shown in Table IV. Equations 1 and 2 can be employed to predict the effluent TVS concentration and the gross treatment efficiency of an anaerobic fermentation system at a given HRT and S_O. Equations 3 and 4 can be used to predict the daily and unit methane production rates, respectively. The volumetric rate of waste utilization can be predicted by Equation 5. The mathematical relationships and design coefficients presented in this table are expressed in terms of VS.

In order to illustrate the applicability of the design approach, an example using the above information is presented for anaerobic fermentation of dairy manure.

Table IV. Summary of expressions comprising the design approach for anaerobic treatment of animal wastes.

Design Relationships	Mathematical Expression	Equation Number
Total VS remaining in the effluent	$S_1 = [S_o/(k \cdot HRT)] + RS_o$	1
Percent reduction of total VS	$E_g = 100\ [S_o - S_1]/S_o$	2
Total daily methane production per liter of reactor at STP	$G_t = 0.35\ M\ [S_o - S_1]/HRT$	3
Methane production per gram BVS added at STP	$G_u = 0.35\ M\ [S_o - S_1]/(1\text{-}R)\ S_o$	4
Volumetric waste utilization rate	$dF/dt = [S_o - S_1]/HRT$	5
Design coefficients for anaerobic treatment of dairy manure		
Substrate removal rate coefficient at 32.5°C	$k = 0.85\ days^{-1}$	
Refractory fraction of VS	R = 0.575 g VS/g VS	
Conversion ratio of COD to VS	M = 1.43 g COD/g VS	

Example:

The problem will be to estimate the size of digester required to treat dairy manure from a 100-head dairy herd. Additionally, it will be beneficial to check the volumetric loading and estimate the percent stabilization and the amount of gas produced. For illustrative purposes, certain assumptions will have to be made. Pertinent design assumptions would include:

1. Waste production characteristics
   85 lb/1000 lb cow/day
   Total solids = 10.6 lb/1000 lb cow/day (12.5%)
   Volatile solids = 8.7 lb/1000 lb cow/day (82% TS)
2. Hydraulic regime of reactor = complete mix
3. Hydraulic retention time = 15 days
4. Influent total solids concentration = 8%
5. Digester operational temperature = 32.5°C
6. Specific gravity of manure = 1.0.

Solution:

1. Compute the daily manure volume and makeup water required for dilution.

   TS = 10.6 lb/cow/day x 100 cows = 1060 lb/day.
   Volume of manure at 12.5% TS = (1060 lb/day)/(8.34 x 0.125) = 1018 gal/day.
   Volume of manure at 8.0% TS = (1060 lb/day)/(8.34 x 0.08) = 1589 gal/day.
   Makeup water = 571 gal/day.

2. Compute the digester volume.
   $HRT = V/Q$
   $V = HRT \cdot Q = 15 \text{ day} \times 1589 \text{ gal/day} \times 0.1337 = 3187 \text{ ft}^3$
3. Compute the volumetric organic loading rate (VOLR)
   $VOLR = S_o/HRT = 65.6 \text{ g VS/l}/(15 \text{ day} \times 16.017) =$
   $0.273 \text{ lb VS/ft}^3\text{/day}$
   where: $S_o = (0.82)\ 8\%\ TS = 6.56\%$ or 65.6 g VS/l
4. Compute the effluent volatile solids concentration.
   $S_1 = S_o/(k \cdot HRT) + R\ S_o = 65.6/(0.85 \times 15) + 0.575(65.6) = 42.9$ g VS/l.
   where: $R = 0.575$ g VS/g VS; $k = 0.85 \text{ days}^{-1}$
   (See Figure 8 for a graphical determination of S_1.)
5. Compute the percent stabilization.
   $E_g = 100\ (S_o\text{-}S_1)/S_o = 100(65.6\text{-}42.9)/65.6 = 34.6\%$.
6. Compute the volumetric waste utilization rate.
   $dF/dt = (S_o\text{-}S_1)/HRT = (65.6\text{-}42.9)/(15 \times 16.017) =$
   $0.095 \text{ lb VS/ft}^3\text{/day}$
7. Compute the volume of methane produced.
   $G_t = 0.350\ M\ (S_o\text{-}S_1)/HRT = 0.350(1.43)(65.6\text{-}42.9)/15$
   $G_t = 0.76 \text{ ft}^3\ CH_4/\text{ft}^3$ reactor/day @ STP
   Also:
   $G_u = 0.350\ M\ (S_o\text{-}S_1)/S_o = 0.350(1.43)(65.6\text{-}42.9)(16.017)/65.6$
   $G_u = 2.77 \text{ ft}^3\ CH_4/\text{lb}$ VS added @ STP
   where: $M = 1.43$ g COD/g VS

SUMMARY

Mathematical models such as the one presented in this study attempt to formulate complex physical and biological phenomena occurring in biological waste treatment systems.[20] Such models are applicable only if they adequately describe the phenomenon under consideration or, in other words, if they fit the data. The approach developed in this study appeared to be able to predict the effluent characteristics of bench-scale anaerobic treatment systems reasonably well. The predictions were generally within about 5% of the actual effluent concentration. This variation is excellent when considering the variations that may have occurred in the experimental program.

The mathematical relationships presented in this study can provide a unified basis for the design of anaerobic fermentation systems. HRT has been suggested in this study as the operational parameter of choice for use in design and control of the treatment process. The operational HRT selected for design directly determines the reactor volume and the waste flow rate of the system.

Based on the information developed from the mathematical relationships, it is suggested that anaerobic systems digesting dairy manure should be designed for and operated at HRTs of at least ten days. Additional consideration should be given to the influent substrate concentration when selecting design and operational criteria, since S_O has an effect on the performance of the treatment system. A summary of the empirical expressions involved in process design and operation was shown in Table IV.

REFERENCES

1. Lawrence, A. W. "Anaerobic Biological Waste Treatment Systems," *Agricultural Wastes: Principles and Guidelines for Practical Solutions,* Proc. Cornell Univ. Conf. Agric. Waste Management (1971).
2. Kotze, J. P., P. G. Thiel and W. H. J. Hattingh. "Anaerobic Digestion—II. The Characteristics and Control of Anaerobic Digestion," *Water Res.* 3(7):459-493 (1969).
3. Kirsch, E. J. and R. M. Sykes. "Anaerobic Digestion in Biological Waste Treatment," *Prog. Ind. Microbiol.* 9:155-237 (1971).
4. Sawyer, C. N. "Anaerobic Units," Proc., Symposium on Advances in Sewage Treatment Design, Sanitary Engineering Division, American Society of Chemical Engineers, New York (1961).
5. Loehr, R. C. "Design of Anaerobic Digestion Systems," *J. Sanitary Eng. Div.,* American Society of Chemical Engineers, 92(SA1):19-29 (1966).
6. Loehr, R. C. *Agricultural Waste Management* (New York: Academic Press, Inc., 1974).
7. O'Rourke, J. T. "Kinetics of Anaerobic Waste Treatment at Reduced Temperatures," Ph.D. Dissertation, Stanford University, Stanford, California (1968).
8. Burd, R. S. "A Study of Sludge Handling and Disposal," Water Pollution Control Research Series Publication No. WP-20-4, Federal Water Pollution Control Administration (1968).
9. American Public Health Association. *Standard Methods for the Examination of Water and Wastewater,* 13th ed. (New York: American Public Health Association, 1971).
10. Prakasam, T. B. S., E. G. Srinath, P. Y. Yang and R. C. Loehr. "Analyzing Physical and Chemical Properties of Liquid Wastes," Special publication SP-0275, *Standardizing Properties and Analytical Methods Related to Animal Waste Research* (St. Joseph, Michigan: American Society of Agricultural Engineers, 1975).
11. Cummings, R., ASTARC, Cornell University, Ithaca, New York, personal communication (1975).
12. Grant, F. and F. Brommenschenkel, Jr. "Liquid Aerobic Composting of Cattle Wastes and Evaluation of By-Products," Report No. 660/2-74-034, U.S. Environmental Protection Agency, Washington, D.C. (1974).
13. Cramer, C. O., J. C. Converse, G. H. Tenpas and D. A. Schlaugh. "The Design of Solid Manure Storage for Dairy Herds," Paper No. 71-910, American Society of Agricultural Engineers (1971).
14. U.S. Environmental Protection Agency. "Development Document for Effluent Limitations Guidelines and New Source Performance Standards for the Feedlots Point Source Category," EPA Report No. 440/1-74-004-a, Washington, D. C. (1974).
15. "Farm Animal-Waste Management," J. R. Miner, Ed., North Central Regional Research Publication No. 206, Iowa Agricultural Experiment Station Special Report 67, Iowa State University, Ames, Iowa (1971).

16. Foree, E. G. and P. L. McCarty. "The Decomposition of Algae in Anaerobic Waters," Technical Report No. 95, Department of Civil Engineering, Stanford University, Stanford, California (1968).
17. Smith, L. W., H. K. Goering and C. H. Gordor. "In Vitro Digestibility of Chemically-Treated Feces," *J. Animal Sci.* 31:1205-1209 (1970).
18. Rankin, R. S. "Digester Capacity Requirements," *Sew. Works J.* 20(5): 478 (1948).
19. McCarty, P. L. "Anaerobic Processes," paper presented at the Birmingham Short Course on Design Aspects of Biological Treatment, Internat'l Assoc. of Water Pollution Research, Birmingham, England (1974).
20. Lawrence, A. W. and P. L. McCarty. "Kinetics of Methane Fermentation in Anaerobic Waste Treatment," Technical Report No. 75, Department of Civil Engineering, Stanford University, Stanford, California (1967).
21. Young, J. C. and P. L. McCarty. "The Anaerobic Filter for Waste Treatment," Technical Report No. 87, Department of Civil Engineering, Stanford University, Stanford, California (1968).
22. Dea, S. J. "Continuous Loading of Completely Mixed, High-Rate Anaerobic Digesters," Ph.D. Dissertation, University of Arizona, Tuscon, Arizona (1966).

GLOSSARY

TS	=	Total solids
VS	=	Volatile solids
TVS	=	Total volatile solids
BVS	=	Biodegradable volatile solids
SRT	=	Mean solids retention time, time
HRT	=	Mean hydraulic retention time, time
VOLR	=	Volumetric organic loading rate, mass/volume-time
V	=	Reactor volume, volume
Q	=	Waste flow rate through the reactor, volume/time
S	=	Substrate concentration, mass/volume
S_o	=	Influent total substrate concentration (TVS), mass/volume
S_1	=	Effluent total substrate concentration (TVS), mass/volume
S_b	=	Biodegradable substrate concentration, mass/volume
$(S_b)_o$	=	Influent biodegradable substrate concentration (BVS), mass/volume
$(S_b)_1$	=	Effluent biodegradable substrate concentration (BVS), mass/volume
S_r	=	Refractory or nonbiodegradable VS concentration, mass/volume
R	=	Refractory fraction or the ratio of the refractory VS concentration to the influent TVS concentration, expressed as a decimal
k	=	Rate of substrate removal, time^{-1}
E_g	=	Gross treatment efficiency, expressed as percent TVS reduction
dF/dt	=	Volumetric rate of substrate utilization, mass/volume-time

G_t = Total daily methane production per reactor volume at STP, volume/volume-time

G_u = Daily unit methane production per mass of BVS added at STP, volume/mass

M = Conversion ratio of COD to VS in order to equate methane production to VS reduction, mass/mass

STP = Standard Temperature ($0^\circ C$) and pressure (1 atm)

28

A COMPARISON OF AN ANAEROBIC DIGESTER AND AN AERATION SYSTEM TREATING PIGGERY WASTE FROM THE SAME SOURCE

P. J. Mills
The North of Scotland College of Agriculture
Aberdeen, Scotland

INTRODUCTION

When work first started in Aberdeen on the anaerobic digestion of piggery wastes, the system was being investigated as a treatment process, with gas production as a by-product. There has been a tendency to ignore the treatment aspect of digestion and dismiss it on the basis of its being an uneconomic energy producer.

This chapter compares the results of an anaerobic and an aeration/separation system operating on the same source of piggery waste. A comparison of effluent quality and costs is given. The costs are those of operating an experimental field-scale plant, but they have been extrapolated to larger but simpler systems to project the place of each process as a treatment system in on-farm situations.

METHODS

Slurry from sows and fattening pigs housed on slats was collected by vacuum tanker and stored in a steel tank (capacity 25 m^3) with a mixer. The contents were agitated by the mixer and slurry transferred, as required, to the loading tank of each system.

The anaerobic digestion system is basically that described by Robertson.[1] The retention time has been reduced to 10 days. Laboratory studies at the Rowett Research Institute have shown that this could be reduced to seven days without instability, but below this there may be problems. The temperature is controlled at 35°C. The electrical power input has been reduced by linking the circulating pump to the thermostat and reducing the time of stirring. The system produces enough gas for theoretical and practical needs at ambient temperatures of 0°C, despite its inefficiencies. A second plant of similar size has recently been commissioned. This plant is much more efficient owing to better insulation and a heat recovery system. Excess gas has not been valued in the costings for the direct comparison.

The aeration system consists of an oxidation ditch aerated by a perforated disc rotor. The slurry is passed through a 1-mm mesh rotary screen separator before being loaded by a peristaltic pump at half hourly intervals. Residence times of 2.5, 5, 10 and 17 days were tested; 2.5 days was found not to be stable, and at 17 days complete nitrification occurred. Only the 5- and 10-day residence times are reported here. The temperature was usually slightly above ambient (7 to 20°C). The limitation of loading rate is probably due to the low operating temperature.

The analyses given for the outputs are of fully mixed effluents. The aeration process also produces an easily handled fiber of approximately 20% dry matter.

RESULTS AND DISCUSSION

The characteristics of the inputs and outputs of the two systems are noted in Tables I and II. The operating parameters of the systems are presented in Table III. The main differences in the results lie in soluble COD and nitrogen levels. The soluble COD levels of aerated effluent suggest that high levels could be applied to land before there is a risk of pollution by seepage. Digester effluent, while not as good, has still half the COD and 20% of the BOD of untreated waste. A subjective assessment of odor indicates

Table I. Analyses of inputs.

	Digester Input Raw Slurry	Aeration Input Screened Slurry
Total Solids (g/l)	43.0	33.2
Suspended Solids (g/l)	34.6	24.8
Dissolved COD (g/l)	14.9	14.9
Total COD (g/l)	74.3	52.5
Organic N (mg/l)	1280	1052
NH_3 N (mg/l)	2490	1635
Oxidized N (mg/l)	0	0
BOD (mg/l)	20100	–

Table II. Analyses of outputs.

	Digester Output	Aeration Output 5[a]	Aeration Output 10[a]
Total Solids (g/l)	25.5	30.5	28.5
Suspended Solids (g/l)	18.1	21.6	21.9
Dissolved COD (g/l)	7.1	1.54	2.31
Total COD (g/l)	36.4	33.6	36.7
Organic N (mg/l)	–	1280	1690
NH_3 N (mg/l)	2240	80	25
Oxidized N (mg/l)	0	270	435
BOD (mg/l)	4250	–	–

[a]Five and ten days hydraulic retention time.

Table III. Operating parameters.

System	Anaerobic	Aerobic	Aerobic
Capacity (m^3)	13.5	3.4 (3.8)[a]	3.4 (3.6)[a]
Throughput (m^3/day)	1.35	0.78	0.35
Residence Time (days)	10	4.9	10.3
Capital Cost ($)	25000	6800	6800
Cost/m^3 Throughput	18500	8750	19500
Electricity ($/yr)	188	270	210

[a]The figures in parentheses indicate the mean volume.

total elimination by aeration and a large reduction by digestion. Most of the sulfur compounds which would be responsible for waste odors appear to be released as gases during digestion, which then dissolve in the water in the condensate traps of the gas holder.

It can be seen that there are considerable losses of nitrogen during aerobic treatment. This would be beneficial if complete treatment prior to discharge to a watercourse were being sought. If all that is required is stabilization before land application, then the loss of fertilizer value becomes significant. Experiments show that oxidized nitrogen is lost rapidly on storage after aerobic treatment. If very high quality effluents are required the major pollutant becomes potassium, the levels of which are unchanged by either process.

There appear to be no benefits in extending aeration to 10 days as this resulted only in a higher oxidized nitrogen level and worse soluble COD. The system was loaded semicontinuously but the level was reduced only once a day, so the volume in the system varied.

The cost figures (Table III) are for running an experimental plant. No labor costs have been included. Owing to its biological stability the digester required only physical attention, clearing of blocked pipes, etc. Owing to its mechanical simplicity the aeration system needed little maintenance but required a higher level of microbiological understanding in the event of process fluctuations.

The operational characteristics of the two systems are quite different. Anaerobic digestion is very stable. No biological upsets have occurred in three years of running, even though loading has been stopped for long periods and the contents exposed to air while mechanical modifications were made.

The stability of digestion has been discussed by Kroeker[2], where it is suggested that operator attention and monitoring requirements are not as important as for domestic plants. This has also been found to be true in comparison with the aeration plant. Variations in loading were readily absorbed by the digester without any need for control, whereas changes in loading of the aeration plant required careful monitoring. Robinson[3] has compared control systems which may lead to steady-state automatic loading aerators, eliminating the need for such detailed monitoring.

Aside from any credit value of excess gas produced by digestion, the running costs are trivial compared to the capital cost. If the excess gas from the existing digester were converted into electrical energy it could supply all the needs of the system. Even so, digestion is still too expensive owing to its high capital cost. This means that, at the scale selected, aeration is a much cheaper way of providing treatment.

PROJECTION

Both the plants used were designed for experimental purposes and were fitted with mechanical equipment much larger than necessary and more complex and expensive monitoring systems. Thus it should be possible to produce larger capacity units by increasing the vessel size alone. The vessel represents approximately 20% of the cost of each system so the capacity of each can be increased by a factor of ten while increasing the cost by only a factor of two. The running cost of the aeration plant increases proportionally with throughput, making the running cost a significant factor. However, with zero running costs the digester would still be more expensive than the aeration process at five days retention time.

Figure 1 shows the projected costs of operating aeration (1) and digestion (2) plants at increasing scales under the same conditions as the experimental units. The annual capital cost has simply been taken as 10% of the total. Grants, tax allowances, etc., would in a real situation make this much more complicated. It can be seen that up to a capacity of 8 m^3 the aeration plant is cheaper to operate.

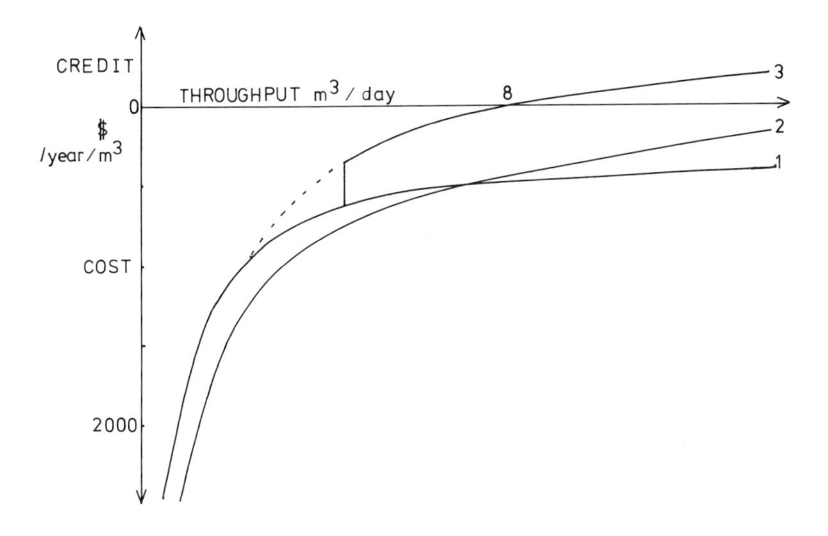

Figure 1. Relative costs of aerobic and anaerobic systems.

A well-insulated large digester using 50% efficient heat recovery system has a theoretical maintenance requirement of about 30% of the gas energy produced. Our second digester operates at about 45%, leaving in excess of about 2.5 kW/m^3 digester throughput. If this credit is added to curve 2, curve 3 results. This shows that, above 2.5 m^3/day, digestion is cheaper. However, increased costs would be involved in using the extra energy so the benefits would be apparent only above 4 m^3/day. Above 8 m^3/day digestion is in credit. Projection of the results from Figure 1 produces the scheme shown in Figure 2.

Zone 1 represents many typical situations of relatively low numbers of stock at a low density. No treatment is required; good management will cover all pollutional risks.

Zone 2a represents units of larger size which have sufficient land not to need treatment. However, above a throughput of about 8 m^3/day, anaerobic digestion could be economic as an energy source.

Zone 2b represents units of the same size which require some form of treatment. The level of treatment will satisfy many situations. Providing that there is a need for the energy produced, the treatment can be a by-product of the gas generation.

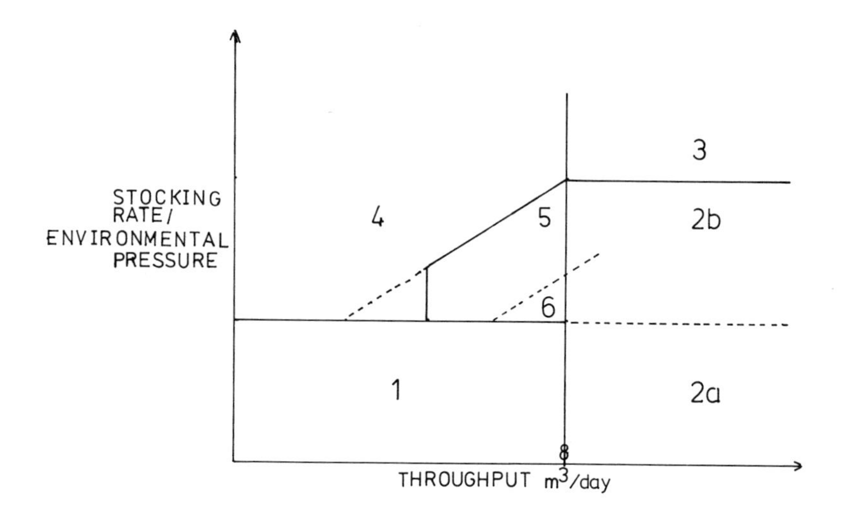

Figure 2. A projected scheme for the selection of treatment systems.

In Zone 3 digestion will not provide a level of treatment sufficient to meet local environmental requirements, so aeration or a two-stage system will be required.

In Zone 4 aeration is the cheapest system to operate. It will provide a level of treatment suitable for most situations except those where direct discharge to a watercourse is required.

In Zone 5, although digestion is not economic from an energy point of view alone, its operating costs are lower than those of the same capacity aeration system. If there is a use for the excess energy, this zone starts at about 2.0 m^3/day throughput, but this scale is probably too small to justify the extra equipment to use the gas, so a throughput of 4.5 m^3/day is more likely.

If there is no use for the gas produced other than maintaining the process, then at the start of Zone 6 (approximately 7.5 m^3/day) digestion would become more economic than aeration.

These figures have been calculated on the basis of the slurry fed to the experimental plant. The solids level indicates a dilution of about 1:1 with water. The economics of digestion improve with solids concentration, as a smaller plant is required and the maintenance energy requirement per unit volume falls with the size of vessel, but the gas yield may increase to 2 to 2.5 volumes per digester volume per day. The size of an aeration plant to handle the reduced volume would also be smaller, but the aeration (power) demand would remain about the same.

STORAGE AND APPLICATION

Effluent from the digester continues to digest at a rate dependent on temperature. This effluent is fed into a lagoon 2 m deep with a residence time of 200 days. This has so far eliminated all settleable suspended solids to produce an effluent of about 300 mg/l total COD. This material could easily be spread by pipeline irrigation owing to its low suspended solids.

On storage of aerated effluent over a winter period, further nitrogen losses occurred and the soluble COD rose by 30%. The oxidized nitrogen decomposed almost immediately. There is approximately 10% less liquid to handle from the aeration process, but there is also a fibrous solid to store and spread. The solid is easy to handle and has a volume of about 20% of the volume of raw slurry. It stacks readily and composts rapidly to produce a material like settled sludge.

The costs of handling and costs benefits of fertilizer value have not been taken into account in the preceding calculations, although both factors appear to be of greater credit to the digestion process.

IMPROVEMENTS

The efficiency of the aeration system could be improved by using a better insulated vessel. This would mean that the system would operate at higher than ambient temperatures, thus permitting higher loading rates, which would probably result in greater process stability. A more efficient aerator also could have been used, lowering the running costs.

The efficiency of the digestion system could be improved by recycling sludge. There is still considerable gas-producing potential in the output with a 10-day retention time. If settled sludge were recycled, gas production might be increased by 25% with only a 10% increase in vessel size. The effluent quality would drop, however. Increased size would lead to great economies of scale with a digestion system with reduced digester area: volume and continuous overflow heat recovery a possibility.

CONCLUSIONS

It has previously been considered that anaerobic digestion was of interest only as an energy-producing process. It can be shown that anaerobic digestion can be economical as a treatment process for units with about 100 breeding sows. Anaerobic digestion does not provide such a high quality effluent as aeration but handles the whole slurry in one stage and preserves all the nutrient content of the raw material.

ACKNOWLEDGMENTS

The anaerobic work described in this paper is carried out as a joint project with the Rowett Research Institute, who are extending the work to cattle and poultry wastes. The aerobic work is carried out as a multidisciplinary project within the North of Scotland College of Agriculture.

REFERENCES

1. Robertson, A. M., G. A. Burnett, P. N. Hobson, S. Bousfield and R. Summers. "Bioengineering's Aspects of Anaerobic Digestion of Piggery Wastes," in *Managing Livestock Wastes* American Society of Agricultural Engineers, St. Joseph, Michigan (1975).
2. Kroeker, E. J., D. D. Schulte, H. M. Lapp and A. B. Sparling. "Anaerobic Digestion Stability and Food-Processing Waste Treatment," 1st International Congress on Engineering and Food, Boston, Massachusetts, August 9-13, 1976.
3. Robinson, K. and D. Fenlon. "A Comparison of pH and Dissolved Oxygen–Controlled Nutrient Addition for the Maintenance of Steady State in a Mixed Continuous Culture," *J. Appl. Bacteriol.* (in press).

29

ANAEROBIC DIGESTION OF POULTRY WASTE WITH AND WITHOUT ACID HYDROLYSIS PRETREATMENT

P. Y. Yang
Department of Agricultural Engineering
University of Hawaii at Manoa
Honolulu, Hawaii

K. K. Chan
Division of Development
The Hong Kong Productivity Center
Hong Kong

INTRODUCTION

To accomplish organic solid or semisolid waste stabilization, both aerobic and anaerobic digestion processes are practiced. Traditionally, the anaerobic process has been used for achieving such stabilization in most of the sewage treatment plants. Because of the difficulty of maintaining an effective anaerobic process, the aerobic process also has been of interest. The difficulty of the anaerobic process includes its low reaction rate and unstable operational performance. In order to improve the anaerobic process, the present study was initiated to investigate the feasibility of increasing the digestion rate and gas production rate by pretreatment of organic waste. Heat and acid hydrolysis were used to break down the complex components in organic waste into small simpler fragments in a relatively short period of time or to take over a large part of the activity performed by hydrolytic enzymes. It was expected that the pretreated organic wastes would be more easily degraded by the microorganism involved in the anaerobic process, especially the methane formers, which play the chief role of stabilization, grow slowly and are more susceptible to a change of environment. A decrease in the activity of methane formers can destroy the equilibrium rate of liquefaction and gasification.

In addition to evaluating the feasibility of improving the waste-stabilization and gas-production rate, the waste-stabilization and gas-production rate constants were evaluated for such modified substrates and compared to those of a conventional unit in which the substrate was not pretreated prior to the anaerobic process. This study was not intended to estimate the costs associated with the pretreatment.

MATERIALS AND METHODS

Organic Waste

Poultry manure was used as organic substrate in the present study. It was obtained from a very small farm (about 50 hens) close to the Asian Institute of Technology, Bangkok, Thailand. A suspension of poultry manure was prepared and allowed to pass through a mesh equivalent to standard U.S. sieve no. 8 in order to screen out large particles such as grit and rice husk. The rather homogeneous filtered suspension was then used for the study.

For the pretreatment of poultry waste, different dilutions of the stock suspension with distilled water were prepared. The acid hydrolysis step involved was followed from Gaudy *et al.*[1] who evaluated the acid hydrolysis of activated sludge in the extended aeration process. The pH was lowered by adding concentrated H_2SO_4 to pH 1.0 and autoclaving the mixture for 5 hr at 15 psi and 121°C. After cooling, the pH was brough back to 7.0 by adding NaOH before adding the mixture to the anaerobic process.

The initial characteristics of acid-hydrolyzed and nonhydrolyzed poultry wastes are shown in Table I. The percentage of hydrolysis of poultry waste is shown in Table II. The average percentage of hydrolysis with different

Table I. Initial characteristics of different batches of poultry waste.

Kind of Poultry Waste (symbol)	Total COD (mg/l)	VSS (mg/l)	Volume of Waste (liters)
Hydrolyzed			
HP1	20,598	9,960	5.5
HP2	14,844	7,040	5.5
HP3	8,927	4,860	5.5
HP4	3,893	1,660	10
Nonhydrolyzed			
NP1	21,977	12,500	5.5
NP2	14,195	7,800	5.5
NP3	10,483	6,540	5.5
NP4	2,567	1,610	4.5

Table II. Poultry waste hydrolysis percentage.

Poultry waste dilution ratio (with distilled water)	0:0	7:3	5:5	3:7
Percentage of hydrolysis	35.3	39.2	38.3	41.0
Average percentage of hydrolysis		38.5		

dilution ratios was 38.5. The value is low when compared to activated sludge which had 49%.[2] The nitrogen content was analyzed and shown to be 13% of the total volatile suspended solid content in the poultry manure. According to McCarty,[3] an 11% nitrogen content of total cell volatile suspended solids would be enough to meet the requirements for the growth of anaerobic microorganisms. Therefore, no supplementary nitrogen was required for this study.

Anaerobic Digester

A simple laboratory digestion apparatus used throughout the experiment is shown in Figure 1. Glass aspirator bottles of six-liter capacity were used as digesters. Each bottle had two openings at the top and one near the bottom. The lower outlet, 1.1-cm diam, was used for sample removal. One of the upper openings, 1.1-cm diam, allowed the addition of chemicals, while another, 0.6-cm diam, functioned as a gas outlet and was connected to a one-liter inverted graduated cylinder by means of glass tubing. Gas was collected by downward displacement of water in the cylinder. The reactor was mounted on a magnetic stirrer with adjustable speed control. COD (total and filtrate), suspended solids (SS), volatile suspended solids (VSS), pH, alkalinity, volatile acids, gas production and temperature were measured during the period of the experiment.

Acclimatization

Effective operation of an anaerobic digestion process cannot be obtained unless the microorganisms involved in the process have been acclimatized to the wastes and to their environments. For the present study, two experimental batches were started: one fed with hydrolyzed poultry waste and the other with raw poultry waste. Actively digesting sludge from the septic tank of a nearby soft-drink bottling plant was used as the seed, making a proportion of 60% waste and 40% seed by volume. The whole acclimatization period was carried out in a 30°C walk-in incubation room. Alkalinity, pH and volatile acids analysis were performed weekly. Acclimatization

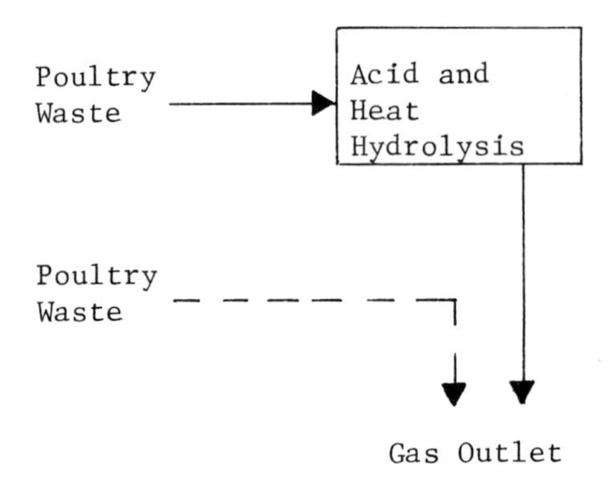

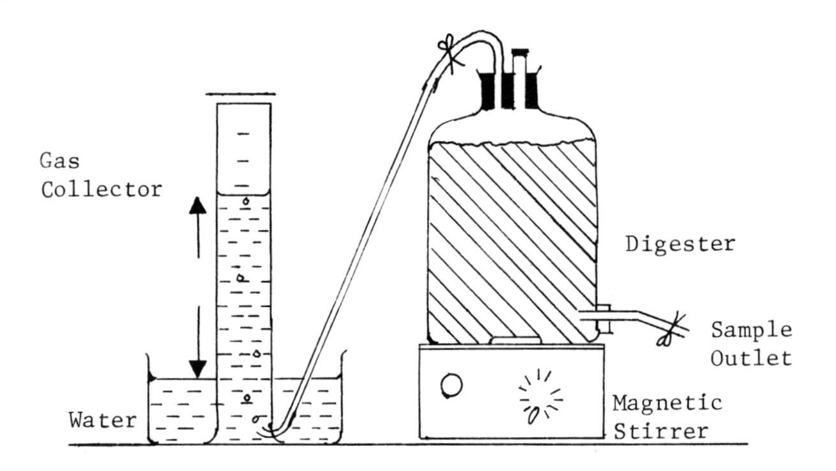

Figure 1. Anaerobic digester and gas collector.

continued until the maximum daily gas production rate was reached and could be used for further study.

Analytical Methods

Determination of COD, SS, VSS and alkalinity followed *Standard Methods*.[4] The pH of samples was measured by a Zeromatic pH meter. Volatile acids were determined by the column partition chromatography method. Methane gas was determined in the residual gas after the total gas produced had been passed through carbonate-free sodium hydroxide, following the method of Malina.[5]

RESULTS AND DISCUSSION

Eight batches of poultry waste with different initial total COD concentrations with and without acid hydrolysis were prepared and subjected to the anaerobic digestion process with daily averaged ambient temperature ranging from 29.7 to 33°C. Their initial characteristics are shown in Table I. With different initial characteristics of the wastes, their performances based on the degrees of digestion and the gas production should be different and will be presented and discussed in the following sections.

Performances of Digestion Process

In Figures 2 and 3, the changes of COD, SS, pH, alkalinity and volatile acid concentrations are shown for the anaerobic digestion of hydrolyzed and nonhydrolyzed wastes, respectively. The changes of these characteristics (except filtrate COD) followed the same pattern in both hydrolyzed and nonhydrolyzed systems. Due to the hydrolytic process or pretreatment of the waste, instead of the accumulation of filtrate COD which occurred in the nonhydrolyzed system, degradation of filtrate COD occurred. Therefore, it took about 17 days to stabilize the total COD content in the hydrolyzed system and 23 days in the nonhydrolyzed system. In addition, the rate of accumulation of volatile acid content in the nonhydrolyzed system was higher than in the hydrolyzed system within the first 8 days of operation. The alkalinity content in the hydrolyzed system was higher than in the nonhydrolyzed system. Therefore, it is expected that the stabilization rate (measured as total COD) is higher in the hydrolyzed system.

In Figure 4 and Table III, it can be seen that the removal rate constant, K_i (measured as total COD), is higher in the hydrolyzed system than in the nonhydrolyzed system. Figure 4 illustrates the evaluation of the constants shown in Table III. The results may be explained in the following way. As the acid hydrolysis provides the partial liquefaction of waste, it also provides more alkalinity, which provides better buffering capacity than with the nonhydrolyzed waste. It is known that the accumulation of volatile acids may delay or inhibit the activity of anaerobic microorganisms.

The methane gas fraction was measured in these two systems. The average value for the hydrolyzed waste was 81.8%, with a range of 71.2 to 90%, and was 80.5%, ranging between 73.5 and 83.5%, for the nonhydrolyzed waste.

Effect of Initial COD Concentration on Total Gas Production

The cumulative gas production with different initial COD concentrations is shown in Figures 5 and 6 for hydrolyzed and nonhydrolyzed systems, respectively. The pattern of the accumulation of gas as a function of the

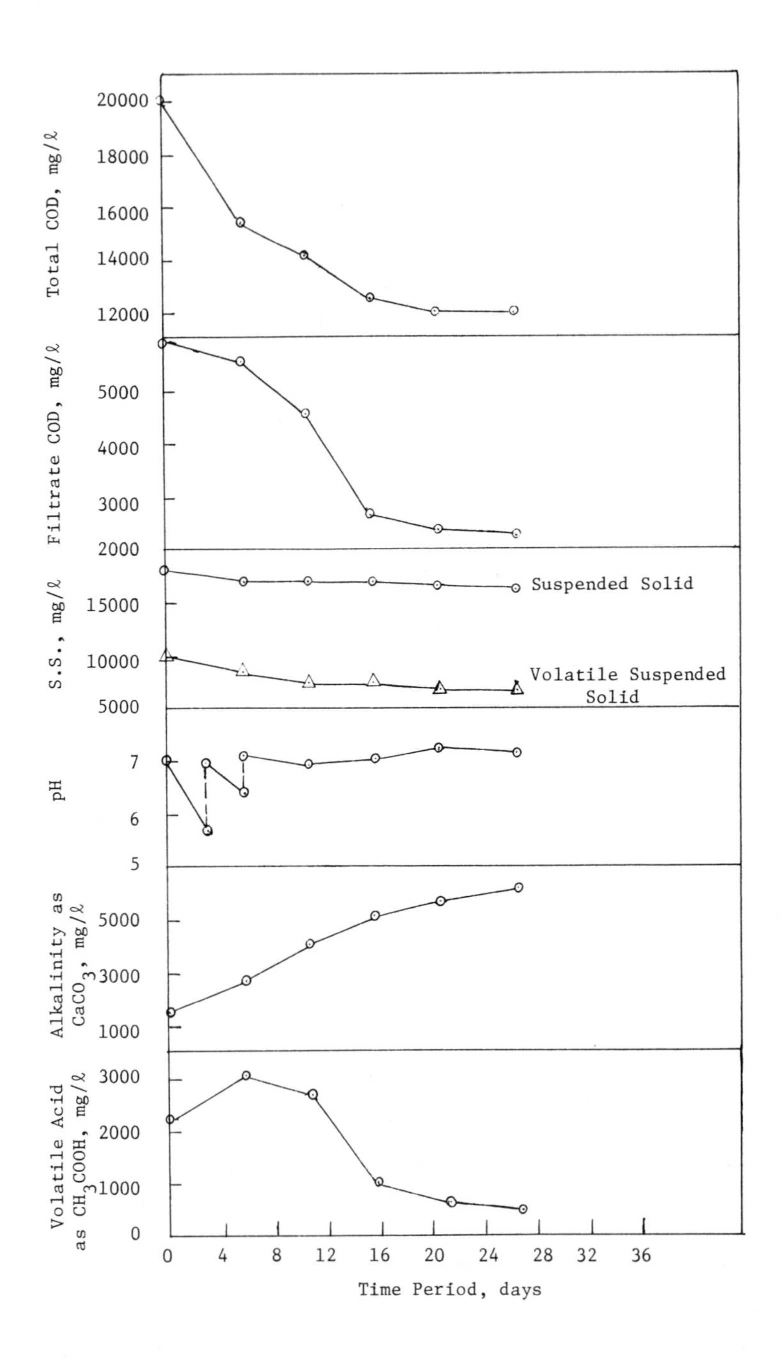

Figure 2. Progress in digestion of hydrolyzed waste.

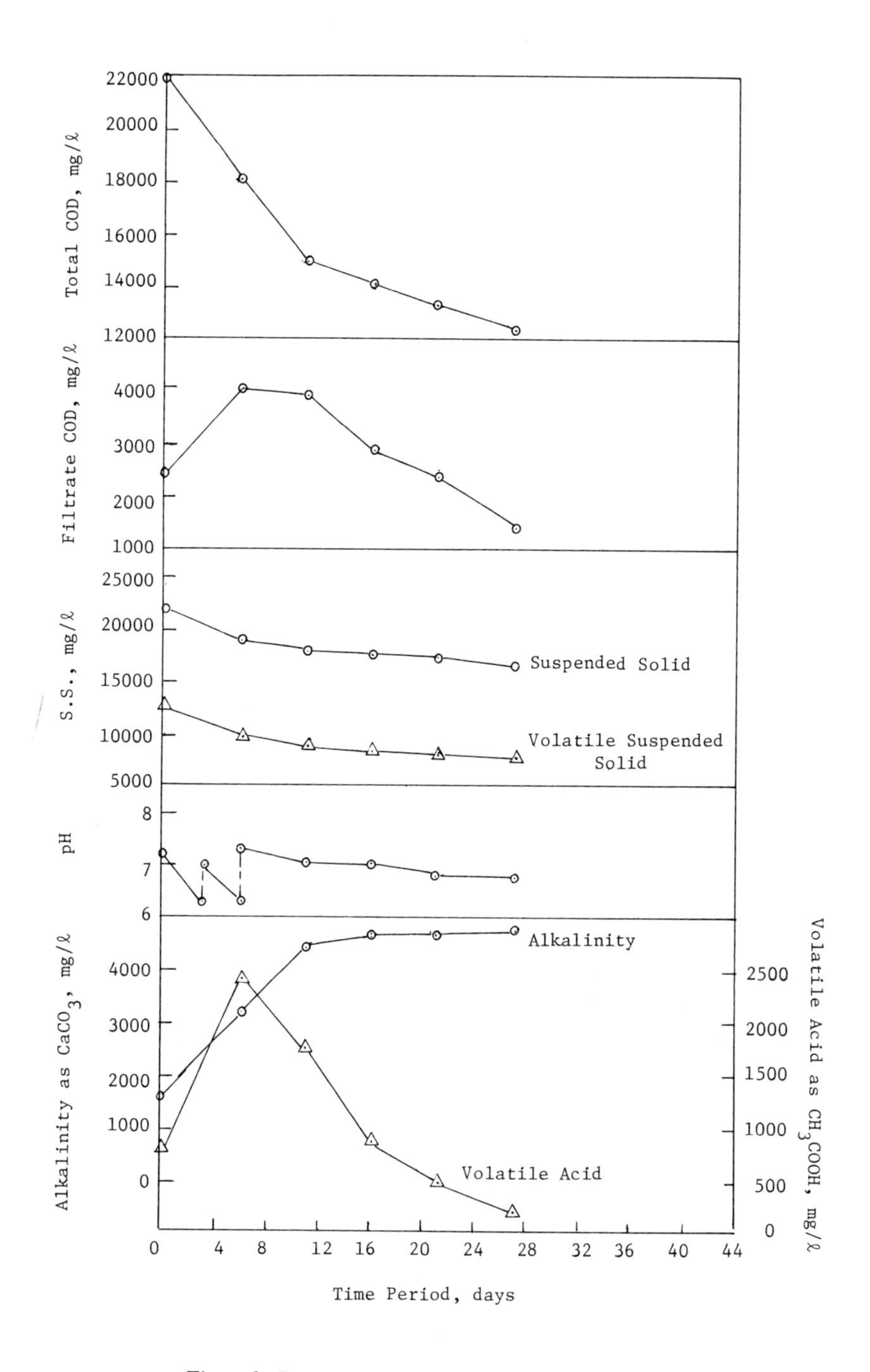

Figure 3. Progress in digestion of nonhydrolyzed waste.

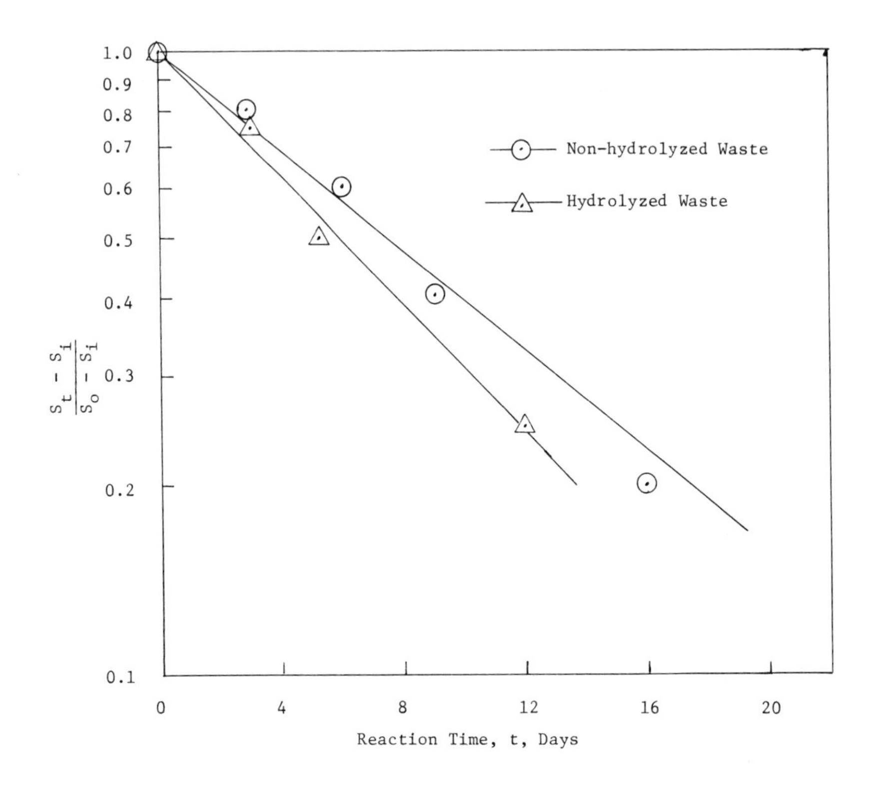

Figure 4. Evaluation of total COD removal rate, K_i.

reaction time is similar to that of the growth pattern of either aerobic or anaerobic organisms. Therefore, the hyperbolic relationship between the initial biodegradable COD and the specific gas production can be established as the following equation:

$$K = \frac{K_m \cdot S_o}{K_s + S_o} \tag{1}$$

where K = specific gas production rate

K_m = maximum gas production rate

K_s = saturation constant which is equal to the substrate concentration at half value of K_m

S_o = biodegradable initial COD concentration (the relationship between total COD and biodegradable COD is shown in Figure 7)

Table III. Comparison of total COD removal rate constants.

Type of Poultry Waste	Initial Total COD (mg/l)	Removal Rate Constants (day^{-1})
Nonhydrolyzed	21,977	0.089
	14,195	0.044
	10,483	0.132
	2,567	0.155
Hydrolyzed	20,598	0.117
	14,844	0.139
	8,927	0.151
	3,893	0.160

$$\ln \frac{S_t - S_i}{S_o - S_i} = K_i t$$

where: S_o = Total COD at time zero,
S_t = Total COD at time t,
S_i = Residual COD, and
K_i = Removal rate constant with consideration of residue, day^{-1}.

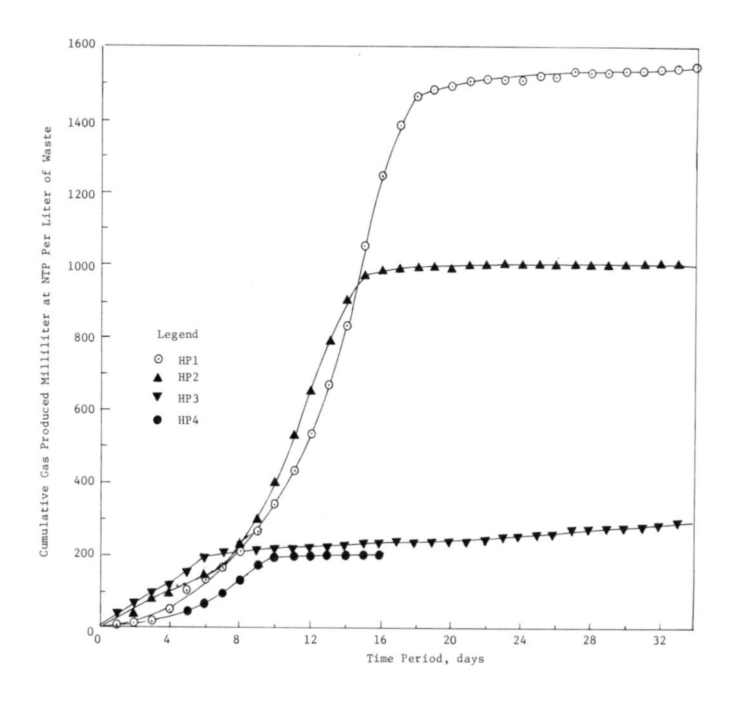

Figure 5. Gas production in hydrolyzed waste.

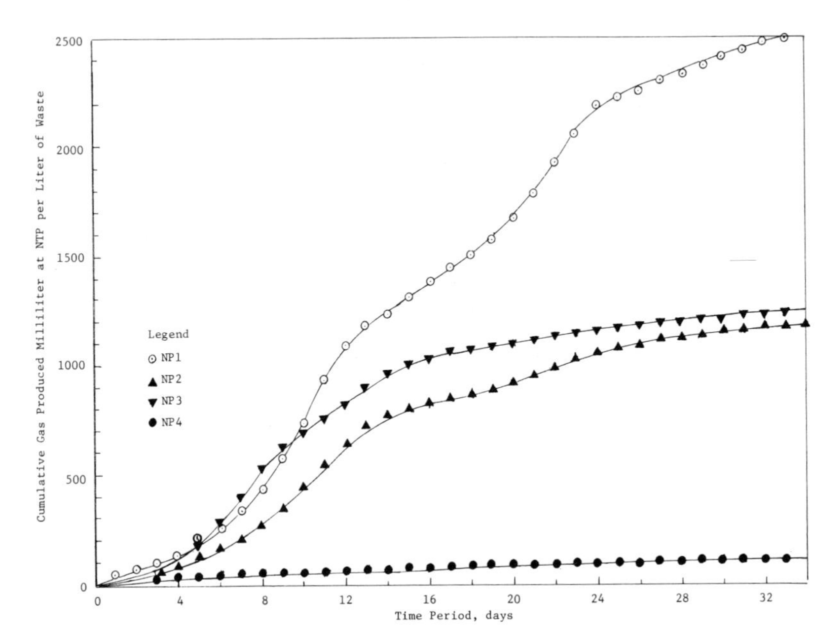

Figure 6. Gas production in nonhydrolyzed waste.

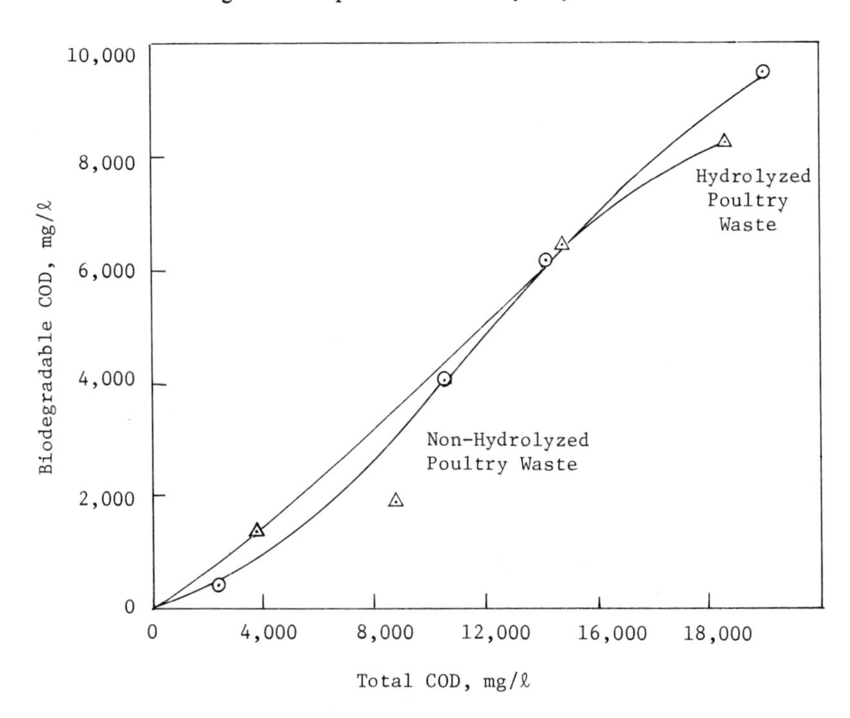

Figure 7. Relationship of initial biodegradable COD and total initial COD.

Equation 1 can be rearranged as follows:

$$\frac{1}{K} = \frac{1}{K_m} + \frac{K_s}{K_m \cdot S_o} \qquad (2)$$

Equations 1 and 2 have been shown in Figures 8, 9 and 10. In Figure 8, the different characteristics between the hydrolyzed and nonhydrolyzed wastes are shown by the degree of the curvature. It also can be concluded that the maximum gas production rate occurs at the biodegradable COD concentration between 8,000 and 10,000 mg/l. The degree of the curvature is dependent on the value of K_s, which was reported to be related to the transport system or binding protein involved in uptake for the growth kinetics of an aerobic heterogenous biological system.[6]

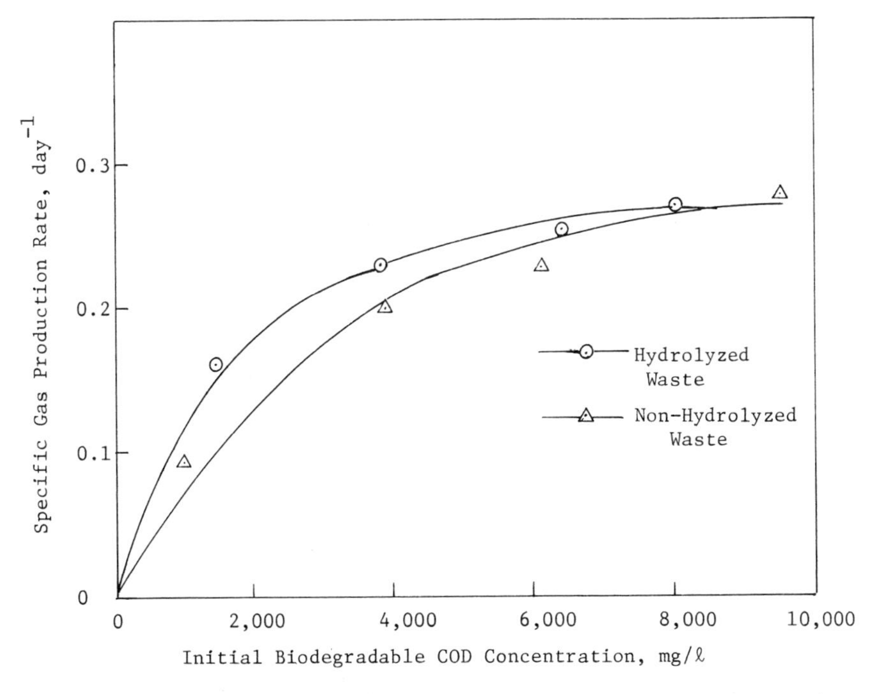

Figure 8. Relationship between initial biodegradable COD and specific gas production rate.

In the present study, the poultry waste was used as the substrate with two different initial forms–hydrolyzed and nonhydrolyzed. In the hydrolyzed system, the value of K_s is less than in the nonhydrolyzed system (Figures 9 and 10). According to the Michaelis-Menten equation,[7] which shows the hyperbolic relationship between enzyme and substrate interaction, the

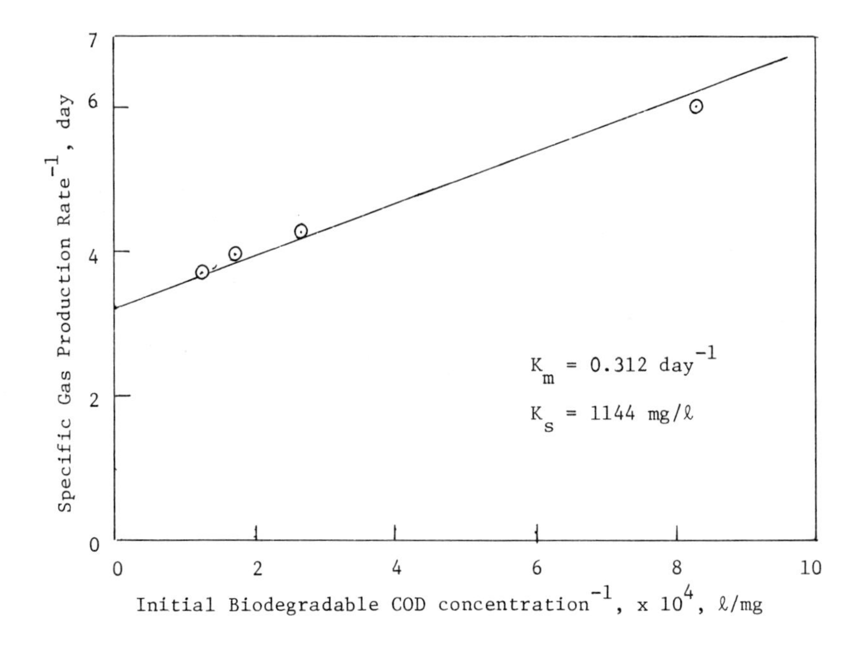

Figure 9. Evaluation of K_m and K_s (hydrolyzed waste).

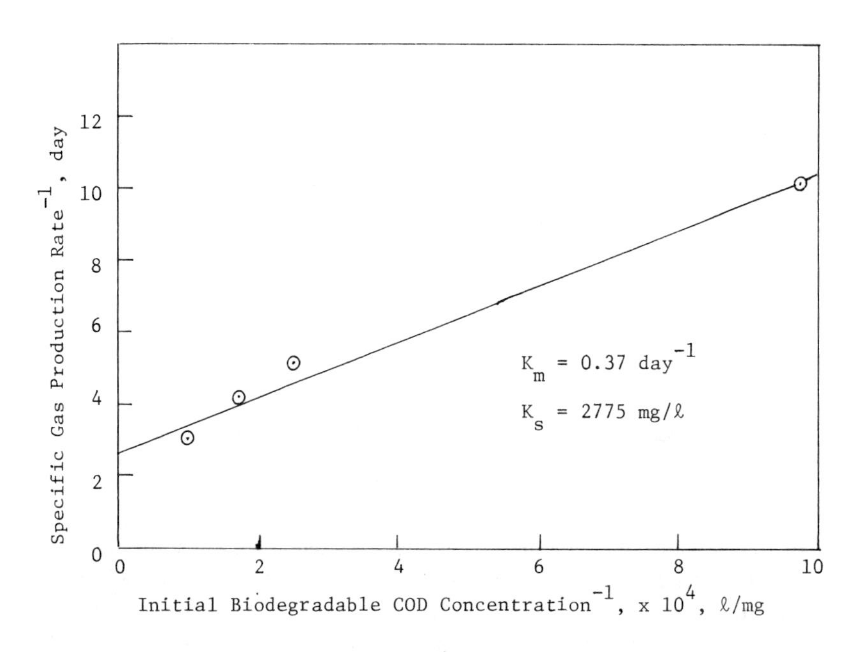

Figure 10. Evaluation of K_m and K_s (nonhydrolyzed waste).

smaller the value of the Michaelis-Menten constant (which is equal to the equilibrium constant, K_s in certain conditions), the greater the affinity of the enzyme for the substrate will be. In other words, the smaller the value of K_s, the greater the reaction rate that can be expected. Therefore, it is expected that the gas production rate in the hydrolyzed system should be greater than in the nonhydrolyzed system within a certain range of substrate concentrations, because K_s in the former system is less than in the latter. The anaerobic study of hydrolyzed and nonhydrolyzed activated sludge produced from a soft-drink wastewater treatment plant provided the same pattern of gas production kinetics as obtained in the present study.[2]

ACKNOWLEDGMENT

This work was supported by the graduate scholarship for K. K. Chan through the Asian Institute of Technology (AIT), Bangkok, Thailand. All the experimental works were conducted in the laboratory of Environmental Engineering Division of AIT.

REFERENCES

1. Gaudy, A. F., Jr., P. Y. Yang and A. W. Obayashi. "Studies on the Total Oxidation of Activated Sludge With and Without Hydrolytic Pretreatment," *J. Water Poll. Control Fed.* 43:40 (1971).
2. Chan, K. K. "Renewable Fuel Recovered From Bio-Conversion of Solid Wastes With and Without Acid Hydrolysis," Masters Thesis, Asian Institute of Technology, Bangkok, Thailand (1975).
3. McCarty, P. L. "Anaerobic Waste Treatment Fundamentals Part One—Chemistry and Microbiology," *Public Works* 107 (1964).
4. American Public Health Association. *Standard Methods for the Examination of Water and Wastewater,* 13th ed. (New York: American Public Health Association, 1971).
5. Malina, J. F., Jr. "Anaerobic Waste Treatment," *Physical, Chemical and Biological Processes* (Austin: University of Texas, 1967), p. 208.
6. Gaudy, A. F., Jr., A. W. Obayashi and E. T. Gaudy. "Control of Growth Rate by Initial Substrate Concentration at Values Below Maximum Rate," *Applied Microbiol.* 22:1041 (1971).
7. Aiba, S., A. E. Humphrey and N. F. Millis. *Biochemical Engineering;* 2nd ed. (New York and London: Academic Press, 1973).

SECTION VI

ENERGY UTILIZATION AND PRODUCTION

30

AN ENERGY AND ECONOMIC ANALYSIS OF CONVENTIONAL AND ORGANIC WHEAT FARMING

G. M. Berardi
Department of Natural Resources
Cornell University
Ithaca, New York

INTRODUCTION

Coal, petroleum products and electricity in the United States have long been relatively cheap and abundantly available. Agriculture has developed (as have other industries) under the stimulus of cheap and readily available energy.[1]

In 1910, farms produced horses, mules, hay, grain and pastures that supplied most of the energy to move machines and implements used in production and short distance transport of agricultural commodities. By 1970 this had been reversed. Most commercial modern-day farms depend on urban industries to build tractors and implements, electric motors and automobiles and trucks that move goods from one place to another on the farm and from farm to market.[2] In 1970, 75% of all inputs were purchased away from the farm, compared to 25% in 1910.[3]

In addition, there has been concern by sociologists over the displacement of agricultural workers, as farms have expanded in size and turned to mechanization.[2] Over the past 5 decades, 5 million fossil fuel tractors have replaced 25 million horses and mules[4] and 9 million farm workers.[1]

Lastly, there has been an increasing effort to have technology maintain soil fertility. In the past, various rotations were used, along with animal and green manuring, to renew soil nutrients.[5,6] Now commercial fertilizers and other technologies are primarily employed for continuous cropping of the land.

The increased use of commercial fertilizers currently faces energy and environmental constraints. In 1968, agriculture ranked third among major industrial consumers of energy in the United States.[1] Within recent years, dependence upon foreign sources of oil and gas has grown,[7] and domestic reserves are rapidly being depleted. With the inevitable rise in the price of energy, domestic fuel supplies could be increased. Yet most of these fuels are finite and thus the energy problem must also be dealt with from the standpoint of energy conservation.[8]

As fossil fuel supplies diminish, choosing crop production schemes less vulnerable to high price increases of such energy-based agricultural inputs may be important. To determine if wheat could be produced economically with minimal amounts of fossil energy, a comparison was made between wheat produced by organic farming (low energy) practices and conventional farming (high energy) practices. The investigation also compared labor inputs and the economic profitability of producing wheat by conventional and organic farming methods in the northeastern United States.

METHODS

Twenty farms growing winter wheat in New York and Pennsylvania were selected for the study. Ten farms used conventional commercial fertilizers and ten used organic fertilizers (animal manures and green manures).* On both types of farms, winter wheat production was only one of several farm enterprises. The main limitation in growing winter wheat reported by both groups of farmers was drainage (or slopes). Growing conditions for winter wheat are similar in New York and Pennsylvania.

The organic farms employed a hay-corn-oats-wheat rotation and regularly applied an average of 24.7 kg of nitrogen (from plant and animal sources) per hectare. The conventional farms employed a similar rotation and applied an average of 40.4 kg of nitrogen (98% was commercial synthetic nitrogen and 2% was of plant and animal origin). Some of the common soil series found on the conventional farms were Ontario, Collamer and Langford, compared to the poorer Volusia and Berks soils of the organic farmers. Winter wheat acreage on the organic farms averaged 7 ha, whereas winter wheat acreages on the conventional farms averaged 13 ha.

Yields, energy inputs, economic profitability and labor inputs were measured in each case and then compared. Process analysis was used in

*The ten conventional farms were obtained from the cost account project in the Department of Agricultural Economics at Cornell University. According to the project supervisor, these farms "are representative of the better farms in New York."[9] The ten organic farms were obtained from the Rodale list of organic farmers and have been managed organically for at least five years.

evaluating the direct and indirect energy requirements of the various production techniques. This provided estimates of specific fossil energy equivalents of inputs (excluding labor) needed for production as well as the establishment of marginal energy costs of production inputs.[10] Several different measures of profitability and income were used in determining the economic viability of the various production techniques, including net cash farm income, as well as economic profitability figures.[9] Estimates of production costs and labor inputs were also established.

Information collected in this study included:

1. soil (series and texture, together with the main limitations reported);
2. total farm organization (crops and acreages);
3. history of cover crops planted; fertilizer, lime and manure (types and amounts) applied to wheat acreage through 1970;
4. rotations used;
5. labor hours to produce the winter wheat crop;
6. size and type of all field implements;
7. land costs;
8. field operations (tillage, planting and harvesting) for the 1975 winter wheat crop year, including those custom-hired;
9. application rate and costs of all fertilizers, pesticides, lime and other materials for winter wheat production;
10. winter wheat varieties, seeding rates and prices of all seeds; and
11. winter wheat yields and prices received.

Data were obtained through personal interviews with the conventional and organic farmers. The data were then analyzed to determine the economic viability of the wheat production, as well as the energy costs and returns.

RESULTS

Economic Profitability

The average profitability for the conventional farmer was $59.50/ha (range: +$260.17 to -$153.49) compared to an average of $14.55/ha for the organic farmer (range: +$237.72 to -$200.18). With such a large variation in values, it was difficult to observe a difference. Nevertheless, when a significance test was performed, no significant difference (5% level) in economic profitability between the two groups of farmers was demonstrated.* (Economic profitability is defined here as revenues minus the total economic production costs.) Note that several of the organic farmers were receiving premiums for their grain, which in turn significantly affected their economic

---

*The Kruskal-Wallis nonparametric test gives the same results—no significant difference in profitability (5% level). The statistically significant difference does not connote actual or practical significance.

profitability. The largest cost for the conventional farmers was certified seed and commercial fertilizer, both of which depend on fossil fuel inputs for production and marketing. The high cost of production for the organic farmers was largely due to land use, the cost being essentially an opportunity cost.

When the cash income is computed using the cash (operating) costs instead of the economic costs, the profitability figures are reversed–$165.22/ha for the conventional group as compared to $244.13/ha for the organic farmers, the difference of which is significant (5% level). These cash costs do not include opportunity or unpaid costs which tend to be higher for the organic farmers.

Minimizing Costs

Despite the agribusiness emphasis on (and research support for) the development and use of fertilizers, chemical pest controls, new equipment and management programs, organic farmers in this study minimized their operating costs and energy consumption while at the same time they increased their cash returns in the following ways:

1. They used older equipment (with perhaps higher maintenance requirements than new machinery) and less equipment. (Use also depends on the tilth and scope of the soil, as well as the previous crop in the rotation.)
2. They used home-grown seed. (The problems associated with home-grown seed, however, should be well noted: storage, seed germination and disease resistance. The latter two, in particular, are likely to be a problem with continued home production.)
3. They used family labor when possible. This is true for conventional farmers as well.
4. They raised some livestock (which requires a larger time commitment than some farmers may be able to make) as a source of manure. (Applying manure and returning crop residues are two major ways in which a nutrient balance is maintained on these organic farms.)
5. They used green manures.
6. They minimized fertilizer use. (Obviously, there will be losses in yields during the transition period, as discussed by Lockeretz;[11] yet minimizing fertilizer use has resulted in a significant savings to the organic wheat farmers and to some of the conventional farmers.)

Energy Use

The conventional farmers' energy inputs for winter wheat production averaged 48% more per ha than the organic farmers' inputs (Table I). The larger energy inputs of the conventional group were mainly due to the use of fertilizer. The average machinery, fuel, electricity and lime use between the

Table I. Per hectare energy inputs in wheat production.[a]

Input	Conventional (kcal x 10^5)	Organic (kcal x 10^5)
Machinery[b]	7.39	7.46
Fuel[c]	5.99	5.22
Nitrogen[d]	5.96	0.45
Phosphorus	1.22	0.14
Potassium	0.87	0.13
Seeds[e]	6.28	5.11
Electricity[f]	0.30	0.27
Total Inputs[g]	28.55	19.73
Total Output[h]	97.82	76.08

[a]Figures are given only for growing and harvesting the crop (storage is not included). All figures in the tables represent averages. Energy Inputs actually refers to the energy equivalents of the various inputs (see Leach *et al.*[12] for more details).

[b]Information on pieces of equipment used, allocation of use to wheat, depreciation schedules, etc., was obtained from individual farmers. The machinery calculation is further explained in Pimentel.[13]

[c]The average fuel use for conventional farmers was 61.65 liter/ha; the average for organic farmers was 53.68 liter/ha. One liter of gasoline = 9725 kcal.[14]

[d]Average nitrogen, phosphorus and potassium use for conventional farmers was 40.40, 38.07 and 39.39 kg/ha respectively, of N, P and K. The organic farmers used an average of 3.01, 4.47 and 5.82 kg/ha of commercial N, P and K. These commercial organic fertilizers contain by-products of other industries, such as cotton hull ashes, bonemeal, cocoa meal, tobacco dust, etc. (The fertilizer energy input in Table I–for organic farmers–represents the marketing costs of the commercial "organic fertilizers" used by two of the farmers.) One kg of nitrogen = 4978 kcal; one kg of phosphorus = 3194 kcal; one kg of potassium = 2203 kcal.[12]

[e]Average seeding rate for conventional farmers was 152.05 kg/ha; for organic farmers, 151.71 kg/ha. In addition to the energy contained in 1 g wheat, 25% more energy was added to account for the effort of producing, handling, transporting and packaging certified seed. For further explanation of this calculation see Pimentel.[10]

[f]For wheat production, electricity is used primarily for seed cleaners and repairs. Electricity requirements for wheat are given as 34.46 kWh/ha.[15] The calculation is as follows: 34.46 kWh x 860 kcal/kWh = 29,635.6 kcal/ha. Note that the average figure is slightly lower for the organic farmers; one of the farmers is Amish and uses no electricity.

[g]Energy estimates for insecticide (for conventional farmers), cover crops (for organic farmers) and lime (for both groups of farmers) are included in the total energy inputs estimates. These inputs account for less than 3% of the total. Wheat is a crop on which few insecticides, herbicides or fungicides are used. For explanation of these calculations see Pimentel.[13]

[h]Total output = yields. The actual per hectare yields are 2961 kg for the conventional farmers and 2303 kg for the organic farmers. Wheat contains about 3300 kcal/kg.[16] The lower yield response on the organic farms was in part due to their poorer soils (Berks, Volusia and Glenelg), as compared to the higher-yielding soils of the conventional group (such as Ontario, Collamer and Langford). Although weather during the 1975 crop year was average for both New York and Pennsylvania's wheat production, the average state wheat yield was 527 kg/ha higher in New York than in Pennsylvania (New York usually averages between 353 and 703 kg/ha higher yields than Pennsylvania). Part of this difference can be attributed to the difference in varieties as well as soils.

two groups was approximately the same. Over one-third of the conventional farmers used fertilizer in amounts exceeding the suggested levels in *Cornell Recommends.*[17] Three of the ten farmers used an average of 45 kg/ha of nitrogen beyond levels recommended in the Cornell publication while averaging a relatively low yield, 2016 kg/ha (range: 1882 to 2150 kg/ha, or 28 to 32 bu/ac).

Increased fertilizer use and mechanization of farming operations have been two major factors in American agricultural production since World War II.[3] Fertilizer use and mechanization economically complement each other in that returns from high yields are needed to help cover increasing capital costs. At the same time, the role of commercial agents in disseminating information on fertilizer use is increasing.[1] To counter these commercial efforts, more effort is needed on the part of land grant colleges and other agricultural institutions to educate farmers and extension agents on proper fertilizer use and alternative fertilization practices.

The organic farmers' labor inputs averaged 21 hr/ha compared to 9 hr/ha for the conventional farmers. The relatively high labor cost for the organic group is largely a result of an Amish farmer's inputs. Exclusion of this farmer from the calculation reduces the labor inputs of the organic farmers to 13 hr/ha. This difference (9 vs 13 hr/ha) is not significant at the 5% level (Kruskal-Wallis test). Note that exclusion or inclusion of this farmer in the other measurements (economic profitability, energy use) does not change their results.

Economic Costs vs Energy Costs

Frequently, agricultural production cost estimates are given only in terms of the economic costs. When trying to evaluate fossil-fuel energy use in food production systems and the environmental impact of such use, these economic cost estimates are inadequate. An estimate of the energy costs associated with various production strategies is also needed (Figure 1).

The economic costs were, on the average, 29%/ha less for conventional wheat production than for organic wheat production. Note that the organic operators were farming on wheat acreage half the size of the conventional farmers. Thus certain economies of scale were operating in favor of the conventional farmers. For example, the machinery cost/ha for the organic farmers was particularly high–$69.23/ha–as compared to $46.04/ha for the conventional farmers.

For the conventional farmers, those with the highest economic costs/ha generally had the highest energy costs as well. However, conventional farmers with low economic costs did not necessarily have low energy costs; in some cases these farmers tended to operate with low overhead costs (thus their "economic" costs were low), with energy-intensive fertilizers and certified seed being their major cost.

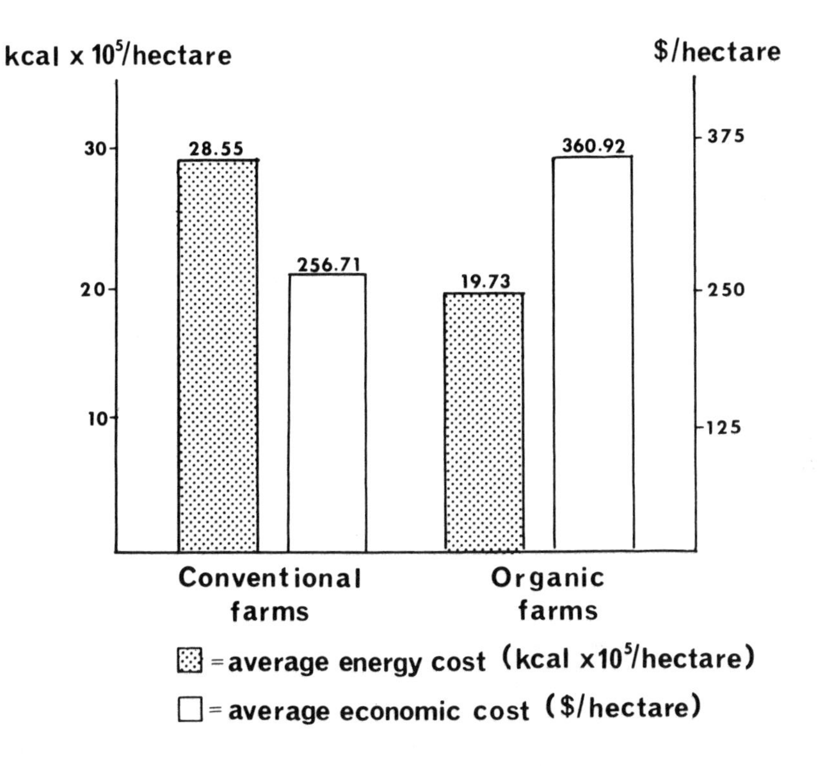

Figure 1. Economic costs vs energy costs: the cost structure of production in comparison.

Alternatively, high economic production costs do not necessarily imply high energy costs for production. This can be illustrated in the case of an Amish farm in Pennsylvania. Although the mode of production by the Amish farmer may appear high in economic costs, over 50% of this calculated cost is unpaid family labor. The Amish farmer operates on low fossil-fuel resources.

DISCUSSION

The results of this preliminary investigation documented that wheat production by conventional farming practices averaged 48% higher energy inputs and only 29% higher yields than wheat produced by organic farming practices. It was primarily the use of nitrogen fertilizer that resulted in the higher total energy inputs for the conventional farmers. Whereas it is well accepted that nitrogen fertilizer raises crop yields, it is also important to recognize that fertilizer use is only one of many factors, such as soil conditions, weather and seed variety, affecting the yield performance of a crop. Given that one-third of the conventional farmers were using levels of fertilizer sufficiently high to cause "fertilizer injury,"[17] with low yields resulting, a decrease in fertilizer use for these farmers is recommended. Decreased fertilizer use would lower not only the total energy inputs of their winter wheat production but the economic costs as well.

The wide range of values obtained in this study for variables such as economic profitability and labor inputs indicates the need for more intensive research with larger samples of organic and conventional farmers for several consecutive crop years.

In addition, with the growing interest in organic farming, organic wastes—both agricultural and municipal—will become increasingly important. Without the addition of such wastes to the soil, organic farmers might run the hazard of creating nutrient imbalances in their crops as well as the soil.

Further investigation of energy costs and production economics of various cropping systems should include the study of larger-scale farming and careful analysis of physical factors affecting production (soil type, weather, etc.), as well as a total input/output analysis of all crops in the rotation. Perhaps the emphasis for such research should be on maximizing agricultural production efficiency rather than maximizing yields.

As discussed above, there was no significant difference in economic profitability between wheat produced by conventional and that produced by organic farming practices in this sample. Extrapolation to the larger sample of conventional and organic farmers in the Northeast is impossible at this point until further research is performed. Nevertheless, perhaps some of the methods practiced by the organic farmers (use of older equipment, less equipment, green manures, etc.) should have stronger support in research and extension work, particularly for small-scale wheat operations. As the results of this preliminary study show, improving cash income and conserving fossil energy need not be mutually exclusive activities.

ACKNOWLEDGMENTS

I thank the following specialists for reading an earlier draft of the manuscript and for their many helpful suggestions: William Lockeretz, Center for the Biology of Natural Systems; and, at Cornell University: Thomas Scott, William Pardee and Douglas Lathwell, Department of Agronomy; George Casler and Pierre Borgoltz, Department of Agricultural Economics; Richard McNeil and Douglas Heimbuch, Department of Natural Resources; and David Pimentel and Elinor Terhune, Department of Entomology. Any errors or omissions are the author's responsibility. This study was supported in part by the Ford Foundation (Resources and the Environment).

REFERENCES

1. National Academy of Sciences. *Agricultural Production Efficiency* (Washington, D.C.: Printing and Publishing Office, 1975).
2. Smith, T. L. "Farm Labour Trends in the United States, 1910-1969," *Int. Labour Rev.* 102:163 (1970).
3. U.S. Department of Agriculture. *Yearbook of Agriculture* (Washington, D.C.: U.S. Government Printing Office, 1975).
4. "You Can Cut Farm Energy Use," *Agway Cooperator* 12:14 (1976).
5. U.S. Department of Agriculture. *Agricultural Statistics* (Washington, D.C.: U.S. Government Printing Office, 1936).
6. U.S. Department of Agriculture. *Agricultural Statistics* (Washington, D.C.: U.S. Government Printing Office, 1968).
7. Ford Foundation. *A Time to Choose* (Cambridge, Massachusetts: Ballinger Publishing Co., 1974).
8. Lincoln, G. A. "Energy Conservation," *Science* 180:161 (1973).
9. Snyder, D. "A Guide for Determining Field and Vegetable Crop Costs and Returns," Department of Agricultural Economics, A.E. Ext. 76-4, Cornell University, Ithaca, New York (1976).
10. Pimentel, D., W. Lynn, W. K. MacReynolds, M. Hewes and S. Ruch. "Workshop on Research Methodologies for Studies of Energy, Food, Man, and the Environment Phase I," Cornell University, Ithaca, New York (1974).
11. Lockeretz, W., R. Klepper, B. Commoner, M. Gertler, S. Fast, D. O'Leary and R. Blobaum. "A Comparison of the Production, Economic Returns, and Energy Intensiveness of Corn Belt Farms that Do and Do Not Use Inorganic Fertilizers and Pesticides," Center for the Biology of Natural Systems - AE-4, St. Louis, Missouri (1975).
12. Leach, G. and M. Slesser. *Energy Equivalents of Network Inputs to Food Producing Processes* (Glasgow: Strathclyde University Press, 1973).
13. Pimentel, D., L. E. Hurd, A. C. Belloti, M. J. Forster, I. N. Oka, O. D. Sholes and R. J. Whitman. "Food Production and the Energy Crisis," *Science* 182:443 (1973).
14. Chemical Rubber Company. *Handbook of Chemistry and Physics* (Cleveland, Ohio: Chemical Rubber Company, 1972).
15. Cervinka, V., W. J. Chancellor, R. J. Coffelt, R. G. Curley and J. B. Dobie. "Energy Requirements for Agriculture in California," California Department of Food and Agriculture, Davis, California (1974).
16. U.S. Department of Agriculture. "Handbook No. 8: Composition of Foods," Consumer and Food Economics Research Division, USDA, Washington, D. C. (1963).
17. Cornell University. "Cornell Recommends for Field Crops," Cornell University, Ithaca, New York (1976).

31

USE OF GINNING WASTE AS AN ENERGY SOURCE

W. F. Lalor, Manager
Systems and Cost Engineering
Cotton Incorporated
Raleigh, North Carolina

M. L. Smith, Associate Professor
Industrial Engineering and Computer Sciences Department
Texas Tech University
Lubbock, Texas

INTRODUCTION

The Ginning Process

The term "seed cotton" is used to describe harvested cotton before it has been ginned. Ginning is the action of separating the fiber from the seed to which it is attached. During the ginning process, seed cotton is dried, cleaned and ginned, in that order. Further cleaning of the lint is done immediately after ginning. The lint is packaged in nominal 480-lb (218-kg) bales. The bale is the unit on which production, performance, heat needs and other inputs and outputs are based.

The average yield per hectare in the United States is about 2.5 bales, with yields of 5.0 to 7.5 bales being common in irrigated areas and yields as low as 1.25 bales being common in some dry-land areas.

Gins vary in ginning rate from 6 to 40 bales per hour, depending on the equipment used. The annual production volume at gins varies from less than 2,000 to more than 30,000 bales, and the operating time per year varies from 200 to 1200 hr. Cotton is ginned as soon as possible after being harvested.

Drying is often done in two stages, between which a cleaning operation is performed. After the second drying stage, further cleaning is done. Cleaning of wet cotton is difficult–so is ginning of wet cotton.[1] On the other hand, ginning at very low moisture content causes fiber damage.[1]

Gin Waste

Ginning waste is of two types. One is the leaf fragments, sticks and other plant parts removed before ginning, and the other is linty material. Some linty material (known as motes) is rejected during actual ginning, and the remainder is lint removed with foreign matter taken out by the lint cleaner after ginning. At some gins the two types of waste are kept separate and the linty material is sold as "motes," while the nonlint material is the true waste. Other gins discard both components as waste.

Objectives

The objective of the study reported here was to collect information needed to design methods of supplying drying heat from the heat released when the waste is burned. Questions posed at the outset were:

1. Can all the drying heat be supplied from the waste?
2. Under what conditions is there a deficit or an excess of heat available?

Previous Studies

Griffin[2] found that the high heat value of ginning waste averaged 18.44 MJ/kg for spindle-picked cotton in the mid-South. The amount of dry waste per bale was 38 kg for the first-harvest cotton and 84 kg for second-harvest cotton. The motes were included in these estimates. California is a cotton-producing area where motes are often kept separate from other waste and are marketed. One gin company manager[3] reports waste production of 63.5 kg/bale and mote production of 11.3 kg/bale (moist basis). Where cotton is harvested with strippers, up to 300 kg of waste is produced per bale during ginning.

The heat requirement for drying cotton can be derived from fuel consumption data published by Holder and McCaskill.[4] These data translate into heat requirements of about 365 MJ/bale when the low heat value and 90%[5] combustion efficiency are assumed. The heat released by burning the waste from one bale would vary from 701 to 1549 MJ. A suitable heat exchanger would thus be one that could extract 23 to 50% of the flue-gas heat.

McCaskill and Wesley[6] estimated that a 30%-efficient heat exchanger would extract enough flue-gas heat for almost all drying needs and designed their system accordingly. They never operated their incinerator, which was an experimental installation, at a commercial cotton gin. The two incinerators from which the data presented here were collected operated at commercial gins and are commercially available, although they are undergoing design changes shown by experience to be needed.

EXPERIMENTS

Our study was started in 1975 at one gin in Arkansas; it was continued in 1976 at that same gin and expanded to include a gin in California.

In 1975, pitot-static probes and thermocouples were permanently installed in the Arkansas gin. Their output was recorded by multipoint recorders and permitted calculation of the mass air-flow rate and heat-flow rate to each dryer. To calculate heat flow, ambient conditions were taken as a baseline and the heat flow was the heat added to the air to raise its temperature above ambient.

The results of the 1975 tests have been published.[7] Heat supplied by the incinerator heat exchanger averaged 121.9 MJ/bale, and the average seed cotton moisture was 15.66 % (w.b.).

Waste production per bale averaged 61.8 kg (dry matter) and ranged from 38 to 84 kg/bale. The waste per bale contained combustion heat of 1004.2 MJ, allowing for 12% noncombustible, soil-derived material. The heat recovery efficiency was thus about 12%. Gas was sometimes used to supplement the recovered incinerator heat. The average gas consumption was 1.4 m^3/bale with a heat value of 58 MJ/bale.

In 1975, therefore, approximately 68% of the annual heat needs was supplied from the incinerator. This included periods when gas was used to supply all the heat because the incinerator was not working, as well as periods when gas was not used or was unavailable for supplying any heat. Except for start-up from cold each morning, virtually all the heat needed was available from the incinerator.

When the experiment was expanded in 1976, the experimental methods were changed. Instead of using permanently placed instruments, we used a hand-held pitot-static tube to which a thermocouple was attached, and we made 20-point scans in each air duct in accordance with standard practice.[8]

The flow rate was calculated for each scan point and summed over the cross section. The effects of temperature, of static pressure in the duct and of barometric pressure (but not of relative humidity) were considered in our calculations of air-flow volume and mass.

Heat-flow rate was the mass air-flow rate multiplied by the heat needed per kg of air to produce the temperature difference between ambient air and the air in the duct. This heat-flow rate was thus directly comparable with the amount of heat flowing to the dryer from the gas burners.

Figure 1 is a schematic drawing of the installation in Arkansas. Air-flow and temperature measurements were made upstream of the temperature control system for each dryer. Figure 2 is a schematic of the California installation. Air-flow and temperature measurements were made in each duct leading to the dryers and at the inlet to the clean-air side of the heat

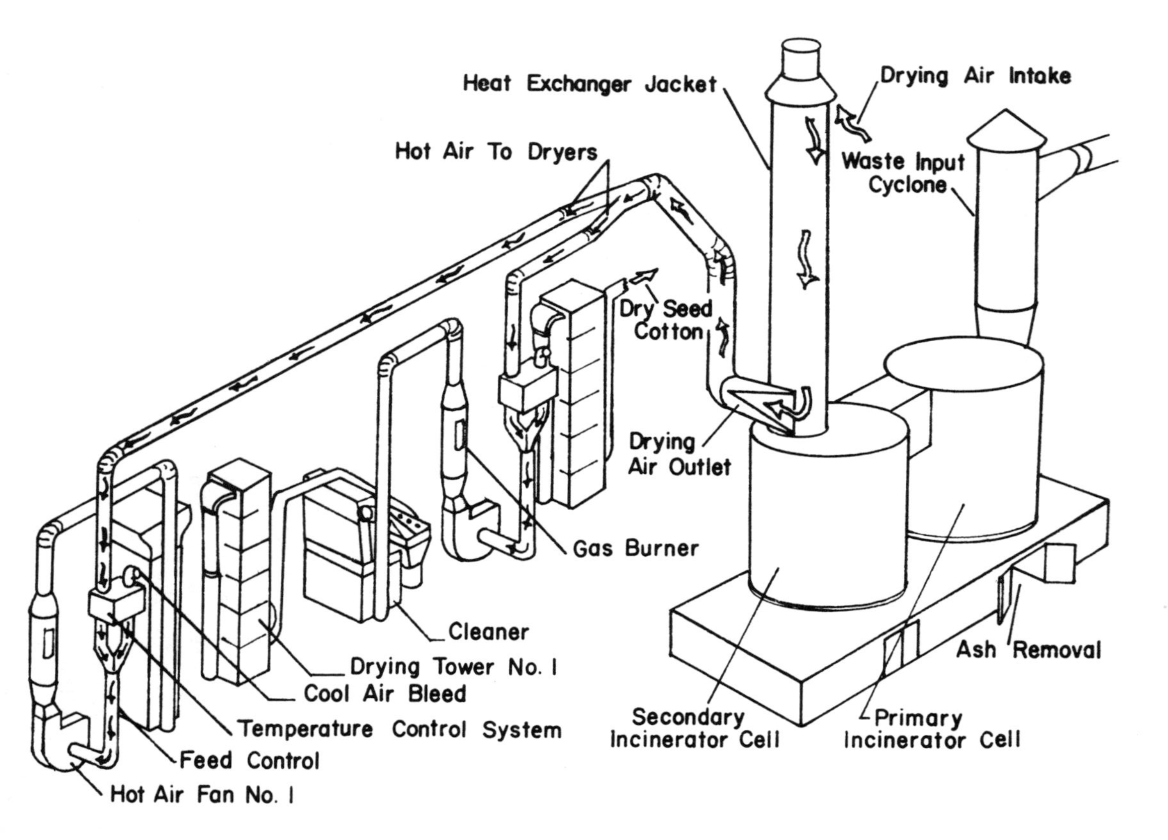

Figure 1. Schematic view of the incinerator in the Arkansas study.

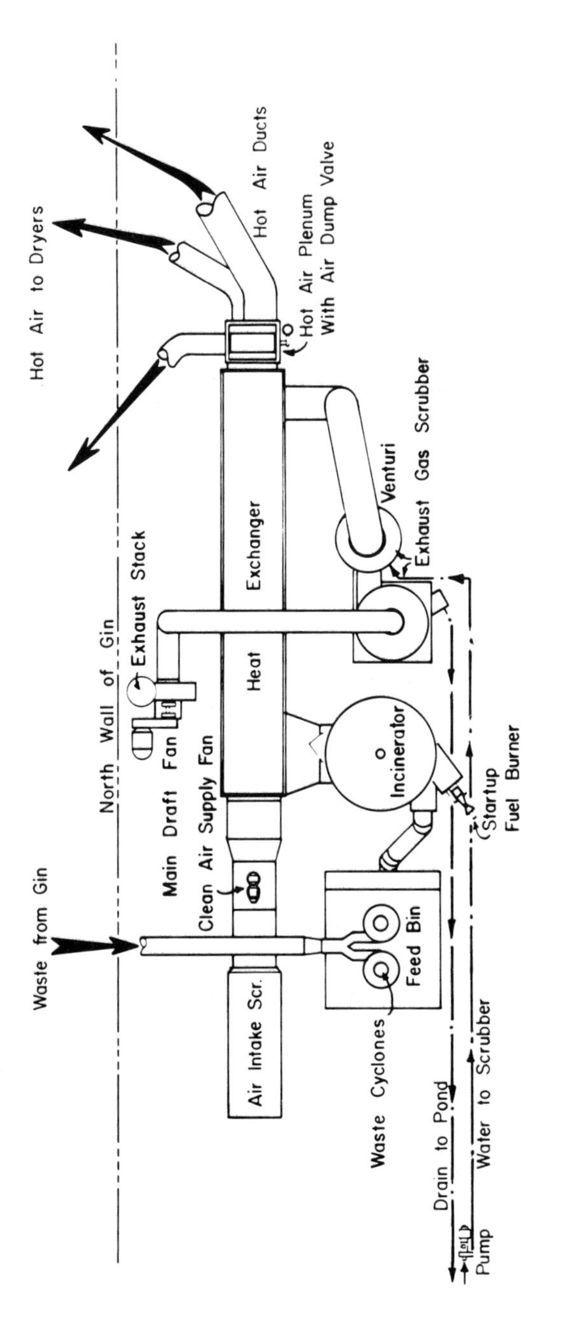

Figure 2. Schematic view of the incinerator in the California study.

exchanger. Hot air not used by the dryers was vented. The amount of air vented was calculated from the flow-continuity principle.

RESULTS

Heat Recovery

Results of the 1976 observations in Arkansas are given in Table I. Gas-derived heat was from natural gas with 41 MJ/m^3 heat content and 90% combustion efficiency. Dryers in a cotton gin are direct fired.

Table I. Heat supplied to Arkansas dryers.

	Drying Heat (MJ/bale)		
Date	From Incinerator	From Gas	Total
10/14/76	71.82	24.94	96.76
10/15/76	0.00	110.98	110.98
10/19/76	85.37	98.55	183.92
11/05/76	109.46	0.00	109.46

The percentage of total heat requirements supplied by the incinerator varied from 100 to 46.4 for days when the incinerator was operating. The incinerator was not operating on October 15 and the supply of natural gas was discontinued on November 5 because of cold weather.

Results of observations in California are shown in Table II. Whenever the incinerator operated, it supplied all the heat needs of the dryers. The heat transmitted from the incinerator to the dryers was derived partly from the heat produced by the start-up burner, which was set to ignite automatically when incinerator temperature dropped below some preset level. This heat is not shown separately in Table II as having been derived from gas. In any case, only a small part of the heat transmitted through the heat exchanger to the dryers is derived from gas burning in the incinerator start-up burner.

Table II. Heat supplied to California dryers.

	Drying Heat (MJ/bale)		
Date	From Incinerator	From Gas	Total
11/09/76	399.37	0.00	399.37
11/11/76	0.00	299.50	299.50
11/17/76	437.22	0.00	437.22
12/06/76	0.00	315.69	315.69
12/09/76	0.00	380.45	380.45

The gas consumed by various processes in the California gin is shown in Table III. The incinerator manufacturer was experimenting with different combustion temperatures in the incinerator. We believe that the gas heat consumption per bale to operate the humidifier and ensure good combustion in the incinerator will be approximately the same as it was on November 9 (64.76 MJ/bale) when the system operates under normal production conditions.

Table III. Gas heat used by gin components (MJ/bale).

	Component			
Date	Humidifier	Incinerator Start-Up	Dryers	Total
11/09/76[a]	48.57	16.19	0.00	64.76
11/11/76	48.57	0.00	299.50	348.07
11/17/76[a]	56.66	97.14	0.00	153.80
12/06/76	48.57	0.00	315.69	364.26
12/09/76	48.57	0.00	380.45	429.02

[a]Incinerator operating.

Energy Savings

The data in Table III make possible an estimate of the energy saved by the California system. If the gas consumption on November 9 is taken to be representative of what can be expected, then comparison with consumption on November 11 and December 6 and 9 shows that a possible savings of 280 to 360 MJ/bale can be expected.

The gas-derived heat consumption avoided by using the system in Arkansas is difficult to estimate but is evidently about 100 MJ/bale. In any case, the total heat used for drying in the Arkansas gin was about half of the 356 MJ/bale to be expected from the results of the survey by Holder and McCaskill.[4] On cool, wet days, especially when seed cotton was damp, we recorded heat consumption in excess of 400 MJ/bale for short periods at this gin. When crop and weather conditions were ideal, we observed drying with heat consumption of 75 MJ/bale.

The drying-heat controls used in cotton gins are less than satisfactory because the heat-input rate is not directly related to the moisture to be removed. The reason for this lies in the difficulty of sensing the moisture content of seed cotton, which is a mixture of materials differing in moisture content by as much as 15% within one sample. The drying heat applied to the crop is largely a result of the ginner's judgment. This judgment is based on the condition of the incoming seed cotton, on the condition of the cleaned seed cotton just before ginning and on the cleanliness of the lint being packaged in the bale.

Resistance-type portable moisture meters are often used to check the lint moisture of seed cotton just before ginning. Some ginners believe the gin does an adequate job with 7 or 8% lint moisture indicated by the meter; others believe the crop should be dried so that only 3% moisture is indicated by the meter.

Seed cotton that is too moist is difficult to clean and sometimes chokes the cleaning equipment. When this happens, production time is lost and severe inconvenience results. On cool, wet days, when cotton may have been rained on in wagons, these choke-ups are common unless the dryer is set to dry the dampest cotton found in a wagon, which may be only a small fraction of the total. This dryer setting might go unchanged for several hours after the damp cotton has been ginned, resulting in a very high heat consumption to dry cotton that is in near-ideal condition coming into the process.

The marked difference between the heat consumed in the two gins (about 200 MJ/bale) is not attributable to the moisture content of the incoming seed cotton. It is principally a result of the belief at the California gin that higher quality lint that has been thoroughly dried and cleaned will bring higher revenues that more than pay for higher drying expenses. Table IV shows the average moisture contents observed in the two locations. The gin stand moisture content reflects the moisture content of the seed. Measured with the portable meter, the lint moisture was about 3% at the gin stand.

Table IV. Moisture content of seed cotton and components.

	Seed Cotton					
	Wagon	Gin Stand	Lint	Seed	Motes	Waste
Arkansas	13.46	11.88	6.19	13.56	-	19.42
California	8.43	6.66	-[a]	8.20	6.47	6.83

[a]Lint was rehumidified after ginning.

Turnout Analysis

Table V shows the turnout analysis which gives the amount of dry waste available per bale for heat generation. The average is shown although we do not believe we have enough data to give a good estimate of the average waste. Besides, the range over which the waste weight varies is of greater importance than the average for most incinerators—seven of the eight systems we know about are direct-on-line. A direct-on-line system takes the fuel at whatever rate it is available—hence the importance of the range. The alternative is to

Table V. Turnout analysis.[a]

Date	Seed Cotton	Lint	Seed	Motes	Waste
		Arkansas[b]			
10/14/76	600	206	319	-	73
10/19/76	570	204	318	-	48
11/05/76	565	204	315	-	45
Mean	578	205	317	-	55
		California			
11/09/76	633	209	359	12	53
11/11/76	630	206	363	13	47
11/17/76	664	206	356	13	88
12/06/76	641	210	355	14	63
12/09/76	681	209	353	12	107
Mean	650	208	357	13	72

[a]Units are kg dry matter per 218-kg bale of moist lint.
[b]The range of waste weights in 1975 was 38 to 84 kg/bale.

use a surge bin and to fuel the incinerator at a constant rate at least over periods as short as two to three hours. Even then, the range is important in designing the surge capacity of the system. The California system had a surge bin; the Arkansas system was direct-on-line.

The waste weight shown in Table V is calculated, not directly measured. All other weights given were directly measured. The weight of waste is the weight of seed cotton less the weights of the other components. The slight error caused by dust loss at various points in the gin is believed to be unimportant. However, ginners often calculate the waste weight in a similar manner but based on wet rather than dry waste. Moisture removed by the dryers is thus counted as waste, and an overestimate of waste weight always results. Using the waste moisture content as an adjustment factor is incorrect.

Percentage Heat Recovery

From Table I, the average heat recovered per bale at the Arkansas gin was 89 MJ. Assuming a heat value of 16.25 MJ/kg (after allowing for soil contamination), the dryers received about 10% of the available heat. This is consistent with 1975 data.

From Table II, the average heat delivered to the dryers in the California gin was 418 MJ/bale and the average waste available was 72 kg. Bomb calorimeter determinations yielded an average heat content of 14.48 MJ/kg. The dryers, therefore, received an average of 40% of the available heat.

Noncombustibles

High soil-contamination levels were found in the California waste—up to about 20% of the sample dry weight. This accounts for the lower heat value than was reported by Griffin.[2] A similar situation existed in Arkansas, but we have no data to show the contamination level. We assumed that about 12% would be soil-derived noncombustible material.

The incinerator in Arkansas was designed to accumulate ash in the primary cell, which was cleaned out daily. But clinkers and "glass" formed in the primary cell and often made clean-out difficult. Ash encrustations often blocked the free entry of combustion air.

The incinerator in California was of a design that required virtually all noncombustible matter to be carried along with the combustion gases, through the heat exchanger, and into a wet-venturi scrubber which removed them from the stack gas. Because the levels of soil contamination were higher than had been expected, some soil-derived noncombustible material dropped out of the gas stream and accumulated in the combustion chamber and in other places where gas velocity slowed due to a widening cross section. The accumulation eventually restricted air flow through the combustion space and this caused the combustion-space temperature to increase to a point at which the incinerator had to be shut down.

Ideally, the incinerator should be designed to cope with high levels of soil contamination, and the manufacturer of the California incinerator now believes that the necessary changes have been made to accomplish this ideal. Another approach would be to remove as much of the undesirable materials as possible from the waste before incineration. Methods of accomplishing this are now being studied. Virtually all the heat-recovering incinerators at cotton gins have been adversely affected by the high noncombustible content of the waste material.

Cost Aspects

When the California incinerator was operating, LP gas consumption of about 15 liters/bale was avoided—this was the gas consumption rate of the dryers when the incinerator was not operating. However, the start-up burner in the incinerator can be expected to consume about 1 liter/bale when the incinerator is operating, resulting in a net saving of about 14 liters/bale. The cost of LP gas varies from place to place and depends on quantities purchased in a single consignment, but 9.5¢/liter is representative. The expense saved would then be about $1.33/bale, or almost $40,000/yr when annual volume is 30,000 bales. These savings would make available the revenue needed to accomplish capital recovery on $160,000 with 10% compound interest in about five years. When applicable tax reductions resulting from investment credit, depreciation allowance and other provisions (such as avoided disposal cost) are considered, additional net revenue would be available. However, labor, repairs, increased insurance (if any) and increased property tax would be expenses to be met from some of the available revenue.

Natural gas was the fuel used in the Arkansas gin, and it cost $6.18/100 m^3 in fall 1976. Gas-derived heat use avoided by using the incinerator in Arkansas is estimated at about 100 MJ/bale or 2.7 m^3 of gas if combustion is assumed 90% efficient. This is a saving of 17¢/bale. For cotton gins in the mid-South as a whole, natural gas costs for drying are about 50¢/bale. As discussed earlier, gas consumption is often a function of the ginner's judgment. The survey referred to[4] resulted in a heat consumption estimate of 365 MJ/bale. The Arkansas gin used 193 MJ/bale in 1974 before the incinerator was installed. If the Arkansas gin could fully eliminate gas consumption, it would therefore hope to save 33¢/bale at 1976 prices for natural gas.

We estimate that the Arkansas incinerator and associate equipment would cost $100,000. Revenue of about $25,000/yr is needed for capital recovery in five years with 10% interest compounded annually. This means that an annual volume of about 75,000 bales would be needed–something which is clearly impossible. Tax savings resulting from various deductions related to the equipment, and savings due to not having to continue paying for waste disposal, would free additional revenue to contribute to capital recovery. Disposal costs alone are often 75¢ to $1.00/bale. When such aspects of the situation as unavilability of natural gas or LP gas are considered, using the incinerator to generate drying heat may be the most economical option. This situation had already arisen before the decision to install the Arkansas incinerator was made. Nevertheless, the conclusion is inescapable that gins at which LP gas is the fuel used are the ones most likely to benefit from installation of a heat-recovering incinerator. Circumstances such as the frequent unavailability of natural gas and/or high alternative disposal costs could justify the system at gins now dependent on natural gas.

POWER GENERATION

In some cotton production areas in the United States, stripper harvesting of cotton is widely used. This harvest method results in gin wastes averaging 318 kg (700 lb)/bale. This amount is approximately five times that obtained from picker-harvested cotton. Stripper-harvested seed cotton usually has a much lower moisture content than does picker-harvested cotton, since the plant is killed by frost or chemicals several weeks before harvest. Often little or no drying of stripper-harvested seed cotton is required. These two factors, presence of relatively large quantities of gin waste and low energy requirements for drying, result in a surplus of gin wastes being available for uses other than drying or for disposal as a nuisance.

A logical alternative is to use the energy from the waste both to dry the seed cotton and to provide mechanical energy for the gin equipment. This alternative has been investigated and will be described in the remainder of this chapter.

All but the oldest gins are powered by electric motors on each individual piece of equipment. A small number of gins use a central power plant such as an internal combustion engine or electric motor that drives a line shaft; each piece of equipment is connected to this line shaft by a belt. Because the gins with individual electric motors are by far the more prevalent type, the power plant investigated here will be one that generates electricity rather than one that serves as a prime mover for the line shaft.

The amount of energy necessary to power a gin varies with the type of cotton ginned, the size and efficiency of the gin and other factors. While the range may vary from 40 to 70 kWh/bale, 65 kWh/bale is a representative figure for stripper-cotton gins. A gin processing 12 bales/hr, therefore, will require a power plant of approximately 800-kW capacity. This size system was selected for the investigation since many gins in stripper cotton areas have capacities near 12 bales/hr.

Three systems were initially considered. These were:

1. direct incineration of gin wastes to produce steam to drive a turbine-generator;
2. biological action to generate methane for powering a dual-fuel engine or a turbine driving a generator; and
3. pyrolysis to produce a combustible gas for powering a dual-fuel engine or a turbine driving a generator.

Methane generation was eliminated as a viable alternative because the digester cost was estimated to be over $500,000; also, the suitability of gin waste for anaerobic digestion is questionable due to the high content of lignin that is not digested by bacteria. A third difficulty involved with methane generation is that the process cannot be started and stopped at will as the gin starts and stops. A means of storing gas will be required, and this will greatly increase the system cost.

The feasibility of pyrolysis of gin waste has been demonstrated by Knight *et al.*[9] The cost of the pyrolysis unit was estimated to be $150,000 to $250,000. While a thorough study of costs of other equipment was not made, it became apparent that total cost of the pyrolysis approach would probably exceed that of the system with direct incineration. Also, the state of the art of pyrolysis is not as advanced as is that of direct incineration.

The Direct Incineration System

An 800-kW power plant with a single-stage turbine may have a relatively low efficiency that will be near 12.5% when boiler, turbine, gear and generator efficiencies are combined. If overall efficiency is assumed to be 12.5%, 100 kg of gin wastes must be burned to provide enough electrical energy to gin one bale. This amount of waste is less than one-half of the waste obtained in ginning a bale of cotton. Therefore, it is apparent that sufficient energy can be obtained from the waste for both electrical and drying energy requirements.

The equipment and costs for an 800-kW power plant are:

1. incinerator–$125,000;
2. waste heat boiler–$40,000;
3. turbine-generator–$85,000;
4. miscellaneous equipment (pumps, water treatment, deaerator, electrical switch gear)–$85,000;
5. building, foundations and installation–$210,000.

A higher cost for the incinerator is used here because of additional heat exchanger equipment not required by the drying heat system described earlier.

Economic Analysis

Annual costs will include fixed costs of depreciation, interest, taxes, maintenance and variable or operating costs that are determined by the number of hours of operation each year. Maintenance cost is assumed to be $5000/yr or 1% of the original cost. It is assumed that an operator is paid $4.00/hr and is paid for 1.2 hr for each hour of gin operation; this will allow time for system maintenance and getting the system on-line at start-up. The cost/bale for several cases is presented in Table VI. It is assumed that the system is depreciated over a 20-yr period, that the interest rate is 10%, and that there is no salvage value. Taxes and insurance are assumed to be 1.75% of initial cost.

Table VI. Annual costs of an 800-kW gin power plant.

Operating hr/yr	Bales/yr	Total Cost	Cost/Bale	Cost/kWh
500	6,000	$81,175	$13.53	0.21
1,000	12,000	$83,575	6.96	0.11
1,500	18,000	$85,975	4.77	0.07
2,000	24,000	$88,375	3.68	0.06

It is obvious from the cost data that a $545,000 investment can best be justified by extensive utilization. The cost/bale figures from Table VI are highly affected by operating hr/yr.

Average costs of electrical energy/kWh for cotton gins in west Texas are given in Table VII. The estimated cost of electrical energy for 1976 to 1977 is less than the most favorable cost from Table VI. However, the increase in costs is about $0.004/kWh/yr. At this rate of increase, within five years the gin power plant will be an economical alternative to purchasing electrical energy.

Table VII. Electrical energy cost to gins in west Texas.[a]

Year	Cost/kWh
1967-68	$0.0302
1968-69	0.0284
1969-70	0.0281
1970-71	0.0268
1971-72	0.0304
1972-73	0.0305
1973-74	0.0283
1974-75	0.0355
1975-76	0.0408
1976-77	0.0440[b]

[a]From Southestern Public Service Southern Division Summary All-Electric Cotton Gin Report.
[b]Estimated.

Discussion

The gin power plant system is not an economically viable alternative with present energy costs; however, there appear to be no technical aspects to prevent operation of such a system. All hardware used in the system evaluation is commercially available.

There are additional considerations that might enhance the economic feasibility of the system. The cost figures in Table VI represent electrical energy costs only and do not reflect any reduction in fossil fuel costs for drying seed cotton or elimination of disposal costs of gin waste. Current costs for drying seed cotton in west Texas are $1.00 to $1.25/bale, and gin waste disposal costs are approximately $1.00/bale. If the $2.00 to $2.25/bale-cost reductions are subtracted from the cost/bale figures in Table VI, the gin power plant is economically feasible at current energy costs at slightly less than an 18,000 bale/yr-volume. This annual volume is high when compared with current volumes but is attainable through extensive utilization of seed cotton storage. If fossil-fuel costs continue to increase as projected, the annual volume for break-even of the gin power plant system will decrease to easily attainable volumes.

It should be noted that in stripper harvesting areas, the gin utilizes less than half the available waste in producing both electricity and heat for drying. Thus an additional 65 to 75 kWh can be generated from the waste of each bale ginned. If this waste is used for electrical energy generation during times other than the cotton ginning season, system output can be more than doubled at a relatively small increase in total annual costs. The difficulty arises when attempting to find a use for the excess energy. Two possibilities are apparent: these are (1) to use the energy for pumping irrigation water in the spring and summer months; or (2) to pump the electrical energy into existing distribution lines. Neither alternative is without some disadvantages. The economic impact of off-season power production from the gin system is substantial and can reduce the cost/kWh by 40 to 60%. Such a reduction will result in the system being not only feasible but profitable.

CONCLUSIONS

The equipment used at the California gin demonstrates that enough heat for drying cotton under almost any circumstances can be recovered from burning gin waste. Approximately 40% of the available heat was supplied to the dryers in the California gin.

At the Arkansas gin, where only about 10% of the heat was recovered for drying, gas was often used as a supplementary heat source. The amount of gas used in this way could be kept at very low levels by careful operation of the gin, but a 10% heat recovery level did not give the flexibility needed to cope with difficult drying situations.

Investment of capital in heat recovering equipment is most easily justified when the combined cost of purchasing fuel and disposing of waste reaches $1.50/bale.

Power and heat can be generated at gins where stripped cotton is ginned. Projected increases in power costs during the next few years lead to the conclusion that studies should be undertaken to identify suitable equipment and assess its impact on gin operation and management. Excess waste would be available for use in generating power for other purposes.

ACKNOWLEDGMENT

We acknowledge the cooperation of Robert G. Curley, George E. Miller and O. D. McCutcheon of the University of California Cooperative Extension Service.

REFERENCES

1. U.S. Department of Agriculture, Agricultural Research Service. *Handbook for Cotton Ginners,* Agriculture Handbook No. 260 (1967).
2. Griffin, A. C., Jr. "Fuel Value and Ash Content of Ginning Wastes," *Transactions of the American Society of Agricultural Engineers* 19(1): 156-58, 167 (1976).
3. Macon Steele, Producers' Cotton Oil Company, Fresno, California, personal communication (1976).
4. Holder, S. H. and O. L. McCaskill. "Costs of Electric Power and Fuel for Dryers in Cotton Gins, Arkansas and Missouri," ERS 138, U.S. Department of Agriculture, Economic Research Service and Agricultural Research Service (1963).
5. McConnell, A. W. McConnell Sales, Birmingham, Alabama, personal communication (1977).
6. McCaskill, O. L. and R. A. Wesley. "Energy from Cotton Gin Waste," *The Cotton Ginners Journal and Yearbook* 44(1):5-14 (1976).
7. Lalor, W. F., J. K. Jones and G. A. Slater. "Performance Test of Heat-Recovering Gin-Waste Incinerator," *Cotton Incorporated Agro-Industrial Report* 3(2) (1976).
8. American Conference of Governmental and Industrial Hygienists. *Industrial Ventilation,* 14th ed. Lansing, Michigan, Section 9 (1976).
9. Knight, J. A., J. W. Taton, M. D. Bowen, A. R. Colcord and L. W. Elston. "Pyrolytic Conversion of Agricultural Wastes to Fuels," Paper No. 74-5017 presented at the 1974 Annual Meeting of the American Society of Agricultural Engineers, Stillwater, Oklahoma, June 1974.

32

GENERATION OF LOW-BTU FUEL GAS FROM AGRICULTURAL RESIDUES EXPERIMENTS WITH A LABORATORY-SCALE GAS PRODUCER

R. O. Williams and B. Horsfield[a]
Department of Agricultural Engineering
University of California
Davis, California

[a]Now with Weyerhaeuser Company
Tacoma, Washington

INTRODUCTION

The theory of gasification, the development of gas-producer technology and the application of gasification to agricultural residues are discussed elsewhere.[1] In a gas producer, low-Btu fuel gas is extracted from crop residue by subjecting the residue to partial combustion in a fixed bed with a limited supply of air. When the reaction temperature is in excess of 1600°F, the resulting "producer gas" may contain up to 30% V/V of carbon monoxide. If steam is added to the incoming air, this reacts with the incandescent carbon to produce hydrogen and carbon monoxide.

In a updraught gas producer, the hot gases flow counter to the fuel; in doing so they pyrolyze a portion of the fuel, and the resulting gas stream has a high tar content. Downdraught producers, on the other hand, have the potential to eliminate tar from the gas and are probably better adapted to crop residue. Pyrolysis products must pass through the reaction zone, where they are broken down, before combining with the exiting gases.

In order to become more familiar with the gasification of agricultural residues, two successive laboratory-scale downdraught gas producers were fabricated at the Agricultural Engineering Department, University of California, Davis.

FIRST GENERATION GAS PRODUCER

A cutaway view of the first producer is shown in Figure 1. The experimental equipment is shown pictorially in Figures 2 and 3.

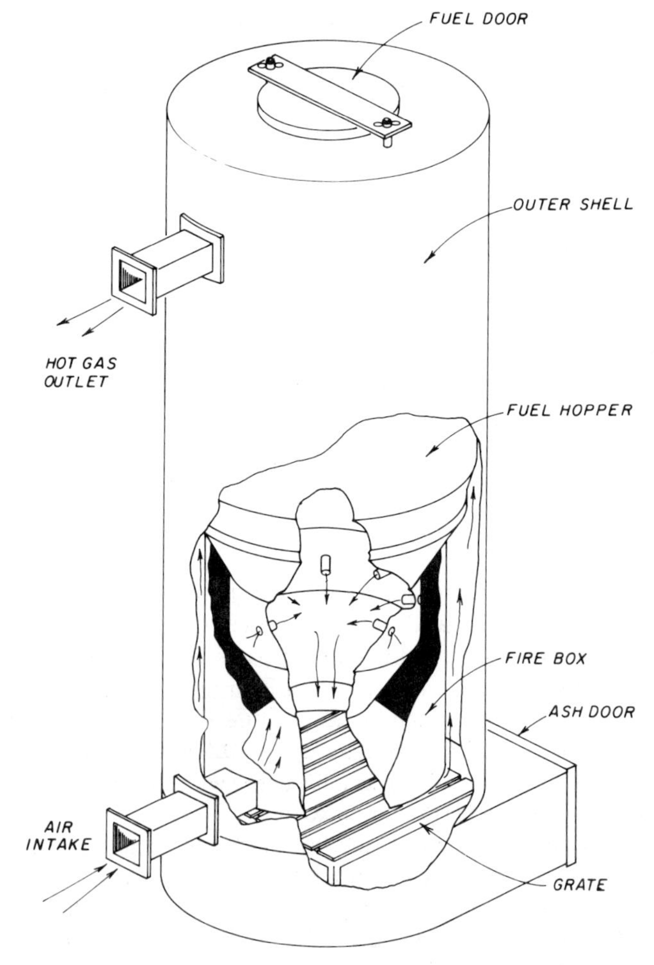

Figure 1. First generation gas producer.
Note: overall height 76-in.; outside diameter 28-in.

Figure 2. Equipment layout showing (right to left) Wisconsin engine, polyphase motor and 10 kW generator.

Figure 3. Equipment layout showing (left to right) gas producer with burner, gas cleaner/cooler and 10 kW generator.

Hot raw gas, generated in the producer, was either flared in a burner or cleaned, cooled and used as fuel for a Wisconsin V460D engine or a 10 kW engine-generator. Engine performance was studied when running on producer gas generated from different agricultural residues.

A switchbox was installed on the inlet manifold to the Wisconsin engine to enable it to be started on gasoline and then converted to a producer gas/air mixture. The flow rates of air and gas into the cylinders were monitored with calibrated orifice meters. The engine was coupled to a 7-hp polyphase motor-generator via a set of V-belts and a friction clutch. When the engine was driving, *i.e.*, under load, the electrical energy output was recorded on a calibrated watt-hour meter and these data were converted to shaft hp delivered by the engine. All tests were performed at constant speed (2100 rpm^{-1}). The air inlet to the engine was constantly regulated to maintain maximum output power.

Maximum gasoline output power was 24.5 bhp. This value was low, since the switchbox adversely affected the engine's volumetric efficiency. During a test when the producer was fueled with tree prunings, the engine delivered between 70 and 80% of maximum gasoline output power. This was unusually high, since the recorded gasoline output power was low. The manifold depression was observed to be 13.7 cm Hg with producer gas as compared with 45 cm Hg when running on gasoline. The volumetric efficiency of the engine was only 60%. The test was conducted for 30 min, during which time output power, manifold depression and gas and air flows throughout the system remained reasonably constant. The pressure recorded across the producer was found to give a good indication of fuel situation in the firebox. This pressure increased steadily with the buildup of char above the grate and a rapid increase (to 20 in.w.g.) occurred just prior to the failure of the engine to run on gas.

Emission tests with a commercial automobile exhaust analyzer, when the engine was under full throttle at 2100 rpm, indicated that hydrocarbons were as low as 25 ppm and CO as low as 0.05% (V/V).

When using walnut shells as fuel, output power, manifold depression, gas and air flows and engine efficiencies were found to be similar to those values obtained for tree prunings.

The results can only be considered to be a rough guide to engine and producer performance because the apparatus leaked air into the gas stream, which caused inaccuracies in recording the air input to the engine. Nevertheless, they do serve to demonstrate the ability of a readily obtainable unmodified engine to run satisfactorily on low Btu gas.

In addition to walnut shells and tree prunings, rice hulls, alfalfa cubes, cotton gin waste and corn cobs were also tested in the producer. With all these fuels it was possible to obtain satisfactory operation of both burner and engine for short periods of time (30 min). Prolonged operation was prevented by stoppages in the flow of fuel through the producer. Wood chips, in particular, tended to bridge in the fuel hopper, and once the material in the fire zone was exhausted, it was not replenished from above. A stirrer was installed in the producer; however, this did not solve the problem. In contrast to tree prunings, free-flowing granular materials (such as nut shells and rice hulls) can be evenly and continuously fed into the fire zone, but these fuels present problems of their own. Rice hulls have a high ash content (20%) and must be fed into and removed from the producer with considerable rapidity. This was not possible in the laboratory batch-filled apparatus. Nut shells, on the other hand, are low in ash content but produce large quantities of tar in the ensuing gas stream. The filtration equipment was at first unable to cope with the tar content of gas derived from nut shells. Problems occurred due to clinker formation when corn cobs were used in the producer. The clinkers impeded the flow of air through the reaction zone. The size, shape and composition of a given fuel are clearly important parameters for the satisfactory operation of a producer.

The tar content of producer gas is an important operational consideration. Relatively small amounts will quickly deposit in the intake manifold and will ultimately seize up the engine. Ideally, tar can be destroyed if it is subjected to high temperatures while passing through the narrow portion of the firebox. Thus with tar-producing fuels the firebox dimensions are critical to ensure that the gas achieves the proper temperature but at the same time is not so restrictive as to cause excessive pressure drop in the producer.

A 10 kW signal corps motor/generator was operated successfully for periods of up to 8 hr on producer gas. A three-way valve, mounted between the carburetor and induction manifold, enabled the engine to be run on either gasoline or producer gas.

It was apparent from the limited experimentation described above and from an extensive literature survey[1] that no serious problems were likely to be encountered when running a gasoline engine on producer gas. A reduction in output power relative to that normally obtained with gasoline was to be expected; however, output power could have been increased by installing a turbo charger and/or by increasing the engine's compression ratio. Of considerably more importance was the design of a gas producer capable of supplying a tar-free gas from agricultural residue fuels. The provision of a continuous flow of fuel through the reaction zone was a prerequisite to the maintenance of a satisfactory supply of producer gas.

SECOND GENERATION GAS PRODUCER

Description of Apparatus

A second downdraught gas producer was constructed and tested. The design of this model was based on a Swedish producer currently being developed for use with automotive transport.[2]

The producer is shown in Figures 4 and 5. Although it is essentially a batch-filled producer, the grate provides for the continuous displacement of

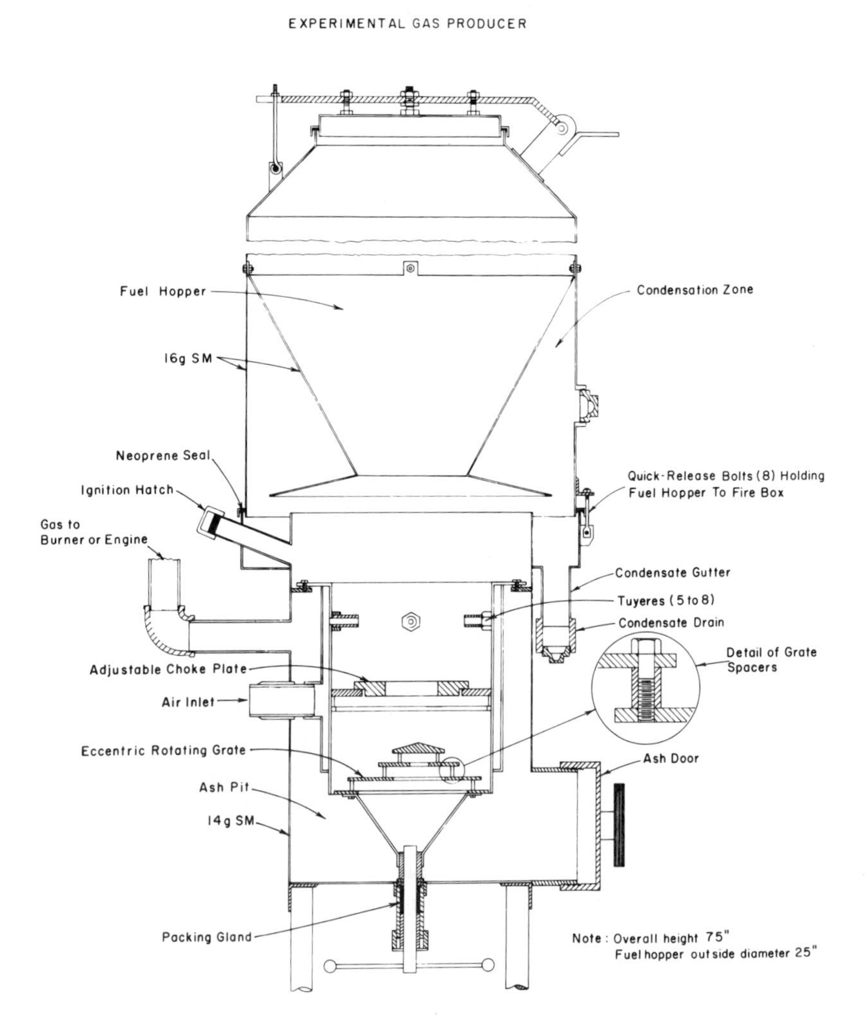

Figure 4. Second generation gas producer.

Figure 5. Second generation gas producer.

solid refuse into the ash pit. It was thus possible to maintain a constant flow of material through the reaction zone.

Five tuyeres, consisting of hollow pieces of stainless steel running thread, were positioned on the circumference of the upper part of the firebox. Each tuyere was screwed into a 3/4-in. nut welded to the firebox inner wall. The running thread was protected by two stainless nuts welded to its exterior. New tuyeres of different length and diameter could be installed as desired.

The restriction in the firebox was designed to be adjustable by making it in the form of a loose 1-in.-thick mild steel ring. This "choke plate" was supported on a larger diameter ring made in two half-pieces which was, in turn, supported in a series of hoops. The entire structure rested on a ring welded to the inside of the firebox. The choke diameter and its position relative to the tuyeres could be altered as desired. A view of the inside of the firebox showing the choke plate and tuyeres is shown in Figure 6.

The eccentric rotating grate consisted of flat, circular steel plate rings mounted one above the other with their edges overlapping. To pass through the grate, solid refuse had to pass horizontally through the space between the plates. This motion was imparted to the refuse by rotation of the grate. The

Figure 6. Inside of firebox showing tuyeres and choke plate.

grate plates were eccentric with the center support so that rotation forced the refuse to move between the plates. In addition, rotation imparted a grinding action to any clinker which may have formed. Clinkers were crushed between the edges of the plates and the sides of the firebox.

At start-up, the firebox was filled with char to a point 6 in. above the tuyeres. The fuel hopper was then filled with fresh fuel. The upper half of the producer could be removed by releasing eight swing-bolts. This provided ready access to the components in the firebox.

Condensable vapors, caused by drying and pyrolysis of the fuel immediately above the tuyeres, migrated upward to the condensation zone. These vapors—a mixture of water, tar, light oil and organic acids—collected in liquid form in the condensate gutter. In this way some of the condensible material could be removed before it entered the producer gas stream.

After charging, the producer was ignited with a standard automobile flare. The flare was fixed inside a protective holder which was pushed through the ignition hatch to a position between the tuyeres in the firebox. The ignition procedure is shown in Figures 7 and 8.

The grate was found to be the most critical part of the producer. It had to perform the two tasks of displacing solid refuse into the ash pit and at the same time providing unhindered passage to the exiting producer gas. The two tasks were not always compatible. A number of modifications to the firebox and grate configuration were tested. These are illustrated in Figures 9 through 15.

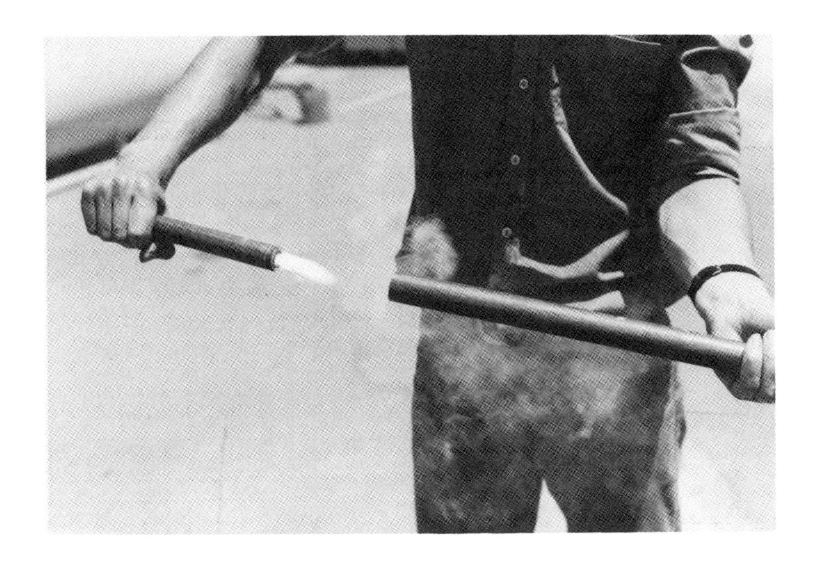

Figure 7. Inserting flare into holder during ignition procedure.

Figure 8. Flare positioned between tuyeres.

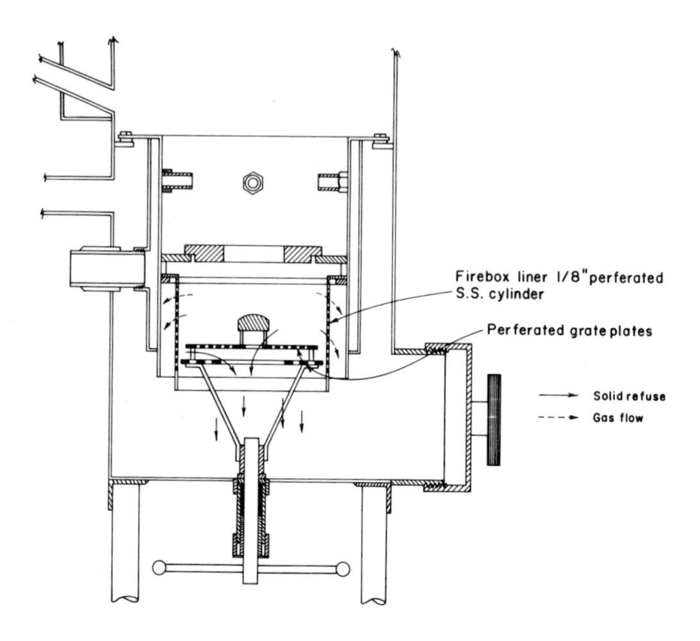

Figure 9. Perforated liner and perforated grate.

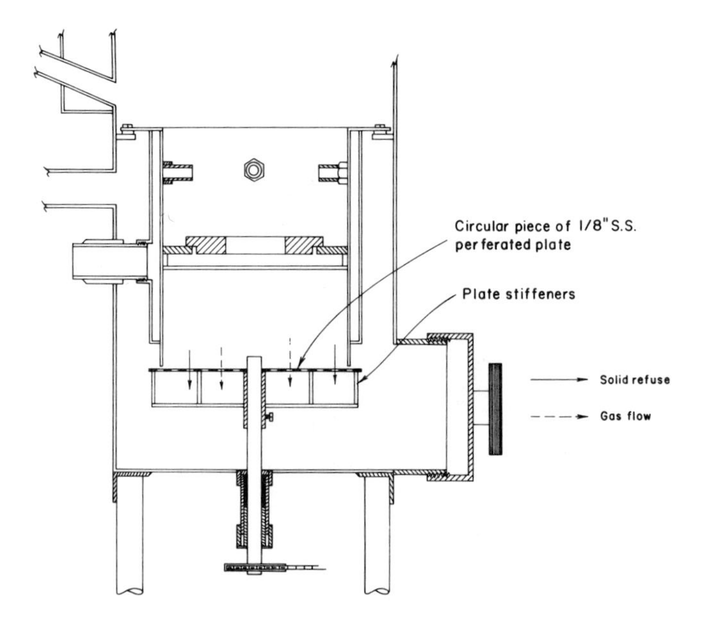

Figure 10. Flat rotating perforated grate.

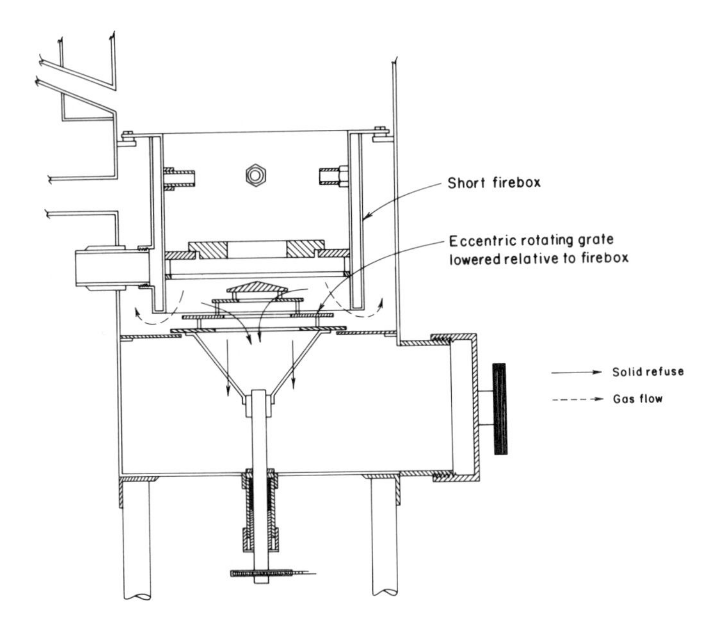

Figure 11. Grate lowered relative to firebox. Gas off-take above grate.

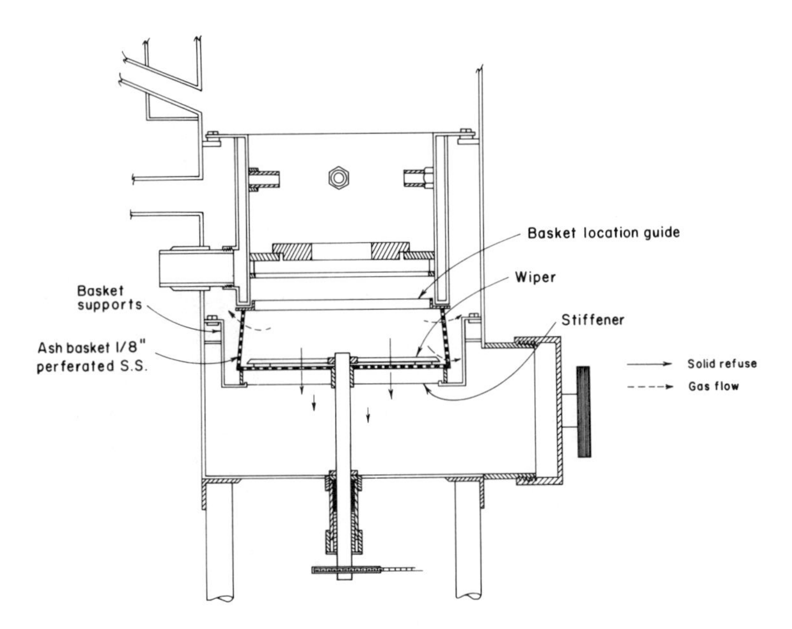

Figure 12. Final design of grate/firebox for use with walnut shells.

Figure 13. Underside of flat, fixed, perforated grate inside firebox.

Figure 14. Underside of flat, rotating, perforated grate.

Figure 15. Final design of grate/firebox for use with walnut shells.

The primary objective of the design shown in Figure 9 was to partially separate the flows of producer gas and solid refuse by providing additional space through which the gas can escape. Producer gas could exit through the perforated walls of the firebox liner, through the perforated grate plates, and between the grate plates as in the original design. The flat rotating perforated grate (Figures 10 and 14) was an attempt to employ the simplest design possible. Solid refuse was broken up by attrition and fell through the perforation into the ash pit. This grate was driven by a 1/15-hp gear-motor coupled to an 18:1 speed reducer.

Figure 13 shows the underside of the firebox with a fixed grate consisting of a circular piece of perforated steel mounted inside it. A wiper was used to scrape the grate surface. The wiper was driven by the system described above.

In the design shown in Figure 11, producer gas was allowed to exit above the grate. In this way the flows of solid refuse and gas were separated above the grate and there was no gas flow into the ash pit.

The grate arrangement which was found to be the most suitable for use with walnut shells is shown in Figures 12 and 15. This consisted of a "basket," fabricated from perforated stainless steel, which extended below the firebox. The base of the basket was a circular piece of perforated stainless steel strengthened by two hoops attached to its underside. The sides of the basket were in the form of a truncated conical membrane, the upper diameter of which matched the bottom of the shortened firebox. A mechanically-driven wiper was used to sweep the circular base; solid refuse fell through the perforations to be collected in the ash pit.

Experimental Procedure

The gas producer was tested with different grate arrangements, different firebox configurations (longer tuyeres, larger diameter chokes, etc.) and different types of fuels. Typical fuels are shown in Figure 16.

After start-up, the producer was run for a period of one to two hours to enable a "char layer" to build up around the inside of the firebox. The formation of this layer was characteristic of all test runs and had the effects of reducing the size of the reaction zone and insulating the firebox walls.

The data reported in this paper were recorded during "steady-state running," *i.e.*, when the following conditions were satisfied:

1. the char layer had built up;
2. a steady temperature had been reached; and
3. the gas was burning with what appeared to be a clean flame.

For each test run, the following data were measured and recorded:

Figure 16. Producer fuel, mulled nutshells (left) and corn cobs (right).

a. air input rate to producer at ambient conditions;
b. pressure differential across the producer;
c. gasification rate;
d. firebox temperature;
e. solid refuse production rate (char);
f. condensate production rate (tar and water); and
g. fixed carbon (FC), ash and volatile and combustible matter content (VCM) of solid refuse.

Items a, b, c and d were measured every 5 minutes; e and f were recorded at the end of a timed run. The analysis for g was conducted by the standard method.[3]

The entire gas producer assembly was mounted on a scale so that gasification rate could be readily determined.

Air input to the producer was measured with a calibrated orifice meter and inclined gauge. The temperature of the input air was measured with a mercury-in-glass thermometer, and the static pressure was recorded on a U-tube manometer connected to the orifice meter upstream pressure tap. The pressure differential across the producer was recorded on a U-tube manometer connected to the air input and gas output parts.

Firebox temperature was measured with a chromel-alumel thermocouple and a compensating direct-reading meter. The thermocouple probe was flexible, which enabled the solid refuse and off-gas temperatures to be measured individually with the same equipment.

The relative quantities of anhydrous tar and water collected in the condensate gutter were determined gravimetrically. Water was evaporated from a weighed sample of condensate, and the sample was reweighed.

The mass flow of steam in the producer gas was obtained by aspirating a sample of hot gas through a cold-finger condenser maintained at 32°F and weighing the condensate. A recommended procedure was followed.[4]

The heating values of the anhydrous tar and the wet fuel were determined with a bomb calorimeter. Some fuel heating values were obtained from the literature.[5] Moisture contents of the various fuels were determined gravimetrically.

The procedures and analyses described above (not including those listed in a through g) were performed only on three or four selected runs, but the results were reasonably constant.

Radiant heat losses were calculated from the following formula:[6]

$$Q_r = 0.171\, F_E\, F_A\, A\left[\left(\frac{T_1}{100}\right)^4 - \left(\frac{T_2}{100}\right)^4\right]$$

where:

Q_r is the radiant heat loss (Btu hr^{-1}).
F_E is the emissivity factor (0.3 for aluminum paint, 0.8 for oxidized metal; a value of 0.6 was assumed).
F_A is the angle factor (assume unity);
A is the area of the firebox outer shell (9 ft^2).
T_1 is 1000°F (1460°R) approximately.
T_2 is 90°F (550°R) approximately.

Depending upon wall temperature and surface condition, the radiant heat loss varied between 15,000 and 60,000 Btu $hr.^{-1}$ A value of 40,000 Btu hr^{-1} was assumed for all runs.

Results and Discussion

Test results obtained from seven runs with the producer are summarized in Table I. The first 12 lines are a record of the data recorded during and after each test run. The remaining part of the table is an energy balance for each run. All energy calculations are based on a temperature of 90°F for the surroundings and for a running time of 1 hr.

Gas analysis equipment was not available, hence the total energy output in the gas (producer gas plus any condensible material) had to be calculated by difference. Heat losses in the solid refuse and the condensate and due to

Table I. Energy balance (producer only).

Run Number	2	5	14	19	20	22	25
Type of Fuel	corn cobs	tree prunings	walnut shell	walnut shell	walnut shell	walnut shell	walnut shell
Firebox and Grate[a] Configuration	Fig. 4	Fig. 9	Fig. 9	Figs. 10& 14	Fig. 13	Fig. 11	Figs. 12& 15
Fuel Consumption (lb hr^{-1})	90	42	37	56	50.5	43	81
Gasified (%)	80	83	91	87	91	88	86
Solid Refuse (%)	20	17	8	11	5	9	12
Condensate (%)	0	0	1	2	4	3	2
Pressure Differential (ins. w.g.)	30	25	17	35	32	10	28
Ash in Refuse (%)	4.5	17.2	6.3	5.3	6.5	3.0	6.5
V.C.M. in Refuse (%)	14.5	18.0	5.3	6.6	7.6	20.7	7.5
F.C. in Refuse (%)	81.0	64.8	88.4	88.1	85.9	76.3	86.0
Firebox Temperature (below choke) (°F)	1600	1800	1420	1550	1600	1600	1600
Potential Heat in Wet Fuel Fired to Producer (Btu hr^{-1})	652,860	300,510	270,118	408,985	368,675	313,921	591,340
Sensible Heat in Air at Input Conditions (Btu hr^{-1})	1,438	690	303	742	554	496	627
Total Energy Input (Btu hr^{-1})	654,298	301,200	270,421	409,727	369,229	314,417	591,967
Heat of Combustion of Solid Refuse (Btu hr^{-1})							
V.C.M.	16,899	8,321	1,015	2,632	1,242	5,187	4,720
F.C.	211,410	67,087	37,941	78,691	31,450	52,773	121,208
Sensible Heat in Solid Refuse (Btu hr^{-1})	6,588	2,613	1,083	2,254	924	1,416	3,557

Heat of Combustion Anhydrous Tar in Condensate (Btu hr^{-1})	0	0	1,396	4,227	7,623	4,868	6,114
Latent and Sensible Heat of Water in Condensate (Btu hr^{-1})	0	0	244	788	1,336	853	1,140
Radiant Heat Loss (Btu hr^{-1})	40,000	40,000	40,000	40,000	40,000	40,000	40,000
Total (Btu hr^{-1})	274,897	118,021	81,679	128,592	82,575	105,097	176,739
Total Energy Output in Gas (Btu hr^{-1})	379,401	183,179	188,742	281,135	286,654	209,320	415,228
Sensible and Latent Heat of Steam in Gas Stream (Btu hr^{-1})	8,352	4,043	3,905	5,591	5,330	4,389	8,080
Sensible Heat in Dry Gas (Btu hr^{-1})	9,882	4,660	4,622	6,618	6,308	5,194	9,561
Total (Btu hr^{-1})	18,234	8,703	8,527	12,209	11,638	9,583	17,641
Net Energy Output in Gas (Btu hr^{-1})	361,167	174,475	180,214	268,926	275,016	199,737	396,587
Conversion Efficiency (%)	55.2	57.9	66.6	65.6	74.5	63.5	67.1

[a] 4-in. choke and 8-in. tuyere diam.

radiation (see calculation) were subtracted from the total energy input. Net energy output in the gas was obtained by further subtracting the latent and sensible heat in the steam and the sensible heat in the dry gas.

Potential Heat in Wet Fuel Fired to the Producer

Dry fuel firing rate is determined from the moisture content of the fuel and the (wet) fuel consumption. Potential heat in the wet fuel is the product of the dry fuel firing rate and the heat content (dry basis) of the fuel.

Sensible Heat in Air at Input Conditions

This value is determined from the measured rate of air input and its temperature and pressure. A value of 0.17 was assumed for c_V. Air density was calculated for the prevailing conditions. Total energy input is the sum of the energy input in the fuel and air only.

Heat Lost in Solid Refuse

Loss of potential heat was determined for the V.C.M. and F.C. portions in the solid refuse. The heating value of the fixed carbon portion was assumed to be 14,500 Btu lb^{-1}. The heating value of the V.C.M. was determined by subtracting the heating value of the F.C. in the fuel from its overall heating value and dividing the result by the percent of V.C.M. in the fuel. Sensible heat loss in the solid refuse is the product of the refuse production rate, the specific heat of the refuse (a value of 0.2 was assumed) and the temperature range (610°F).

Heat Lost in Condensate

The potential heat loss in the anhydrous tar is the product of the tar production rate and its heating value (10,200 Btu lb^{-1}), and the heat loss in the water is the product of its production rate and 1,050 Btu lb^{-1}.

Heat Loss Due to Steam in Gas

This value was obtained from the mass flow rate of steam and assumed values of 1050 Btu lb^{-1} and 110 Btu lb^{-1} for the latent and sensible heats of the condensing water.

Sensible Heat in Dry Gas

The mass flow of dry gas was determined by subtracting the steam flow from the total gas flow; c_p was assumed to be 0.25, and the temperature of the stack was 700°F.

Run #2

When the producer was fueled with large-size material (+0.5 in.), such as corn cobs or cubed cereal straw, the rotating eccentric grate was used very successfully. It was possible to pass sufficient air through the fuel bed to give a high enough reaction temperature and a tar-free gas. In this run a high fuel consumption rate was achieved, but the serious loss of energy in the fixed-carbon fraction of the solid refuse caused the conversion efficiency to be low.

Run #5 and Run #14

In these two tests tree prunings and mulled walnut shells were fired to the producer at a similar rate. The same firebox/grate configuration was used in each case. With prunings, the fixed carbon content of the refuse was lower and no condensate was obtained. Nevertheless, more solid refuse was obtained with the prunings and the conversion efficiency was lower than that achieved with the walnut shells.

Run #19 and Run #20

A major problem associated with the gasification of mulled walnut shells (tyler mesh range 4 to 14) is the ease with which this material fluidizes. If the reduction reaction is to be maintained in equilibrium, sufficient air must be available to maintain the oxidation reaction at the desired temperature (about 2000°F). Fixed beds of granular material offer a high resistance to the passage of air or gas due to their low voidance. A high pressure differential has to be maintained across the bed (32 to 35 in. w.g.) if sufficient air is to be passed through it. For a given area of grate, there is a maximum gas flow above which the gas velocity exceeds the critical value for fluidization to occur (120 ft min^{-1} in the case of mulled walnut shells). When fluidization occurs in downdraught flow, the high pressure causes the entire bed of fuel to be displaced into the ash pit.

The types of firebox and grate used in these two runs were designed primarily for use with walnut shells and to obviate the problem of fluidization. The stationary grate used in conjunction with a wiper (Run #20) was more effective, and a higher conversion efficiency was obtained with this design.

Run #22

In this test producer gas was allowed to exit from the fuel bed above the grate. The design was not entirely successful, and the producer had to be run with a low air input to prevent large quantities of unburned fuel being blown into the ash pit. As a result, the fuel consumption rate was low.

Table II. Energy balance (producer only).

Run Number	3D	6D	9D	13D	14D	18D	16D
Proportions of Raw and Mulled Walnut Shells Constituting Fuel	100% raw	80% raw 20% mulled	60% raw 40% mulled	50% raw 50% mulled	40% raw 60% mulled	20% raw 80% mulled	100% mulled
Firebox and Grate Configuration	Figs. 12 & 15	Figs. 12 & 15	Figs. 12 & 15	Figs. 12 & 15	Figs. 12 & 15	Figs. 12 & 15	Figs. 12 & 15
Fuel Consumption (lb hr^{-1})	44.0	48.0	45.0	54.2	55.2	62.6	82.6
Fuel Gasified (%)	97.2	93.3	93.4	92.5	88.8	87.9	89.6
Solid Refuse (%)	1.4	4.2	4.4	5.5	9.4	7.9	9.7
Condensate (%)	1.4	2.5	2.2	2.0	1.8	4.2	0.7
Pressure Drop Across Producer (ins. w.g.)	20	20	20	20	20	20	20
Ash in Refuse (%)	18.8	28.4	23.6	16.0	15.5	10.7	3.6
F. C. in Refuse (%)	69.9	56.1	61.5	73.8	77.4	83.6	88.7
V.C.M. in Refuse (%)	11.3	15.5	14.9	10.2	7.1	5.7	7.7
Firebox Temperature (below choke (°F)	1600	1600	1600	1600	1600	1500	1450
Potential Heat in Wet Fuel Fired to Producer (Btu hr^{-1})	338,492	369,264	335,587	404,198	411,654	466,839	615,989
Sensible Heat of Air Blast (Btu hr^{-1})	1,026	1,026	1,026	1,026	1,026	1,026	1,026
Total Energy Input (Btu hr^{-1})	339,518	370,290	336,613	405,222	412,680	467,865	617,015

Heat of Combustion of Solid Refuse (Btu hr^{-1})							
F.C.	6,118	16,071	16,774	30,305	55,322	56,951	97,896
V.C.M.	441	1,981	1,813	1,869	2,264	1,733	3,792
Sensible Heat of Solid Refuse (Btu hr^{-1})	74	241	229	345	601	573	929
Heat of Combustion of Anhydrous Tar in Condensate (Btu hr^{-1})	2,463	4,798	3,837	4,201	3,851	10,190	2,241
Latent and Sensible Heat of Water in Condensate (Btu hr^{-1})	420	818	654	717	657	1,738	382
Radiant Heat Loss (Btu hr^{-1})	40,000	40,000	40,000	40,000	40,000	40,000	40,000
Total Heat Loss (Btu hr^{-1})	49,516	63,909	63,307	77,437	102,695	111,185	145,240
Total Energy Output in Gas (Btu hr^{-1})	290,002	306,381	273,306	327,785	309,985	356,680	471,775
Sensible and Latent Heat of Steam in Gas (Btu hr^{-1})	2,917	3,055	2,779	3,315	3,241	3,638	4,893
Sensible Heat of Dry Gas (Btu hr^{-1})	6,008	6,291	5,724	6,827	6,675	7,493	10,079
Total (Btu hr^{-1})	8,925	9,346	8,503	10,142	9,916	11,131	14,972
Net Energy Output in Gas (Btu hr^{-1})	281,077	297,037	264,803	317,643	300,079	345,549	456,803
Conversion Efficiency (%)	82.8	80.2	78.7	78.4	72.7	73.8	74.0

Run #25

Fluidization was completely avoided with the grate used in this run. Solid refuse was reduced in size as a result of wiper action which also prevented the perforated plate from blocking. This design was tested on three occasions. The results of early runs were highly satisfactory and, after a few simple mechanical problems were corrected, the producer was run continuously for 7.5 hr. Air input was maintained at 20 ft^3 min^{-1} (ambient conditions) throughout the run. The gas burned with a clean flame. No damage to any of the internal parts of the producer was observed at the end of the run.

A further series of tests were conducted to study the gasification of fuels of controlled size distribution. The conversion efficiencies which could be achieved with different size distributions of fuel were investigated. Mixtures of mulled shell and raw shell (approximately 0.5-in. material) were prepared such that the relative amounts of the two types of shell varied from 100% w/w raw shell to 100% w/w mulled shell. Gasification tests were run on each mixture. The grate/firebox configuration was that shown in Figures 12 and 15. Air input rate and producer pressure drop were maintained constant at 18 ft^3 min^{-1} and 20 in. w.g., respectively. The grate wiper was rotated at whatever speed was necessary to keep the pressure differential at the desired level.

The results are given in Table II. It can be seen from the data that as the fuel mixture approached 100% mulled shell, it became progressively more difficult to pass sufficient air through the bed of fuel. To maintain the air flow at the desired rate, the bed had to be agitated more violently by turning the wiper at a higher speed. As a result, fuel consumption rates were increased, but the solid refuse production rate and the fixed carbon content in the solid refuse also increased, leading to lower conversion efficiencies.

GENERAL CONCLUSIONS

- In a downdraught gas producer, the grate/firebox configuration is of paramount importance in providing for the continuous supply of good quality gas.
- Different designs of grate/firebox are likely to be used with different fuels.
- Conversion efficiencies in excess of 70% can be readily achieved, and 80% is possible with an ideally sized material.
- As the size of the fuel decreases, gasification in downdraught mode becomes increasingly more difficult.
- Mixing fuels of different sizes could alleviate problems caused by the high resistance to air/gas flow which occur with granular fuels.

ACKNOWLEDGMENTS

The authors wish to thank the funding bodies who have made this research possible. Thanks are also due to John R. Goss and Jim Hewell of the Department of Agricultural Engineering, University of California, Davis, California.

This research was supported in part by funds from USDA-ARS together with RANN (later ERDA) under Contract No. 12-14-1001-603 and in part by the State of California Energy Resources, Conservation and Development Commission under Contract No. 4-0138.

REFERENCES

1. Horsfield, B. and R. O. Williams. "Energy for Agriculture and the Gasification of Crop Residues," paper to be published in *Energy Sources.*
2. Nördstrom, O. "Aktuelle Arbeiten auf dem Geibeit der Ersatztreibstoffe in Schweden Diesel gas betrieb, Entwicklung der Holzgasgeneratoren and Reinigee," *Motor lastwagen L'autocamion* 45 (1960).
3. Engelder, C. J. *Gas, Oil and Fuel Analysis* (New York: Wiley, 1931).
4. "Methods for Determination of Velocity, Volume, Dust and Mist Content of Gases," Bulletin WP-50, 7th ed., Western Precipitation Division, Joy Manufacturing Company.
5. The Power Gas Corporation Ltd. "P-G Waste Fuel. Producer Gas Plant."
6. Faires, V. M. *Thermodynamics* (New York: Macmillan, 1957).

33

USE OF CROP RESIDUES TO SUPPORT A MUNICIPAL ELECTRICAL UTILITY

R. K. Koelsch, S. J. Clark, W. H. Johnson, G. H. Larson
Department of Agricultural Engineering
Kansas State University
Manhattan, Kansas

INTRODUCTION

Present farming practices for many crops harvest only a small portion of the energy captured by plants. The remainder of the plant, which represents a sizable quantity of energy, is left in the field. Considering this fact, a potential energy source in the form of crop residue is available for use. Operating under this premise, the engineering colleges of Kansas State University and Kansas University in conjunction with the electric utility at Pratt, Kansas, are considering the use of crop residues as a possible fuel for electrical generation.

The project, sponsored by the Ozark Regional Commission, is concerned with the removal of crop residue and the conversion process of crop residues and manure to a useful energy form for electrical generation. Presently, direct combustion, anaerobic digestion and wet oxidation represent alternative conversion processes that are being considered. However, this chapter is concerned primarily with the problems and costs of residue removal.

AGRICULTURAL BACKGROUND

The major emphasis of our study of crop residues involved wheat straw. Wheat represents the primary crop in Pratt and surrounding counties. By far, the largest number of acres is planted to wheat, as illustrated by Table I. The only other crop to be produced in sizable quantities is grain sorghum. Corn is not well adapted to the area except when irrigated. If crop residues are going to be used as energy, wheat straw will need to be our primary source of residue in the Pratt area as well as in most of Kansas. Irrigated grain sorghum stover and corn stover represent a growing potential energy source due to the expansion of irrigation in the area. However, the potential of these crop residues is still small due to the limited number of irrigated acres (Table II).

Table I. Crop production, 1970-1975 Average.[1]

	Wheat		Corn for Grain		Grain Sorghum	
	Ha Harvested	Yield (t/ha[a])	Ha Harvested	Yield (t/ha)	Ha Harvested	Yield (t/ha)
Pratt County	67,380	2.04	2,266	6.53	16,300	2.30
Surrounding Four Counties	241,400	2.06	3,059	6.06	37,900	2.53
State	4,156,000	2.18	591,000	5.47	1,486,000	3.09

[a]m. ton/ha.

Table II. Irrigation, 1975.[1]

	Wheat		Corn for Grain		Grain Sorghum	
	Ha Harvested	Yield (t/ha)	Ha Harvested	Yield (t/ha)	Ha Harvested	Yield (t/ha)
Pratt County	0	0	3,240	8.77	970	3.87
Surrounding Four Counties	6,270	2.30	5,140	8.42	7,100	4.69
State	177,000	2.70	425,000	6.52	174,000	4.28

One final observation of agriculture in the Pratt area reveals a wide variation in wheat growing practices. To the west of Pratt, the wheat is produced primarily on summer fallowed land, while continuous wheat is the predominant practice to the east of Pratt. This particular factor will have tremendous influence on wind erosion soil losses. The practice of summer fallowing also reduces the amount of land in wheat production, which will affect residue transportation costs.

STRAW COLLECTION FIELD TESTS

Cost of collection of straw, yields and problems of different collection methods are areas of major concern. Field studies were made on three different collection systems. The first method involved a conventional rectangular baler and an automatic stack wagon. The rectangular bales, dropped by the baler in the field, were collected and moved to the field's edge in groups of 120 bales by the stack wagon. Secondly, a large round baler was used to produce straw packages 1.5 m (5 ft) long and 1.8 m (6 ft) in diameter. The bales were collected and moved individually to the edge of the field by a second tractor with a rear-mounted pickup attachment. Finally, a large package 2.720 m.ton (3-ton) hay stacker was used in the studies. It collected the stacks and moved them to the edge of the field. These three methods were used to collect a total of 50.3 m.ton (55.5 tons) of straw from 43.7 ha (108 ac) of wheat subble during 1976 (an average of 1.150 m.ton/ha or 0.51 tons/ac).

During all the tests conducted in 1975 and 1976, straw removal consisted of only the residue that passes through the combine. The straw spreaders or choppers on the combine were disconnected from the power train, allowing the straw to collect in a windrow. This portion amounts to between 20 and 50% of the total residue. Approximately 1-2/3 kg of wheat straw are available per kg of grain. For corn and grain sorghum, 1 kg of residue per kg of grain is produced.[2] The remainder of the standing stubble was left in place to lessen the problems caused by residue removal.

From the data collected during the field tests (Table III), several comparisons can be made. Of the available straw, the stacker, round baler and conventional baler collected 33, 25 and 15%, respectively. The round baler and conventional baler have higher field capacities than the stacker. However, a separate machine is needed to transport the round and rectangular bales to the edge of the field. The actual labor requirements to collect and transport the straw to the field's edge for the stacks, round bales and rectangular bales are, respectively, 0.487, 0.548 and 0.713 man-hr per m.ton (0.442, 0.506 and 0.647 man-hr/ton).

The cost of collection of the straw also varies with the method of collection. In Table IV the cost per ton based on custom rates paid to the operators involved in this project are listed. These values indicate a definite advantage for the large package stacker. However, the custom rates paid to the round baler were 20 to 30% above average custom rates in the area. This decrease could make the round baler more competitive. Custom rates paid for the conventional baler were about average for rates charged in this region. This alternative is not very competitive from the standpoint of high collection cost. Table IV lists average custom rates for south central Kansas and the expected cost per ton at these rates. Note that the custom rates for transportation of rectangular and round bales to the edge of the field were not available.

Table III. Field performance tests for straw packaging methods studied during 1976 (based on dry weight).

	Loose Hay	Round Bales		Conventional Baler	
	Stacker	Baler	Hauler	Baler	Bale Stacker
Ha Harvested	15.4	15.7	-	12.5	-
Straw Collected (t/ha)	1.460	1.23	-	0.689	-
(kg straw/kg grain)	0.55	0.44	-	0.25	-
Field Capacity (m.ton/hr)	2.05	3.78	3.41	3.61	2.30
Bale Weight (kg/bale)	1060.0	232.0	-	14.3	-
Bale Density (kg/m^3)	38.4	69.2	-	82.0	-
Cost ($/m.ton)	18.8	30.1	4.3	19.0	15.3

Table IV. Average custom rates for south central Kansas.[3]

	$/Bale or Stack	$/m.ton of Straw
Baling Square Bales	0.23	16.10
Stacking Square Bales	?	?
Baling Large Round Bales	5.40	23.10
Transporting Large Round Bales to Edge of Field	?	?
Stacking Straw	21.75	20.50

Table V. Energy value of some organic agricultural wastes.[4]

Type	kcal/kg Dry Weight
Cotton Stalk	4440
Corn Stalk	4390
Soybean Stalk	4330
Sunflower Stalk	4720
Wheat Straw	4220

From the point of view of collection costs and characteristics, both the round baler and the large package stacker systems have a definite edge over the square-baling method. However, such factors as transportation and changing custom rates charged if the electric utility chooses to contract the collection work need to be considered.

STORAGE AND TRANSPORTATION OF THE STRAW

The majority of the straw would probably be stored on the farmer's land from which it was collected. It would later be transported to the power plant at a rate roughly equal to the demand of the plant. This would minimize equipment requirements and land for storage at the power plant. This type of system may require storage on the farmers' land up to 12 months.

The packaged straw will need to withstand the weather during this period. Observations of the straw collected during the summer of 1976 were made over a 6-month period. The round bales withstood the elements during this period with very little loss. As of December 1976, the round bales were still a transportable package. The loose straw stacks were very susceptible to the weather. Several stacks appeared to have lost as much as 50% of their original amount of straw. The loose stacks of straw would be almost impossible to transport on an open trailer similar to present equipment used for alfalfa stacks. Rectangular bales seemed to hold together well during the 6-month observation period. However, the stacks of rectangular bales had partially collapsed by winter. Although the bales were still in good shape, much labor would be required to load the bales for transport to the power plant.

Since average custom rates for transportation are not available, costs were based on conversations with individuals who do custom hauling of alfalfa. Using this information[5] the cost per ton for transportation of straw stacks can be expressed in the following function:

$$3.77 + 0.333x = \text{dollars per m.ton of straw}$$

where x = maximum driving distance of the area from which the straw is removed. The cost of transport of round bales[6] can be expressed by the following function:

$$1.72 + 0.347x = \text{dollars per m.ton of straw}$$

The storage and transportation characteristics of round bales seem to be the best of the three different straw packages. Rectangular bales suffer from large labor requirements necessary for loading the bales for transportation. The low density and lack of binding on loose straw stacks cause problems during storage and transportation.

TOTAL COST

The total cost of round bales and stacks will be considered using average custom cost for collection and the above data for transportation cost. The conventional rectangular baler was not deemed practical at this point due to its high collection cost and large labor requirements for transportation. The total costs for round bale and stack delivery to the power plant are summarized in Table VI for various maximum collection radii.

Table VI. Cost of straw delivery to power plant (based on dry weight).

Collection Radius	Payment for Residue ($/m.ton)	Collection Cost ($/m.ton)	Transport Cost ($/m.ton)	Total Cost ($/m.ton)	Total Cost ($/10^6 kcal)
16.1 km					
Round Baler	11.0	23.2	5.2	39.4	9.34
Stacker	11.0	20.5	7.1	38.6	9.15
32.2 km					
Round Baler	11.0	23.2	8.7	42.9	10.2
Stacker	11.0	20.5	10.5	42.0	9.95
48.3 km					
Round Baler	11.0	23.2	12.1	46.3	11.0
Stacker	11.0	20.5	13.8	45.3	10.7
64.4 km					
Round Baler	11.0	23.2	15.6	49.8	11.8
Stacker	11.0	20.5	17.1	48.6	11.5

To achieve a better perspective on these costs, one should compare these values to the cost of coal. Low-sulfur Wyoming coal with a higher heating value of 4720 kcal/kg (8500 Btu/lb) can be delivered at a cost of $34.2/m.ton (AMAX Coal Company estimates of small load delivery charges) or $7.25/$10^6$ kcal.[7] From the viewpoint of the power plant, crop residue is not yet economically competitive.

NUTRIENT VALUE OF WHEAT STRAW

There is some concern that the wheat straw has some value by returning it to the soil to improve its physical properties and nutrient content. In consultation with Hobbs[8] it was estimated that there are about 7 kg nitrogen per 10 kg of wheat straw. Also, there are approximately 1.67 kg of phosphate and 22.5 kg of potash per 1000 kg of straw. When the straw is burned, all the nitrogen is lost and a small percentage of the phosphate is destroyed. The remaining elements can be recovered in the ash after burning.

It takes about 13,800 kcal/kg (24,800 Btu/lb) to produce 1 lb of nitrogen fertilizer. Thus, the replacement of the nitrogen in the straw represents 2.3% of the total energy available in the straw. It should be pointed out that when the straw is returned to the soil, the nitrogen in the straw does not become available immediately for plant production, but instead gradually becomes available over a period of time as the straw decomposes.

SOIL LOSSES FROM EROSION

Pratt and surrounding counties are characterized by nearly level to gently rolling or sloping plains. The northern third of Pratt County consists of mostly to moderately sandy soils that are highly susceptible to wind erosion.[9] Sandhills and dunes are characteristic of the northwestern part of the county. Silt and clay loams are predominant in the remainder of the county. The climate of Pratt County is subhumid continental. Precipitation occurs primarily during late spring and summer. Annual precipitation amounts to about 61 cm (24 in.). Surface winds are moderate to occasionally strong during all seasons. March and April represent the windiest months with average hourly speeds in excess of 24 km/hr (15 mph).

Wind erosion represents a serious problem of the Pratt area due to a combination of sandy soils, high winds and low rainfall. During winter and early spring, wind can cause blowouts of growing wheat. The quantity of soil erosion losses due to wind is heavily dependent upon the amount of crop residue or growing plants on the soil's surface. For this reason, additional wind erosion soil losses resulting from removal of straw will be considered closely. The rate of water erosion also may be affected by residue removal, but it is not nearly as serious as wind erosion in this region.

The Soil Conservation Service has established two guidelines for determining maximum allowable soil erosion losses. First, the "soil-loss tolerance" is used to denote a maximum rate of soil erosion from wind and water that will permit a high level of crop productivity to be sustained economically and indefinitely. For most Pratt soils, the tolerance is set at 11.21 m.ton/ha (5 tons/ac) per year. A few of the soil types have tolerance set as low as 6.73 m.ton/ha (3 tons/ac) per year. The second factor to consider is the crop tolerance to wind erosion. The crop tolerance for corn and grain sorghum is set at 4.48 m.ton/ha (2 tons/ac) per year, while wheat will withstand soil losses equal to or greater than the soil-loss tolerance.[3] Despite the fact that values for corn and grain sorghum are very low, one must remember that these crops are not exposed to the peak wind erosion periods of winter and early spring due to the timing of their growing season. Under all circumstances, it will be beneficial to keep wind erosion losses less than 5 tons/ac per year.

Annual soil losses due to wind erosion were determined by a procedure already employed by the Soil Conservation Service. This method defines annual wind erosion soil losses as a function of the following variables: (a) soil erodibility; (b) surface roughness of the soil; (c) climate; (d) unsheltered distance across a field along the prevailing wind direction; and (e) vegetative cover during the peak wind period (February and March in Pratt County).

Assumptions were made based on average conditions for surface roughness, climate, field width and tillage practice. From these assumptions, average soil losses due to wind erosion were determined for each soil erodibility group at three different field widths of 30.5, 305 and 792.5 m (100, 1000 and 2600 ft).

The different soil erodibility groups are categorized into nine wind erosion groups (WEG). Our studies centered around six of these groups in Pratt County that are used for grain crops (Table VII). As indicated by Figure 1, the high WEG soil numbers are less susceptible to wind erosion. Field widths for wheat production in Pratt County also vary considerably. For WEG equal to 5, 6 and 7, field widths of 305 to 792.5 m (1000 to 2600 ft) are predominant. Wind erosion losses in this range will not vary much. For WEG equal to 2, 3 and 4L, strip cropping is an established practice for many farmers, which means field widths in the range of 30.5 m (100 ft). However, wider fields are numerous even in these easily erodible soils. Neither practice is dominant. The collected information of our model studies are shown in Figures 1, 2 and 3 for field widths of 30.5 and 792.5 m (100 and 2600 ft).

Some general conclusions can be drawn from Figure 1. For WEG equal to 5, 6 or 7, approximately 1.0 and 1.3 m.ton/ha (0.45 and 0.6 tons/ac) of equivalent flat small grain residue are needed to hold soil loss due to wind erosion to a maximum of 11.21 m.ton/ha (5 tons/ac) per year on the longer fields. This amount of residue must be present on the land during February and March. At the same soil loss level, WEG equal to 2 requires 1.6 m.ton/ha (0.7 tons/ac) of equivalent flat small grain residue, while WEG equal to 3 and 4L need 1.1 m.ton/ha (0.5 tons/ac) of flat small grain residue under strip-cropping conditions (30.5 m). If strip cropping practices are not followed, considerably more residue is needed.

To give these graphs more meaning, the term "equivalent flat small grain residue" can be converted to other residue forms. The Soil Conservation Service makes a distinction between flat and standing stubble, height of stubble and type of stubble (small grain, corn or grain sorghum). During an average year, the Soil Conservation Service estimates that 0.45 m.ton/ha

Table VII. Descriptions of wind erodibility groups (WEG) and percentage of land area in Pratt County belonging to various wind erosion groups.[2,9]

WEG	Predominant Soil Textural Class	Dry Soil Aggregates (0.84 mm %)	Land Area (%)
1	Very fine, fine and medium sands; dune sands	1, 2, 3, 5 and 7	0
2	Loamy sands; loamy fine sands	10	20.0
3	Very fine sandy loams; fine sandy loams; sandy loams	25	35.7
4	Clays; silty clays; noncalcareous clay loams and silty clay loams with more than 35% clay content	25	0
4L	Calcareous loams and silt loams; calcareous clay loams and silty clay loams with less than 35% clay content	25	9.4
5	Noncalcareous loams and silty loams with less than 20% clay content; sandy clay loams; sandy clay	40	11.9
6	Noncalcareous loams and silt loams with more than 20% clay content; noncalcareous clay loams with less than 35% clay content	45	19.7
7	Silts; noncalcareous silty clay loams with less than 35% clay content	50	1.9
8	Very wet or stony; usually not subject to wind erosion	-	0
	Other	-	1.4

(0.2 ton/ac) of growing wheat (0.90 m.ton/ha of equivalent flat small grain residue) are present during February and March plus whatever residual straw left over from the previous crop.

Wheat represents the largest potential source of residue for energy. Both continuous cropping and summer fallowed wheat are grown in the Pratt area. The cropping program, continuous or summer fallowed wheat and the tillage practice between harvest and winter will have a large effect on wind erosion soil losses. Different tillage tools bury different amounts of residue, as indicated by Table VIII. The tillage practices vary greatly due to a number of conditions. The more common tillage practices for continuous wheat listed in Table IX will be considered.

The cropping practice of wheat followed by summer fallow almost completely eliminates any possibility of residue removal. Proper management of residues prior to the first winter will keep wind erosion losses at a very low level the first year. However, tillage requirements for weed control and seed

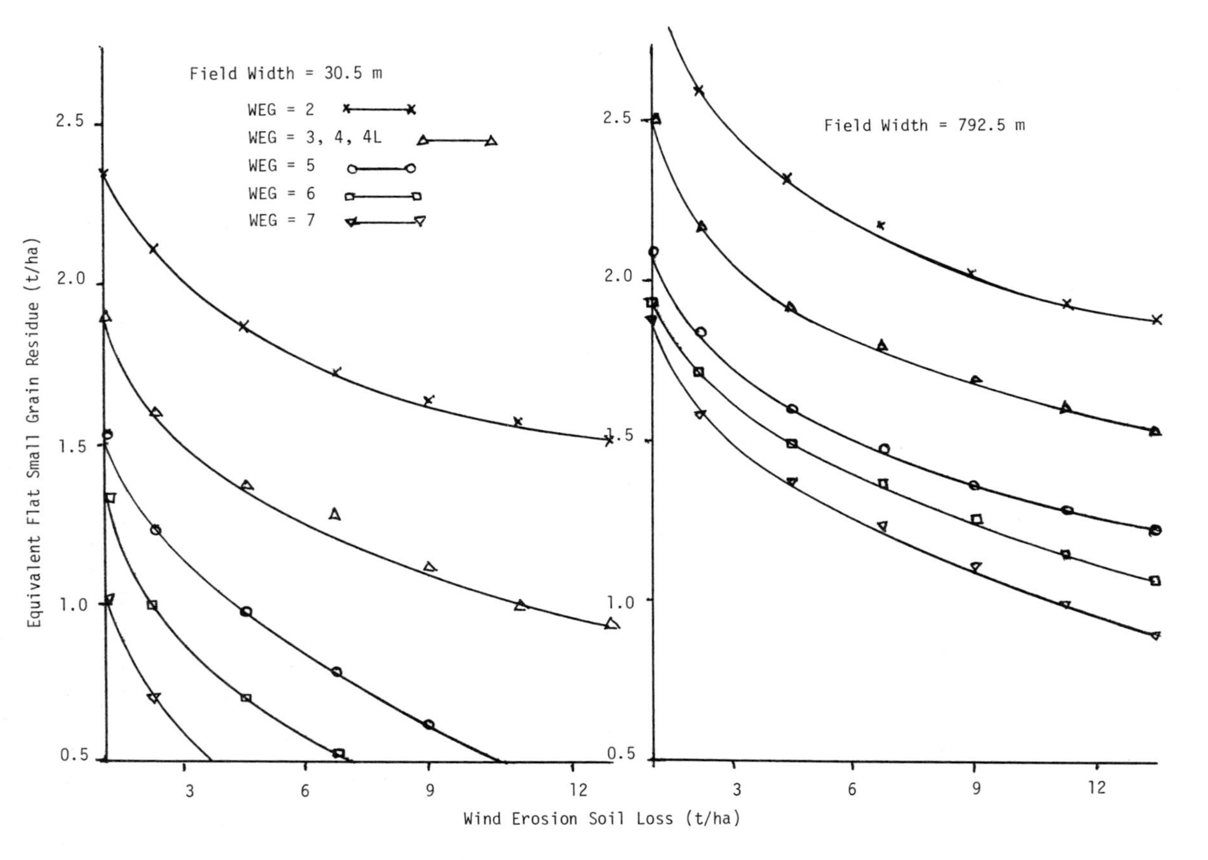

Figure 1. Soil loss vs residue level.

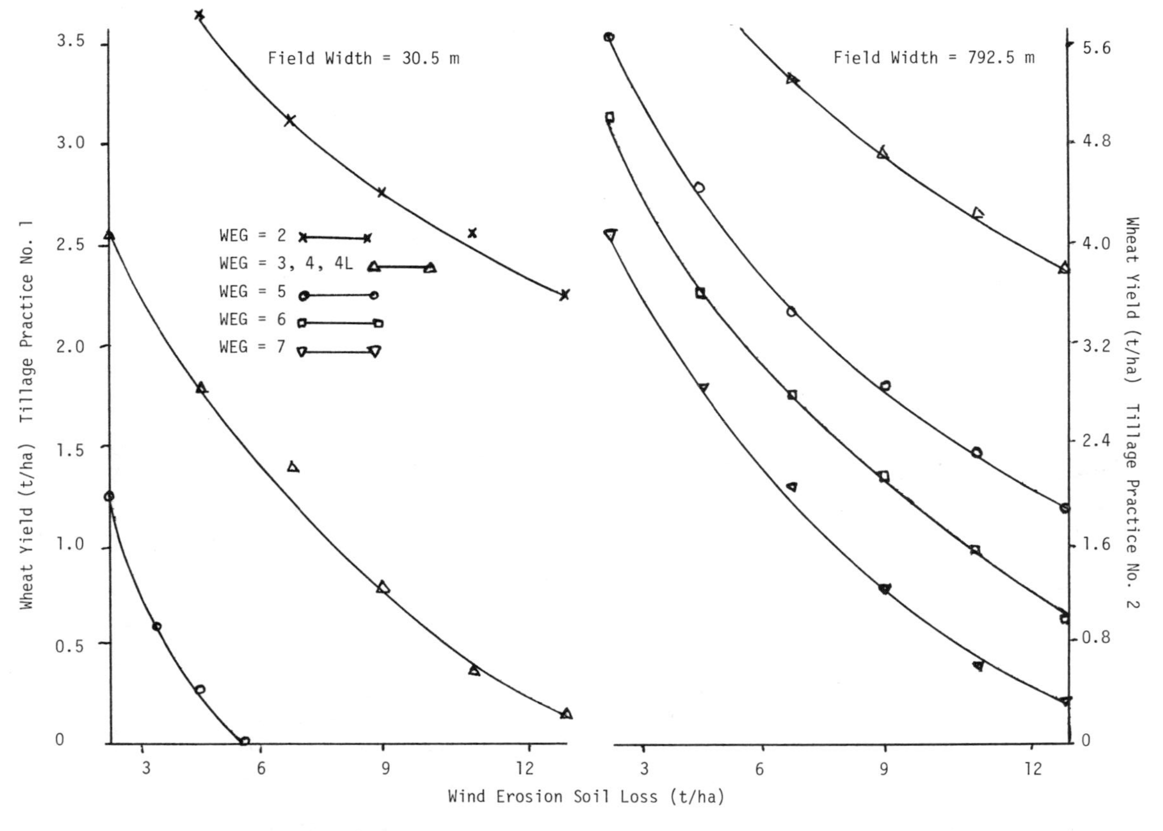

Figure 2. Soil loss vs wheat yield. Crop residue removed: 33%.

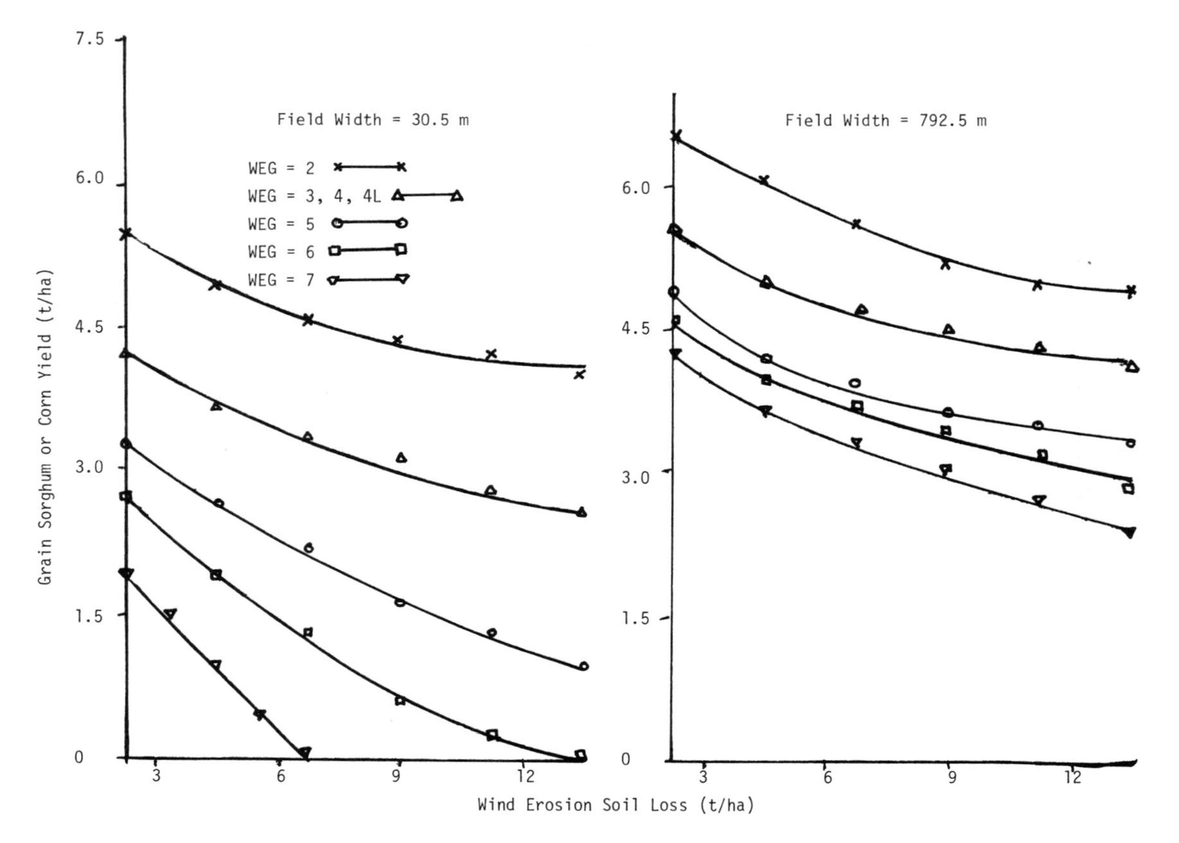

Figure 3. Soil loss vs grain sorghum or corn yield. Crop residue removed: 33%. Tillage operation prior to winter: none. Stubble height = 40.6 cm.

Table VIII. Residue reduction due to tillage equipment.[10]

Tillage Machine	% of Residue Reduction by Each Tillage Operation
One-way disk (61 to 66 cm disks)	50
Tandem or offset disk	50
Field cultivator (41 to 46 cm sweeps)	20
Chisel plow (5 cm chisel 30.5 cm apart)	25
Blades (91.4 cm or wider)	10
Sweeps (61 to 91.4 cm)	15
Rodweeders–plain rod	10
Rodweeders–with semichisels or shovels	15

Table IX. Common tillage practices for wheat in Pratt area.[11]

Continuous Wheat	
1. Harvest a. Disking: once b. Chiseling: once c. Field Cultivator: twice Fall planting	2. Harvest a. Disking: twice b. Chiseling: once c. Field Cultivator: once Fall planting
% Residue Buried: 76% Condition of Residue: Flat	% Residue Buried: 85% Condition of Residue: Flat

bed preparation buries most all of the residue by the second winter. In fact, the Soil Conservation Service assumes no residue remains on the surface by the second winter. This leaves only the growing wheat as a means of controlling wind erosion during the winter following summer fallow. The 0.90 m.ton/ha (0.4 ton/ac) of equivalent flat small grain residue in the form of growing wheat is simply not enough to control the wind erosion, as seen in Figure 1.

Continuous wheat offers a better opportunity for the use of residue as an energy source. During each winter, growing wheat and some residue from the previous crop will be present for controlling wind erosion. In the more stable soils, some of the residue could be removed following harvest and still avoid serious wind erosion problems. Figure 2 provides an idea of the soil losses due to wind that will result at different crop yields.

Soils of WEG = 6 and 7 should be able to withstand 33% residue removal for average wheat yields without serious soil losses. A decision to remove one-third of the residue from soils of WEG = 5 will depend on the tillage practice and the yield of the wheat. Soils of WEG = 2, 3, 4 and 4L would probably not be able to withstand removal of any of the crop residue. The soils from which one-third of the residues could be removed represent 33.5% of the land area of Pratt, Kansas. The amount of continuous wheat grown on this land is unknown.

Dryland grain sorghum represents a possible source of crop residues. Grain sorghum is harvested during fall. Normally, no tillage follows harvest until the following spring. Thus, all of the residues are left on the soil surface. However, removal of one-third of residues may cause severe wind erosion problems as indicated by Figure 3. Assuming that dryland grain sorghum averages about 2.51 m.ton/ha (40 bu/ac), wind erosion soil losses will average more than 11.21 m.ton/ha (5 ton/ac) per year for all soil types where strip cropping is not practiced. Even in the soils where strip cropping is practiced, wind erosion still represents a serious hazard. Under present practices, the removal of dryland grain sorghum will create a serious wind erosion potential. Thus, dryland grain sorghum does not represent a viable source of residues for use as an energy source.

Irrigated grain sorghum and corn represent a better source of residues for energy. High-irrigated corn and grain sorghum yields indicate a very high residue level providing more residue for use in controlling wind erosion and as an energy source. Tillage prior to winter on irrigated row crops will vary between no tillage and one disking. If no tillage is practiced prior to winter, one-third of the residues could be removed under nearly all conditions with very few serious wind erosion problems resulting. Residue removal should be avoided if a disking occurs prior to winter. A disk will not only bury about half of the residue but will also flatten the residue, making it less effective in controlling soil loss. Since the primary purpose of a disking is to begin incorporating excess residue into the soil, this practice may be unnecessary if some of the residue is removed. One final observation reveals that if no tillage occurs prior to winter, more than one-third of the residue could be removed from high-yielding crops grown on the more stable soils without causing serious wind erosion problems. Overall, wind erosion does not appear to be a major obstacle to partial removal of residues from irrigated corn and grain sorghum.

One must realize that all of the previous recommendations apply to average or normal conditions. Changes in weather from year to year can have a major effect on wind erosion. Also, our recommendations were based on generalized field conditions and tillage practices. These generalized conditions were only for exploring the potential for the removal of crop residues. They cannot be applied to an individual situation without first evaluating this particular situation. If a program of crop residue removal is to be undertaken, recommendations will need to be set up for individual farmers based on their own operation. The farmer must also be allowed to make a decision each year as to whether or not it would be wise to allow partial removal of crop residues.

Tillage programs have a tremendous effect upon wind erosion potential. Present tillage programs are designed primarily for weed control and seed bed preparation, with wind erosion control only a secondary consideration. Greater use of tillage tools that bury less residue and herbicides for weed control could increase the availability of crop residue for energy. This is particularly true for corn and grain sorghum, which presently have herbicides compatible with these crops and planting equipment capable of handling high residue levels. However, these areas are not well developed for wheat production. A farmer must bury a large part of his wheat stubble by planting time to allow present planting equipment to do its job. Research in a minimum tillage program for wheat is needed.

FARMER ATTITUDES

The importance of farmer support of the use of residue as energy cannot be overemphasized. Since farmers control the supply of residue, their acceptance of such a program is a necessity. A survey of Pratt area farmers was conducted to gain an insight as to their acceptance of partial residue removal. Review of the 154 responses to the survey revealed several trends. Of the farmers answering the survey, 60% indicated that they were unwilling to allow any of their crop residue removed. The 40% who favored this program were willing to allow partial removal of residues from only part of their land. These people often indicated problems with erosion on part of their land. Some farmers also wanted to try this program for a few years before they would allow removal of the straw from all of the harvested acres. The farmers replying to the survey, who owned 25,218 ha (62,312 ac) of wheat, indicated a willingness to allow straw removal on only 18.4% of the land. A willingness to allow residue removal on irrigated corn and grain sorghum was expressed by farmers on 29% of this land. These values may be poor averages due to the lower number of acres of irrigated corn and milo involved in our survey (5,091 ha or 12,580 ac).

The survey also brought out the major reasons of farmers for not wanting the residue taken off the land. The greatest concern involved the need of residue to control erosion. One hundred surveyed farmers indicated this as a major obstacle. Retaining of the residue's nutrients on the land was also a major reason given by farmers for not supporting our program. The low price offered and the inconveniences of working with custom operators were much smaller concerns of those farmers. Only a small number of farmers indicated that their residues are already being used as cattle feed.

Acceptance of residue removal by farmers seems to be another major obstacle. The fears of increased erosion problems are quite justifiable. The loss of the nutrients of crop residue seems to be less justified and more information provided to the farmers could probably quiet this fear.

CONCLUSIONS

The use of crop residue offers a large potential source of energy. The exact quantity of available crop residue is heavily dependent on farmer attitudes and field conditions. Considering these factors, several conclusions can be drawn:

1. Presently, the cost of wheat straw delivered to the Pratt electric utility is not economically competitive with coal.

2. Wind erosion and farmer attitudes represent major limitations controlling the available residue. Our survey indicates that approximately 20% of the wheat acres would be made available for partial residue removal initially. Approximately 35% of the land area of Pratt, Kansas, has soil types which will tolerate partial residue removal from wheat stubble.

3. The available quantity of straw may not be sufficient to provide the energy needs of the electric utility at Pratt, Kansas. A mixture of coal and crop residue may be necessary for meeting the energy needs. The utility has an average load of about 10 MW.

4. Fields will need to be modeled individually to indicate how much residue can be removed before erosion becomes an overriding concern.

5. An educational program for the farmers will probably be necessary to improve support for a residue removal program. Promotion of proven tillage practices that make more efficient use of available crop residue for controlling erosion must be the main thrust.

6. Development of a minimum tillage program for wheat is essential for increasing the availability of wheat straw for use as energy.

Although we cannot overlook the problems associated with the removal of crop residue for its energy value, these problems are surmountable. Thus, consideration should be given to the value of crop residue as an energy source.

REFERENCES

1. Kansas State Board of Agriculture. "Farm Facts," 1970/1971, 1971/1972, 1972/1973, 1973/1974, 1974/1975, 1975/1976.

2. U.S. Department of Agriculture. "Estimating Soil Loss Resulting from Water and Wind Erosion in the Midwest," Soil Conservation Service, Midwest RTSC, Lincoln, Nebraska (July 1973).
3. Schlender, J.R. and L. Figurski. "Custom Rates for Harvesting and Haying Operations," Cooperative Extension Service Publication No. MF-254, Kansas State University (September 1976).
4. Barth, C. L. and D. T. Hill. "Energy and Nutrient Conservation in Swine Waste Management," ASAE 75-4040, presented at 1975 annual meeting of the American Society of Agricultural Engineers, June 22-25, 1975.
5. Reynolds, Mr., Agricultural Sales, Inc., P.O. Box 589, McPherson, Kansas, personal interview.
6. Saloga, E., personal communication (April 1976).
7. Lester, T. W. "The Direct Firing of Wheat Straw in a Municipal Electrical Generating Station," part of final report to Ozark Regional Commission under ORC Contract No. TA-76-29(NEG) (January 27, 1977).
8. Hobbs, J.A. Department of Agronomy, Kansas State University, personal interview.
9. U.S. Department of Agriculture. "Soil Survey: Pratt County, Kansas," USDA Soil Conservation Service and Kansas Agricultural Experiment Station (September 1968).
10. U.S. Department of Agriculture, Agricultural Research Service. "How to Control Wind Erosion," Agriculture Information Bulletin No. 354, ARS, USDA.
11. Clark, S. J. and W. H. Johnson. "Evaluating Crop Productions Systems by Energy Used," SAE Paper No. 740647.

ADDITIONAL REFERENCES

Rider, A. R. and W. Bowers. "Hay Handling and Harvesting," *Agric. Eng.* (August 1974).

Schrock, M. D., D. L. Figurski and K. L. McReynolds. "Hay Handling Systems," Cooperative Extension Service Publication C-537, Kansas State University (October 1975).

Sproule, J., Pratt Soil Conservation Service Office, personal interview (June 1976).

SECTION VII

ANIMAL WASTE MANAGEMENT

EFFECTS OF POULTRY WASTE EFFLUENT AND INSECTICIDES ON CORN PRODUCTION

R. O. Hegg
Department of Agricultural Engineering
H. D. Skipper
Department of Agronomy and Soils
Clemson University
Clemson, South Carolina

INTRODUCTION

The highest income from an agricultural animal product in South Carolina is from eggs produced by approximately six million layers. A very high percentage of these layers are housed in cages above the floor. Normally the manure droppings are allowed to accumulate for several months and then are transported to the field. During the warm-weather months, there are more problems with handling this waste than during the colder weather because of higher moisture content, more odors and increased fly problems. There is a need to develop management techniques that will combine economics, non-offensiveness and nutrient conservation, along with proper land utilization for these manures.

Several studies have been conducted using various sources and treatments of poultry manure on soil for growing corn. Shorthall and Liebhardt[1] put poultry manure on a sandy loam soil in a single application at rates as high as 5000 kg N/ha from poultry manure plus litter and 224 kg N/ha as commercial fertilizer. The poultry manure significantly reduced yield at the high application rates, and it was determined that soil salinity was considered to be the most important cause of yield reduction. Stuedemann *et al.*[2] used broiler litter and NH_4NO_3 on fescue pasture to determine the effect of grass tetany, fat necrosis and nitrate toxicity on beef cattle. The broiler litter was applied at rates of nearly 800 kg N/ha/yr in five applications/yr for six yr. The authors concluded that grass tetany was a problem with the broiler litter, but this could be alleviated if the application rate were kept below 9 m.ton/ha/yr (322 kg N/ha/yr).

The use of poultry manure and its effect on corn was studied by MacMillan, *et al.*[3] who used stored oxidation-ditch liquid, oxidation-ditch liquid, diffused air-treated liquid and raw manure. These were applied at rates up to 1000 mg N/g dry soil on two different types of silt loam soil, one acid (pH 4.2) and the other neutral (pH 7.1). The acid soil showed increased corn yield as the N application increased, while the neutral soil had decreased corn yield. This was attributed to nitrite toxicity which was measured to be as high as 250 mg NO_2-N/g in the neutral soil, while the acid soil had no measurable level of nitrite.

Fertilization of range land with poultry manure showed a great deal of promise in experiments conducted in California.[4] Forage quality and palatability increased, but the initial abundance of legumes was decreased by increased rates of poultry manure. The increased yield of forage from the manure lasted three years after the application.

Wengel and Kolega[5] studied the effect of poultry manure in relation to soil-water quality and corn yield and concluded that maximum manure application rates should be 30 ton/ac dry matter (67 ton/ha). The manure depressed corn yields with high application rates (2000 to 5000 kg N/ha) compared to commercial fertilizer applied at 280 kg N/ha. High levels of nitrate (over 1000 ppm NO_3-N) were found in the soil after heavy application of poultry manure, resulting in nitrate toxicity of the corn.

Soils and crops grown on soils are not adversely affected by application of arsenic-containing poultry waste litter.[6] The normal application rates of litter on cropland resulted in less than 2 ppm arsenic even after the soil had been treated for 20 years with litter containing arsenicals.

Ethylene was evolved in the surface layers of grassland after application of 550 m.ton/ha of cow waste slurry.[7] The concentrations evolved were high enough to limit plant growth. The conditions to produce the ethylene include significant concentrations of readily decomposable organic matter and restricted oxygen levels in the soil.

The primary objective of this research was to determine the effect of poultry waste effluent and matching fertilizer solutions on corn seedlings under growth-chamber conditions. The digesters were maintained under anaerobic conditions, because this probably more closely approximated actual farm operations. Nitrogen content of the effluent was considered to be an important parameter affecting plant growth. During the progress of the research, it was determined that insecticides were used at the poultry house under warm-weather conditions to control the fly population. Since this poultry house was the source of manure for our studies, additional experiments were conducted to determine the effect of the insecticides on the *E. coli* population in the manure and on plant growth. The Poultry Science Department was not using arsenic, feed additives or antibiotics in their feed program.

EXPERIMENTAL PROCEDURE

Poultry waste effluent from different methods of biological treatment was used as a "fertilizer" on young corn plants. Plexiglass columns 15 cm in diameter and 150 cm high (12 L) were used to simulate mixed and nonmixed anaerobic lagoons. Loading rates and solids retention times (SRT), 10 days, were chosen to give a range of actual field conditions. The digesters were mixed by means of an aerator located at the bottom of the column. Rates of mixing ranged from vigorous to none (except at time of sampling). An aliquot was pumped from the columns, and poultry waste and water were added back 5 days/week to provide a semicontinuous loading schedule. Samples of the effluent were analyzed weekly for total nitrogen, nitrate-N, ammonium-N, phosphate, pH, total solids, volatile solids (VS) and fixed solids.

The bioassay portion of the experiments consisted of adding poultry waste effluent or the synthetic fertilizer solution (50 ml/application) to young corn plants. These synthetic fertilizer solutions were made up to match the N, P, K, Na, Ca and Mg concentrations and the pH of the effluent. Compounds used were K_2HPO_4, KH_2PO_4, $MgSO_4 \cdot 7H_2O$, NaCl, $Ca(H_2PO_4)_2 \cdot H_2O$, $CaCl_2$, NH_4OH, $(NH_4)_2SO_4$ and KCl. The corn plants (thinned to 10 plants/pot) were from 1 to 3 weeks old and were grown in 1000 g of Appling sandy loam soil (pH 6.1). Four replications and a completely randomized block design were used. The plants were grown in a growth chamber that had a light intensity of 2800 foot-candles, a 12-hr day at 26°C, a night temperature of 18°C and 50 to 60% relative humidity. Control pots receiving water only were also grown in the growth chamber. The pots were brought to an approximate field capacity (0.1 bar) daily by weighing. The pots received from 1 to 5 applications of effluent or fertilizer solution per week, depending on the experiment, with a total duration of from 2 to 5 weeks. Approximately 10 days after the final application, the plants were harvested by cutting off the tops and weighing them. *E. coli* counts were made on Tergitol-7 agar from samples of the effluent following the procedures described by Janzen, *et al.*[8] Appropriate statistical analyses were performed on the bioassay data to determine treatment effects.

RESULTS AND DISCUSSION

Experiment I

Experiment I consisted of four mixed anaerobic digesters that were fed poultry waste at loading rates of 50, 100, 150 and 200 lb VS/1000 ft^3-day (0.8, 1.6, 2.4 and 3.2 g/m^3 - day). These digesters were fed 5 days/week for 7 weeks under laboratory temperature conditions (20 to 22°C). Figure 1 shows the effect of effluent or synthetic fertilizer solution (expressed on a nitrogen basis) on 1-week-old corn seedlings 1 week after application. Injury symptoms were noted within 24 hr on the plants receiving effluent at the three higher loading rates. Symptoms were a massive browning-out (discoloration) of the older leaves causing death of many plants. The yield for the fertilizer solutions tended to increase as the loading rate increased, while the opposite was true for the effluent-treated plants. The plants treated with water only had higher yields than the plants receiving effluent at the three higher loading rates, which suggests some very toxic compound(s). Young corn plants could be expected to be more susceptible to toxic conditions than older plants and were used in the experiment for that reason. Response of older plants was determined in Experiment II.

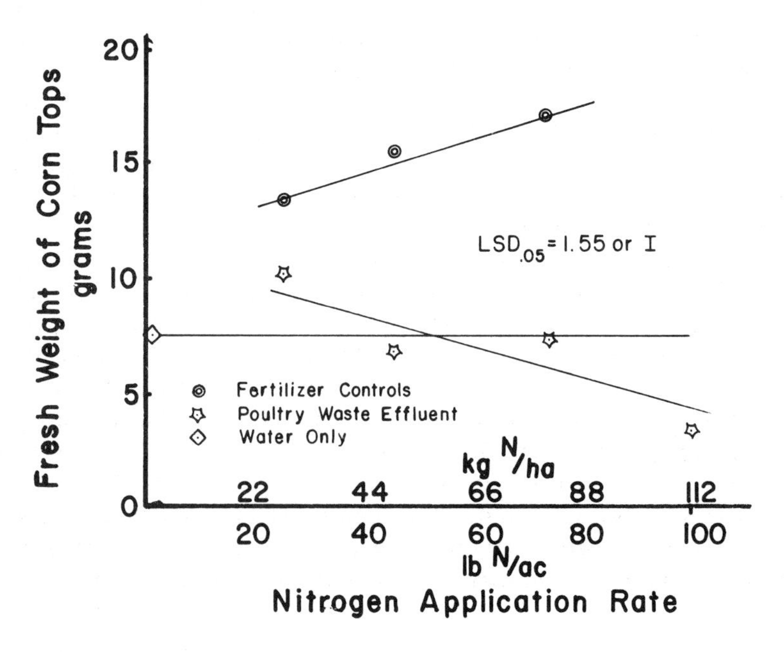

Figure 1. Experiment I, corn response to poultry waste effluent and fertilizer control solutions.

The *E. coli* population of the effluent ranged from thousands to ten thousands/ml. This was a relatively high level compared to subsequent experiments, which will be covered later in this chapter.

The reasons for the toxic effect on the plants from the poultry effluent were not determined. Various possible reasons are: (1) excessive salts;[1] (2) a phytotoxic organic constituent, such as a phenol or dimethoate (an organophosphate insecticide); (3) the high energy value of the effluent causing denitrification; or (4) a combination of the above items. The results of the experiment certainly pointed out some areas where further work was needed.

Experiment II

In this experiment, four digesters were fed poultry waste at rates of 25, 50, 100 and 200 lb of VS/1000 ft^3-day (0.4, 0.8, 1.6 and 3.2 kg of VS/m^3-day). The digesters were continually mixed (anaerobic) and fed 5 days/week. The digesters were operated for 4 weeks before the effluent was used in bioassay studies.

The environmental conditions for this bioassay study were the same as for Experiment I, except corn seedlings of 1, 2 and 3 weeks of age were used and the results of all three ages receiving the same application rate were averaged together. This was done because all three ages of corn responded similarly to treatment. The effluent was applied once a week for 3 consecutive weeks, rather than just once as in Experiment I. Foliage fresh weights from the corn seedlings were collected 7 days after the third treatment.

The *E. coli* population was very low (zero to hundreds/ml) compared to Experiment I. Viable *E. coli* populations did not exist in the digesters except for about one week of the experiment. It was not determined what caused the *E. coli* die-off, but results of Experiments III and IV will discuss possible reasons.

There were no drastic phytotoxicities observed in the effluent-treated corn, although it produced significantly less foliage weight than the corresponding corn treated with fertilizer control solutions (Figure 2). The fertilizer solutions were made up to equal total-N in the effluent rather than the ammonium-N concentration of the effluent. This may account for the higher weight of corn that received the fertilizer solution, because all nitrogen was in the ammonium form. Apparently the nutrients in the waste effluent were not as readily available as those in the fertilizer solutions. This could have been due to denitrification in the soil. The ammonium-N concentration was lower in Experiment II (500 ppm) than in Experiment I (2000 ppm) for the corresponding digesters. There is some indication that the lower ammonium concentrations may have been due to the low *E. coli* populations, since *E. coli* are important in the conversion of organic-N to ammonium-N.[9]

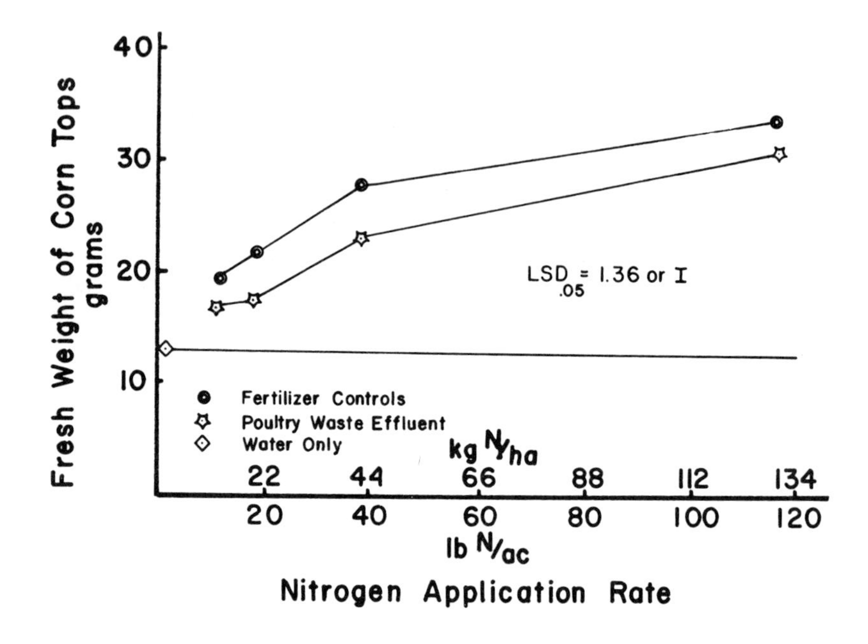

Figure 2. Experiment II, corn response to poultry waste effluent and fertilizer control solutions.

Experiment III

Experiment III involved two digesters that were continually mixed by means of low aeration rates (only for mixing purposes and not to maintain a dissolved oxygen content) and two digesters that were not mixed, except at time of feeding. One mixed and one nonmixed digester were fed at the rate of 30 lb VS/1000 ft^3-day (0.053 kg VS/m^3-day), and the other two digesters were fed at three times this rate. This experiment ran for 15 weeks, with effluent from weeks 10, 11, 13 and 14 being used in the bioassay analysis. Table I shows the mean values of several constituents in the poultry waste effluent. Total-N concentrations of the effluent ranged as high as 400 ppm at the low feeding rate and up to 1100 ppm at the high feeding rate. The total N as ammonium-N was approximately 75 to 80%.

The results of the first of three bioassays conducted during Experiment III were similar to those from Experiment II in which the growth from the fertilizer solutions was significantly higher at the 5% confidence level than for the effluent-grown corn, although the effluent-grown corn produced higher yields than the pots that received water only. Figure 3 includes results from all three bioassays. The fertilizer solutions were made up on the basis of ammonium-N rather than total-N, which is probably more

Table I. Average concentrations (mg/l) of the nutrients in the poultry waste effluent applied to corn seedlings in Experiment III.

Parameter	Mixed		Nonmixed	
Loading Rate (kg VS/m^3-day)	0.053	0.106	0.053	0.106
pH	7.6	7.6	7.4	7.2
Total-N	166	941	378	1092
NH_4-N	79	727	290	834
P	98	178	105	293
K	355	888	350	852
Na	109	289	120	261
Ca	93	156	140	227
Mg	43	39	49	95
Fe	2	4	3	4
Cu	0.4	0.6	0.6	0.7
Zn	0.6	1.6	0.9	1.8
Mn	0.6	1.7	0.9	1.6
Total Solids	2100	5100	2400	5400
Volatile Solids	1500	3100	1500	3300

correct, because in this way the effluent and fertilizer solutions are matched more equally with respect to available nitrogen.

The second bioassay in Experiment III consisted of 1X (50 ml) and 2X (100 ml) applications of poultry waste effluent from the mixed and non-mixed high-rate digesters once a week for two successive weeks. Again there was a significantly higher yield from the fertilizer solutions compared to the effluent at the 5% confidence level. The 2X application rate of fertilizer solution increased the corn response over the single application rate, while the 2X application of nonmixed effluent actually decreased the corn growth. The effluent from the nonmixed digester had a higher N content than that from the mixed digester (potential for nitrogen loss by nitrification-denitrification under mixed conditions) so the 2X application rate resulted in an application of 146 kg N/ha. This produced a yield of less than the water controls (Figure 3), so it was quite evident that the effluent contained phytotoxic substances which were more injurious with a 2X application rate. The reasons for this were probably similar to those expressed in the discussion of Experiment I.

The third bioassay of Experiment III was designed to determine the effect on corn growth if applications of effluent or fertilizer solutions were made either three or five times per week. This resulted in the highest application

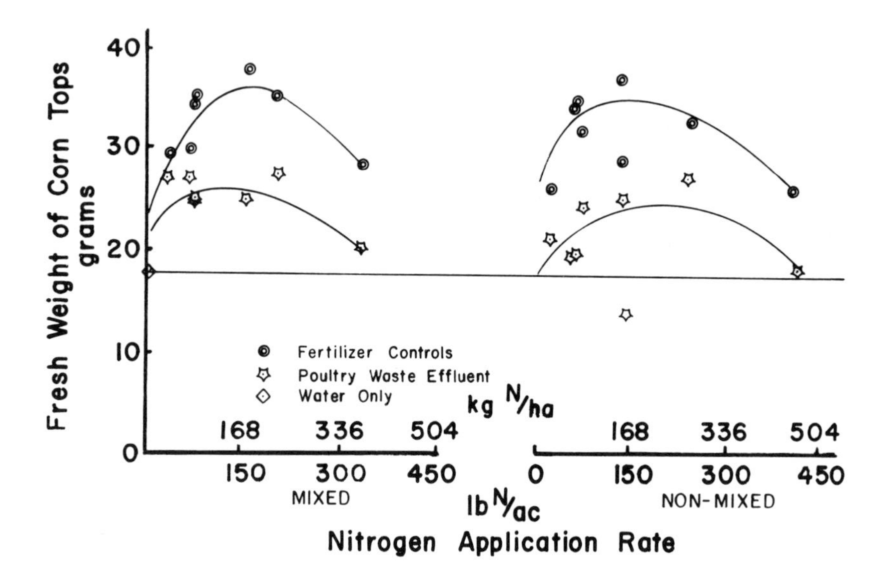

Figure 3. Experiment III, corn response to poultry waste effluent and fertilizer control solutions from three bioassays.

rates of 380 kg N/ha from the mixed reactors and 470 kg N/ha from the non-mixed reactors and caused a further decrease in yield (Figure 3).

E. coli counts were made 3 to 4 times per week on the effluent samples. During the first several weeks of the experiment the coliform counts were quite high, as in Experiment I; however, the counts suddenly dropped to zero. There appeared to be no logical explanation for this occurrence. A subsequent check at the poultry farm, which was the source of waste for the digesters, revealed that upon the initiation of warmer weather (February), fly-control methods were started through the use of two insecticides, dichlorvos (2,2-dichlorvinyl dimethyl phosphate) and dimethoate (0,0-dimethyl S-(N-methylcarbamoylmethyl) phosphorodithioate). The use of the insecticides coincided with the drop in *E. coli* population. Ballington *et al.*[10] evaluated the effect of these insecticides on *E. coli* populations and reported that 10,000 ppmv (volume basis) of these insecticides caused a complete die-off of *E. coli* similar to that experienced in Experiments II and III. Since dimethoate was used directly on the manure and was more injurious to the *E. coli*, it was chosen for further study.

Experiment IV

Experiment IV was designed to evaluate the effect of dimethoate on *E. coli* and corn plants when the dimethoate was added regularly to the digesters. In this experiment the loading rate was 90 lb VS/1000 ft^3-day (0.16 kg VS/m^3-day) for two nonmixed digesters. Fertilizer control solution results were extrapolated from Figure 3. The rate of dimethoate addition was chosen to match the conditions in Experiment III when complete die-off of *E. coli* occurred. This rate appeared to be approximately 5000 ppmv, so dimethoate was added several times a week to the digester with a resulting concentration of 2000 to 3000 ppmv. The degradation rate of dimethoate would account for the concentration being less than the desired rate.

Dimethoate in the poultry effluent, applied once or three times per week for two weeks, did effectively eliminate *E. coli*, while the untreated digester had *E. coli* populations around 10^6/ml. The dimethoate-treated effluent did drastically reduce the corn growth (Figure 4).

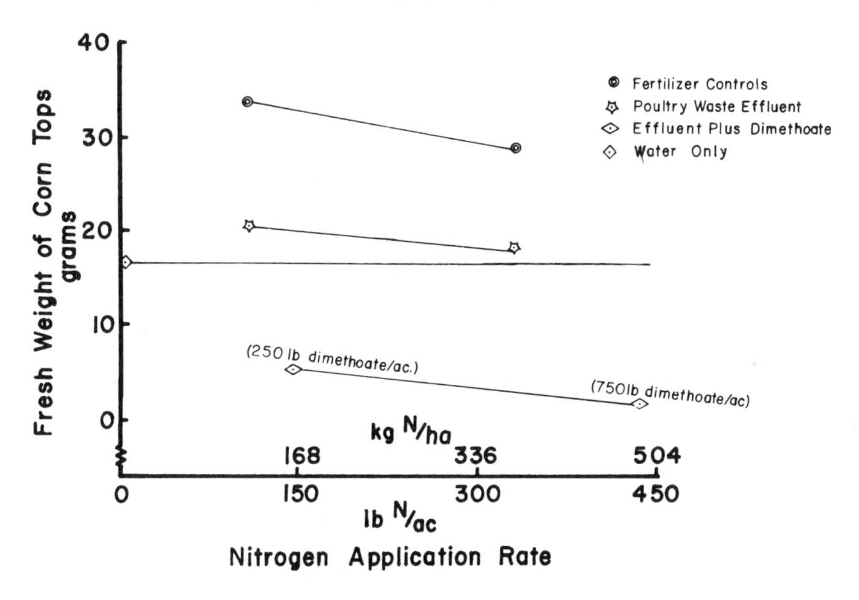

Figure 4. Experiment IV, corn response to poultry waste effluent with and without addition of dimethoate.

This response is similar to that of Experiment I (Figure 1), where corn growth was less than the water control. At the time of the bioassay study in Experiment I there was no explanation for the phytotoxic effect on the corn seedlings. In Experiment IV it was concluded that the dimethoate was the most likely factor involved in the phytotoxic effect.

This rate (250 and 750 lb dimethoate/ac) corresponds to several applications and is representative of an actual situation where poultry manure has to be disposed of regularly. The recommended application rate[11] of dimethoate to poultry manure is 1 gal/100 ft^2 of 1% active ingredient solution

(10,000 ppmv), which is equal to approximately 38 lb of dimethoate per acre per application.

Figure 5 shows the effect of dimethoate only and poultry effluent with dimethoate on corn growth. It is obvious that dimethoate decreases the growth of corn below that of the water control pots. This curve is a typical phytotoxic response curve. Dimethoate inhibits utilization of the nutrients in the effluent (Figures 4 and 5) at the rates shown. Extrapolation of the data would indicate some potential crop injury with low rates of dimethoate application. Based on a half-life in soil of 2 to 4 days,[12] it is suggested that a similar time period be allowed for degradation of the dimethoate before land application.

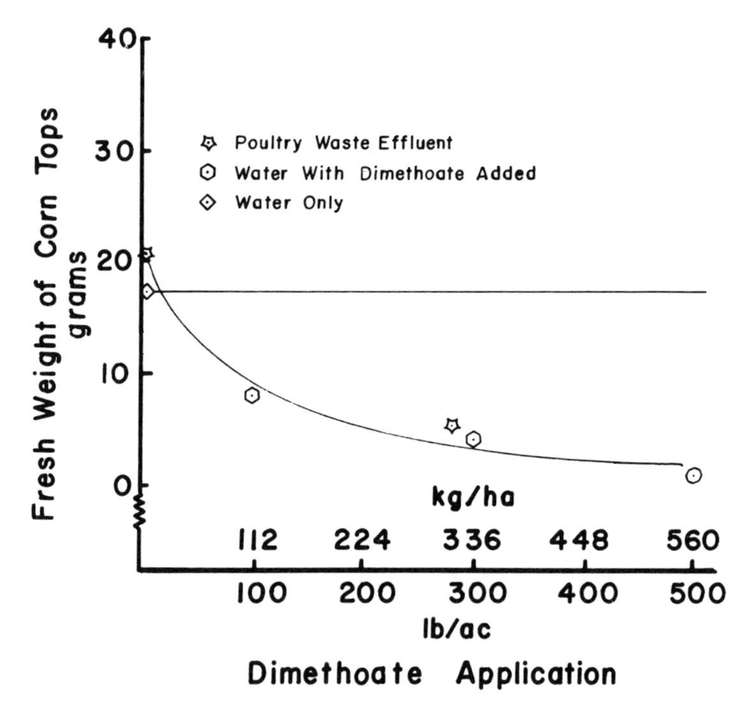

Figure 5. Experiment IV, corn response vs dimethoate application rate.

CONCLUSIONS

The following conclusions are based on poultry waste effluent being applied to 1- to 3-week-old corn seedlings over a 1- to 2-week period:

- Application of poultry waste effluent over a 1- to 2-week period onto corn seedlings produced maximum growth when the nitrogen levels were 125 to 200 lb N/ac (140 to 224 kg N/ha).

• Matching fertilizer control solutions consistently produced significantly higher yields than the poultry waste effluent.

• Mixed and nonmixed poultry waste effluent produced similar results when applied to corn plants.

• Dimethoate (an insecticide) eliminated the *E. coli* population in poultry waste at concentrations of 2000 to 3000 ppmv.

• Poultry manure effluent containing dimethoate (>25 lb/ac) could cause corn injury.

ACKNOWLEDGMENTS

This paper is contribution No. 1459 from the South Carolina Agricultural Experimental Station. The cooperation of the Clemson University Department of Experimental Statistics and Agricultural Chemical Services is greatly appreciated. The technical assistance of R. E. Gantt, W. F. Krebs, J. L. Allen and S. V. Malcolm of Clemson University made it possible to conduct the experiments.

REFERENCES

1. Shorthall, J. G. and W. C. Liebhardt. "Yield and Growth of Corn as Affected by Poultry Manure," *J. Environ. Qual.* 4(2):186-191 (1975).
2. Stuedemann, J. A., S. R. Wilkinson, D. J. Williams, H. Ciordia, J. V. Ernst, W. A. Jackson and J. B. Jones, Jr. "Long-Term Broiler Litter Fertilization of Tall Fescue Pastures and Health and Performance of Beef Cows," Proc. 3rd International Symposium on Livestock Waste, American Society of Agricultural Engineers, St. Joseph, Michigan, pp. 264-268 (1975).
3. MacMillan, K. A., T. W. Scott and T. W. Bateman. "Corn Response and Soil Nitrogen Transformations Following Varied Applications of Poultry Manure Treated to Minimize Odor," *Can. J. Soil Sci.* 55:29-34 (1975).
4. McKell, C. M., V. W. Brown, R. H. Adolph and C. Duncan. "Fertilization of Annual Rangeland with Chicken Manure," *J. Range Management* 23:336-340 (1970).
5. Wengel, R. W. and J. J. Kolega. "Land Disposal of Poultry Manure in Relations to Soil Water Quality and Sillage Corn Yield," ASAE paper No. 72-957, St. Joseph, Michigan (1972).
6. Morrison, J. L. "Distribution of Arsenic from Poultry Litter in Broiler Chickens, Soil, and Crops," *J. Agric. Food Chem.* 17:1288-1290 (1969).
7. Burford, J. R. "Ethylene in Grassland Soil Treated with Animal Excreta," *J. Environ. Qual.* 4(1):55-57 (1975).
8. Janzen, J. J., A. B. Bodine and L. J. Luszcz. "A Survey of Effects of Animal Wastes on Stream Pollution from Selected Dairy Farms," *J. Dairy Sci.* 57(2):260-263 (1974).
9. Cabes, L. J., A. R. Colmer, H. T. Barr and B. A. Tower. "The Bacterial Population of an Indoor Poultry Lagoon," *J. Poultry Sci.* 48:54-63 (1969).

10. Ballington, P. E., H. D. Skipper, R. O. Hegg, W. F. Krebs and R. E. Gantt. "Effects of Insecticides on *E. coli* in Poultry Waste Digesters," abstract from Agronomy Society of South Carolina meeting, Columbia, South Carolina, 1976.
11. *Agricultural Chemicals Handbook,* Clemson University, Clemson, South Carolina.
12. Bohn, W. R. "The Disappearance of Dimethoate from Soil," *J. Econ. Entomol.* 57:798-799 (1964).

35

PRACTICAL APPLICATION OF AEROBIC TREATMENT AND LAND APPLICATION OF POULTRY MANURE

A. C. Anthonisen
Monteco Environmental Management Associates, Inc.
Montgomery, New York

D. H. Wagner
Mountain Pride Farms, Inc.
Woodbourne, New York

INTRODUCTION

Mountain Pride Farms, Inc., located in Woodbourne, New York, produces approximately 30 tons of raw poultry manure each day. Problems associated with odors resulting from years of accumulated manure required that Mountain Pride Farms look at alternative methods of manure management. It was necessary that the manure management system meet not only the requirements of Mountain Pride Farms, but also the requirements of the New York State Department of Environmental Conservation. A liquid aeration system using in-house oxidation ditches was ultimately selected to handle the manure produced. This chapter will present the experiences of Mountain Pride Farms, Inc., during the development and operation of the oxidation ditch system. The project has been funded by appropriations made by the Board of Directors from the Mountain Pride Farms budget, and at no time were funds made available through state or federal grant agencies.

Background

Construction of Mountain Pride Farms began in 1968 on a 74-ac parcel of land located in an agricultural district in Sullivan County, New York. The farm is situated next to the Neversink River as shown in Figure 1. Each of

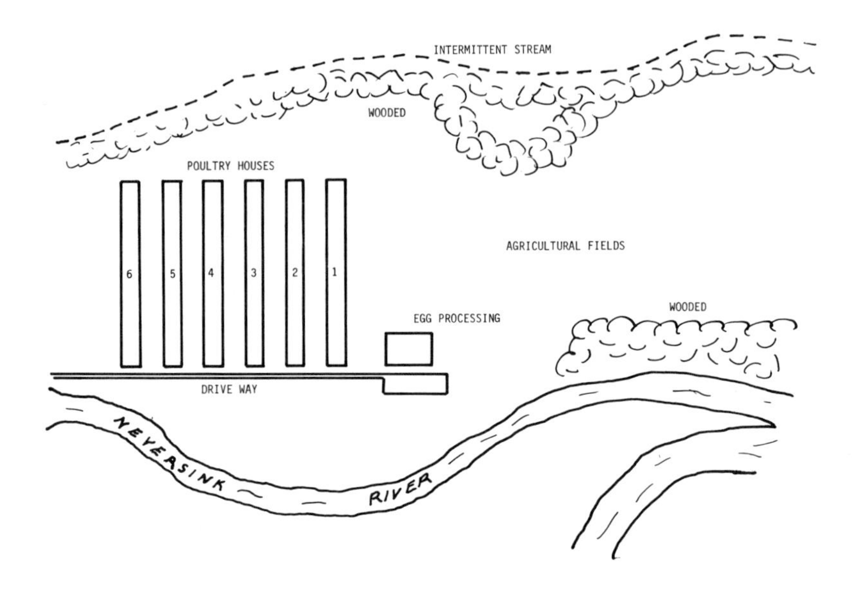

Figure 1. Site plan of Mountain Pride Farms.

the six poultry houses contains 40,000 laying hens and is 50 ft wide and 500 ft long with an 8 ft deep basement for manure storage. Each building is sloped so that one end is approximately 1.5 ft lower than the other. Five buildings contain stair-step cages with a Hart cup watering system, while the sixth building used a flat deck system with trough waterers. The first two buildings used outside feed bins, and the remaining four used feed bins that were constructed within the building and supported by columns that extended into the basement.

After a few years of operation, it became apparent that the accumulated manure was creating a nuisance odor throughout the area. Neighboring residents began meeting on a regular basis to discuss the situation and demonstrated their dissatisfaction by picketing the New Paltz Regional Office of the New York State Department of Environmental Conservation (NYSDEC). The Town Board enacted rules and regulations that specified how, where and when the poultry manure could be disposed of and transported on town roads, and also required that a permit be obtained prior to any hauling of manure. Local newspapers followed the events, and soon the New York State DEC requested to meet with Mountain Pride Farms, Inc. On April 2, 1973, the DEC indicated that the odor from Mountain Pride Farms contributed to a contravention of the standards of air pollution and was in violation of Article 19 of the Environmental Conservation Law. On May 11, 1973, Mountain Pride Farms, Inc. submitted to DEC the following project plan as shown in Figure 2.

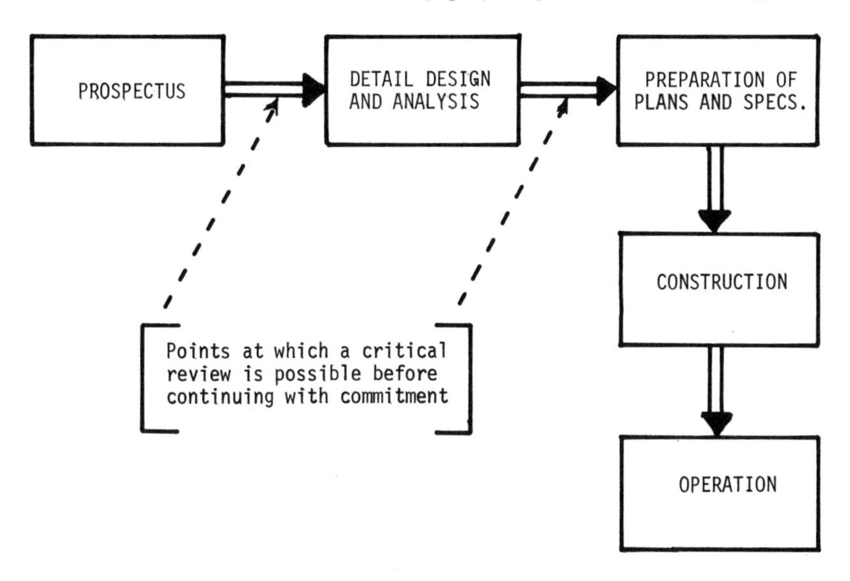

Figure 2. Project plan.

The plan described procedures that would be followed to develop a manure management system at Mountain Pride Farms for effectively controlling odors. A prospectus was submitted the following August summarizing alternatives for the development of a liquid aeration system for use at Mountain Pride Farms, and on August 29, 1973, Mountain Pride Farms signed a Consent Order that required the submission of a detailed design and analysis and a time schedule for the remaining phases of the project plan.

The detailed design and analysis and construction schedules were approved by the DEC on October 21, 1974, and construction of oxidation ditches began in January 1975. Operation of the first oxidation-ditch system began in March 1975.

CONSTRUCTION

Figure 3 presents a plan view of the oxidation-ditch system as designed for each building. Prior to construction, all existing manure was removed from the buildings using bulldozers, farm tractors with blades, a backhoe and dump trucks. Two separate ditches were then constructed within each building. The curved end walls are 12-in.-wide poured concrete and provided for a

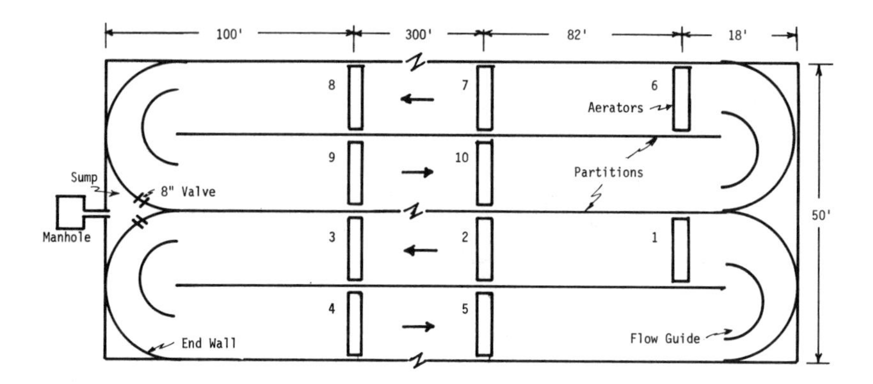

Figure 3. Plan of oxidation ditches.

maximum liquid depth of 54 in. at the building's south end and 36 in. at the building's north end. The ditch partitions were constructed of treated 2 x 6 tongue-and-groove lumber and were attached to the 6 x 6 wooden posts supporting the cage system. The flow guides at the end of each ditch were 8 in. concrete blocks. An 8 in. gate valve was placed in the end wall of each ditch, and both ditches were connected to an outside manhole. Subsequently, the manholes for each of the six buildings will be interconnected. All construction was performed on a contract basis with a local building contractor.

Aeration was supplied by five 5-hp MONTAIR brush aerators for each ditch as is shown in Figure 3. The MONTAIR aerator uses a rotor with a counter-auger design and is mounted on floats for operation at varying liquid depths. Power is transmitted directly to the rotor from a low-speed motor using sprockets and roller chain. Rotor speeds may be varied by interchanging the sprockets to change the sprocket ratio. The rotors normally operated at 285 rpm with a 6-in. immersion depth. A 200 amp service was provided to supply the necessary power for each building.

OPERATION

The operation of the oxidation ditches began according to the schedule in Table I.

Building 2 was filled to a depth of 12 in. at the building's shallow end, which provided approximately 1 ft^3 of water per bird. The ditch was started without the benefit of seed organisms and went through a period of excessive foaming. Foaming was not a problem when adequate seeding was provided during start-up.

Table I. Schedule of operation.

	Number of Birds	Date Started
Building #2		
Ditch #1	18,680	March 19, 1975
Ditch #2	22,345	March 19, 1975
Building #3		
Ditch #1	20,473	August 25, 1975
Ditch #2	22,391	August 25, 1975
Building #6		
Ditch #1	18,760	Jarıuary 12, 1976
Ditch #2	19,590	January 12, 1976

Dissolved oxygen, oxygen uptake rate, pH, solids concentration and power consumption were monitored on a routine basis for the purpose of process control. Funds were not allocated for the purpose of obtaining research-type data.

Building 2 was allowed to operate as a "batch"-type process where the solids concentration was permitted to accumulate to a maximum level while still maintaining ditch movement and odor control. The rate of solids accumulation is presented in Figure 4. Dissolved oxygen levels throughout

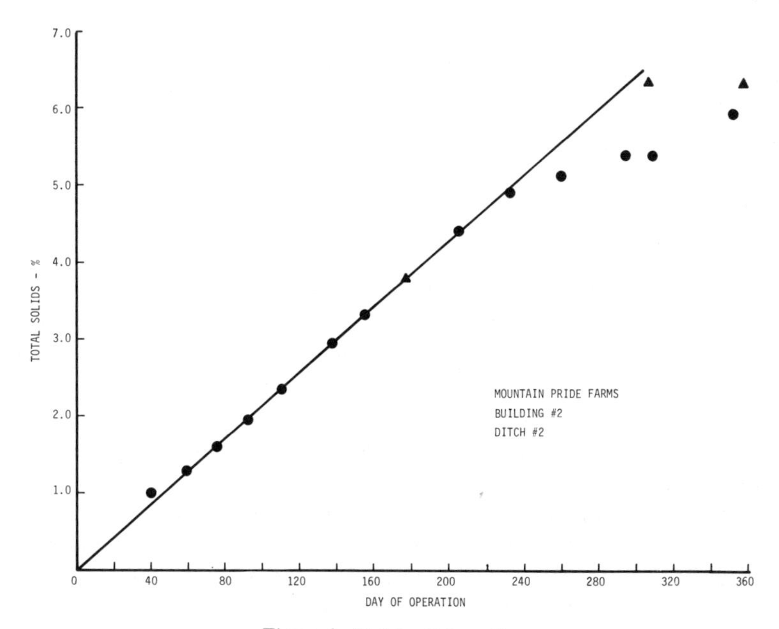

Figure 4. Total solids vs time.

the time shown varied between 0.2 and 3.5 mg/l. Microscopic observations revealed the presence of free-swimming and stalked protozoa, and on occasion rotifers were also observed. These organisms are indicators of a well-stabilized aerated sludge. During these conditions, the oxygen uptake rate ranged between 25 and 60 mg/l/hr.

The aerated sludge in building 6 appeared to be nitrifying. The ammonia concentrations were less than 50 mg/l and the pH was in the range of 7.5 to 7.8, as compared to 8.0 to 9.0 for building 2 which had high ammonia residuals in the range of 2000 to 3000 mg/l. The sludge would rise during quiescent anaerobic conditions, further suggesting that denitrification of the nitrified sludge was occurring. The fact that nitrification was probably occurring indicates that the aeration capacity of the system was not only adequate for odor control, but was also sufficient for satisfying the BOD (biochemical oxygen demand) and NOD (nitrogenous oxygen demand). Odor control was very good at all times.

After building 2 had operated for one year under batch conditions, the oxidation ditch mixed liquor (ODML) was analyzed as summarized in Table II.

Table II. Analyses of ODML: after one year of operation, building #2.[a]

Parameter	Ditch #1 (mg/l)	Ditch #2 (mg/l)
NH_3-N	2,620	3,200
TKN	7,060	7,630
Calcium (as Ca)	2,290	2,080
Phosphate (as P)	4,260	4,500
Potassium (as K)	5,600	4,930
Total Solids	61,700	63,430
Volatile Solids	34,740	38,140
pH	8.2	8.9

[a]All analyses were performed by Envirotest Laboratories, Newburgh, New York.

Based upon these and other analyses, it was determined that the nutrient value of the ODML would be useful for crop fertilization. In terms of N, P_2O_5 and K_2O, the nutrient value for each 1000 gal of ODML was determined using the average value of the analyses for ditch 1 and ditch 2 of building 2, as shown in Table III.

Table IV shows the mass balance calculated to verify that these nutrients were in fact present within the system, based upon input from the birds.

These figures suggest that the amount of nutrients represented to be present within the ODML is reasonable, and that its nutrient value based on 1976-1977 prices is in the range of $50/1000 gal.

Table III. Value of ODML (building #2–March 17, 1976).

	Concentration (mg/l)	Concentration (lb/1000 gal)	1976 Cost ($/lb)	Value ($/1000 gal)
N	7,345	61	0.20	12.20
P_2O_5	20,060	167	0.17	28.39
K_2O	12,688	106	0.09	9.54
			TOTAL	$50.13

Table IV. Mass balance (building #2–365 days).[a]

	Expected Input (g)	ODML (g)	Loss
N	33,782,000	13,902,000	59%
P_2O_5	45,260,000	37,968,000	13%
K_2O	23,360,000	24,016,000	-6.6%

[a]N–2.4 g/bird-day; P_2O_5–3.1 g/bird-day; K_2O–1.6 g/bird-day; 1% mortality/mo.

Using the same input values that were used to calculate the mass balance, and using a total solids input of 27.4 g/bird-day, a comparison that may be made between raw manure and ODML is shown in Table V.

Table V. Nutrients as a percentage of total solids.

	24-hr Sample[a] (raw manure)	ODML (after one yr)
N	8.73%	11.74%
P_2O_5	11.27%	32.06%
K_2O	5.82%	20.28%

[a]From A. G. Hashimoto, *Characterization of White Leghorn Manure.* Proc. Agric. Waste Management Conf., Cornell University, Ithaca, New York (1974), pp. 141-152.

This indicates, as would be expected, that during one year of aerobic treatment, N, P_2O_5 and K_2O concentrations increased as other solids were being volatilized. The comparison also indicates that the primary value of either the raw poultry manure or the ODML is the concentration of P_2O_5.

Based upon this information, the following projection was made to determine the total yearly value of the ODML that could be produced from all the birds (240,000) at Mountain Pride Farms. For the purpose of being economically conservative, it was assumed that 75% of the input N, 25% of the input P_2O_5 and 25% of the input K_2O would be lost. In practice it would not be expected that either the P_2O_5 or K_2O would be lost from the system.

Total Manure Value (240,000 birds)

	N	$ 22,322.91 (assume 75% loss)
	P_2O_5	76,264.10 (assume 25% loss)
	K_2O	20,838.77 (assume 25% loss)
	Total	$119,425.77
Using:	N	2.4 g/bird per day; $0.20/lb
	P_2O_5	3.1 g/bird per day; $0.17/lb
	K_2O	1.6 g/bird per day; $0.09/lb

These figures suggest that the yearly value of the poultry manure would be approximately $120,000 using only the above nutrients to assess its value. The presence of calcium and other essential growth factors would tend to increase the value of the ODML as a soil conditioner and nutrient source. In view of these considerations, a decision was made to make the necessary investments for marketing the ODML to area crop farmers.

MARKETING OF THE ODML

Marketing of the ODML required a degree of salesmanship. Many of the crop farmers had had past experiences with using poultry manure as a fertilizer that were not very good. In some cases, fields were used as disposal sites rather than as fields with a crop to be fertilized. Equipment would get bogged down, manure would be indiscriminately dumped and fields would receive physical damage by the heavy equipment. Overapplication had caused "burning," and odors brought complaints from the neighbors. It was necessary to convince the farmer that the ODML was not the traditional poultry manure, but rather a homogeneous nuisance-free material that could be evenly distributed on his field at rates that would satisfy his needs. To market the ODML, it was also required to provide the service of spreading it on the field. A number of farmers within a 50-mile radius of Mountain Pride Farms were willing to use the ODML to supplement their fertilizer requirements. The crops were mostly corn but also included hay and greens.

Procedures

The procedures used to transport and spread the ODML involved the following equipment:

- high-velocity manure pump (2500 gal/min)
- 40-hp tractor w/PTO for manure pump
- two 6000-gal liquid tank trailers
- liquid manure spreader (1500 gal)
- 85-hp tractor to haul spreader

The ODML in building # 2 was pumped directly from the outside manhole, which was connected to both ditches, to one of the 6000-gal tank trucks. A 4-in. hose from the pump was attached to the trailer by the truck driver, and pumping was begun by starting the tractor's PTO. The filling operation required about 15 to 20 min.

The ODML was then hauled to the field being spread. Hauling time varied from 30 to 90 min one way, depending on the location of the field. At the field, the full tanker was exchanged for one that had just been emptied. To do this, a two-speed load-bearing landing gear was installed on each trailer to support the weight when fully loaded. The truck then returned to Mountain Pride Farms to be refilled and returned to the field. The ODML was shuttled to each location in this manner.

A 1200 Series David Brown tractor was used to haul the liquid spreader. This tractor provided a wide range of operating speeds, thereby permitting different loading rates to the field. The spray from the spreader was approximately 40 to 45 ft wide. When changing locations, the tractor and spreader were either hauled by truck or driven to each successive site.

Cost of Marketing

Approximately 500,000 gal of ODML were marketed during the spring of 1976, and from this experience it was possible to make the analyses shown in Table VI for marketing the total volume of ODML that might be produced from all 240,000 birds at Mountain Pride Farms. For the purpose of this analysis, it was assumed that oxidation ditches in each building would produce results similar to those observed in building #2. This projection also assumes that suitable cropland will be available throughout the 100 days estimated as required for spreading the ODML.

The cost of trucking presented is based upon a contract price negotiated with a local trucker. The actual value of the ODML was obtained by using the total projected value of $120,000 previously determined and dividing by an expected volume of 3,300,000 gal/yr at 6% total solids.

Table VI. Analysis of marketing (3,300,000 gal/yr at 6.0% TS).

Capital Costs		Annual Costs (100 days at 12 hr/day)	
Trailers (2)	$10,000	Trucking at $25/hr	$30,000
Tractor	$10,000	Labor at $10/hr	$12,000
2nd Tractor	$ 2,500	Maintenance (10% equip.)	$ 3,200
Spreader	$ 7,000		
Pump	$ 2,500		$45,200
		Annual Capital Cost	$ 5,209
TOTAL	$32,000		
			$50,409
Annual Capital Cost (at 10% and 10 yr)	$5,209	Actual Cost	$15.28/1000 gal
		Actual Value of ODML	$36.36/1000 gal

The cost to the farmers using the ODML ranged between $20 and $30, depending on the distance traveled. The crop responses were favorable, and the farmers are using ODML as a fertilizer supplement again in 1977.

COST OF AERATION

Table VII shows the cost of installing and operating the oxidation ditch system based on 1977 prices received from the contractors involved. The costs do not include consulting fees required for design and supervision of the overall project. The cost of electricity was based on actual meter readings for one year of operation.

Table VII. Cost of aeration system.

Installation (one building– 40,000 birds)			
Construction		$22,000	
Electrical		$ 5,000	
Equipment		$35,000	
	TOTAL	$62,000	
Cost per bird		$ 1.55	
Operation (one building– 800,000 dozen)			¢/doz
Labor (0.25 man/bldg)		$ 3,500	0.44
Parts (5% equipment)		$ 1,750	0.22
Electricity (3.37¢/kWh)		$ 7,872	0.98
	TOTAL	$13,122	1.64

The installation cost presented may vary for other buildings containing a similar number of birds, depending on the type of structure and methods selected for construction. The operating cost as presented in terms of ¢/dozen was based upon a production rate of 20 dozen/bird-yr.

SUMMARY

There still exists, for some poultry farms, a need for an effective manure management system. Mountain Pride Farms, Inc., producing raw manure at a rate of approximately 30 tons per day, was faced with this need. Circumstances required that an effective system for odor control be installed at Mountain Pride Farms to meet the requirements of a New York State Department of Environmental Conservation Consent Order.

The management system selected included the use of oxidation ditches. Previous research and demonstration by Cornell University provided the necessary guidelines for implementing the system at Mountain Pride Farms.

This chapter has summarized the planning, coordination, design, construction and operation involved in the development of the oxidation ditches at Mountain Pride Farms and covers a time period of approximately four years. The primary objective of the system has been to maintain odor control. Comparisons between those buildings with the oxidation ditch system and those without clearly have demonstrated that this objective has been accomplished.

A secondary objective of utilizing the oxidation ditch mixed liquor (ODML) has also been demonstrated through a successful marketing program to area crop farmers. The marketing of the ODML is consistent with the view that poultry manure is a resource to be utilized rather than a waste requiring disposal. The resource value of aerated poultry manure, in terms of a feed supplement, is presently being investigated at Cornell University. It is hoped that, through the results of these studies, more economically effective methods for using the nutrients synthesized during the aeration process will be demonstrated.

ECONOMIC COMPARISON OF THE OXIDATION DITCH AND HIGH-RISE MANURE DRYING AS POULTRY WASTE MANAGEMENT ALTERNATIVES

J. H. Martin
Department of Agricultural Engineering
R. C. Loehr
Environmental Studies Program
College of Agriculture and Life Sciences
Cornell University, Ithaca, New York

INTRODUCTION

The introduction of the cage system of management into the poultry industry resulted in a significant increase in the efficiency of egg production. However, the adoption of new management techniques produced changes in the physical characteristics of poultry manure resulting in odor problems which have been well documented. Increased efficiency also provided the basis for the subsequent concentration of egg production activities on fewer but larger farms. The higher concentrations of birds increased the pollution potential of this industry to both surface and ground waters.

Numerous research studies have been conducted to evaluate processes and develop systems which will alleviate the problems of odor and water pollution potential associated with these wastes. In laboratory- and pilot-plant-scale studies, two processes—aerobic biological stabilization and drying—have been identified as capable of abating these problems. The technical feasibility of two systems (the oxidation ditch and high-rise undercage drying) has been demonstrated under commercial conditions.[1,2] Results of these studies have permitted the development of process design approaches for both systems.[3]

The decision to employ a particular waste management system should be based on the evaluation of several factors. Among these should be the costs of alternative systems. While cost information relative to the oxidation ditch and high-rise undercage drying has not been totally lacking, available information has focused on operating costs. Moreover, differences in factors such as size of operation and level of technology development have made comparisons on an equal basis impossible.

The objective of this chapter is to present an analysis of the costs associated with use of the oxidation ditch and high-rise undercage drying as poultry waste management alternatives. Also included is an estimate of labor requirements for ultimate disposal.

SYSTEM DESCRIPTIONS

High-Rise Undercage Drying

The high-rise undercage drying system utilizes unheated ventilation air circulated above the surface of accumulated manure to remove moisture. Drying occurs in an undercage collection and storage area which provides storage capacity for up to three years. A cross-section view of a typical high-rise house with undercage manure drying is presented in Figure 1. An important aspect of this system is the ventilation airflow pattern. Air enters the building at the eaves, passes down through the cages and over the manure, and is then exhausted from the building.

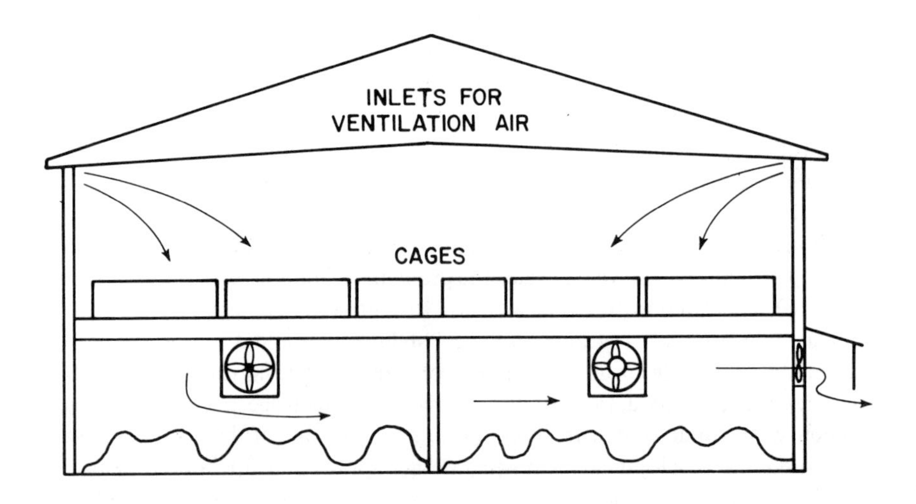

Figure 1. Cross section of a high-rise, undercage drying system for poultry manure.

The performance of the drying process is sensitive to a number of variables. Included are waste characteristics, bird density, vapor pressure differential and drying-air velocity. Drying-air velocity serves as the principal process design parameter for these systems. While the normal house ventilation system provides a degree of air movement over the accumulated manure, velocity varies greatly with both location and ventilation requirements. Thus, independent forced-air circulation using fans located in the manure storage area (Figure 1) is necessary to provide for drying of the manure. These fans are positioned to provide a race track shaped airflow pattern in the manure storage area (Figure 2).

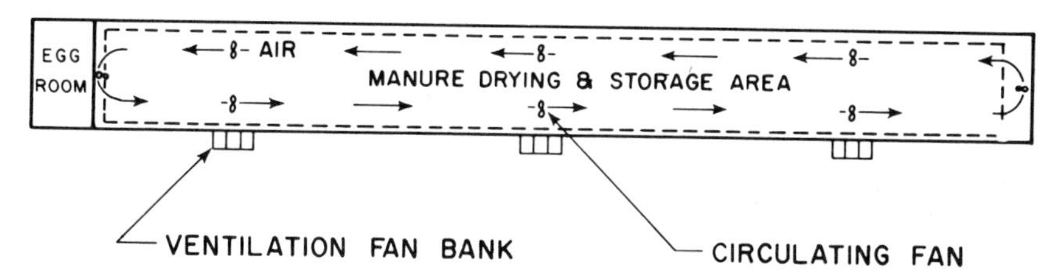

Figure 2. Plan view of a typical high-rise undercage drying system for poultry waste.

The high-rise building design may provide manure storage capacity for up to three years. However, yearly removal is more common. The drying process permits handling of poultry manure as a solid. The high-rise building design, while not a prerequisite for undercage drying, is necessary to provide clearance for manure removal equipment such as front-end loaders.

The Oxidation Ditch

The oxidation ditch incorporated into a confinement building is probably the most commonly used aerobic biological stabilization system for animal wastes. The oxidation ditch has two components—the aeration basin or ditch and the aeration unit. The ditch is typically a rectangular channel which forms a closed loop. An advantage of the oxidation ditch in comparison to other aeration systems is the ability to shape the ditch to conform to the cage layout. This eliminates the need for equipment to transport the wastes to the aeration unit. A cross-sectional view of a typical undercage oxidation ditch for laying hens is presented in Figure 3. Either brush- or cage-type horizontal surface aerators are used to provide both oxygen transfer and mixing.

An aeration system such as the oxidation ditch provides both odor control and reduction of water pollution potential by aerobically stabilizing biologically available carbonaceous compounds. Nitrogen may be removed either by ammonia stripping or via a nitrification-denitrification sequence. It has been

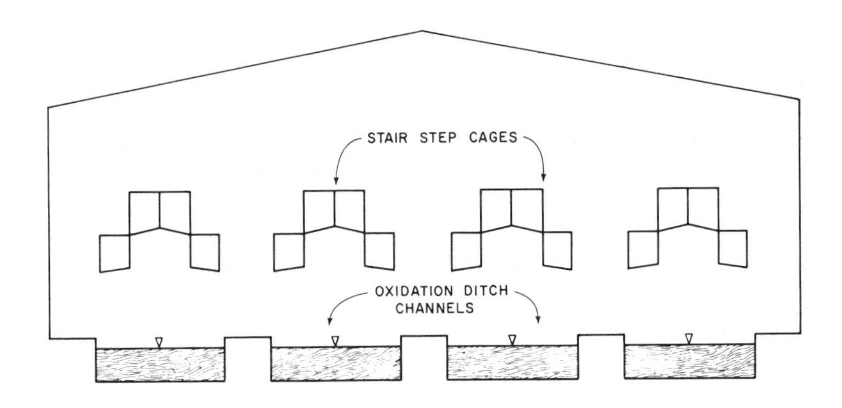

Figure 3. Cross section of undercage oxidation ditches for poultry waste.

demonstrated that minimum aeration can provide both odor control and stabilization of carbonaceous compounds.[1,4] Oxygen requirements for this mode of operation can be equated to the exerted carbonaceous oxygen demand measured as chemical oxygen demand (COD) removed. Increasing oxygen transfer beyond this minimum level will provide oxygen to meet the nitrogenous oxygen demand.

An oxidation ditch for poultry wastes can be operated as either a continuous-flow or batch reactor. The advantage of continuous-flow operation lies in the ability to control mixed liquor total solids (MLTS) concentrations. In a batch system, the MLTS concentration continually increases due to the accumulation of residual solids which are slowly biodegradable. Research[5] has shown that the efficiency of oxygen transfer, the ratio of oxygen transfer under process conditions to that in tap water (alpha), decreases as MLTS concentration increases beyond 20 g/l (Figure 4). While continuous-flow systems permit maintenance of low MLTS concentrations, the volume of stabilized waste requiring ultimate disposal is increased significantly. Techniques for removal and concentration of residual solids would alleviate this problem. However, practical approaches are presently lacking. Therefore, the batch mode of operation with the inherent inefficiency in oxygen transfer appears to be the most practical alternative.

WASTE MANAGEMENT SYSTEM DESIGNS FOR EACH ALTERNATIVE

In order to provide an equal basis for an economic comparison of the oxidation ditch and high-rise undercage drying, a system design was developed for each alternative. The rationale for this approach, instead of basing cost analyses on existing systems, was to eliminate differences

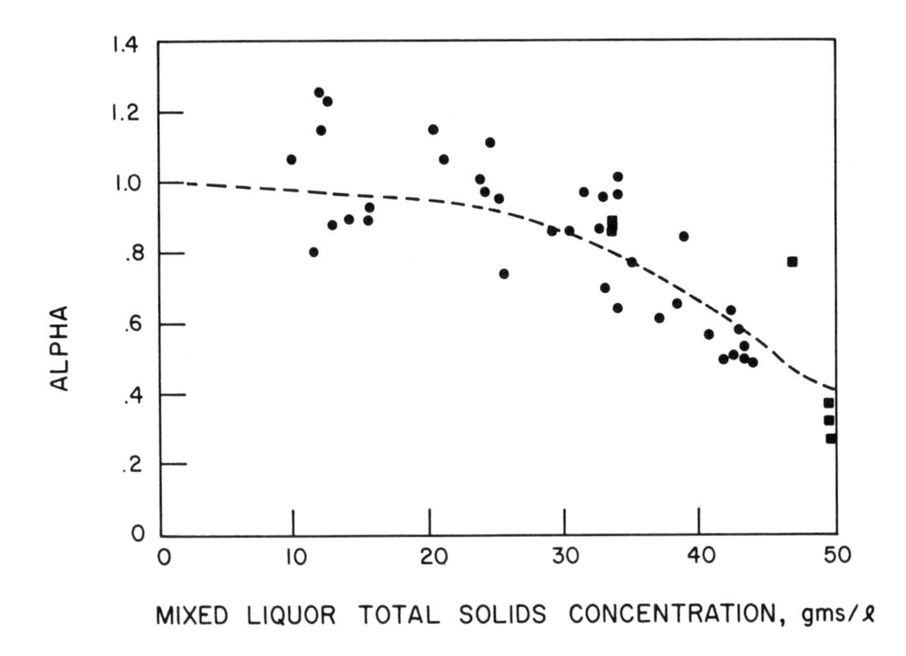

Figure 4. Relationship between alpha and mixed liquor total solids concentration in aerated poultry waste.[5]

such as number of birds, waste characteristics and bird density, which varies with type of cage system and number of birds per cage. Differences in these factors can indirectly affect waste management costs. Common criteria used in the design of both waste management systems are summarized in Table I.

Table I. Common criteria for oxidation ditch and high-rise drying systems design.

Number of Birds	30,000
Type of Cage System	Full Stairstep, 4 Rows
Management Practices	4 Birds per 31- x 46-cm (12- x 18-in.) cage
Building	12.8 m x 152.4 m (42 ft x 500 ft)
Building Structural Design	Timber Column

It was assumed that costs for cages, feeding equipment, ventilation fans, etc., would be equal for both waste management systems. Thus, these items were excluded from consideration since the costs of interest are those due to the different waste management systems.

Process design methodologies presented by Martin and Loehr[3] were used for the design of both waste management systems. The objective in the design of both systems was odor control, with the degree of waste stabilization dependent on that factor. The following is a brief description of the waste management designs.

High-Rise Undercage Drying

This poultry waste management system necessitates the construction of a two-story structure as opposed to a conventional single-story poultry house. It also requires construction of a floor system to support the cages and to provide aisles between the cage rows. A concrete floor in the manure storage area was included in the design to facilitate manure removal. Although many high-rise houses have been constructed with a compacted earthen or cinder base, experience indicates that the use of a concrete floor is desirable.

The average drying-air velocity to provide odor control and permit handling of manure as a solid for the criteria presented in Table 1 was determined to be 0.34 m/sec (71 ft/min). This requires drying air circulating fans with capacities of 4.7 m^3/sec (10,000 ft^3/min) spaced at 30 m (100 ft) intervals. A 91 cm, 0.373 kW (36 in., 0.5 hp) fan will provide an airflow of 4.7 m^3/sec. For the 152-m (500-ft) building under consideration, a total of eight 0.373 kW fans are required. To provide airflow in a race track-shaped pattern similar to that shown in Figure 2, two 61 cm, 0.186 kW (24 in., 0.25 hp) fans were also included to provide cross airflow at each end of the building. This design is similar to that of the system evaluated by Sobel[2] with the exception of the number of 0.373 kW fans, which is increased by two due to greater building length.

Based on reported data,[2] the anticipated quantity of dried manure from this system (30,000 birds) should be 1314 m^3 (46,400 ft^3) per yr. This is based on a density of 32 kg/m^3 (20 lb/ft^3). Density as accumulated should be higher but decreases due to handling.

Undercage Oxidation Ditches

Oxidation ditches can be easily incorporated into conventional single-story poultry houses replacing manure collection pits. The only difference between the dropping pits and oxidation ditch channels is the construction of the semicircular connecting loops at each end of an oxidation ditch. Thus, structural changes from conventional laying-house design for the use of oxidation ditch waste management systems are minimal.

The aeration system for this cost analysis was designed as a batch system with a system volume of 30 liters/bird (8 gal/bird). Operation to a maximum

MLTS concentration of 60 g/l should provide storage for a time period of 100 days.

The carbonaceous oxygen demand for this system has a maximum estimated value of 466 g O_2/1000 bird-hr, or 13.98 kg O_2/hr for 30,000 hens. To meet this oxygen demand at a MLTS concentration of 60 g/l, an aeration capacity of 35 kg O_2/hr is necessary due to the decrease in oxygen transfer efficiency with increase in MLTS concentration. A comparison of changes in carbonaceous oxygen demand and aeration capacity requirements to meet that demand with time in a batch system is presented in Figure 5.

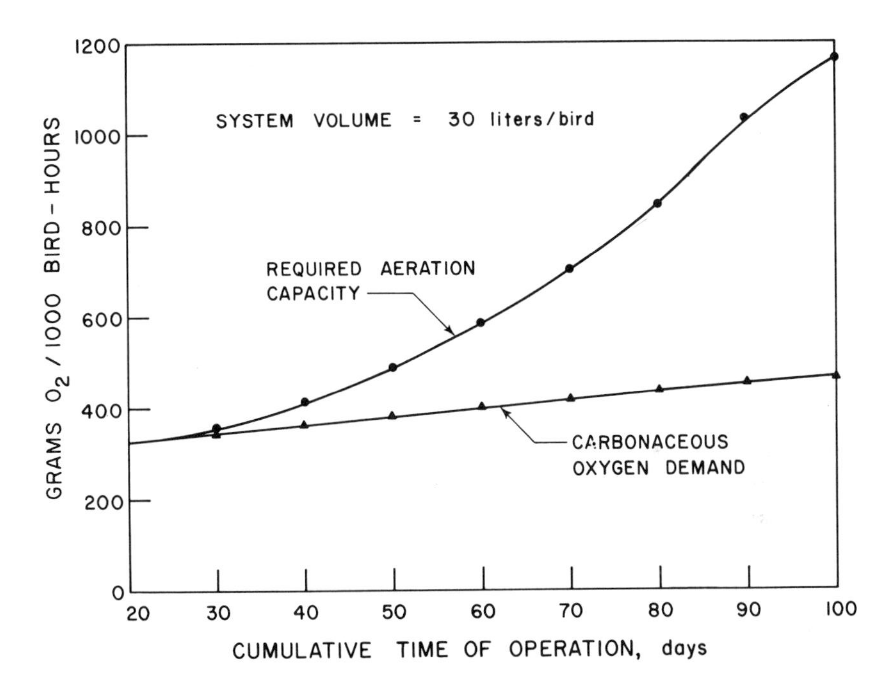

Figure 5. Comparison of carbonaceous oxygen demand and required aeration capacity for odor control in aeration of poultry waste.

The ultimate disposal requirements for this system are based on three and six-tenths 100-day batch cycles/yr. The volume of stabilized waste from 30,000 hens should be 3240 m^3 (114,404 ft^3) per yr.

Cost Analyses

Each waste management system was divided into four components to simplify cost analyses. They are: (1) facilities costs related to the waste management system; (2) fixed and operating costs for stabilization equipment; (3) fixed costs for handling and disposal of manure; and (4) value of the stabilized waste in terms of plant nutrient content. Annual capital costs were determined using an amortization rate of 9% assuming a 20-yr life for structural components and a 10-yr life for equipment with no salvage value. Taxes and insurance were based on a rate of 3.5%/yr of the investment cost. Maintenance costs were assumed to be 1 and 2% of the respective investment costs for structural components and equipment. A value of $0.035/kWh was used for the cost of electrical power. Costs for equipment, such as aeration units, fans, and manure handling equipment, were obtained from manufacturers or their representatives.

Facilities Costs

Both oxidation ditches and high-rise undercage drying systems increase poultry housing costs above that for a conventional cage-type poultry house. These additional costs should be included in determining the total cost for each waste management system. Since the high-rise drying system is an integral part of the structure, total structural costs for a conventional house with manure collection pits, a conventional house with undercage oxidation ditches, and a high-rise house were compared. Cost estimates for each, based upon the common criteria presented in Table I, were obtained from the building department of Agway, Inc., a northeastern agricultural cooperative.[6] Total structural costs for each building and the waste management component of these costs for oxidation ditches and high-rise drying are presented in Table II.

Table II. Estimates of the waste management component of structural costs for oxidation ditches and high-rise drying.

	Structural Costs ($)	Waste Management Component of Structural Costs ($)
Conventional Poultry House with Manure Collection Pits	144,000	–
Conventional Poultry House with Oxidation Ditches	146,500	2,500
High-Rise Poultry House	172,500	28,500

Based upon the waste management component of structural costs, total annual facilities costs including annual capital costs, taxes and insurance, and repairs and maintenance for each waste management alternative are presented in Table III.

Table III. Annual waste management facilities costs for oxidation ditches and high-rise drying for a 30,000-bird operation.

	Undercage Oxidation Ditches	High-Rise Undercage Drying
Annual Capital Cost ($)[a]	274	3125
Taxes and Insurance ($)[b]	10	109
Repairs and Maintenance ($)[c]	3	31
Annual Facilities Cost ($)	287	3265

[a]Waste management component of structural costs amortized at 9% over 20 years.
[b]3.5% of annual capital cost.
[c]1% of initial investment.

Stabilization Costs

Stabilization costs were defined as annual equipment and operating costs associated with the operation of aeration units and drying-air circulation fans. It was first attempted to analyze annual equipment and operating costs for oxidation ditch aeration units independently. Oxygen-transfer capacities and power requirements were obtained from reported research results or from manufacturers' brochures when independently developed data were not available. However, it was found that operating costs in terms of cost/kg of oxygen transfer capacity varied significantly between aeration units evaluated (Table IV). Analysis of annual equipment costs also revealed wide variation (Table V). However, it was found that units with high operating costs had low annual equipment costs. Expressing annual equipment costs in terms of cost/kg of oxygen-transfer capacity and combining this value with operating

Table IV. Estimated operating costs for oxidation ditch aeration units.

Manufacturer	Capacity ($g\ O_2$/hr)	Power Requirements[a] (kW)	$g\ O_2$/kWh	Cost/kg O_2[b] ($)
1. 1.8-m rotor	4857	2.94	1652	0.021
2.4-m rotor	6476	3.94	1644	0.021
2.	1244	1.93	644	0.105
3. 1.8-m rotor	3360	3.68	913	0.040
4. 3.0-m rotor	3110	2.98	1044	0.034

[a]Calculated from net power requirements assuming maximum motor efficiency of 75%.
[b]Based upon an electrical energy cost of $0.035/kWh.

Table V. Estimated annual equipment costs for oxidation ditch aeration units.

Manufacturer	Initial Cost ($)	Capital Cost[a] ($)	Taxes and Insurance[b] ($)	Maintenance and Repairs[c] ($)	Total Annual Equipment Cost ($)
1. 1.8-m rotor	8170	1274	286	163	1723
2.4-m rotor	8550	1333	299	171	1803
2.	1270	198	44	25	267
3. 1.8-m rotor	2610	407	91	52	550
4. 3.0-m rotor	3500	546	122	70	738

[a]Amortized at 9%/yr over an estimated useful life of 10 yr.
[b]Estimated at the rate of 3.5% of initial cost/yr.
[c]Estimated at the rate of 2% of initial cost/yr.

costs, revealed that there was little difference between four of the five units evaluated (Table VI). Excluding unit B, the estimated average cost/kg of oxygen transferred is $ 0.058. Based upon a maximum required oxygen-transfer capacity of 35 kg O_2/hr for the oxidation ditch system design under consideration, the total annual cost for oxygen transfer was estimated to be $17,539.

As previously noted the 30,000 bird high-rise system design requires eight 0.373 kW and two 0.186 kW drying-air circulating fans. A study of the air moving efficiencies of agricultural propeller-type fans such as those used to circulate drying air in a high-rise manure drying system has shown a wide variation in efficiency between manufacturers.[7] Analysis of both annual equipment and operating costs were based on a maximum reported air flow efficiency of 0.010 m^3/sec/W (21 ft^3/sec/W). For the ten fans required, the

Table VI. Summary of total costs for oxidation ditch aeration unit oxygen transfer.

Manufacturer	Operating Cost/kg O_2[a] ($)	Annual Equipment Cost/kg O_2 ($)	Cost/kg O_2 ($)
1.	0.021	0.040	0.061
	0.021	0.032	0.053
2.	0.105	0.046	0.151
3.	0.040	0.019	0.059
4.	0.034	0.027	0.061

[a]Based on 24-hr 360-day operation.

total annual equipment cost was found to be \$446/yr based on initial costs obtained from vendors. The annual operating cost was calculated to be \$1285 at \$ 0.035/kWh.

Handling and Disposal Equipment Costs

The total cost for handling and disposal of manure is perhaps the most difficult component of total waste management costs to quantify due to the comparatively high labor component in comparison to other aspects of these waste management systems. Trade-offs exist between investment and other fixed costs which are related to equipment capacities and operating and labor costs. Another variable is transport distance, which is a major factor in labor and operating costs. In the interest of simplicity, only fixed costs for handling and disposal equipment were considered. Labor and operating costs were evaluated indirectly in terms of number of loads of waste/yr from each system requiring ultimate disposal.

The high-rise drying alternative requires both a tractor-mounted loader for manure removal from the building and a box-type manure spreader for transport and disposal. Only fixed costs directly related to the front-end loader were considered. It was assumed that 75% of the annual operating time of the loader would be for manure handling. Fixed and operating costs associated with the tractor were omitted since manure handling should represent only a small fraction of the annual tractor operating time. Manure spreader costs were based on an intermediate-size unit with a capacity of 7.6 m^3 (268 ft^3).

Only a closed-tank-type liquid manure spreader with a capacity of 11.4 m^3 (3000 gal) was considered in the estimate of handling and disposal and equipment costs for the oxidation ditch alternative. It was assumed that the aerated slurry could be removed from the oxidation ditches using gravity. A summary of the fixed handling and disposal equipment costs associated with each alternative are presented in Table VII. Also included are estimates of the number of loads of manure requiring ultimate disposal annually based on anticipated waste volumes noted earlier and spreader capacities specified for each system.

Plant Nutrient Value

In situations where plant nutrients can be utilized for field crop production or marketed, the ability of a waste management system to conserve these nutrients must be considered as a credit. However, it must be recognized that the value of these nutrients is realized only when these nutrients are actually utilized in place of purchased fertilizer inputs. Poultry manure as produced

Table VII. Comparison of annual manure handling and disposal equipment costs.

	Oxidation Ditches	High-rise Drying
Initial Cost ($)	6159[a]	4915[b]
Annual Cost ($)	960	766
Taxes and Insurance ($)	216	172
Repairs and Maintenance ($)	123	98
Total Annual Equipment Cost ($)	1299	1036
Number of Loads/yr	284	173

[a]11.4 m^3 liquid manure spreader loaded by gravity.
[b]7.6 m^3 solid manure spreader and 75% of a tractor mounted front-end loader.

contains nitrogen, phosphorus and potassium, as well as calcium, in significant quantities.

Neither high-rise undercage drying systems nor oxidation ditches operated for odor control appear to be effective systems for nitrogen conservation. A nitrogen loss of 53% based on mass balance results from a full-scale high-rise system evaluation has been reported.[2] The amount of nitrogen remaining as ammonia was 37.5%. Since opportunities for volatilization during and following spreading are sizable, the potential for plant utilization of this ammoniacal nitrogen appears minimal. Thus an assumption of a 71% nitrogen loss appears reasonable. Nitrogen losses of this magnitude have also been observed in oxidation ditches where aeration was limited to odor control requirements.[1]

The transformations of phosphorus and potassium as well as calcium in either system have not been clearly delineated. Since these elements do not possess volatile forms, losses should not occur. Possible chemical transformations rendering these elements unavailable to plants should be equal in both systems. Thus, it appears that the plant nutrient values of the manure from high-rise undercage drying systems and of that from oxidation ditches are equal. Since no difference in plant nutrient value appears to exist between the two systems, this factor was excluded from further consideration.

DISCUSSION

A summary of the component costs for oxidation ditches and high-rise undercage drying is presented in Table VIII. These data suggest that the use of oxidation ditches as batch reactors for poultry wastes has a higher cost. However, it should be recognized that opportunities exist for reducing the stabilization costs, which is the major cost component for oxidation ditches (Table VIII). The simplest approach would be to utilize cyclic aerator

Table VIII. Summary of waste management component costs for oxidation ditches and high-rise undercage drying.

	Undercage Oxidation Ditches	High-rise Undercage Drying
Facilities Costs ($)	287	3,265
Stabilization Costs ($)	17,539	1,731
Manure Handling and Disposal Equipment Costs ($)	1,299	1,036
Total Annual Waste Management Costs ($)	19,125	6,032

operation adjusted with time to match changes in aeration requirements as shown in Figure 5. This practice would reduce the time of aeration unit operation, and thus operating costs, by 33%. The total reduction in stabilization costs would be less due to constant fixed costs.

A more effective alternative would be oxidation ditch operation as continuous-flow reactors at MLTS concentrations of less than 20 g/l. This approach would serve to overcome the high stabilization costs due to the inefficiency of aeration poultry manure slurries at high MLTS concentrations. This would reduce oxygen transfer capacity requirements, thus reducing both fixed and operating costs. A 60% reduction of stabilization costs would be possible. However, it should be recognized that this alternative awaits the development of a practical liquid-solids separation process. Otherwise, the reduction in stabilization cost will merely be shifted into ultimate disposal costs.

A summary of the unit costs for three alternative methods of oxidation ditch operation and high-rise undercage drying is presented in Table IX. It should be noted that the costs for the continuous-flow mode of oxidation ditch operation do not include liquid-solids separation costs. Thus, actual costs may be somewhat higher than the stated cost due to added fixed and operating costs. However, if concentration of solids to values exceeding 6% can be achieved, reduced ultimate disposal costs will, to some degree, offset the added costs of liquid-solids separation.

SUMMARY AND CONCLUSIONS

The results of cost analyses of the oxidation ditch and high-rise undercage drying as poultry waste management alternatives show that drying represents

Table IX. Summary of poultry waste management unit costs.[a]

System	Cost/ 1000 Hens/Yr ($)	Cost/ Dozen Eggs[b] ($)
Oxidation Ditch		
Batch		
Continuous Aerator Operation	638	0.032
Cyclic Aerator Operation	540	0.027
Continuous Flow	287	0.014
High-Rise Undercage Drying	201	0.010

[a]Excludes labor and operating costs for ultimate waste disposal.
[b]Assumes 20 dozen eggs/hen-yr.

a least-cost alternative for odor control. Total cost for high-rise drying excluding labor and operating costs for ultimate disposal was estimated to be $0.01 per dozen eggs. Comparative cost for the batch-reactor oxidation ditch alternative was $0.032 per dozen eggs. High waste stabilization costs for this type of oxidation ditch result from the inefficiency of oxygen transfer in poultry manure slurries with high mixed liquor total solids concentrations. Maximizing oxygen transfer efficiency would reduce costs to a level approaching $0.014 per dozen eggs.

The results of this study should provide a basis for a rational economic decision concerning the selection of a management system for poultry wastes. However, it should not be used independently but rather in conjunction with other factors, such as operating characteristics and the ease of integration into an overall management system.

REFERENCES

1. Martin, J. H. and R. C. Loehr. "Demonstration of Aeration Systems for Poultry Wastes," Environmental Protection Technology Series Report No. EPA 600/2-76-186, U.S. Environmental Protection Agency, Washington, D. C. (1976).
2. Sobel, A. T. "The High-Rise System of Manure Management," AWM 76-01, Department of Agricultural Engineering, Cornell University, Ithaca, New York (1976).
3. Martin, J. H. and R. C. Loehr. "A Process Design Manual for Poultry Waste Management Alternatives," U.S. Environmental Protection Technology Series Report (in press, 1977).
4. Martin, J. H. and R. C. Loehr. "Aerobic Treatment of Poultry Wastes," *J. Agric. Eng. Res.* 21:157-167 (1976).

5. Baker, D. R., R. C. Loehr and A. C. Anthonisen. "Oxygen Transfer at High Solids Concentrations," *J. Environ. Eng. Div.*, American Society of Civil Engineers, 101:759-774 (1975).
6. Mellor, C., Production Manager, Building Department, Agway, Inc., Syracuse, New York, personal communication (1977).
7. Albright, L. D. "Air Moving Efficiencies of Ventilating Fans," ASAE Paper No. NA75-034, American Society of Agricultural Engineers, St. Joseph, Michigan (1975).

MANAGEMENT OF LAYING HEN MANURE BY MOISTURE REMOVAL–RESULTS OF SEVERAL RESEARCH INVESTIGATIONS

A. T. Sobel and D. C. Ludington
Agricultural Engineering Department
New York State College of Agriculture and Life Sciences
Cornell University
Ithaca, New York

INTRODUCTION

Management of the manure from laying hens has for many producers been very difficult. The manure as produced by caged laying hens is high in moisture, difficult to handle and, if allowed to accumulate, produces offensive odors. A management scheme which removes and applies the manure to the land on a daily basis has had the minimum difficulties. However, there are problems associated with daily spreading, such as getting onto the land due to soft ground and runoff during snow melt, and a more desirable solution to this management dilemma is sought.

Moisture removal presents itself as a means of solving some of the problems associated with laying hen manure. The removal of sufficient moisture from manure will result in a material that can be handled as a solid, has low odor and can be stored. The incorporation of moisture removal into a management program requires knowledge of the parameters of such a system. The research results presented attempt to characterize some of the operational parameters of a management system for laying hen manure utilizing moisture removal. Investigations were made on various methods to promote drying under the caged birds, on the use of unheated air for drying, on the direct measurement of manure moisture and on a procedure for calculating the component changes through a management system.

The major portion of this research was conducted under cooperative agreement with the Livestock Engineering and Farm Structures Research Branch, AERD, Agricultural Research Service, U.S. Department of Agriculture.

The research reported was conducted at facilities located at the New York State College of Agriculture and Life Sciences, Cornell University, Ithaca, New York. Experiments of a pilot scale level were conducted at the Agricultural Waste Management Laboratory (AWML), financed jointly by funds from the Office of Water Programs, EPA, and the College of Agriculture and Life Sciences.

UNDERCAGE DRYING–MECHANICAL DEVICES–NO FORCED AIR

Theoretical Considerations

There are many factors which contribute to the drying of the manure. Such factors as air velocity, initial moisture content of manure which is affected by breed and strain of bird and ration being fed, time of exposure, configuration of undercage device and environmental conditions including temperature and humidity affect the amount of moisture that is removed from the manure by evaporation. An undercage device for promoting drying holds a percentage of the manure produced and results in a moisture loss from the held manure and from the passed manure. The moisture lost from the two portions of manure will be a function of the above factors. The final moisture content of the system (Table I) will be a result of mixing these two portions of manure.

Table I. Parameters of produced, held, passed and mixed manure.

	Wet Weight	Weight of Total Solids	Moisture Content Dry Basis Unit/ Unit Total Solids	Fraction of Total Solids
Produced by bird	W_O	S	M_O	1.0
Held on device	W_H	S_H	M_H	H
Passed through device	W_P	S_P	M_P	P
Mixture of held + passed	W_F	S	M_F	1.0

Moisture Content of Mixture, $M_F = HM_H + PM_P$

$= HM_H + (1 - H)M_P$

Moisture Removal Rate, $\frac{M}{\theta} = \frac{M_o - M\theta}{\theta}$

Where θ = time.

Example:

If $M_P = M_o = 3.00$,

$M_F = HM_H + (1 - H)3$

Values of M for: M_H =

H	3.00	2.00	1.00
0.0	3.00	3.00	3.00
0.25	3.00	2.75	2.50
0.50	3.00	2.50	2.00
0.75	3.00	2.25	1.50
1.00	3.00	2.00	1.00

Note: This example is hypothetical, for if 100% of the manure were held, very little if any moisture loss would occur. Therefore, for H = 100%, the moisture content of the manure held would not be as low as 1.00. Values below the dotted lines are highly improbable.

Experimental Application

In an actual system, $M_H = f(H)$. Two examples of possible systems for holding manure for drying were previously reported.[1] One system had metal fins oriented at an angle with respect to the horizontal. The angle was varied; consequently, the amount of manure that was held was a function of the angle. The moisture content of the held manure was a function of the amount of manure held. The data from this system in the format of the above theory are presented in Table II. The moisture content of the passed manure was not measured but was assumed to be 300 (75).[a] The data are presented in graph form in Figure 1 on Log-Log coordinates and fit to an equation of the form $Y = AX^B$.

For this system the reduction in the final moisture content was not large, with the greatest reduction occurring when a large amount of manure was held. When 80% of the manure was held the resulting moisture content was 204 (67) even though the moisture content of the manure held was only 180 (64). More moisture was evaporated with a large amount of manure held than with a small amount of manure held and reduced to a low moisture content.

[a]Moisture contents are presented as [percent dry basis (percent wet basis)] unless noted.

Table II. Moisture contents of manure for varied angle system.

Angle	Percent Held, H x 100	Moisture Content			
		M_H[a]	$(1\text{-}H)M_P$[b]	HM_H	M_F[a]
45	100 (assumed)	2.205 (69)	0	2.205	2.205 (69)
55	80.0	1.801 (64)	0.600	1.441	2.041 (67)
	45.0	0.923 (48)	1.650	0.415	2.065 (67)
	32.5	1.000 (50)	2.025	0.325	2.350 (70)
	46.0	0.821 (45)	1.620	0.378	1.998 (67)
65	38.5	1.429 (59)	1.845	0.554	2.399 (71)
	43.5	1.688 (63)	1.695	0.734	2.429 (71)
	20.0	1.110 (53)	2.400	0.222	2.622 (72)
	28.0	2.135 (68)	2.160	0.598	2.758 (73)
75	8.5	0.789 (44)	2.745	0.067	2.812 (74)
	6.0	1.179 (54)	2.820	0.070	2.890 (74)
	2.5	0.560 (36)	2.925	0.014	2.939 (75)
	2.0	0.348 (26)	2.940	0.007	2.947 (75)
80	1.0	0.151 (13)	2.970	0.002	2.972 (75)
	0.25	0.357 (26)	2.993	0.001	2.994 (75)
	0.50	0.107 (15)	2.985	0.001	2.986 (75)

[a]Moisture contents are shown as (Percent Dry Basis)/100 (Percent Wet Basis).
[b]M_P assumed to be 3.00(75).

A second example of an undercage device to promote drying that illustrates the combining of two portions of manure is that of vertical fins.

Metal fins 2.5 in. high were placed under two adjoining cages of birds, three birds per cage (0.5 sq ft/bird). Air circulation was normal ventilation of 3 to 4 cfm per bird. Two fin spacings were tested, 3/4 in. and 1 in. A plastic-covered board suspended below the fins collected the manure that passed through the fins. The fins were scraped at one- to five-day intervals and the manure separated as to manure held and manure passed. The data from this system in the format of the above theory are presented in Table III. The data are presented in graph form in Figure 2 on Log-Log coordinates and fit to an equation of the form $Y = AX^B$.

For this system the moisture content of the manure passed was also a function of the system of the manure held, $M_P = f(H)$. Consequently, the final moisture content was influenced by drying of both the manure held and the manure passed. As with the varied angle fins, the greatest contribution to drying occurred when a high percentage was held and the moisture content reduction was low. This indicates a larger removal of water by evaporation than when a small quantity of manure was dried to a low moisture content.

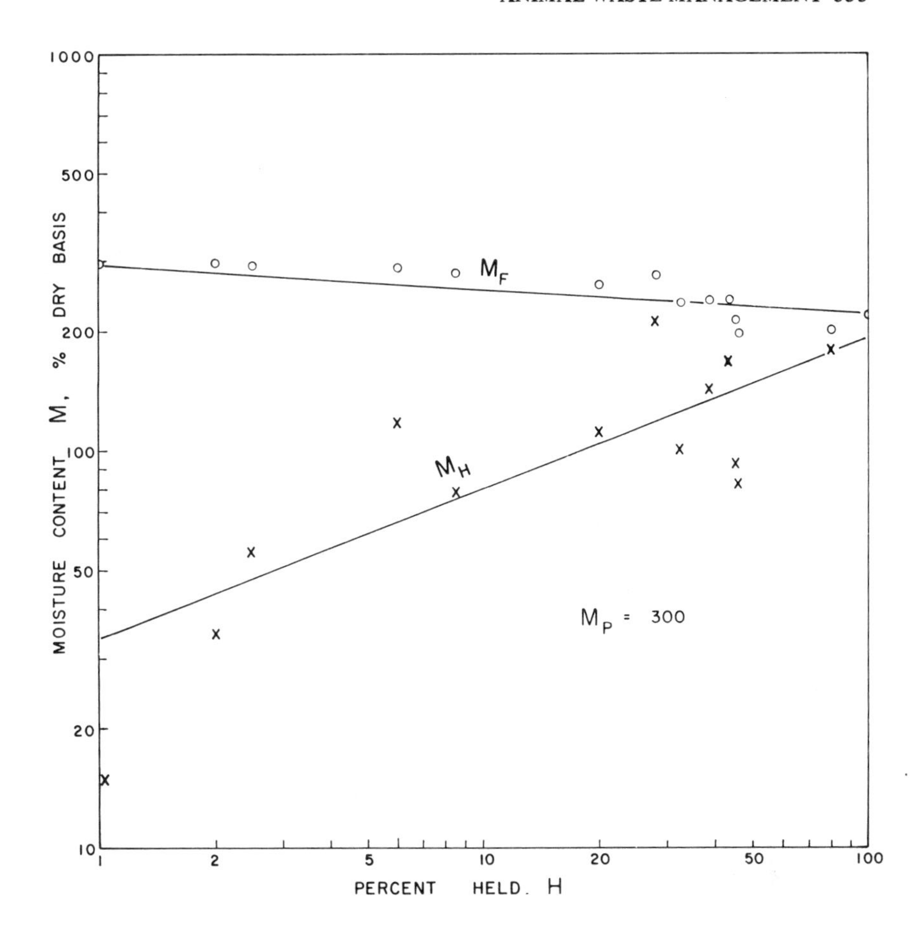

Figure 1. Percent held vs moisture content for metal fins at varied angles.

The 1-in. fins held a small quantity of manure (18%), while the 3/4-in. fins held nearly one-half of the manure produced (46%). The 1-in. fins had a high M_P, 282(73), and therefore a high M_F, 256(72). The 3/4-in. fins allowed some drying of the manure passed and M_P was 176(63); therefore, M_F was lower, 164(62).

The moisture content of the held manure and consequently the final moisture content was also related to the time of exposure. For the vertical fins the moisture content of the manure held was a function of time. The moisture removal rates are presented in Table IV.

The moisture removal rate can be presented as several relationships. The moisture removed per day per unit weight of total solids present decreases as the number of days of accumulation increases. The moisture removed per

Table III. Moisture contents of manure for vertical fins system.

Fin Spacing (in.)	Days Collection	Percent Held 100 H	Moisture Content[a]				
			M_H	M_P	HM_H	$(1-H)M_P$	M_F
1	1	25.64	1.969(66)	2.957(75)	0.505	2.199	2.704(73)
	2	16.86	0.947(49)	2.764(73)	0.160	2.298	2.458(71)
	2	18.23	1.432(59)	3.088(75)	0.261	2.525	2.786(74)
	3	13.32	1.351(57)	3.243(76)	0.180	2.811	2.991(75)
	5	17.06	1.123(53)	2.028(67)	0.192	1.682	1.874(65)
	Average	18.22	1.364(58)	2.816(74)			2.563(72)
3/4	1	30.19	1.741(63)	2.111(68)	0.526	1.474	2.000(67)
	2	46.68	1.777(54)	1.493(60)	0.549	0.796	1.345(57)
	3	50.67	1.650(62)	2.466(71)	0.836	1.216	2.052(67)
	4	32.25	2.398(71)	2.095(68)	0.773	1.419	2.192(69)
	4	54.11	1.221(55)	1.333(57)	0.661	0.612	1.273(56)
	5	64.84	0.916(48)	1.055(51)	0.594	0.371	0.965(49)
	Average	46.45	1.517(60)	1.759(64)			1.638(62)

[a]Moisture Contents are shown as (Percent Dry Basis)/100 (Percent Wet Basis).

unit weight of total solids produced per day is more constant but also decreases as the number of days of accumulation increases. This indicates the effect of manure being covered with fresh manure and reducing the ability to dry further. Additional exposure of the manure under these conditions would not increase the drying rate or result in the lower moisture content. For the vertical fin system, based on the average moisture removal data, the M_F is as follows:

$$M_F = \frac{\text{Moisture Produced/Day-Moisture Evaporated/Day}}{\text{Total Solids Produced/Day}}$$

Moisture Produced/Day = 3(117) = 351 g/day @ 300(75)

$$1 \text{ in: } M_F = \frac{351 - 50}{117} = 2.57 \text{ or } 257(72)$$

$$3/4 \text{ in.: } M_F = \frac{351 - 170}{117} = 1.55 \text{ or } 155(61)$$

These final moisture contents are in agreement with the average final moisture contents obtained.

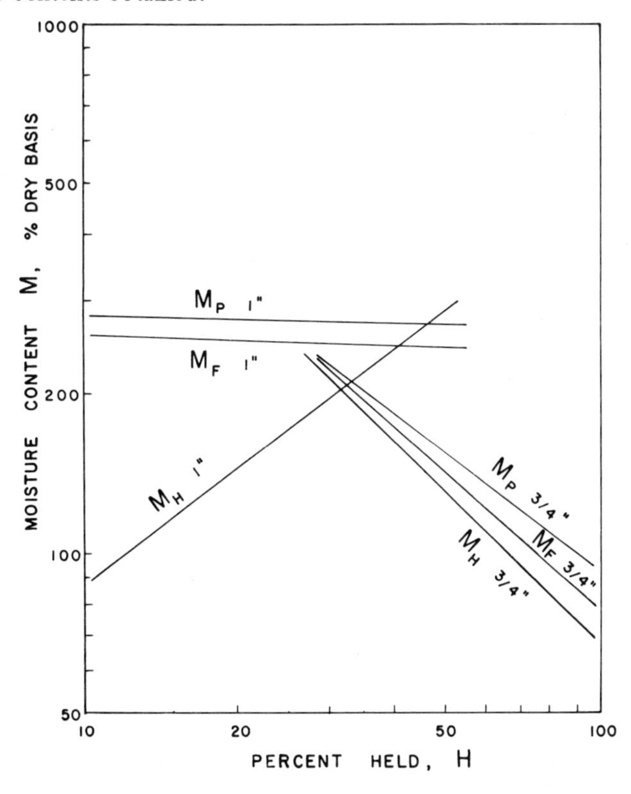

Figure 2. Percent held vs moisture content for vertical metal fins.

Table IV. Moisture removal rate for vertical fin system.

Fin Spacing (in.)	Days Collection	M_H	M_F	Total Solids (g)			Total/Day	g/g TS/Day		g/Day		g/g TS Produced
				Held	Passed	Total		Held	Final	Held	Final	
1	1	1.969	2.704	28.96	83.97	112.95	112.95	1.031	0.296	29.86	33.4	33.43
	2	0.947	2.458	41.49	204.64	246.13	123.06	1.026	0.271	42.57	66.7	33.34
	2	1.432	2.786	45.30	203.21	248.51	124.25	0.784	0.107	35.52	26.6	13.29
	3	1.351	2.991	43.69	284.21	327.90	109.30	0.550	0.003	24.03	0.98	0.33
	5	1.123	1.874	93.00	452.18	545.18	109.04	0.375	0.225	34.88	123	24.53
							Average	0.753	0.180		50.1	20.98
3/4	1	1.741	2.000	37.54	86.81	124.35	124.35	1.259	1.000	47.26	124	124.35
	2	1.177	1.345	103.23	117.92	221.15	110.57	0.911	0.827	94.04	183	91.44
	3	1.650	2.052	169.07	164.62	333.69	111.23	0.450	0.316	76.08	105	35.15
	4	2.398	2.192	99.65	209.30	308.95	77.24	0.150	0.202	14.95	62.4	15.60
	4	1.221	1.273	261.94	222.11	484.05	121.01	0.445	0.432	116.56	209	52.28
	5	0.916	0.965	539.23	292.34	831.57	116.31	0.417	0.407	224.85	338	67.42
							Average	0.605	0.531		170	64.42
				Averages of 1 and 3/4			117.21[b]	0.672	0.371			

[a]M_o assumed to be 3.0.
[b]19.5 g TS/bird/day for 6 birds.

Miscellaneous Devices

Several other devices were given preliminary tests to determine the possible application as undercage drying devices.

Wires

The idea of using wires to hold a portion of the manure was investigated by placing wires beneath caged laying hens. Plastic-coated electrical wire (NO. 14 TW solid) approximately 1/8 in. diameter was spaced 3/4, 7/8 and 1 in. apart under twelve cages (12 in. x 18 in.) of three birds each (3 ft of cage row, bird density, 2 birds/sq ft of manure collection area). The wire was placed perpendicular to the length of the cage row and approximately one foot above the dropping board. Air movement around the manure was only that associated with ventilation provided for the birds. The wires were under the birds for 48 days, during which time they were not disturbed. Observations were made during this time and samples were taken at the end of this time.

Within one week the manure started accumulating on the wires. The manure soon bridged the wires, and after two weeks essentially 100% of the manure was held on the wires. At 48 days the wires were sagging and the manure had an average moisture content of 163(62). The manure was removed from the wires with difficulty.

Metal Strips

In conjunction with the study on wires, galvanized metal strips were placed under caged birds. Strips 3/4-, 1-1/4-, and 2-in. wide were placed 3/4 in. apart with each size strip under four cages (one foot of cage row). Bird density was the same as with wires. Ater three weeks of operation, a modification was made and the 2-in. strips were twisted and swivels were added to the ends so the strips were free to turn.

At the end of the test period (49 days) the manure had bridged all the strips including those with the swivels. The average moisture content of the manure on the strips at this time was 256(72). The strips with the swivel would turn only if encouraged to move by starting to turn them. No difference in manure accumulation or moisture content was noted between strip widths.

Ridges

Observations of the manure deposited under the caged laying hens indicate that the major portion is deposited at the center line of the cage. This is a result of the position of the animal during feeding. A device designed to take advantage of this observation was constructed. The device consisted of four ridges parallel with the cage row located at the center-line of each cage row. The ridges were constructed of galvanized sheet metal. The sides of the ridge were at 45° with the horizontal. The base of the ridge was 13 in. wide and therefore occupied approximately two-thirds of the manure collection area. The manure collection area was increased by a factor of one-third with this device. This resulted in an effective decrease in the bird density. A mechanized scraping device was utilized to clean the ridges. The manure from the ridges fell by gravity to a dropping board and was removed by a mechanical cleaner.

The manure from this system was monitored for 40 days. During this time there was considerable difficulty in cleaning the manure from the ridges due to insufficient drying and resulting sticking. The average final moisture content of the manure from this system was 203(67).

Conclusions

The final moisture content of the manure from these systems is summarized in Table V. In all cases M_F is above that required for handling manure as a solid. The lowest final moisture content obtained was 163(62). From investigations with various solid handling systems a moisture content below 100(50) is required to allow the manure to be handled as a solid. However, manure at 100(50) is not storable without odor production. The small contribution to the final moisture content by the low moisture content held manure at the low percentages held is apparent in the low values of HM_H.

Table V. Final moisture contents for various undercage drying systems with no forced air.

System Description	Final Moisture Content
Vertical Fins–1-in. spacing	256(72)
Vertical Fins–3/4-in. spacing	163(62)
Wires–3/4-, 7/8-, 1-in. spacing	163(62)
Metal Strips	245(71)
Ridge	203(67)

The application of intermediate drying devices to undercage drying appears to be limited under environmental conditions present in the Northeast. Unless other factors which enhance drying, such as increased air velocity or reduced bird density, are included with the drying device, the final manure moisture content will not be acceptable for proper handling and storage.

A PILOT SCALE UNHEATED AIR BELT DRYER FOR LAYING-HEN MANURE

Initial investigations in the laboratory[2] indicated two significant properties of laying-hen manure related to thermal (evaporative) moisture removal or drying. These are:

1. Laying-hen manure has a low equilibrium moisture content.
2. Drying is enhanced by utilizing a thin layer.

To investigate the possibilities of utilizing these properties for large-scale commercial drying of laying-hen manure, a prototype dryer was constructed. The basic principle of the dryer was to form the manure into a thin layer, place it on a wire mesh belt, and circulate unheated air around the manure. The purpose of the belt was to expose the manure to the air and to provide transport for the manure.

Description and Operation of Dryer

The dryer constructed was centered around a 20-in.-wide belt. The belt was galvanized wire mesh with an approximate mesh of 3.5 openings per inch. The belt from center to center of rollers was 28.75 ft. The effective length was 27.25 ft and the effective width was 1.5 ft. The effective drying area was therefore 40.8 ft^2. The framework of the dryer was angle iron which was covered with plywood. Air was moved by a centrifugal fan and the air could be recirculated or mixed with room air. Figure 3 illustrates the basic configuration of the dryer.

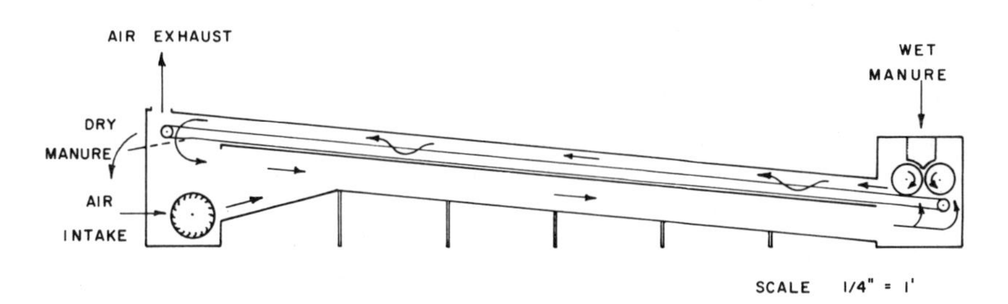

Figure 3. Basic configuration of belt dryer.

A unique part of the dryer was the device used to form the manure into a thin layer. This device consisted of two 14-in.-diam horizontal drums with a 0.25-in. clearance between. The drums rotated such that the manure was forced through the narrow opening. The wet manure fed very well through the opening as there was sufficient friction between the manure and the surface of the 14-in. drums. The thin layer formed would adhere to one or the other drum and was scraped off in such a manner that it would drop onto the belt moving beneath. The speed of the drums and the speed of the belt were controlled independently. The density of the manure on the belt could be controlled by the relative speed of the drums and the belt.

Ideally, the belt could be run at such a speed that the manure would dry during its passage through the dryer. The manure would leave the other end of the belt at a moisture content that would depend on the initial moisture content, the belt speed, the temperature and humidity of the drying air, and the thickness and density of the manure on the belt. However, for the unit constructed the belt was not long enough for continuous operation. Therefore, the unit was run as a batch operation. The belt was loaded and then stopped until the manure was at the desired moisture content. Information on the drying time was then projected to determine the size of belt required. When the manure was dry, the belt would be run such that the dry manure would fall off the end of the belt and slide down a chute. The dryer was then reloaded for the next batch operation.

Instrumentation was provided for temperature, relative humidity and air velocity. Moisture contents of the manure before, during and after drying were determined by oven drying. Samples were taken directly from the belt through access ports, or a sample was placed in a wire basket on the belt and weighed periodically. Weights of ingoing wet manure and outgoing dried manure were recorded.

Experimental Results

After preliminary operation of the dryer, several detailed runs were made to determine the operating characteristics of the dryer. Table VI summarizes the parameters of these runs. Drying curves for Runs 1 and 3 are shown in Figures 4 and 5. Run 1 ran overnight (17.2 hr), and the manure first contacted by incoming air (air entering at the blower first came in contact with the manure at the rollers) appeared to be dry in 14 hr. Manure located near the air outlet and exposed to air moistened with water from passing over the belt of manure did not dry as fast and was still losing moisture at 17.2 hr. The manure dried at a constant rate down to a moisture content of 70(41).

During Run 3 samples were taken through the access ports to compare with the basket method of determining the drying curve. The basket method gave a more uniform drying curve as indicated in Figure 5. The run also went overnight, and the manure at the air inlet appeared to be dry in 15 hr. The manure dried at a constant rate down to a moisture content of 43(30). The relative humidity for Run 3 inside the dryer is shown in Figure 6. The higher humidity near the outlet is apparent.

Table VI. Summary of drying runs.

Run Number	1	2	3	4
Date	05/08/67	05/09/67	12/15/67	04/12/68
Weight (lb)				
Initial	35.0	40.2	20.8	48.6
Final	13.0	16.0	6.0	21.5
Weight Reduction (%)	62.8	60.2	71.1	55.8
Moisture Content[a]				
Initial	210(67.8)	227(69.4)	285(74.0)	151(60.1)
Final	15.2(13.2)	30.0(23.1)	11.1(10.0)	10.8(9.8)
Water Removed (lb/ lb Total Solids)	1.95	1.97	2.74	1.40
Time on Dryer (hr)	17.2	8.5	23	24
Approximate Time to Dry to 25(20)(hr)	13	-	12	11
Relative Humidity of Inlet Air (%)	31	31	19	15
Belt Coverage[b] (%)	53	60	31	73

[a]Moisture contents are presented as dry basis (wet basis).

[b]Based on thickness of 5/16 in. and density of 63 lb/ft^3; 100% coverage would be 66.8 lb of wet manure.

The amount of moisture picked up by the air passing over the manure was calculated by considering the change in absolute humidity between incoming air and air having passed over the wet manure. For Run 3 the difference in absolute humidity as a function of time taken from Figure 6 is presented in Figure 7.

$$\text{Moisture picked up by air} = [\,\Sigma (h_o - h_i)\,\theta\,]\,Q$$

where:
- h_i = absolute humidity of air at inlet at time θ, lb water/lb dry air
- h_o = absolute humidity of air at outlet at time θ
- Q = air-flow rate, lb dry air/hr
- θ = time, hr.

For Run 3 from Figure 7:

$$\Sigma \text{ area under curve} = 64 \times 10^{-4}\ \frac{\text{lb water}}{\text{lb dry air}}\ \text{hr}$$

$$Q = 300\ \text{ft/min} \times 60 \times 2\ \text{ft}^2 = 36{,}000\ \text{ft}^3/\text{hr}$$

$$\frac{36{,}000}{13.35} = 2{,}696\ \text{lb dry air/hr}$$

$$\text{Moisture picked up by air} = (64 \times 10^{-4})(2{,}696) = 17.3\ \text{lb water}$$

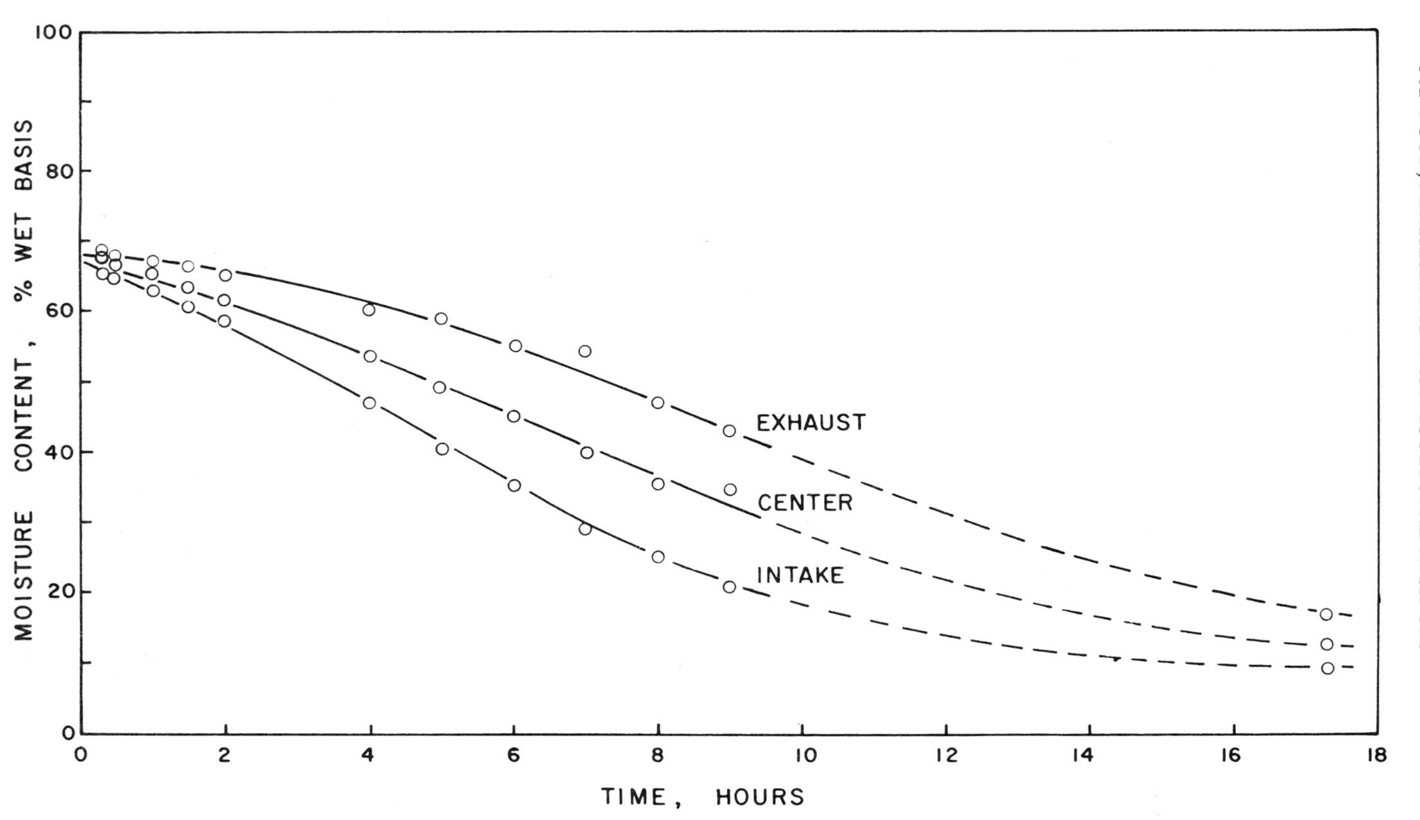

Figure 4. Drying relationship for Run 1.

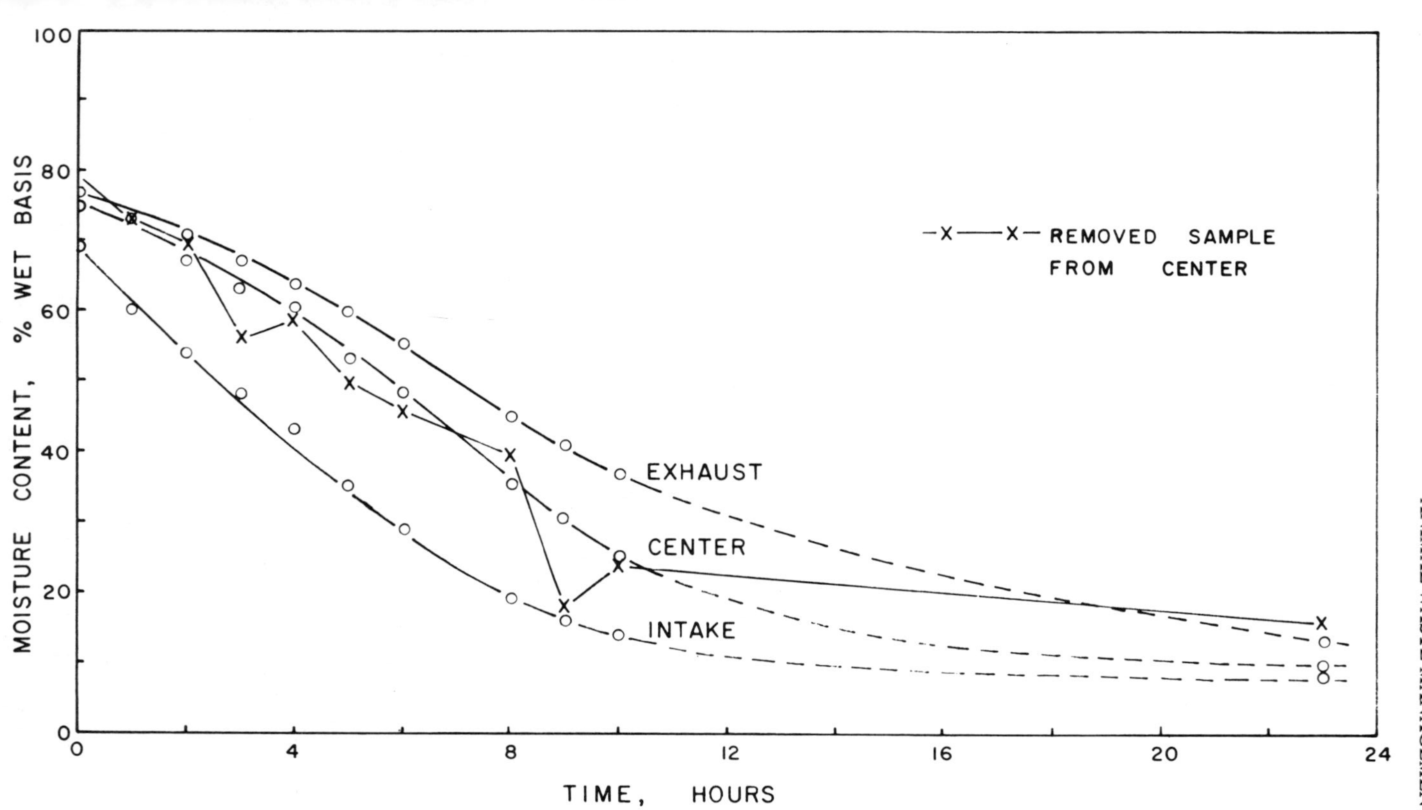

Figure 5. Drying relationship for Run 3.

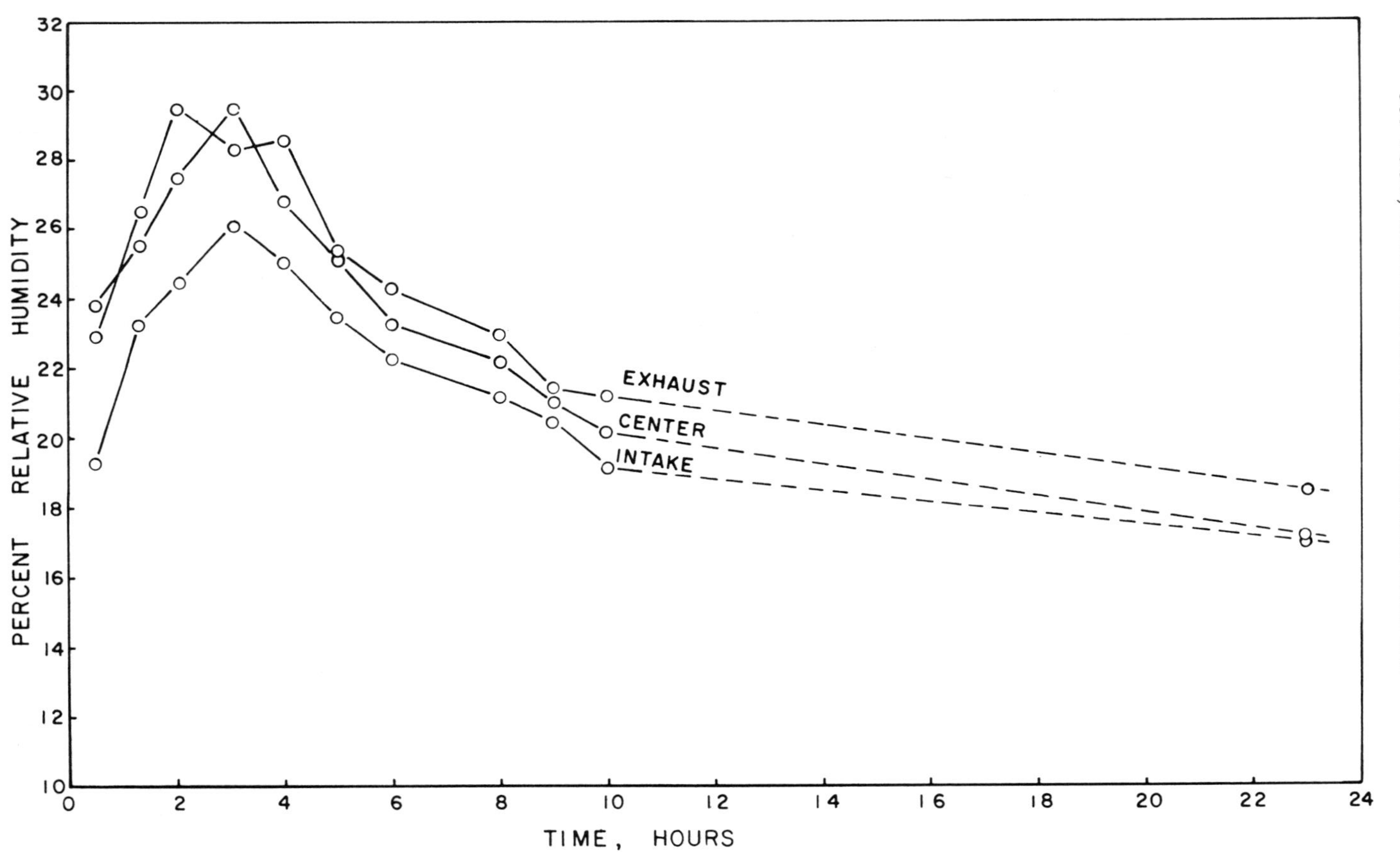

Figure 6. Relative humidity within dryer during Run 3.

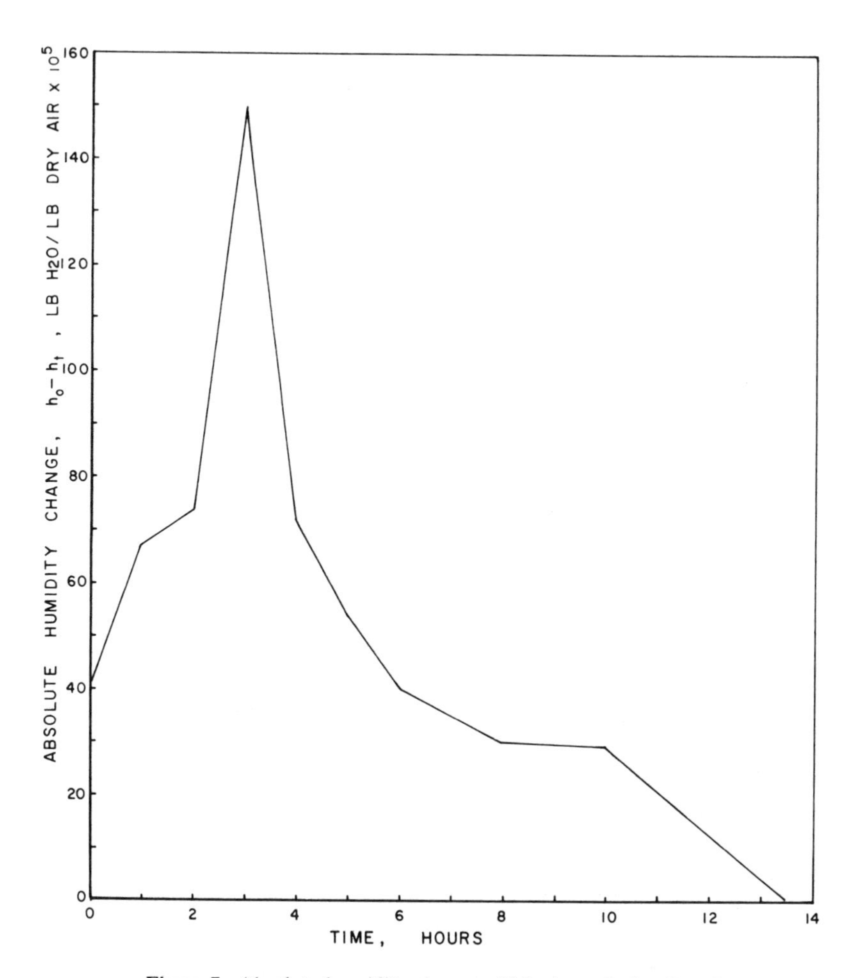

Figure 7. Absolute humidity change within dryer during Run 3.

Based on moisture loss of manure:

6 lb dry manure at 16% moisture, wet basis = 5.04 lb total solids

Initial moisture, dry basis,–final = 370 - 19 = 351

Moisture loss = 3.51 (5.04) = 17.7 lb water

Effective utilization of the surface area of the belt requires coverages approaching 100%. Complete coverage, however, would not allow adequate air circulation. The size of belt required per bird was determined using the following considerations:

Thickness of Drying Layer	5/16 in.
Coverage of Belt	67%
Moisture Content of Fresh Manure	300 (75)
Manure Production per Bird	0.09 lb solids/day
Density of Manure	63 lb/ft^3

Table VII. Size of belt related to drying time.

Drying Time (hr)	Runs/Day	Ft^2 of Belt/1000 Birds	Birds/Ft^2 of Belt
3	8	41	24
4	6	55	18
6	4	82	12
8	3	109	9
10	2.4	136	7
12	2	164	6
24	1	328	3

Conclusions

The drying time is a function of manure configuration, amount of water to be removed, *i.e.,* initial and final moisture contents, environmental conditions (temperature, humidity, air velocity) and drying efficiency of the overall system. For the conditions in Table VII the manure configuration is a constant. For a particular dryer the velocity is a constant. Any variations in these constants would be included in the drying efficiency.

During drying with unheated air, the heat for evaporation comes from the sensible heat of the air. This results in a depression of the dry bulb temperature and an increase in the absolute humidity.

The time to dry to 25(20) was approximately 12 hr. This would allow two drying cycles per day. Therefore, a dryer as computed in Table VII would handle 6 birds/ft^2 of belt. For most single-deck or stair-step cage systems the bird density is approximately 2 birds/ft^2 of manure collection area. The use of the belt dryer makes it possible to dry manure in one-third the area.

DIRECT MEASUREMENT OF MANURE MOISTURE BY ELECTRICAL RESISTANCE

Various devices are used for the direct measurement of soil moisture. The gypsum or plaster block buried in the soil will absorb or lose water in proportion to the water in surrounding soil. Fine-mesh stainless steel electrodes in the block allow measurement of electrical resistance which is proportional to water in the block. A wet block has low resistance, while a dry block has high resistance. Similarly, the moisture content of wood is measured by the electrical resistance between two pointed probes.

Experimental Procedure and Results

The adaption of several of these devices to the measurement of manure moisture was investigated. The following devices were tested:

1. two platinum measuring electrodes spaced 3/4 in. center to center;
2. plaster soil moisture block (2-1/2 x 1-3/8 x 1/2 in.);
3. fiberglass cloth soil moisture grids (1-3/8 x 1-5/16 in.); and
4. laboratory-constructed stainless steel probe consisting of two 1/4-in.-diam rods separated by a plexiglass holder; spacing set at 3/4 in. center to center.

Manure partially dried using the undercage slot-outlet drying system was utilized for the moisture tests. Table VIII presents the conditions of the test and the resulting resistance values. Resistance was measured with an AC Wheatstone Bridge at 1 KHz. The values for calibration chosen from Table VIII are presented in Table IX. Percent available moisture was calculated based on an initial moisture content of 300 (75) and an equilibrium moisture content of 17 (14.5).

$$\text{Percent Available Moisture} = 100\,\frac{(M - M_E)}{(M_o - M_E)} = 100\,\frac{(M - 17)}{283}$$

Figure 8 is the calibration curve obtained which relates dry basis moisture content to resistance. Figure 9 relates wet basis moisture content and available moisture to resistance.

The information in Table VIII indicates the time required for the plaster block to recover from wet moisture manure to dry moisture manure. When removed from manure at 262 (72.4) and placed in manure at 18 (15.3), 15 days were required to reach an equilibrium with the dry manure. The fiberglass grids also required time to recover but the recovery time was only three days for the same moisture conditions. At high moisture contents the plaster block appeared to be acted upon by the microorganisms in the manure.

All the resistance devices except the platinum probes presented approximately the same calibration curve This curve was nonlinear on log-log coordinates. A large range of resistance, 10 K - 700 K, occurred for moisture contents near equilibrium, 20 (16.7) - 17 (14.5).

The platinum probes with limited data presented a linear relationship which indicated a higher resistance at a specific moisture content than the other devices. These differences are probably a result of the platinum probes having very small contact surface with the manure.

Table VIII. Conditions of test and resulting data.

Identification (month, day year)	Moisture Content[a]	Platinum Probes	Plaster Block	Resistance, Ohms				
				Fiberglass Cloth Grids				Stainless Steel Probes
				1	2	3	Ave.	
Manure from Slot Outlet Undercage Drying System								
50572	44.9(31.0)	48,000	19,000	-	-	-	-	-
50872	41.2(29.2)	35,500	830	-	-	-	-	-
50972	38.9(28.0)	48,000	630	-	-	-	-	-
51072	36.6(26.8)	81,000	560	-	-	-	-	-
51172	36.6(26.8)	29,000	520	27,000	38,500	26,500	30,080	-
Manure from Slot Outlet System Stored in Plastic Bag Approximately One Year								
51272	18.0(15.3)	-	2,600	450,000	460,000	610,000	507,000	-
51572	17.6(15.0)	-	35,000	510,000	460,000	620,000	530,000	-
51672	17.1(14.6)	-	46,000	530,000	460,000	730,000	570,000	375,000
51772	17.2(14.7)	-	49.000	560,000	550,000	760,000	620,000	420,000
Moisture Added to 51772 Manure								
51872	85.2(46.0)	4,000	170	290	550	490	443	87
Additional Moisture Added to 51872								
51972	248(71.3)	61	39	4.6	4.4	3.5	4.2	3.8
52272	262(72.4)	78	60	2.3	2.4	2.0	2.2	2.7
Manure of 51172–Additional Air Drying								
52372	17.6(15.0)	-	190	5,300	8,300	20,500	11,400	385,000
52472	17.9(15.2)	-	330	9,200	14,300	28,000	17,200	375,000

52572	18.0(15.3)	-	790	11,000	18,000	33,000	21,000	415,000
52672	19.0(16.0)	-	2,100	12,900	20,500	35,300	22,900	410,000
53072	16.9(14.5)	-	6,300	16,200	27,000	41,500	28,200	335,000
53172	(17.9 assumed)	-	7,500	16,300	28,500	-	22,400	225,000
60172	(17.9 assumed)	-	8,400	18,200	32,200	-	25,200	280,000
60272	(17.9 assumed)	-	10,100	23,600	39.500	-	31,500	305,000
60572	(17.9 assumed)	-	12,700	24,000	40,000	82,000	49.000	180,000
60672	(17.9 assumed)	-	14,200	25,500	45,000	84,000	51,500	160,000
Manure from Slot Outlet Undercage Drying System								
60772	26.6(21.0)	-	2,700	2,100	3,050	10,000	5,200	12,200
60872	23.9(19.3)	-	2,800	2,500	4,000	-	3,200	15,700
60972	(23.5 assumed)	-	2,900	2,800	4,400	15,800	7,700	12,600
61272	(23.5 assumed)	-	4,100	4,250	5,900	19,900	10,000	19,400
61372	23.1(18.8)	-	3,900	3,700	4.950	16,000	8,200	11,600
Moisture Added to 61372 Manure								
61572	35.1(26.0)	-	1,280	540	600	1,890	1,010	1,630
61972	30.9(23.6)	-	410	700	880	2,300	1,290	2,230
62172	29.0(22.5)	-	310	730	960	2,300	1,330	2,500

[a]Moisture content is expressed as percent dry basis (percent wet basis).

Table IX. Calibration values–manure moisture by electrical resistance.

Identification	Moisture Content	% Available Water[a]	Platinum Probes	Resistance, Ohms Plaster Block	Fiberglass Grids (average)	Stainless Steel
51172	36.6(26.8)	6.9	29,000	520	30,800	-
51772	17.2(14.7)	0	-	49,000	620,000	420,000
51872	85.2(46.0)	24.1	4,000	170	443	87
52272	262.0(72.4)	86.6	78	60	2.2	2.7
60672	17.9(15.2)	0.2	-	14,200	51,500	160,000
60872	23.9(19.3)	2.4	-	2,800	3,200	15,700
61372	23.1(18.8)	2.2	-	3,900	8,200	11,600
62172	29.0(22.5)	4.2	-	310	1,330	2,500

[a]Based on initial moisture of 300 (75) and an equilibrium moisture content of 17 (14.5).

Conclusions

With a small number of experimental conditions investigated and without checking all commercially available equipment, there appears to be the possibility of obtaining approximate moisture contents of partially dried manure by means of electrical resistance methods. Further investigation and calibration would be required.

CHANGES IN COMPONENTS DURING DRYING, TREATMENT AND STORAGE

Changes in a biologically active system are difficult to evaluate due to the continual loss of volatile solids. A fixed reference is needed throughout the process in order to determine the absolute changes which have occurred. During a biological process the possibility of the concentration of a component referenced to total solids remaining unchanged is a reality even though losses are occurring. This can occur if the total solids and the component being investigated both are reduced in magnitude. As an example, if the TKN at the start of a process was 6.0% of the total solids and at the end of the process was 6.0%, the tendency is to conclude that no loss of nitrogen has occurred. However, if the fixed solid content was observed to be 25% of the total solids at the start and 40% at the end of the process, loss of nitrogen did occur and can be calculated.

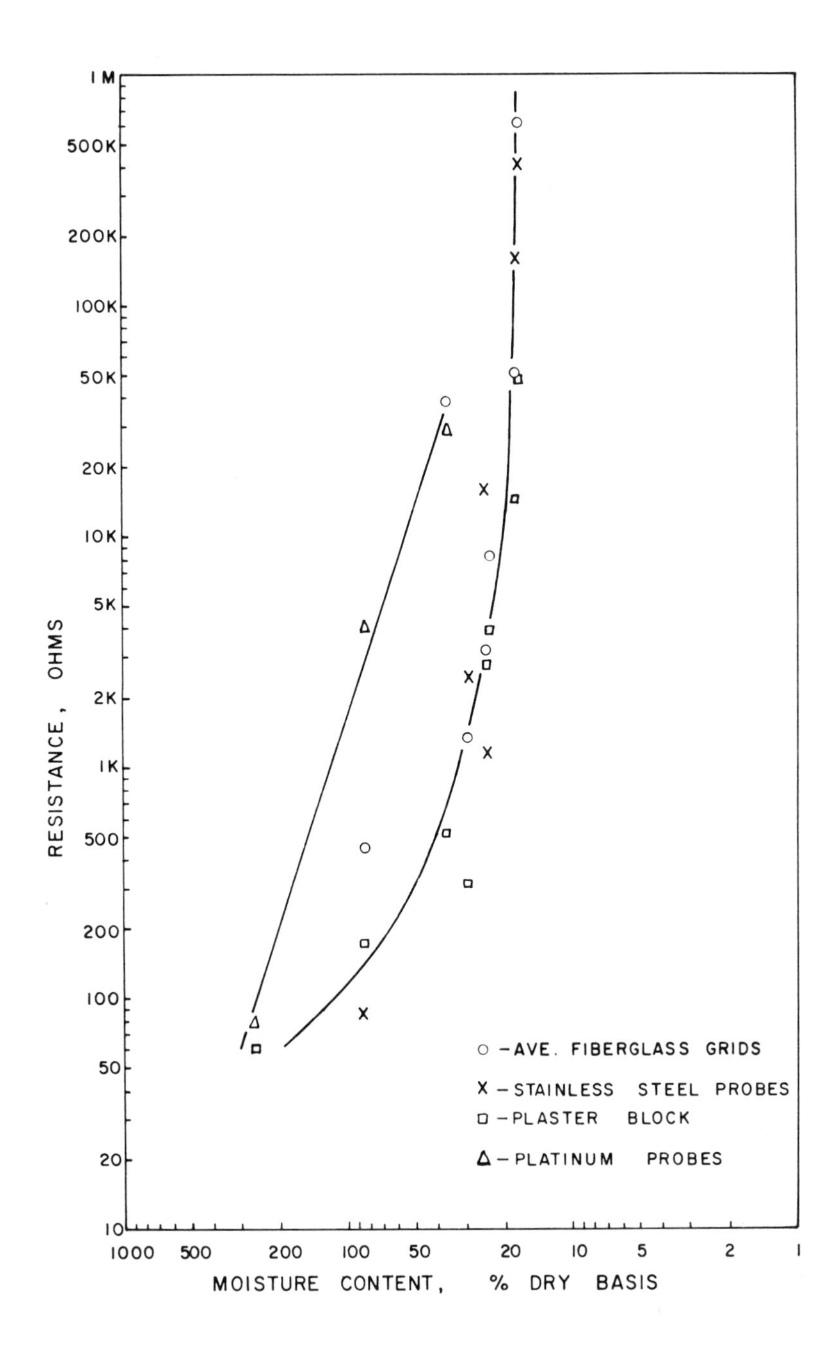

Figure 8. Calibration curve–resistance vs dry basis moisture.

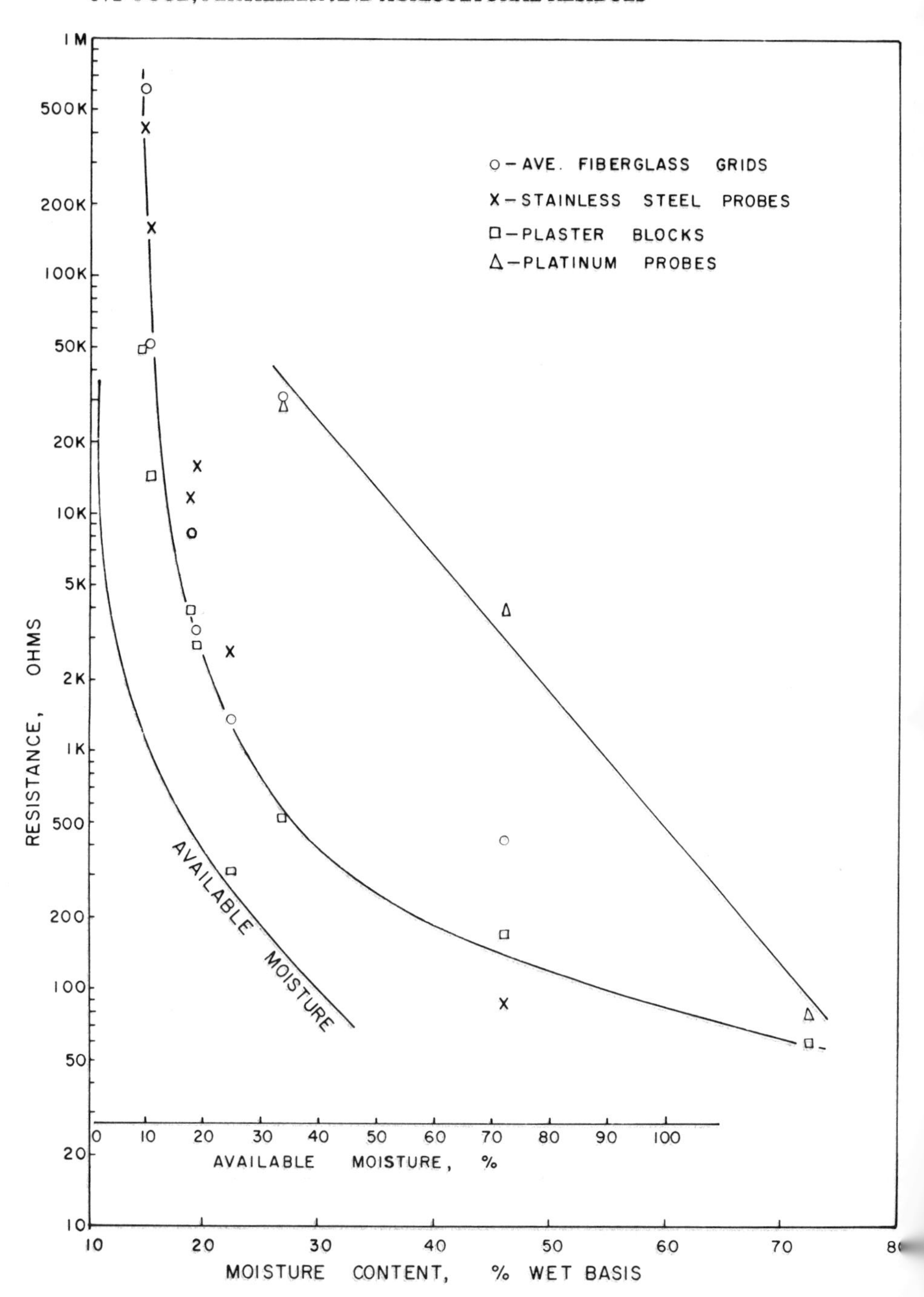

Figure 9. Calibration curve–resistance vs wet basis moisture content and available moisture.

	Nitrogen–TKN		Total Solids
	Total Solids Basis	Fixed Solids Basis	Fixed Solids Basis
Start	0.06	0.06/0.25 = 0.24	1/0.25 = 4.0
End	0.06	0.06/0.40 = 0.15	1/0.40 = 2.5
Change (%)	0.00	-37.5	-37.5

Therefore, in the example the nitrogen loss was 37.5%. This basis for calculation assumes that the mass of fixed solids remains constant throughout the process. There are some references to problems with this assumption due to changes occurring within the fixed solids. Determination of fixed solids at a temperature different from the 600°C used for the studies reported and at different times may possibly provide some insight into the problem.

Based on the assumption that the fixed solids remain constant, the loss and the percentage remaining of a component can be calculated knowing the concentrations measured and computed in the conventional manner. The relationships for these calculations are shown in Table X.

An example utilizing data from a storage study is presented in Table XI. When possible the wet weight should be obtained and the changes calculated by measurement. For some situations this is not possible, and the fixed solids basis can be utilized.

Table XII summarizes the changes for examples of various moisture removal systems. The losses of organic nitrogen are particularly of interest, as this is the long-term nitrogen available to plants. The organic-nitrogen remaining for several systems is presented in Figure 10. For the direct storage of manure after predrying, the organic nitrogen loss was approximately the same during a 100-day storage period for varying moisture contents at the start of storage. When the manure was treated and then stored, significantly greater loss occurred.

"As produced" manure is directly from the caged laying hen. This would approximately represent a system that cleaned once or twice a day. In actual practice, however, some losses would occur due to bacterial activity in the residual manure on the dropping boards. All systems are compared with "as produced" values.

Conclusions

Based on the assumption that the fixed solids remain constant throughout the movement of a waste within a manure management system, these fixed solids can be used as a tracer to calculate changes occurring to a component of the waste.

Table X. Relationships for calculations based on conventional measurements.[a]

Component	Actual Weight	Concentrations: Wet Weight Basis[b]	Concentrations: Total Solids Basis[c]	Concentrations: Fixed Solids Basis[d]	Fixed Solids Basis: Change from Time 0 to	Fixed Solids Basis: Remaining[e]
Wet Weight	W	1	$\frac{W}{WS} = \frac{1}{S}$	$\frac{W}{WSA} = \frac{1}{SA}$	$\frac{1}{A_oS_o} - \frac{1}{A_\theta S_\theta}$	$\frac{A_oS_o}{A_\theta S_\theta}$
Moisture	W(1-S)	$\frac{W(1-S)}{W} = 1-S$	$\frac{W(1-S)}{WS} = (\frac{1}{S}) -1$	$\frac{W(1-S)}{WSA} = \frac{1}{AS} - \frac{1}{A}$	$\frac{1}{A_oS_o} - \frac{1}{A_\theta S_\theta}$ $- \frac{1}{A_o} - \frac{1}{A_\theta}$	$\frac{A_oS_o(1-S_\theta)}{A_\theta S_\theta (1-S_o)}$
Total Solids	WS	$\frac{WS}{W} = S$	$\frac{WS}{WS} = 1$	$\frac{WS}{WSA} = \frac{1}{A}$	$\frac{1}{A_o} - \frac{1}{A_\theta}$	$\frac{A_o}{A_\theta}$
Volatile Solids	WS(1-A)	S(1-A)	$\frac{WS(1-A)}{WS} = 1-A$	$\frac{WS(1-A)}{WSA} = (\frac{1}{A}) -1$	$\frac{1}{A_o} - \frac{1}{A_\theta}$	$\frac{\frac{1}{A_\theta} -1}{\frac{1}{A_o} -1}$
Fixed Solids	WSA	SA	$\frac{WSA}{WS} = A$	$\frac{WSA}{WSA} = 1$	0	1
Nitrogen[b]	WSN	SN	$\frac{WSN}{WS} = N$	$\frac{WSN}{WSA} = \frac{N}{A}$	$\frac{N_o}{A_o} - \frac{N_\theta}{A_\theta}$	$\frac{N_\theta A_\theta}{N_oA_\theta}$

[a]Total solids on a wet weight basis. Fixed solids on a dry weight basis.
[b]Units/unit wet weight, dimensionless.
[c]Units/unit total solids, dimensionless.
[d]Units/unit fixed solids, dimensionless.
[e]Percent loss = (1-remaining) 100.
[f]Other components such as potassium, calcium, etc., measured on a total solids basis, would have the same relationship as nitrogen.

Table XI. Example of changes calculated by fixed solids basis compared with losses calculated by total solids basis and by measurement.[a]

	In Storage			Out Storage		
		Calc. from			Calc. from Fixed Solids Basis	
Component	Measured	Measured, kg	Units/Units FS	Measured	Measured, kg	Units/Units FS
Wet Weight, W	182.95 kg	182.95	5.13	149.14 kg	149.14	4.09
Moisture, (1-S)	0.286	52.32	1.47	0.172	25.65	0.70
Total Solids, S	0.714	130.63	3.66	0.828	123.49	3.39
Volatile Solids (1-A)	0.727	94.97	2.66	0.705	87.06	2.39
Fixed Solids, A	0.273	35.66	1.00	0.295	36.43	1.00

Percent Change During Storage	Total Solids Basis	Fixed Solids Basis	By Measurement
Wet Weight	-14.3	-20.3	-18.5
Moisture	-50.0	-52.4	-51.0
Total Solids	0	- 7.4	- 5.5
Volatile Solids	- 3.0	-10.2	- 8.3
Fixed Solids	+ 8.0	0[b]	+ 2.2

[a]Data from a manure storage study.

[b]Assumes no loss in fixed solids.

Table XII. Summary of changes for various moisture removal systems.

	Wet Basis		Moisture		Total Solids		Volatile Solids		TKN		Organic-N		NH_3-N	
	FSB[a]	△[b]	FSB	△	FSB	△	FSB	△	FSB	△	FSB	△	FSB	△
0. "As Produced"	16.67	0	12.50	0	4.17	0	3.17	0	0.313	0	0.291	0	0.022	0
1. Dry to 30%	5.88	-65	1.94	-85	3.94	- 6	2.94	- 7	0.251	-20	0.214	-26	0.037	+ 68
	5.13	-69	1.47	-88	3.66	-12	2.66	-16	0.226	-28	0.209	-28	0.017	- 23
2. Predry to 50%; Store Inside	6.30	-62	3.29	-74	3.00	-28	2.00	-37	0.232	-26	0.147	-49	0.084	+282
3. Predry to 30%; Store Inside	4.18	-75	0.72	-94	3.46	-17	2.46	-22	0.215	-31	0.183	-37	0.032	+ 45
	4.59	-72	1.18	-91	3.43	-18	2.43	-23	0.228	-27	0.172	-41	0.056	+154
4. Predry to 15%; Store Inside	4.49	-73	0.42	-97	4.08	- 2	3.08	- 3	0.221	-29	0.190	-35	0.031	+ 41
5. Predry to 30%; Treat;[c] Store Outside	6.46	-61	3.50	-72	2.96	-29	1.96	-38	0.118	-62	0.055	-81	0.063	+186

[a] FSB = fixed solids basis.
[b] △ = percent loss,- (or gain, +) from "As Produced."
[c] Treatment for this example was composting with wood shavings.

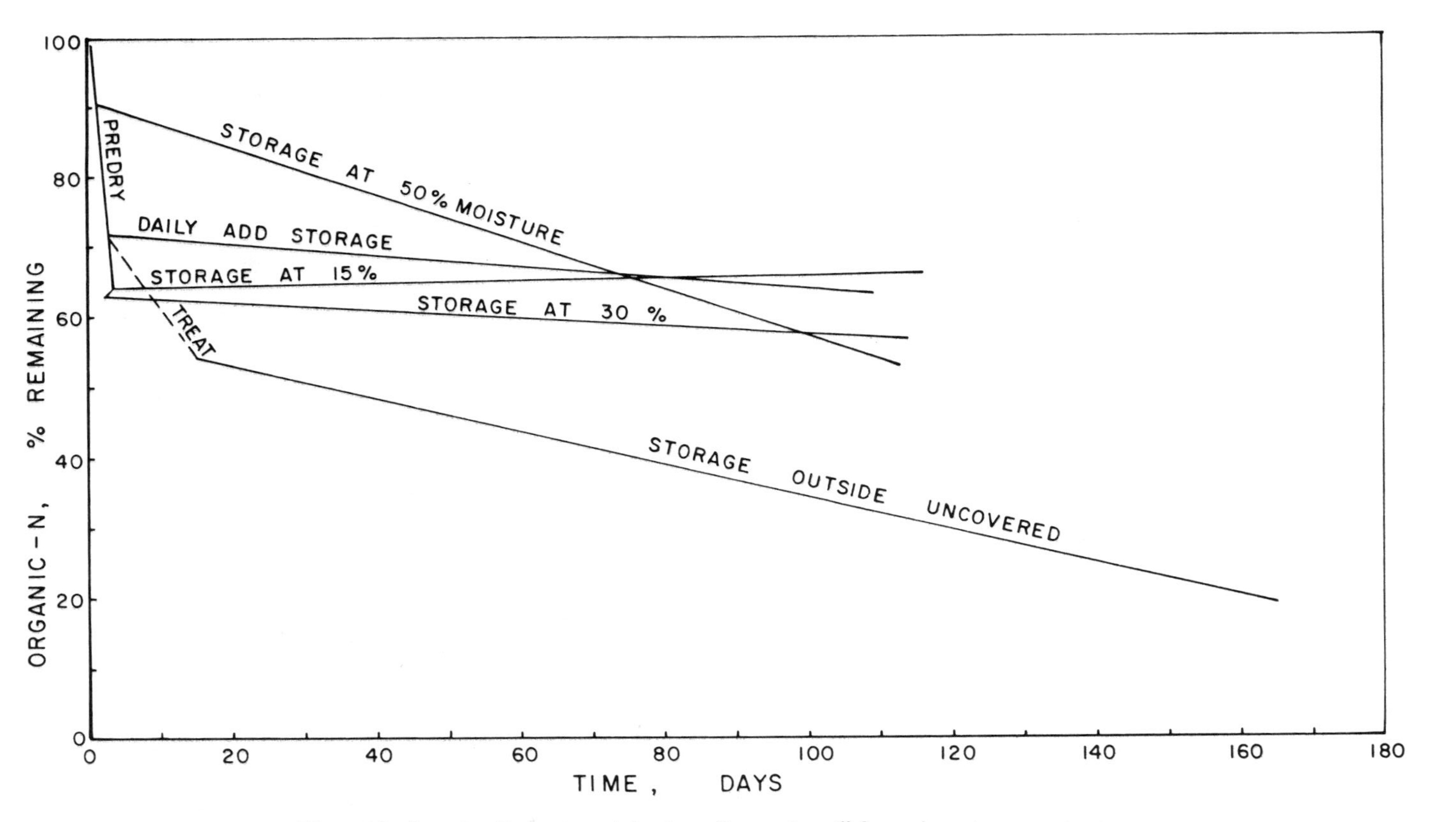

Figure 10. Organic nitrogen remaining from "as produced" for various storage systems.

SUMMARY

Undercage Drying–Mechanical Devices–No Forced Air

Research was conducted on the use of mechanical undercage devices to promote drying of droppings. A theoretical approach was directed towards determining the resulting moisture content when a percentage of the manure is partially dried and then mixed with the remaining manure. This theory was applied to data from studies with metal fins, wires, metal strips and metal ridges. In all the cases studied the final moisture content was above that required for handling manure as a solid. The lowest final moisture content contained was 62%, wet basis.

A Pilot Scale Belt Dryer

A pilot scale belt dryer using unheated air was designed and constructed. The dryer was capable of handling the manure from 120 birds with a drying time of 24 hr. The final moisture content was as low as 10%, wet basis.

Direct Measurement of Manure Moisture by Electrical Resistance

Several methods of measuring the moisture content of manure were investigated. The direct measurement of manure moisture by electrical resistance is possible.

Changes in Components During Drying, Treatment and Storage

Various manure management systems studied utilize moisture removal as a means of controlling odor, increasing ease of handling and providing storage as a part of the overall system. As the manure passes through the different phases of a system, losses and gains occur in components such as wet weight, total solids and forms of nitrogen. The assumption is made that the fixed solids do not change. This allows reference to a common base throughout the phases of the system.

A summary of the changes for various systems is detailed. The losses of organic nitrogen are particularly of interest as this is the long-term nitrogen available to plants. For the direct storage of manure after predrying, the organic nitrogen loss was approximately the same during a 100-day storage period for varying moisture contents at the start of storage. When the manure was treated and then stored, significantly greater loss occurred.

ACKNOWLEDGMENTS

The authors wish to acknowledge George Hoffman and Bradley Gormel for their assistance in conducting the various tests and in handling of data. Also, the senior author would like to thank the many people who worked on the unheated air-belt dryer, including Stuart Neagle, George Hoffman and Paul Roske.

REFERENCES

1. Sobel, A. T. "Undercage Drying of Laying Hen Manure," Proc., Cornell Waste Management Conference, February 1972.
2. Sobel, A. T. "Removal of Water from Animal Manure," Proc., Cornell Waste Management Conference, January 1969.

38

STORABILITY OF PARTIALLY DRIED LAYING HEN MANURE

D. C. Ludington and A. T. Sobel
Department of Agricultural Engineering
Cornell University
Ithaca, New York

INTRODUCTION

Removing water from poultry manure reduces weight, tends to reduce bulk and is known to reduce odor production and thus reduce the potential for air pollution. To utilize manure properly and at the same time improve environmental quality, manure must be stored perhaps for several months. How dry must poultry manure be before the manure can be successfully stored? Manure which is too dry will be dusty and difficult to handle. If the manure is too moist, bacterial and fungal growths will take place, producing odorous compounds; moreover, the manure is sticky and difficult to handle. These studies were conducted to investigate the storage of partially dried poultry manure and the changes that occur during storage. The first study was carried out in plastic bags (bag study) and the second in bins (bin study).

BAG STUDY

Procedure

To measure the storability of manure, samples were partially dried to three moisture contents (15, 30 and 50% w.b.) and stored in small plastic bags for up to 112 days. The manure used was dried with unheated air to the three moisture contents and then placed in the bags. Twenty-one bags, each containing approximately 300 g, were prepared for each moisture content. The bags were not sealed; instead, the neck of the bag was tied with a string around a capillary tube to permit an exchange of gases with the atmosphere. After the bags were filled, a number was randomly assigned to each bag. At the same time, numbers were randomly selected to identify the bags that would be analyzed on each of the designated days: 0, 7, 14, 28, 56, 84 and 112. The bags were stored in a random order on top of a cabinet where they would not be disturbed or exposed to direct sunlight. Ambient air temperature was between 65 and 75°F.

On each designated day the three bags previously identified for each moisture content were analyzed. The bags were moved to a Varian Aerograph Series 200 gas chromatograph with as little disturbance as possible. The gases within the bag were analyzed for CO_2, one of the products of microbial respiration. This was done to provide a quantitative measure of the biological activity within the manure in each bag.

Following CO_2 measurement, the bags were weighed and the content analyzed for moisture content, volatile solids, ammonia-N, organic N and odor. Samples of 14 g or more were placed in evaporating dishes, weighed and heated at 103°C overnight to determine the moisture content. After being cooled and weighed again, the samples were placed in a muffle furnace at 600°C overnight for determination of volatile solids. Determinations of the organic and ammonia nitrogen content of the manure samples were made using standard macroKjeldahl techniques (*Standard Methods,* 1965).

The manure in the bags at one moisture content was not exactly the same when the bags were filled on day zero because of the nonhomogeneity of the manure. Due to this fact, comparing the results of an analysis for a particular parameter in a particular bag on a particular day with the results from another bag on another day could have led to false conclusions. Also, during storage, biological activity in the manure consumed volatile solids which are a part of the total solids. Tracing the changes in a parameter with time when the base values are changing leads to erroneous conclusions. This same difficulty arises when comparing biological processes which are progressing at different rates. The approach chosen to eliminate this problem was to use a fixed-solids base for all analysis.

Results

Figure 1 shows the changes in moisture content (based on wet weight) for each of the three basic moisture contents during the 112-day storage. Each value plotted is the average of the three values—one value for each of the bags analyzed. The 15% moisture manure appeared to lose water slowly during the entire storage period. The 30% manure showed an initial moisture increase after one week. Since water is one product of respiration, this increase could be the result of biological activity. After the first week, the

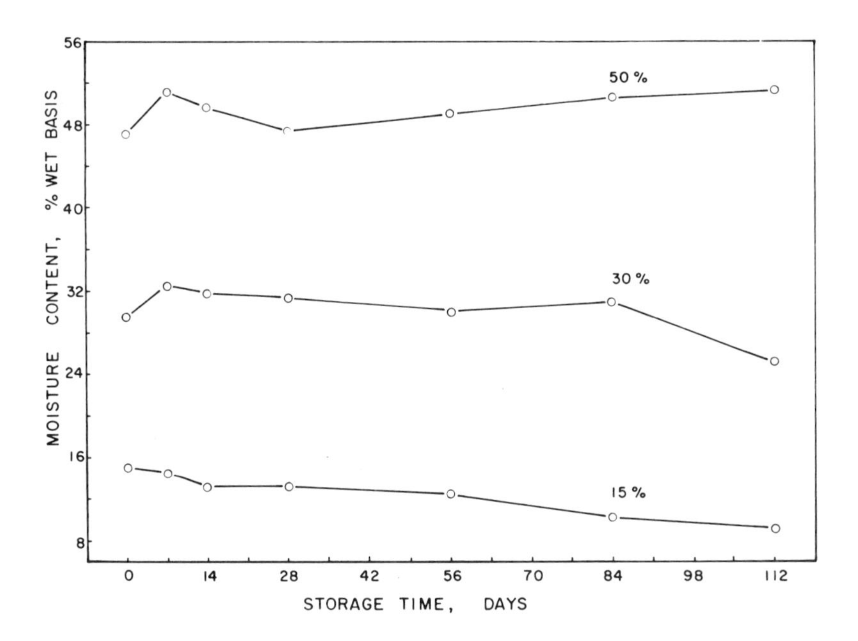

Figure 1. Moisture content, wet basis, vs storage time.

bags in general gradually lost water for the remainder of the study. A slight gain occurred at the end of 12 weeks, but it was followed by a decline near the end of storage. The 50% manure also showed an increase for one week followed by a decline, and then a slight increase occurred during the next three months. This measured increase may have been due to the water produced by biological action in combination with a loss in volatile solids.

Figure 2 represents the changes in moisture contents for the manure at the three levels of moisture but based on fixed solids. This method of expressing concentrations shows the greatest variation from those based on wet weight (Figure 1) at the high moisture contents. This is to be expected because the greatest biological activity occurs here, resulting in greater losses of total solids and thus causing the greatest deviation in the concentration values which are based on total solids.

The concentration of CO_2 within each bag was measured to provide an indication of the level of biological activity. Figure 3 shows the results of the CO_2 analysis. The rapid increase in CO_2 concentration during the first few weeks in the 30 and 50% moisture manure indicates a high rate of microbial activity which produces water. This adds validity to the explanation for the increase in moisture content during the initial stage of storage. The level of CO_2 in the 15% moisture bags increased slightly and then decreased to nearly zero, indicating little biological activity. The level of CO_2 maintained in the

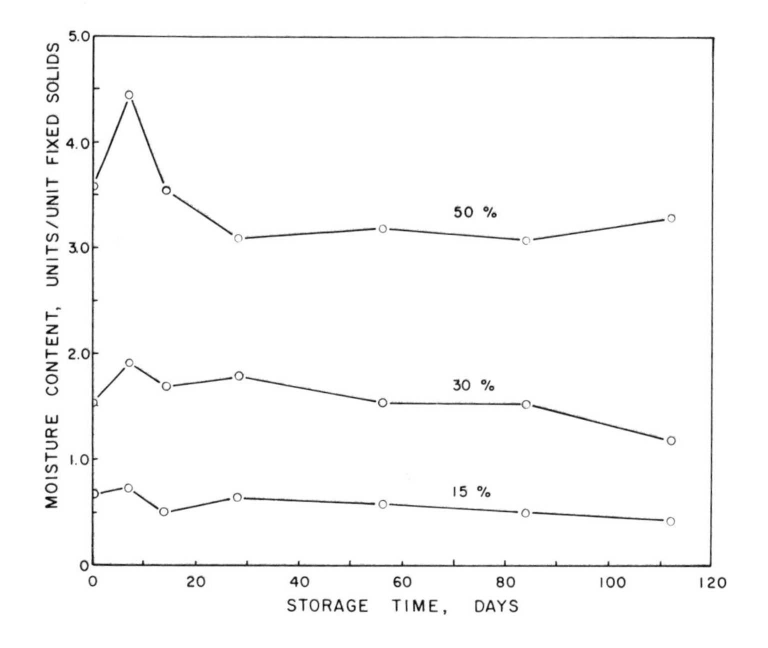

Figure 2. Moisture content, fixed solids basis, vs storage time.

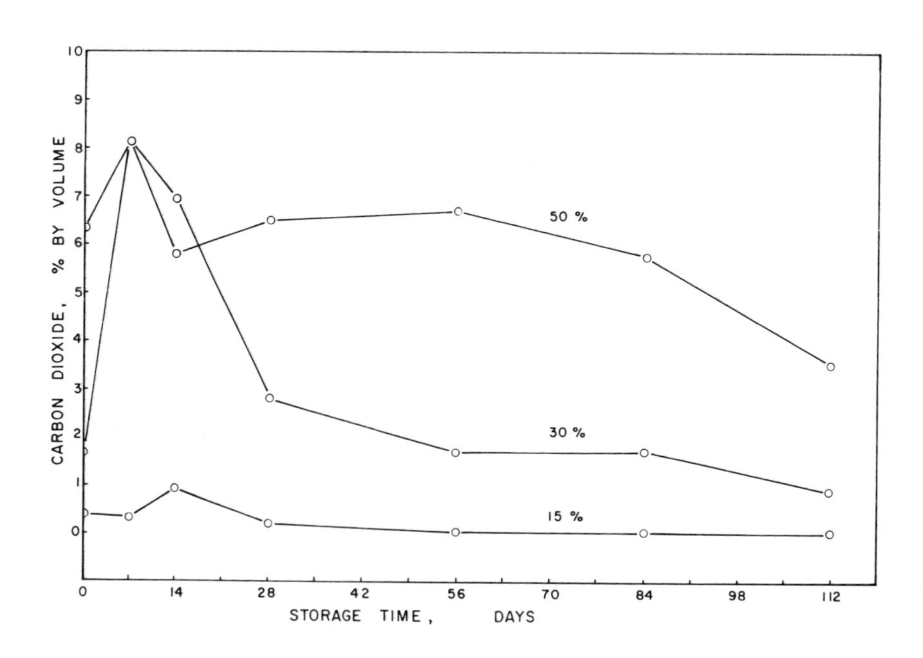

Figure 3. Carbon dioxide concentration in bags vs storage time.

30 and 50% moisture bag is a good indicator of the activity because the drop in CO_2 concentration in the 30% bags shows that a fairly rapid exchange of CO_2 with the outside atmosphere was occurring. For the 50% bags to maintain a high level of CO_2 required a high biological activity. This conclusion will be further validated during the discussion of losses in volatile solids.

Changes in the concentration of volatile solids (VS) with time are shown in Figure 4. The linear regression lines are included. The VS in the 15% manure fluctuated with time for no known reason. The regression showed a slight increase in concentration with time which is highly improbable. In 50 and 30% manure there was a general decline in VS, with a greater slope for the 50% manure. For the 50% and, to a lesser degree, for the 30%, an increase in VS was measured in day 7. This phenomenon has been measured in other studies but no explanation is known.

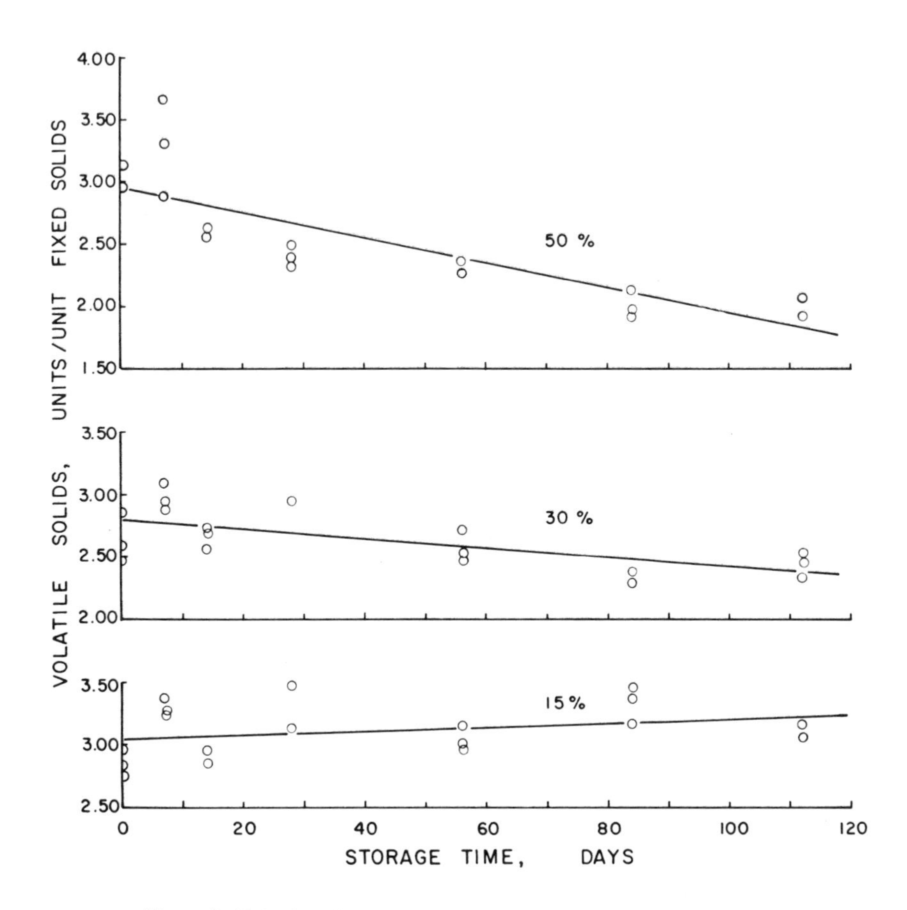

Figure 4. Volatile solids concentration, fixed solids, vs storage time.

Figure 5 shows the concentration of the three forms of nitrogen found in the manure. The slopes for these regression lines were taken from Table I. The TKN in the manure at the end of 100 days of storage was nearly the same concentration for all three moistures, even though the initial concentrations (θ = 0) were considerably further apart. The reason for the variation in TKN at θ = 0 is due to losses of TKN during the predrying process required to attain the three different moisture levels. Based on these data, the

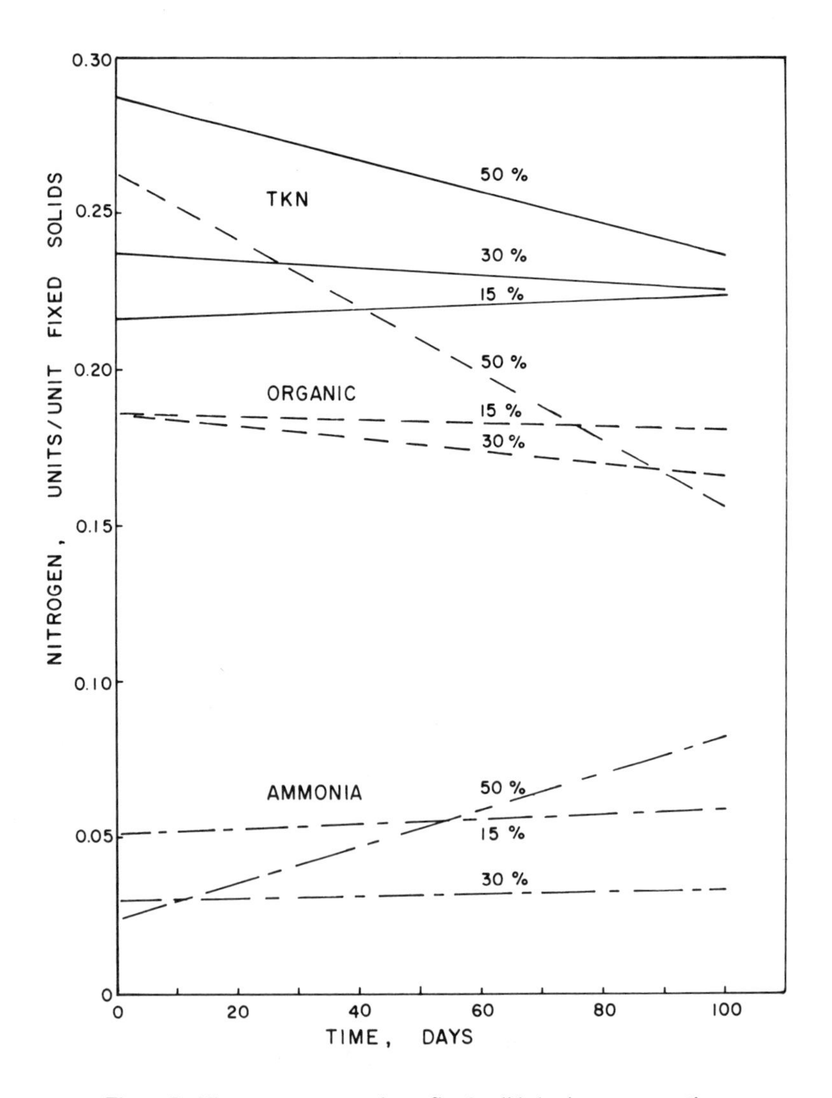

Figure 5. Nitrogen concentrations, fixed solids basis, vs storage time.

moisture content at time of storage has little influence on the TKN content of the manure at the end of 100-day storage. These data should not be extrapolated beyond the 100 days shown.

An increase in ammonia was measured at all three moisture levels with a significant increase in the 50% manure. Again, this moist manure supported greater microbial activity and thus higher rates of ammonification. Nitrogen for the ammonia came from the organic-N component. The percentage changes and absolute changes for organic-N are shown in Table I.

Figure 6 shows the percentage remaining for four components (wet weight, VS, TKN and organic-N) vs storage time at the three different moisture contents. The slope of the straight lines are from the linear regressions of actual data collected based on fixed solids. The intercepts were all placed at 100% remaining at $\theta = 0$ because the changes with storage time are being investigated.

If conserving organic matter (VS), TKN and organic-N are important, then manure should be stored as dry as possible. The higher levels of moisture promote higher losses. The losses in TKN and organic-N were similar in magnitude for the 15 and 30% moistures. At the 50% level, the losses of organic-N were more than twice the TKN losses for the same period of time. This indicates that the rate of ammonification increased sharply between 30 and 50% moisture, and that the ammonia did not volatilize but remained in the 50% moisture bag and elevated the TKN.

Table I presents the changes that occurred during 100 days of storage. These values are the slopes of the linear regression lines computed from the data.

Table I. Change in eight parameters during 100-day storage.

	15		30		50	
Parameters	**%[a]**	**Abs.[b]**	**%[a]**	**Abs.[b]**	**%[a]**	**Abs.[b]**
Wet Weight	-2.1	0.088	-15.0	-0.79	-23.	1.7
Moisture	-35	0.24	-27.	-0.42	-20.	0.7
Total Solids	+4.0	0.15	-10.	-0.37	-26.	1.0
Volatile Solids	+5.0	0.15	-15.	-0.37	-35.	1.0
Fixed Solids	0.0	0.0	0.0	0.0	0.0	0.0
Ammonia-N	+10	0.0029	+18.	+0.008	+30.2	+0.058
Organic-N	+1.8	0.0048	-11.	-0.020	-37.	-0.11
TKN	+3.8	0.0077	-5.5	0.012	-16.	-0.051

[a]Slope of linear regression of % remaining x 100.
[b]Slope of linear regression of parameter concentrations (units/unit fixed solids x 100).

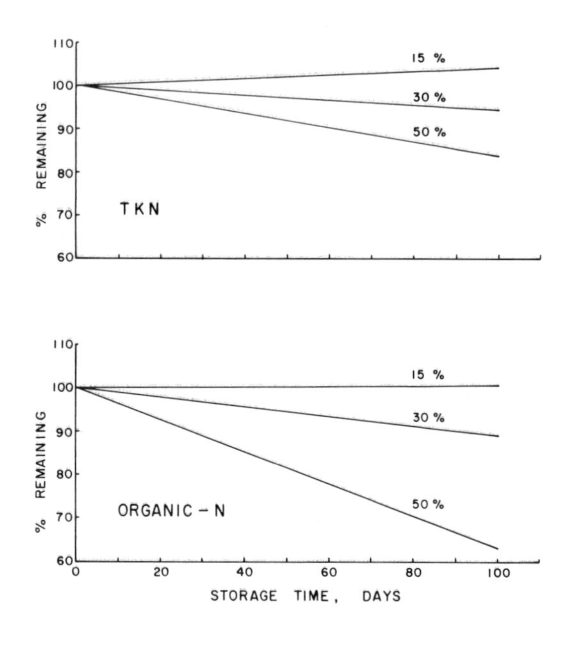

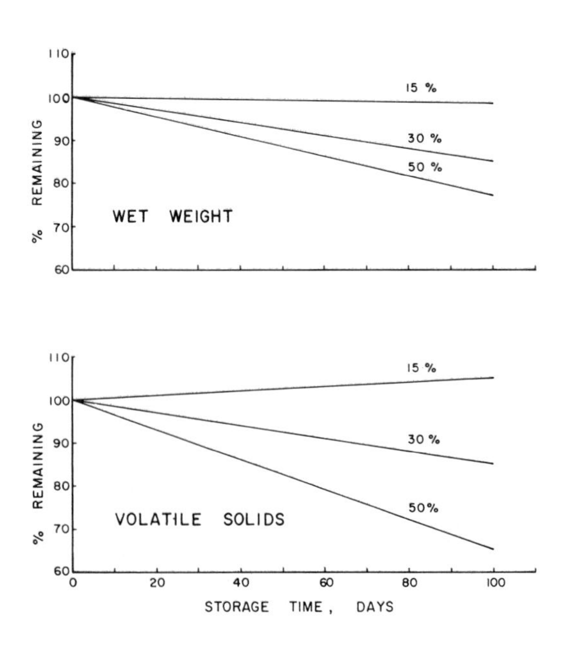

Figure 6. Percent remaining of wet weight, volatile solids, TKN and organic nitrogen vs storage time.

The wet weight decreased with time for all three moisture contents. The 50% manure showed the highest loss of wet weight, 23% for 100 days. Change of wet weight is a combination of the change in moisture and total solids (volatile solids). Percentage change in moisture in the 15% manure was higher than for the 50% manure, but this does not mean that the absolute losses were highest. The 50% manure lost 0.7 units of moisture per unit of fixed solids each day, while the 15% bags lost 0.24. This higher loss is reasonable because the higher absolute humidity maintained above the 50% manure would exchange a greater mass of water vapor with the outside atmosphere for each unit of gas exchanged.

Total and volatile solids, organic-N and TKN showed a slight increase with time in the 15% bags. This is highly improbable and thus must be considered an indication of the magnitude of the error for the analysis. Values for these same components for the 30 and 50% bags all show a negative percentage change and a decline in concentration based on fixed solids.

Table II shows the composite opinion of at least three different people on the offensiveness of the bagged manure. The 15% bags were never offensive but tended to be slightly dusty. For the first few weeks of storage, the 30% bags were offensive in odor. After two weeks, there was little offensive odor except for ammonia, which was strong throughout. After two weeks, no odor was detectable from the outside of the bags. The 30% bags were of

Table II. Results of odor panel tests, bag study.

	Description of Odor by Panel % Moisture		
Time, Days	15	30	50
0	dry, dusty, feed not offensive	NH_3, bad stinks	glue, sweet sickish
7	not offensive slight, NH_3	sharp offensive	sharp, strong moldy
14	almost odor free	mildly offensive NH_3 detected	sharp, moldy
28	faint offensive odor	soggy, moldy NH_3	strong mold manure, bad
56	dusty-dry	NH_3, something else offensive	moldy sickening
84	almost none	strong NH_3	moldy, NH_3 sickening
112	no chemical odor	strong NH_3	moldy, NH_3 sickening

good consistency for handling throughout the storage period. At the beginning of the storage period, the odor was described as sickish sweet, with ammonia and mold smell increasing as time proceeded. The growth of mold increased until at the end of two months over 90% of the surface area of the manure was covered. Throughout the period this manure remained sticky and difficult to handle.

BIN STUDY

Procedure

To simulate the conditions in a large pile, two plywood bins (45.7 cm x 45.7 cm) were made in 30.5-cm sections so that sections could be added as the depth of manure increased. To reduce heat loss from the bins in an attempt to simulate conditions in a large pile, 5 cm of insulation were cemented to the outside of the bins. The bottom of each bin was closed to prevent convection currents within the bins. Poultry manure with a moisture content of about 29% was added to the bins at two rates: approximately 2.5 cm (1 in.) per day for Bin 1 and 10 cm (4 in.) per day for Bin 4. (Bin number corresponds to increment added in inches.) Manure was added 5 days per week for 4 weeks. Thermocouples were placed in the bins at 30 cm intervals to monitor temperature.

The moisture content of the manure being added and its bulk density were measured as the manure was added. The moisture content of the manure at the surface of the pile and the depth of the manure were measured prior to adding more manure. Odors emanating from the bins were evaluated. A continuous record of temperatures was kept. When the manure was removed from the bins (70 days after the last addition of manure for Bin 1 and 72 days for Bin 4), the depth, weight and moisture and nitrogen content were measured.

Tables III and IV present the data regarding the manure added to the two bins.

Results

At the conclusion of loading, Bin 1 was 51.3 cm deep and Bin 4 was 195 cm deep. Bin 1 should have been 50.8 cm deep if no settling had occurred. In Bin 4, the depth should have been 203 cm if the correct amount were added and no settling occurred.

Odors during loading are noted in Table IV. For Bin 1 the odors were never more than "very faint." In contrast to this, Bin 4 had odors ranging from "steamy, musty, definite ammonia" to "definite-peculiar manure." The major reason for this difference may be the temperature which developed in

Table III. Partially dried manure added to bins.

Parameter (units)	Bin 1	Bin 4
Manure added, (cm/day)	2.5	10.0
(cm^3/day)	5,250	21,000
(loading days)	20	20
Total wet weight, [g(m)][a]	47,640	182,950
Moisture content, [%(m)]	28.6	28.6
Water, [g(c)]	13,620	52,820
Total solids [g(c)]	34,020	130,630
Fixed solids (ash), [%(m)]	27.3	27.3
Fixed solids (ash), [g(c)]	9,290	35,660
Volatile solids [g(c)]	24,730	94,970

[a](m) = measured; (c) = computed.

the two bins. Bin 1 (Figure 7) had a maximum temperature of 106°F and was above 100° for less than 2 days. Bin 4 (Figure 8) had temperatures above 100°F for nearly 50 days and a maximum temperature of about 142°F. The higher temperature in Bin 4 would increase the volatilization of odorous compounds and water plus a possible increase in the production of the odorous compounds. The bin temperatures were, in general, responsive to ambient temperatures with little lag time. This would indicate moderately rapid heat transfer or high thermal conductivity for the manure-bin system.

The manure at the surface of Bin 4 dried less than that in Bin 1 during loading (Table IV). In Bin 4 the surface moisture content decreased only 2.1 percentage points, while in Bin 1 a decrease of 4.9 percentage points was measured. Data presented in Table VIII show that the actual water loss from the surface of both bins during loading was approximately the same (3050 g vs 2770 g). Because there was more water to be lost from Bin 4 than from Bin 1, the expected decline in percentage of moisture would be less.

In order to estimate the water lost while loading the bins and that lost during storage, the moisture content at the end of loading was computed. For Bin 1, each 2.5-cm layer was assumed to have dried to a uniform moisture content between loadings, and this resulting moisture content was the same as that measured at the surface prior to loading the new layer of manure. Based on this assumption, the average moisture content of Bin 1 at the end of loading was 23.7%. In Bin 4, the moisture content of the 10-cm layer was assumed to be the average between the moisture content of the manure when added (assuming no drying at the bottom of the new layer) and the moisture content at the surface just prior to adding new manure. Table V presents the results of measurements and computations made regarding the manure at the end of the loading period (day 28). Further assumptions were made:

Table IV. Data taken while loading the bins.

Day of Test	Loading No.	Moisture Content "in"	Bin 1			Bin 4		
			Moisture Content % (surface)	Depth[a] (cm)	Observed Odors	Moisture Content % (surface)	Depth[a] (cm)	Observed Odors
3	3	23.8	14.9	8.6	Very Faint	20.4	32.5	
6	4	16.2	20.2	–	Very Faint	20.5	–	Detectable
8	6	27.5	28.3	17.8	Very Faint	26.0	69.1	
10	8	33.5	20.9	19.0	Very Faint	30.5	71.4	
13	9	26.9	23.1	20.6	Very Faint	25.3	79.2	Steamy
15	11	28.4	26.6	27.2	Very Faint	28.9	99.8	Musty
17	13	29.0	31.4	33.0	Very Faint	27.8	119	NH_3
20	14	35.4	26.4	34.0	Very Faint	33.3	128	Moldy
22	16	31.2	26.9	39.9	Very Faint	26.9	149	Very Faint
24	18	34.2	17.9	44.5	Very Faint	24.9	166	Definite Manure
28	20	–	–	49.3	Very Faint	–	186	Definite Manure
29	–	–	–	51.3	Very Faint		195	Peculiar
Average	–	28.6	23.7	–	–	26.5		

[a]Depth taken before adding new manure.

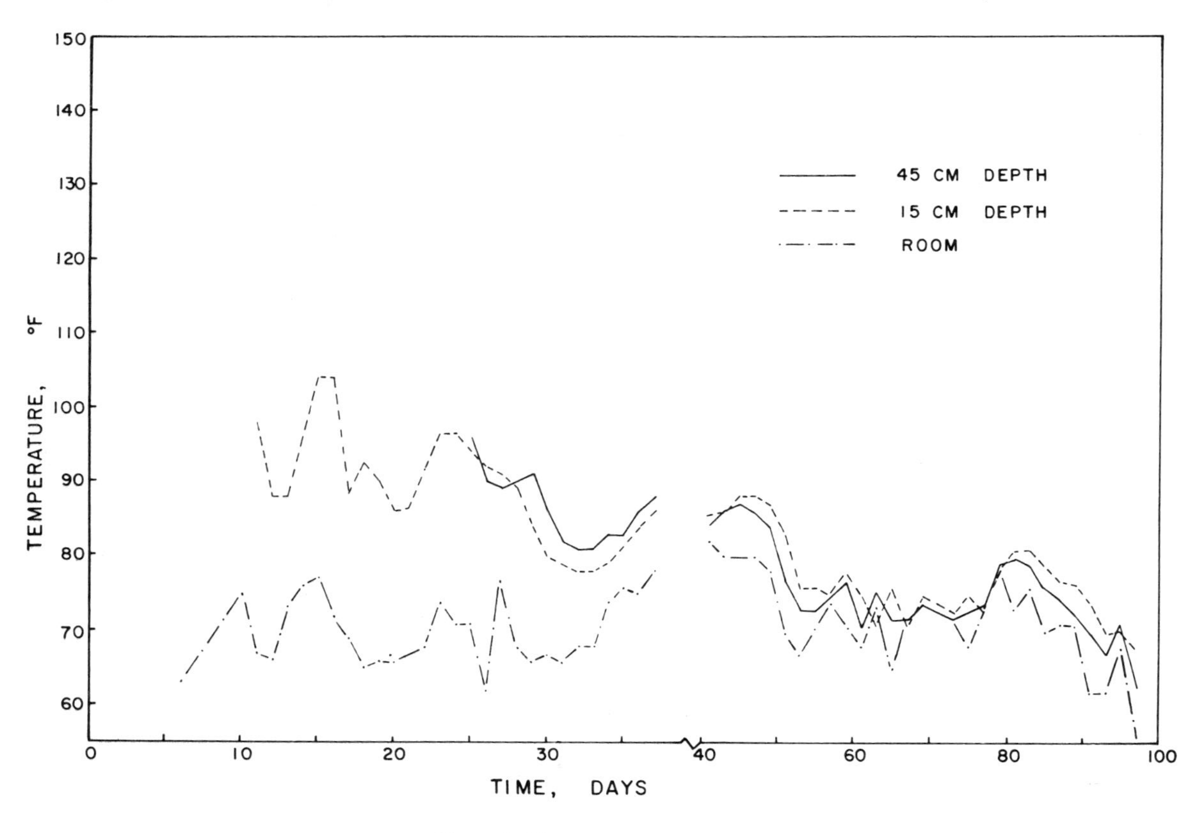

Figure 7. Temperatures within Bin 1.

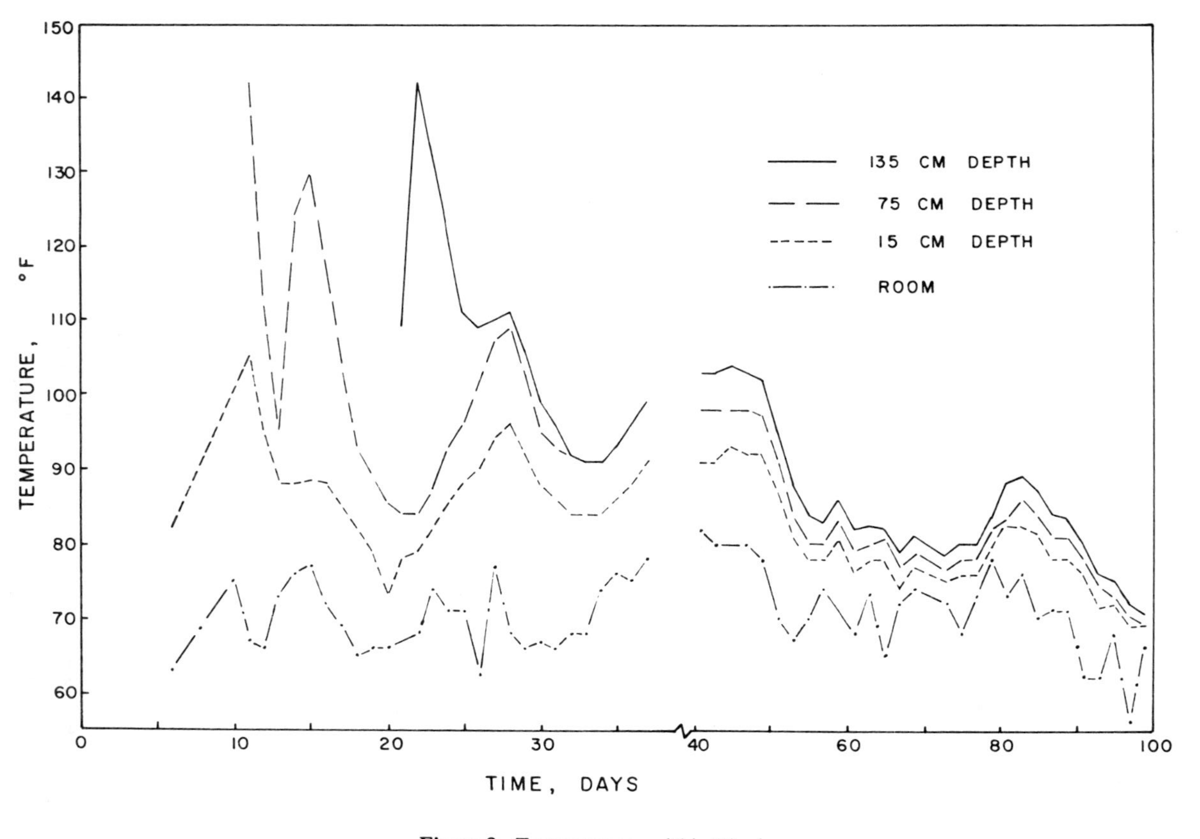

Figure 8. Temperatures within Bin 4.

that no moisture was lost from a layer after that layer had been covered with the succeeding layer until loading was complete, and that no volatile solids were destroyed by biological activity during loading. These are gross and simplifying assumptions which were made in order to compute the moisture loss from the "pile" during loading and the 70-day storage period.

Table V. Condition of poultry manure at end of loading.

Parameters, Units	Bin 1	Bin 4
Depth of manure, cm (m)[a]	51.3	195
Time, days	28	28
Wet weight, g (c) [a]	44,590	180,180
Moisture content, % (m)[b]	23.7	27.5
Water, g (c)	10,570	49,550
Total solids, g (n)[a]	34,020	130,630
Fixed solids (ash), % (n)	27.3	27.3
Fixed solids (ash), g (n)	9,290	25,660
Volatile solids, g (n)	24,730	94,970

[a](m) = measured ; (c) = computed; (n) = no change assumed.
[b]See text for rationale for computing these values.

After 97 days the bins were emptied. As they were being dismantled, the weight, moisture content, nitrogen content and odors were measured. Table VI summarizes the conditions at the end of the storage period. The only assumption made in preparing this table was that the weight of fixed solids did not change during the test.

Table VI. Condition of poultry manure at end of storage.

Parameters, Units	Bin 1	Bin 4
Time, (days)	97	97
Depth, [cm (m)[a]]	48.5	185
Settling, [cm/day (m)]	0.067	0.075
Wet weight, [g (m)]	37,680	149,140
Moisture content, [% (m)]	16.6	17.2
Water, [g (c)][a]	6,250	25,650
Total solids, [g (c)]	31,430	123,490
Fixed solids (ash), [% (m)]	28.9	29.5
Fixed solids (ash), [g (n)][a]	9,290	35,690
Fixed solids (ash), [% (c)]	29.6	28.9
Volatile solids, [g (c)]	22,140	87,800

[a](m) = measured; (c) = computed; (n) = no change assumed.

The bulk density of the manure was measured as the manure was added to the bins and as the stored manure was emptied from the bins. Table VII shows the changes in density during storage. The reduction in bulk density can be explained by the loss of water from the manure.

Table VII. Bulk densities of partially dried poultry manure.

Bin 1		Bin 4	
Bulk Density	**(g/cm^3)**	**Bulk Density**	**(g/cm^3)**
Added to bin (avg.)	0.45	Added to bin (avg.)	0.43
Removed from bin section		Removed from bin section	
Bottom - 7.5 cm	0.38	Bottom - 30 cm	0.39
7.5 - 17.5	0.31	30 - 60	0.38
17.5 - 27.5	0.36	60 - 90	0.38
27.5 - 37.5	0.44	90 - 120	0.38
37.5 - 47.5 (top)	0.37	120 - 150	0.38
		150 - 160 (top)	0.39

The loss of water and volatile solids during loading and storage can be seen in Table VIII. Bin 1 lost about 10% more water during loading than Bin 4 because the loading rate was lower. Bin 4 lost nearly 90% of the total water loss during storage because of the prolonged self heating within the manure. At the end of 90 days, both bins had lost over 50% of the original water. Bin 4 experienced a smaller percentage loss in VS than Bin 1, but nearly three times the VS were biologically consumed in Bin 4.

Table VIII. Loss of water and volatile solids during loading and storage

	Bin 1		Bin 4	
Parameter	**(g)**	**(%)**	**(g)**	**(%)**
Water lost during loading	3,050	22	2,770	5
Water lost during storage	4,320	32	23,900	46
Total water lost	7,370	54	26,670	51
Volatile solids loss	2,590	10	7,170	7.6

As the bins were emptied at the completion of the storage period, the odors released by the manure were recorded. The word "sharper" was used to describe the odor near the middle of Bin 1. Ammonia was the dominant odor for Bin 4. In no case was the odor putrid or characteristic of anaerobic conditions. The manure in both bins was granular with no indication of caking. Thus, the handling characteristics were not altered by storing partially dried manure for 90 days. The ammonia and organic nitrogen were measured as the manure was added to the bins and when removed following storage. Table IX presents the results of these measurements.

Table IX. Ammonia and organic nitrogen.

	Manure Added % of Dry Solids			Manure Removed % of Dry Solids	
Depth (cm)	Organic-N	NH_3-N	Depth (cm)	Organic-N	NH_3-N
Bin 1					
47	6.0	0.29	48	5.3	0.74
36	4.1	0.51	30	5.6	0.92
21	5.4	0.60	20	4.2	0.92
12	7.2	0.46	10	5.3	0.90
			0	5.2	0.83
Average	5.7	0.46		5.1	0.86
Bin 4					
180	6.0	0.29	185	2.1	0.54
132	4.1	0.51	167	2.9	1.2
83	5.4	0.60	137	6.1	1.0
48	7.2	0.46	107	5.6	0.90
			76	7.1	1.1
			46	7.1	0.72
			15	6.0	1.3
Average	5.7	0.46		5.3	0.96

Table X presents a mass balance for the two bins from loading to unloading. Change in dry solids and nitrogen are given. The loss in organic-N is most important because the change signifies a potential loss of nitrogen for crop production. The assumption is made that little of the nitrogen in the ammonia form will be available for crop utilization due to rapid volatilization of the ammonia. An 18% decline in organic-N was measured during loading and storage in Bin 1, but only a 12% decline was seen for Bin 4. The relationship between the two bins with regard to organic-N is nearly the same as that for the loss of volatile solids.

If the same quantity of manure were stored under the two conditions tested (2.5 cm/day and 10 cm/day), the manure stored at 2.5 cm/day would lose 33% more volatile solids and 50% more organic-N, based on the results of this research. There apparently was more biological activity per unit of

Table X. Quantitative analysis of organic and ammonia nitrogen.[a]

	Bin 1		Bin 4	
Parameter	(g)	(%)	(g)	(%)
Material Added				
Dry Solids	34,020	-	130,630	-
TKN	2,110	6.2	8,100	6.2
Organic-N	1,940	5.7	7,450	5.7
Ammonia	160	0.46	600	0.46
Material Removed				
Dry Solids	31,430	-	123,490	-
TKN	1,850	5.9	7,660	6.2
Organic-N	1,600	5.1	6,540	5.3
Ammonia	270	0.86	1,190	0.86
Loss of TKN	260	12	440	5
Loss of Organic-N	340	18	910	12
Gain of Ammonia	114	73	585	97

[a]Numbers in a given category may not total due to rounding errors.

manure in the 2.5 cm/day storage. The 2.5-cm layer may be more aerobic due to thinner layers, therefore encouraging greater activity and biodegradation of organic matter.

CONCLUSIONS

Rapid piling of manure in thick layers was simulated in the plastic bags. To maximize conservation of organic nitrogen and maintain odor control, the manure should be dried to 15%. This requires more time and energy than drying to a higher moisture content. A more moist manure can be stored with a slight increase in organic-N losses and more odor. The moisture content should not be greater than 30% for generally successful storage. Conservation of nitrogen, odors and handleability were all termed acceptable at 30% moisture.

Poultry manure which has been partially dried to about 30% moisture can be stored without any adverse effects when the piles are built in layers (10 cm was the maximum layer thickness tested). No putrid odors developed, and the material remained granular. About 50% of the water in the manure was lost during loading and storage. Manure stored in piles built in thin layers (2.5 cm) will suffer greater losses of organic matter and organic nitrogen.

MANURE RESIDUE AS A SUBSTRATE FOR PROTEIN PRODUCTION VIA *HERMETIA ILLUCENS* LARVAE

C. V. Booram, Jr.
Department of Agricultural Engineering

G. L. Newton and O. M. Hale
Animal Science Department
Georgia Coastal Plain Experiment Station
Tifton, Georgia

R. W. Barker
Department of Entomology
Oklahoma State University
Stillwater, Oklahoma

BACKGROUND

Hermetia illucens, better known as the soldier fly, exists as some 1500 different species in the United States. The larvae are aquatic and have been found growing in animal manures, carrion, nests of rodents, wet feed, decayed fruit, vegetables and garbage. Environmentally, the characteristic home of the soldier fly in the southeastern United States is wet poultry droppings in caged layer units.

The objectives of this study were: (1) to evaluate *H. illucens* larvae as a possible nutrient source for swine rations; (2) to characterize the larvae meal chemically; (3) to consider the differences in composition of resulting feces when soybean meal or *H. illucens* larvae are used as a dietary source of protein; and (4) to consider amounts of waste material utilized by *H. illucens.*

LITERATURE

The search for alternate sources of protein has found scientists looking at many possibilities. Fly pupae is one possibility that has been explored by several scientists. Teotia and Miller[1] reported that housefly pupae have a protein content of 61% and amino acid characteristics comparable to those of meat and bone meal or fish meal.

Termites are also among the insects that have been used extensively as human food. Dried termites contain 36 to 46% protein and 36 to 44% fat. Dried locusts have a crude protein content ranging from 51 to 75% and a fat content ranging from 4 to 18%.[2]

The soldier fly is of the suborder Brachycera that resembles a wasp in appearance. Tingle *et al.*[3] reported on research on the life cycle of *H. illucens,* including mating activity, pupation and emergence. They reared *H. illucens* from eggs collected from poultry droppings as well as from eggs laid by females in the greenhouse. They concluded that rearing of the soldier fly is possible and that the immature stages may be maintained for stocking purposes by a reduction in the available food supply or by holding at low temperatures.

MATERIALS AND METHODS

During the summer of 1974, *H. illucens* larvae were harvested from anaerobic manure storage pits under a totally slotted floor at the Beef Cattle Research Center, Georgia Coastal Plain Experiment Station, Tifton, Georgia. The larvae were actively growing in the surface layer of manure in the storage pit and were also found crawling up the walls of the pit.

Larvae were harvested three different times from the pits by dipping the larvae out of the tank with a dipping net. The material that was harvested contained a mixture of soldier fly larvae, manure and hair. The larvae and hair were separated by placing the mixture in a reciprocating separator that allowed the larvae to be collected and the hair removed. The separated larvae were then washed with potable water to remove traces of fecal material and dried in a forced-air oven for 24 hr at 85°C. Hair remaining with the larvae after drying was removed using air separation. The clean, dried larvae were ground using a hammer mill, after which an antioxidant was added to retard deterioration and the meal was frozen until it was utilized experimentally.

Two 20% protein diets were prepared using *H. illucens* larvae as the protein source for one diet and soybean meal for the second diet (see Newton *et al.*[4] for details of the digestion trial). The two prepared diets were fed to weanling pigs.

Approximate analysis of the larvae was conducted using AOAC Methods, and a spectrographic analysis was conducted on the larvae ration, the soybean meal ration, the larvae ration feces, the soybean meal ration feces and the *H. illucens* larvae meal. To evaluate the effect of *H. illucens* larvae on the manure substrate, fresh beef cattle feces and urine were mixed to form a slurry with a moisture content of about 76%. A 2270-g sample of this slurry was placed in each of six plastic pans. Two pans each were inoculated with 100 and 300 g of washed *H. illucens* larvae, with the remaining two serving as controls. The pans were loosely covered and not disturbed until 14 days later, when they were again weighed and sampled. Nitrogen content and dry matter were determined initially and after 14 days.

DISCUSSION AND RESULTS

The larvae found growing on the substrate of anaerobic beef waste were about 2.5 cm long and light brown in color. The coloration was probably due to the darkness of the substrate. The dried *H. illucens* larvae contained 42% crude protein and 35% crude fat. A detailed analysis of the amino acid content can be found in Hale[5] and Newton *et al.*[4] Analysis indicated that the protein quality is similar to that of meat scraps and soybean meal. Thus, based on the characteristics of the protein, it would seem to be an acceptable protein replacement. Refeeding of the larvae to weanling pigs indicated that although the difference in apparent nitrogen digestibility was small, it was significant and was in favor of the soybean meal diet. Table I illustrates some mean values of different nutrients found in fecal material when *H. illucens* larvae and soybean meal are used as the protein source.

Table I. Spectrographic analysis of diets.

Element	Larvae Diet	Feces	Soy Diet	Feces
Phosphorus (%)	0.92	3.25	1.07	3.16
Calcium (%)	2.71	7.90	0.92	2.58[b]
Magnesium (%)	0.18	0.75	0.14	0.58[b]
Potassium (%)	0.49	0.39	0.90	0.73
Sodium (ppm)	2217.0	1886.2	2212.00	3878.3
Manganese (ppm)	144.0	503.3	71.00	258.3[b]
Copper (ppm)	7.0	22.8	7.0	30.2
Boron (ppm)	2.0	0	12.00	2.2
Zinc (ppm)	217.0	a	168.00	a
Iron (ppm)	622.0	4168.3	435.00	4648.3
Strontium (ppm)	21.0	103.0	12.00	53.8[b]
Barium (ppm)	15.0	50.0	11.00	33.0[b]

[a] Zinc samples were contaminated during the experiment by the cages used for the weanling pigs.

[b] Treatments statistically different at the $\alpha = 0.05$ level.

A spectrographic analysis of the *H. illucens* larvae meal (ground larvae) was made and is given in Table II. The calcium content of the larvae meal is high but is characteristic of the *H. illucens* larvae. The outer part of the *H. illucens* larvae secretes a heavy deposit of calcium carbonate. This deposit forms on all parts of the body and is thick and dense.[6] An examination of the larvae used in this study revealed that they were encased in a dense crust.

Table II. Mineral content of *Hermetia illucens* larvae (moisture free basis) and proximate[a] analysis.

Element	Concentration
P (%)	1.51
K (%)	0.69
Ca (%)	5.00
Mg (%)	0.39
Mn (ppm)	246
Fe (ppm)	1370
B (ppm)	0
Cu (ppm)	6
Zn (ppm)	108
Al (ppm)	97
Sr (ppm)	53
Ba (ppm)	33
Na (ppm)	1325
Crude Protein (N x 6.25) (%)	42.1
Ether Extract (%)	34.8
Crude Fiber (%)	7.0
Ash (%)	14.6
Nitrogen Free Extract (%)	1.4
Moisture (%)	7.0

[a]Dry matter basis except moisture.

The results of inoculating manure slurry with soldier fly larvae are shown in Table III. The pans not inoculated with soldier fly larvae maintained a solid crust on the surface of the manure. The pans containing 100 g of soldier fly larvae maintained a slight liquid appearance with a moderate amount of larvae activity. The 300-g larvae treatment was completely fluid and the larval activity was intense.

The control treatment contained numerous housefly larvae, while the two treatments seeded with soldier fly larvae contained few housefly larvae. All of the treatments lost 20 to 22% of the dry matter that was put into the system. This was lower than expected since prior work by Hale[5] indicated that 35 to 36% of the dry matter was lost when *H. illucens* naturally infested

Table III. Summary of soldier fly larvae manure digestion trial.

Characteristic	Control[a]	L 100[b]	L 300[c]	I Manure[d]	Hale[5]
Moisture Content	75.5	75.1	75.7	76.2	70.1
Dry Matter Content (g)	426.4	430.3	422.8	540.0	-
N in Manure (g)	13.8	13.2	13.8	18.9	-
N Lost (%)	27.0	30.7	27.0	-	-
pH	6.51	6.74	6.95	6.30	-
Dry Matter Lost (%)	21.0	20.3	21.7	-	35.5

[a]0 g of larvae added.
[b]100 g of larvae added.
[c]300 g of larvae added.
[d]Slurry used for trial.

swine manure over a 14-day period. The difference is partially explained by the fact that this study was conducted in late fall and temperatures were not optimal.

The ability to produce *H. illucens* eggs in captivity and rear the larvae on a manure substrate is critical if the protein in the larvae is to be utilized. If the *H. illucens* can reproduce under artificial conditions, then it should be possible to utilize *H. illucens* as a mechanism to treat livestock wastes and to produce a source of protein with extremely low energy inputs.

The larvae are large enough so that they can be separated with little difficulty and will dry to a storable state within 24 hours. Currently laboratory work is under way to investigate the possibility of mass rearing of *H. illucens.* With this basic information, an environmental system will be developed that will allow the waste to be seeded with *H. illucens* eggs and growth to a harvestable size. The benefits are twofold: production of protein and treatment of livestock wastes.

REFERENCES

1. Teotia, J. S. and B. F. Miller. "Nutritive Content of House Fly Pupae and Manure Residue," *Brit. Poultry Sci.* 15:177-182 (1974).
2. De Foliart, G. R. "Insects as a Source of Protein," *Bull. Entomol. Soc. Am.* 21:161-163 (1975).
3. Tingle, F. C., E. R. Mitchell and W. W. Copeland. "The Soldier Fly *Hermetia illucens* in Poultry Houses in North Central Florida," *J. Georgia Entomol. Soc.* 10:179-183 (1975).
4. Newton, G. L., C. V. Booram, R. W. Barker and O. M. Hale. "Dried *Hermetia illucens* Larvae as a Supplement for Swine," *J. Animal Sci.* 44(3):395-400 (1977).

5. Hale, O. M. "Dried *Hermetia illucens* Larvae (Diptera: Stratiomyidae) as a Feed Additive for Poultry," *J. Georgia Entomol. Soc.* 8(1):16-20 (1973).
6. Johannsen, O. A. "Stratiomyidae Larvae and Puparia of the North Eastern States," *J. New York. Entomol. Soc.* 30(4):141-153 (1922).

LAND APPLICATION OF SWINE WASTE RESIDUE FOR INTEGRATED CROP PRODUCTION

C. V. Booram, Jr.
Department of Agricultural Engineering
Georgia Coastal Plain Experiment Station
Tifton, Georgia

INTRODUCTION

Livestock waste management systems produce residues, and some of these residues must ultimately be applied to the soil whether the manure is treated by an anaerobic lagoon or energy is produced by anaerobic digestion.

Observation of the management habits of livestock producers indicates that nutrients present in animal wastes are not being effectively utilized because such utilization adds additional labor expense to the already busy management schedule. If all inputs and outputs from a swine-production system are to be used efficiently, it will be necessary to utilize the nutrients produced via the production of swine.

The objective of this research was to evaluate the ability of agricultural land to serve as a receptor for the application of anaerobic swine residues when the land is to be used for feed-grain production.

LITERATURE

Land application of wastes began long before the complex technology of today's treatment systems was developed. Livestock waste differs from municipal sludge only in the quantity of heavy metals contained in the waste. A review of some of the existing literature will provide a valid basis for comparison of livestock wastes and municipal wastes.

Hinesly *et al.*[1] found that furrow irrigation of anaerobically digested sludge significantly increased corn yields during the 1968 to 1972 growing seasons. During the 4-yr period, the amounts shown in Table I were applied. Data were gathered on macroelements and microelements in the corn ear leaf at tasseling and in the grain taken when harvested. In the ear leaf at tasseling, N, P, Ca, Mn, Zn, Cd and B increased significantly and Mg decreased with

Table I. Total plant macronutrients, total plant essential micronutrients, minor elements and additional total minor elements (lb/ac).[1]

Total Plant Macronutrients		Total Plant Essential Micronutrients		Minor Elements[a]		Total Minor Elements[b]	
Total-N	13,140	Fe	13,140	Na	540	Pb	410
NH_4-N	5,770	Zn	1,850	Cr	1,150	Hg	0.14
P	7,540	Cu	480	Co	1.1	Cd	130
K	1,140	Mn	160	Se	1.4	Sn	16.4
Ca	8,920	Mo	0.4	Ni	116		
Mg	2,580	B	16				
S	1,010	Cl	2,020				

[a]Essential for animals but not for plants.
[b]Not considered essential for either plants or animals.

sludge treatment. Grain produced on the same plants showed fewer significant differences with only K, Zn and Cd increases with sludge treatment. A major concern with sludge applications is the impact of continuous sludge disposal in terms of the possible toxic effects that trace metals may have on the environment. Dean[2] reported that crop yields were depressed when sewage sludge containing zinc was applied to sewage cropland in France.

There are many examples of metal toxicity in agriculture.[3] Toxic amounts of Cu, Zn and Ni have accumulated in soils from fungicides, unneeded fertilizers and sewage sludge. Most toxicities have occurred under intensive agricultural practices, such as in orchards, vineyards or vegetable fields. Correction is often quite expensive.

Spotswood and Raymer[4] discussed some aspects of sludge disposal on agricultural land. The limiting factor in the disposal of sewage sludge, even when it is from domestic sources, is the toxic metal accumulation in the soil. Heavy metals, once added to the soil, will remain there almost indefinitely.

Metal toxicity from application of sludge and effluent has been observed in England,[5] though generally it has occurred in unmanaged situations most favorable for toxicity: low pH, sandy soils and sludges with high metal contents.

Toxic effects of zinc have often been reported. Previous evidence[6] indicated that liming reduced the harmful effects of zinc on crops. Analysis of crops grown in soils contaminated with zinc shows that the element is readily taken up and translocated in the plant. In general, it is not possible to diagnose heavy metal toxicity from the observation of the symptoms.

Many other cases could be documented, but because of the concentrations of Zn, Cu and Ni in livestock wastes it seems highly unlikely that toxicities will develop unless feeding practices change drastically.

SITE DESCRIPTION AND PROCEDURES

The application site for the excess lagoon liquid encompasses about six acres of poorly drained Clarion-Webster complex soil with a maximum slope of 4%. A part of the 6-ac site has 12 plots measuring 40 ft by 60 ft that were tile drained so that drainage from each plot could be sampled during and after application of swine wastes.

The lagoon serving the confinement building has been described previously by Willrich,[7] Smith[8,9] and Koelliker *et al.*[10] The lagoon was constructed during the fall of 1962 and was put into operation in 1963. Willrich, Smith and Koelliker *et al.*[7-10] describe many of the changes that have been made in the system since 1962. The lagoon has a volume of 100,000 ft^3, and excess liquid in the lagoon has been applied to adjoining land by a sprinkler irrigation system. The waste-handling system uses hydraulic transport techniques that recycle anaerobically-treated lagoon effluent to transport the wastes out of the confinement-finishing unit.

During the study, the seedbed was prepared by plowing, disking and harrowing before planting. After seedbed preparation, the plots were planted with Pioneer 3388 corn seed at a rate of 30,000 plants/ac, and Atrazine was applied to the plots for preemergence weed control.

Soil samples were collected preceding the 1972 season, in October 1972 and in April 1973 after the second irrigation season. These samples were analyzed for nitrogen, carbon, pH, electrical conductivity and exchangeable potassium. Electrical conductivity determinations were made as described by Bower and Wilcox[11] and nitrogen determinations as described by Chapman and Pratt;[12] pH was determined by mixing 50 ml of dry soil with an equal volume of distilled water.

Water samples were collected during the growing season from the anaerobic lagoon on a weekly basis during 1972 and 1973. Water quality measurements were made according to *Standard Methods.*[13] Porous cups also were installed in one block of the irrigated plots at depths of 6, 12 and 24 in. Contents of the cups were analyzed for nitrate-N, ammonia-N and total phosphorus. Finally, effluent from the tile drains located about 4 ft below the soil surface was sampled.

Treatment 1. Control treatment.
Treatment 2. Apply 2 in. at soil tension of 400 mb.
Treatment 3. Apply 1 in. of lagoon effluent per week.
Treatment 4. Apply 2 in. of lagoon effluent per week.

WATER-QUALITY CHARACTERISTICS

Water quality was measured at three different stages during the disposal path: (1) the water quality of the lagoon liquid; (2) the water quality of liquid from porous-cup samples; and (3) the water quality of liquid from tile drains located in each plot. Table II provides some average values of the water-quality parameters during 1972 and 1973.

Table II. Average values of water quality data, Unit K, anaerobic lagoon–1972 and 1973 growing season.

Parameter	1972	1973
pH	7.3	-
Conductivity (millimhos/cm)	5.0	-
Kjeldahl-N (mg/l)	462	581
Ammonia-N (mg/l)	434	307
Chloride (mg/l)	343	254
Chemical Oxygen Demand (mg/l)	1177	1839
Total Phosphorus (mg/l)	69	79

Analysis of the water for selected samples for N, P, K, Mg, Ca, Na, Fe, Mn, B, Cu, Zn, Pb and Ni is illustrated in Table III. These values, placed on a lb/ac basis, illustrate the quantity of material applied from swine wastes (Table IV). The net result is that, under normal rates of nutrient application, the quantity of metals applied, such as Zn, Cu, Pb and Ni, is very small.

The porous-cup method of sampling soil water provides an index of soil waste-treatment effects with depth. Table V summarizes some data for 1972 and 1973. Nitrate nitrogen is mobile, and this is illustrated by the relatively high values at greater depths in the soil profile. Table VI summarizes the removal of nitrogen and phosphorus (on the basis of concentration values) for the porous cups.

Theoretically, the use of a tile drain to monitor water quality is ideal. On a practical basis, however, it is difficult to put on an input-output basis without continuous measurement of flow and sampling. In this case, instrumentation was not available. Discussions of treatment effectiveness are based on concentration differences.

Table III. Elemental analysis (ppm)–Unit K lagoon, Swine Nutrition Station, Iowa State University, Ames, Iowa.

	1971[14]		1972		1973		
	Aug. 16	Oct. 5	June 22	Aug. 28	June 27	Aug. 28	Average
Aluminum	10	1.2	-	-	-	-	5.6
Boron	0.43	0.63	1.0	0.55	0.81	0.78	0.70
Calcium	53	103	59	43	79	-	67.4
Chromium	-	-	3.1	-	0.09	0.12	1.1
Copper	0.16	0.24	1.7	0.04	0.05	0.08	0.38
Iron	1.8	4.2	4.6	0.81	5.2	6.82	3.91
Potassium	303	347	-	256	275	369.0	310
Magnesium	73	70	52	16.3	19.6	58.0	48.2
Manganese	0.35	0.37	1.6	0.39	0.5	0.27	0.58
Molybdenum	0.02	0.02	-	-	-	-	0.02
Sodium	150	203	104	112	18	142	121.5
Nickel	-	-	0.8	0.02	0.018	0.023	0.22
Lead	-	-	2.9	0.03	0.08	0.11	0.78
Zinc	0.22	0.67	0.8	0.10	2.25	3.40	1.24

Table IV. Nutrients applied to plots by treatment, 1972 and 1973.

	Application Rate (lb/ac) Treatment					
Nutrient	2 (1972)	2 (1973)	3 (1972)	3 (1973)	4 (1972)	4 (1973)
N	361	541	1104	1518	2208	3247
P	57	70	175	195	351	418
K	166	299	506	840	1013	1796
Mg	28.4	36.1	86.6	101.2	173	216.5
Ca	42.7	73.4	130	205.9	261	440.3
Na	90.4	74.2	276	208.2	552	445.3
Fe	2.30	5.66	7.02	15.9	14.10	33.43
Mn	0.83	0.36	2.55	1.00	5.11	2.14
B	0.63	0.74	1.92	2.07	3.83	4.43
Cu	0.73	0.06	2.22	0.16	4.44	0.34
Zn	0.39	2.62	1.20	7.36	2.40	15.74
Pb	1.22	0.09	3.73	0.25	7.46	0.53
Ni	0.38	0.11	1.18	0.32	2.35	0.68

Table V. Summary of total phosphorus, ammonia nitrogen and nitrate plus nitrate-nitrogen concentrations in porous cup samples from 6-, 12- and 24- in. soil depths–1972, 1973.

	Treatment					
Parameter	**2**	**(2)**	**3**	**(3)**	**4**	**(4)**
6-in. depth						
Total P	-	(6.8)	6.0	(8.1)	9.5	(11.2)
Nitrate + nitrite N	-	(9.8)	74.3	(92.9)	43.5	(121.1)
Ammonia N	-	(5.6)	0	(6.0)	0	(10.2)
12-in. depth						
Total P	-	(6.2)	3.2	(6.9)	2.4	(9.0)
Nitrate + nitrite N	-	(2.8)	-	(74.5)	-	(57.6)
Ammonia N	-	(0)	0	(57.6)	0	(1.8)
24-in. depth						
Total P	-	(6.8)	2.0	(6.3)	2.3	(3.6)
Nitrate + nitrite N	-	(16.8)	80.0	(27.7)	81.0	(29.2)
Ammonia N	-	(5.6)	0	(2.1)	0	(1.0)

Table VI. Percentage removal of phosphorus and nitrogen in the surface 6 in. of soil.

	1972	**1973**
Element	**% Removed**	**% Removed**
Nitrogen		
Treatment 2	-	97
Treatment 3	84	80
Treatment 4	91	78
Phosphorus		
Treatment 2	-	91
Treatment 3	94	89
Treatment 4	87	86

Actually, the percentages of removal in undisturbed soil profiles probably are higher than those reported. This is because tile-drainage systems cause the soil treatment system to be more artificial and permit direct channeling of water into the tile-drainage system. Thus, values reported in this chapter are lower than those that could possibly be expected.

Tables VII and VIII summarize the mean concentrations of the drainage for several parameters for the years 1972 and 1973. Tables IX and X illustrate the percentage removal of nitrogen, phosphorus and COD on a concentration basis.

Table VII. Mean concentrations of parameters measured in tile drainage (1972).

Treatment	Conductivity (μmhos/cm)	pH	Kjeldahl -N (mg/l)	NH_3-N (mg/l)	NO_3+ NO_2-N (mg/l)	Cl (mg/l)	COD (mg/l)	Total P (mg/l)
1	1739	7.1	0	0	0.85	230.7	23.3	0.69
2	1845	7.3	7.3	4.2	39.3	223.3	50.9	2.90
3	1703	7.1	6.6	4.1	62.8	247.9	60.0	3.13
4	1864	7.0	12.1	6.5	65.3	227.3	156.5	5.00

Table VIII. Mean concentrations of parameters measured in tile drainage (1973).

Treatment	pH	Kjeldahl -N (mg/l)	NH_3-N (mg/l)	NO_3+ NO_2-N (mg/l)	COD (mg/l)	Total P (mg/l)
1	-	-	-	-	-	-
2	7.7	5.0	0	11.5	647.5	23.0
3	7.2	23.4	4.7	38.9	319.7	12.1
4	7.2	51.4	26.4	48.0	711.7	23.7

Table IX. Percent removal of nitrogen, COD and total phosphorus through 4 ft of soil profile (1972).

	Treatment			
Parameter	1	2	3	4
Nitrogen	-	91	86	85
Total Phosphorus	-	96	95	93
COD	-	96	95	87

Table X. Percent removal of nitrogen, COD and total phosphorus through 4 ft of soil profile (1973).

	Treatment			
Parameter	1	2	3	4
Nitrogen	-	97	89	83
Total Phosphorus	-	71	85	70
COD	-	65	83	61

PLANT TISSUE

To check for undesirable increases in nutrient concentrations or nutrient imbalances caused by the application of lagoon effluent, tissue samples were taken at maturity from both the leaves and grain of the planted corn crop.

Leaf samples opposite and just below the primary ear leaf were collected at random from each plot. They were then wiped free of salt and soil, rinsed twice in distilled water, dried at 65°C, and ground with a Thomas-Wiley mill.

The grain samples were shelled, dried, ground and prepared for analysis. Both the leaf and grain tissue were analyzed at the Ohio Agricultural Research and Development Center's Spectrographic Laboratory for P, K, Ca, Mg, Na, Mn, Fe, B, Cu, Zn, Al, Sr, Ba and Mo. Table XI illustrates the treatment means for each of the treatments and those means that were significantly different from the control treatment during 1972. Concentrations of P, Na, Fe, Zn, Al and Mg showed differences from the control treatment means. During 1973, P, Mg, N, Na, Mn, Al, Sr and Zn showed differences for the control treatment means (Table XII).

Table XI. Effect of anaerobic lagoon effluent upon the leaf tissue of corn[a]–1972 (tile-drained plots).

Element	Treatment			
	1	2	3	4
Phosphorus (%)	0.51	0.61	0.67[b]	0.71[b]
Magnesium (%)	0.33	0.29[b]	0.24[b]	0.26[b]
Sodium (%)	0.03	0.07[b]	0.11[b]	0.11[b]
Iron (ppm)	166	187	210[b]	251[b]
Zinc (ppm)	31	36	45	60[b]
Aluminum (ppm)	148	192	218[b]	270[b]

[a]Treatment means for corn leaves opposite and below the primary ear shoot at physiological maturity of corn grain.
[b]Significant at 0.10 level.

Statistical analysis of the grain-tissue data revealed significant differences for potassium and magnesium for the 1972 growing season (Table XIII). Both decreased with increasing amounts of swine wastes applied. The differences were not statistically significant during 1973, although the same trend was indicated.

At the time this research began, very little information had been published on the content of nutrients in a growing crop when high rates of swine wastes were used as the nutrient source.

Table XII. Effect of anaerobic lagoon effluent upon the leaf tissue of corn[a]–1973 (tile-drained plots).

	Treatment			
Element	**1**	**2**	**3**	**4**
Phosphorus (%)	0.50	0.68[b]	0.68[b]	0.72[b]
Magnesium (%)	0.31	0.26[b]	0.20[b]	0.26[b]
Nitrogen (%)	2.72	2.97	3.16[b]	3.48[b]
Sodium (%)	0.023	0.090[b]	0.103[b]	0.107[b]
Manganese (ppm)	44.0	44.0	50.3	83.3[b]
Aluminum (ppm)	64.7	101.0[b]	102.3[b]	125.7[b]
Strontium (ppm)	26.0	26.0	25.0	28.3[b]
Zinc (ppm)	29.7	34.7	36.0	51.7[b]

[a]Treatment means for corn leaves opposite and below the primary ear shoot at physiological maturity of corn plant.
[b]Significant at 0.10 level.

Table XIII. Effect of anaerobic lagoon effluent upon corn grain tissue[a]–1972 (tile-drailed plots).

	Treatment			
Element	**1**	**2**	**3**	**4**
Potassium (ppm)	0.68	0.62[b]	0.61[b]	0.60[b]
Magnesium (ppm)	0.13	0.10	0.10	0.06[b]

[a]Treatment means for corn grain at physiological maturity.
[b]Significant at the 0.10 level.

The results of this research indicate that anaerobic lagoon liquid did not significantly affect the yield during 1973 (Table XIV), although there was a decrease in yield with increasing amounts of lagoon effluent applied. The 1973 data indicated that the yield depression was statistically significant (Table XV).

Even though high levels of nitrogen were applied, the concentrations of metals were very low as compared with municipal wastes. The long-term effects of applying swine wastes to a site would seem minimal under current swine nutritional concepts. It also seems unlikely that large quantities of heavy metals will be applied to the soil. The 1973 data indicated that the yield depression was statistically significant, but the leaves were severely burned when the anaerobic lagoon liquid was applied to the plots. This burning probably was due to a combination of the high ammonia concentration and high ambient air temperature. Plots receiving the maximum amount

Table XIV. Corn yield, plant population and inches of lagoon effluent applied to tile-drained plots–1972.

Treatment	Yield (bu/ac)	Plant Population (plants/ac)	Lagoon Effluent Applied (in.)
1	148	20,959	0
2	150	21,983	3.7
3	140	21,418	11.3
4	137	18,800	22.6

Table XV. Corn yield, plant population and inches of lagoon effluent applied to tile-drained plots–1973.

Treatment	Yield (bu/ac)	Plant Population (plants/ac)	Lagoon Effluent Applied (in.)
1	143	16,790	0
2	109[a]	16,509	4.1
3	124	15,758	11.5
4	90[a]	15,238	24.6

[a]Significant at 0.10 level.

of anaerobic lagoon effluent matured much earlier than the check plots. Tassels dropped off, and the corn plants were shorter than in the check plots. Characteristically, the ears were quite short and stubby on the plots receiving the maximum amount of anaerobic lagoon liquid.

SOIL ANALYSIS

Soil samples were collected in June 1972, April 1973 and October 1973. During the initial sampling there were not any statistical differences among any of the plots for the measures of pH, conductivity, carbon and nitrogen (Table XVI).

Analysis of samples collected in April 1973 after the first growing season revealed that the soil conductivity for the 12- to 24-in. depth was the only parameter that showed any statistically significant changes according to the analysis-of-variance test (Table XVII).

Soil samples collected and analyzed after the second growing season indicated that total nitrogen (0 to 6 in.), conductivity (0 to 6 in., 6 to 12 in. and 12 to 24 in.) and exchangeable potassium (0 to 6 in.) were statistically different from the control treatments (Table XVIII).

Table XVI. Means of pH, conductivity (μmhos/cm), carbon (%) and nitrogen (%)–June 1972 (tile-drained plots).

	Plot Depth (in.)	1	2	3	4
pH	0- 6	6.2	6.2	6.4	6.4
	6-12	6.2	6.4	6.6	6.3
	12-24	6.4	6.4	6.6	6.6
Conductivity (μmhos/cm)	0- 6	320	362	390	374
	6-12	340	321	369	283
	12-24	318	388	394	348
Carbon (%)	0- 6	4.19	4.07	4.16	3.87
	6-12	2.86	2.87	3.09	2.97
	12-24	1.32	1.20	1.78	1.29
Nitrogen (%)	0- 6	0.27	0.29	0.29	0.24
	6-12	0.21	0.20	0.23	0.18
	12-24	0.08	0.09	0.10	0.09

Table XVII. Means of pH, conductivity (μmhos/cm), carbon (%), nitrogen (%) and exchangeable potassium (ppm)–April 1973 (tile-drained plots).

	Plot Depth (in.)	1	2	3	4
pH	0- 6	6.3	6.4	6.5	6.3
	6-12	5.9	5.9	6.3	6.0
	12-24	6.2	6.3	6.2	6.2
Conductivity (μmhos/cm)	0- 6	165	181	204	202
	6-12	145	158	188	199
	12-24	118	151	206[a]	235[a]
Carbon (%)	0- 6	4.08	3.85	4.55	4.14
	6-12	3.88	3.64	3.78	3.21
	12-24	1.93	2.52	1.84	2.32
Nitrogen (%)	0- 6	0.31	0.29	0.35	0.30
	6-12	0.25	0.25	0.26	0.25
	12-24	0.14	0.13	0.12	0.12
Exchangeable Potassium (ppm)	0- 6	282	297	350	385
	6-12	198	277	237	316
	12-24	89	67	37	103

[a]Significant at 0.10 level.

Table XVIII. Means of pH, conductivity (μmhos/cm), carbon (%), nitrogen (%) and exchangeable potassium (ppm)–October 1973 (tile-drained plots).

	Plot Depth (in.)	1	2	3	4
pH	0- 6	6.4	6.1	6.2	6.2
	6-12	6.2	5.9	6.3	6.3
	12-24	6.3	6.1	6.4	6.4
Conductivity (μmhos/cm)	0- 6	144	332	833[a]	863[a]
	6-12	130	252	595[a]	651[a]
	12-24	95	188	460[a]	473[a]
Carbon (%)	0- 6	3.89	3.63	4.69	3.77
	6-12	3.73	3.45	4.28	3.11
	12-24	1.98	1.66	2.16	1.79
Nitrogen (%)	0- 6	0.24	0.28	0.33[a]	0.25
	6-12	0.23	0.22	0.21	0.20
	12-24	0.12	0.09	0.12	0.09
Exchangeable Potassium (ppm)	0- 6	348	476	639[a]	610[a]
	6-12	291	361	392	523
	12-24	105	82	95	240

[a]Significant at 0.10 level.

If the conductivity test is used as an index of salt accumulation, the data collected in June 1972 and April 1973 indicate that the rainfall in Iowa is adequate to flush accumulation of salt through the soil profile. The conductivity values for October 1973 indicate that the wastes applied during the growing season had not been flushed through the soil (Table XIX).

Table XIX. Summary of tile drainage conductivity data (0- to 6-in. level) (μmhos/cm), 1972-1973.

Date Effluent Applied (in.)	Treatment			
	1	2	3	4
June 1972	320[a] (–)[b]	362 (–)	390 (–)	374 (–)
April 1973	165 (0)	181 (3.7)	204 (11.3)	202 (22.6)
October 1973	144 (0)	332 (4.1)	833 (11.5)	863 (24.6)

[a]μhos/cm.

[b]in. of lagoon effluent applied.

The plots for this experiment received from 57 to 108 inches of anaerobic lagoon effluents before the site was used to produce corn. The application of this material did not cause any detrimental effects as measured by yield or conductivity. This would seem to indicate that the assimilative capacity of the soil is high and the moisture regeneration during the winter months is sufficient to flush salts applied through the soil profile and out into the tile drainage system.

DISCUSSION AND CONCLUSIONS

At the time this research was conducted, very little had been published about the nutrient content of plants when animal wastes were applied to a growing crop. Research conducted during 1972 and 1973 at Iowa State University was directed to the measurement of nutrient changes in plant tissue. The results of this research indicate that anaerobic lagoon liquid did not significantly affect the yield in 1972, although yield did decrease slightly with increasing amounts of lagoon effluent applied.

During 1972 and 1973, the leaf-tissue concentration of phosphorus, sodium, aluminum and zinc increased with increasing amounts of lagoon effluent applied, and magnesium concentrations decreased with increasing amounts of anaerobic lagoon effluent applied.

The analysis of the grain-tissue data for 1972 indicated that differences in potassium and magnesium contents were statistically significant. The potassium and magnesium content decreased with increasing amounts of lagoon effluent applied. Potassium and magnesium were not found to be statistically significant in 1973.

Plants need only a small quantity of the metals for satisfactory growth. Thus, when 22 inches of anaerobic lagoon effluent (the equivalent of 2000 to 3000 lb N/ac) was applied to a growing crop of corn, the percentage removal in the plant tissue of N, P, K, Fe, Cu and Zn was very poor, and an accumulation of metals in the soil could be expected. This accumulation is likely to be insignificant when compared with industrial wastes. Even if accumulation occurs, the effects probably will be insignificant over a long period of time.

The data show that, if the crop yield is not important, large quantities of swine wastes could be applied to a disposal site on a short-term basis without any problems.

Thus, the choice seems to be one of acceptable management; *i.e.,* whether to use a small area for short-duration, intensive disposal or to increase the area sufficiently so that it capitalizes on nutrient utilization while serving for long-term disposal.

For the study reported here, the first set of soil samples was collected in June of 1972. The pH for the surface layer of soil ranged from 6.2 to 6.4, and the electrical conductivity of the soil was essentially the same at all depths. A second set of soil samples was collected in April 1973, and both pH and electrical conductivity had decreased with respect to the samples collected in June 1972. The electrical conductivity of samples collected in October 1973 increased above those observed during June 1972 and April 1973. Even though extremely large amounts of lagoon effluent were applied from 1968 to 1971, the electrical conductivity was normal after application was stopped for one year. This supports what Swanson *et al.*[15] reported when large quantities of feedlot runoff were applied to the soil in eastern Nebraska.

Exchangeable potassium was first monitored in the samples collected during April 1973, and increases in concentration were observed when samples were collected in October 1973.

In summation, the two years of data suggest that the application of animal wastes to soil will cause increases in conductivity and exchangeable potassium and a slight decrease in pH. Application rates, based on nitrogen, greater than 450 to 500 lb/ac, seem to cause greater increases in conductivity and exchangeable potassium.

The anaerobic lagoon liquid used in this study typically had a pH ranging from 7.0 to 7.3, a nitrogen content of 400 to 500 mg/l, a chloride content of 250 to 300 mg/l and a phosphorus content of 70 to 80 mg/l. On the basis of analyses of porous-cup samples, the soil to which this effluent was applied showed a capacity to remove 80% of the nitrogen and 85% of the phosphorus.

Analyses of the tile drainage revealed that removal of phosphorus, nitrogen and COD in the soil profile was slightly better in 1972 than in 1973. It is believed, however, that more direct channeling into the tile drains in 1973 is responsible for the difference. Climatic events during the 1972-73 winter and spring are believed to have caused soil conditions that allowed channeling into the tile drains to occur. If so, this could account for the slightly better removal of nitrogen and COD in the soil profile in 1972.

Crop production on an application site is an option for removing nutrients from the soil. Removal of the entire plant is an effective nutrient-removal technique for major plant nutrients, and micronutrients are removed from the soil. As more and more wastes are applied to the soil, the ability of plants to remove plant nutrients and micronutrient metals is reduced. Logically, this means that metals can increase in the soil. If swine-waste-disposal applications to the soil are limited by the requirements of nitrogen, phosphorus and potassium, a buildup of micronutrients is not likely to occur.

Salinity is not seen as a problem for swine-waste-disposal systems in Iowa. As an example, the tile-drained plots used for the research (1972 and 1973) were used from 1968 to 1970 for disposal of swine wastes, and an accumulated maximum of 108 in. lagoon effluent has been applied to the plots.

Analysis of soil samples collected before the beginning of irrigation in 1972 indicated that the conductivity of the soil was about 0.3 μmhos/cm after the site had been idle for one year. The conductivity test is a measure of soil salinity, and there is a close correlation between electrical conductivity and dissolved salts. The test is simple, and equipment is available in most soils laboratories. Thus, this test is logically used to verify that salts are not accumulating. Salt in the plots at the Swine Nutrition Research Station evidently has leached through the soil profile and into the tile-drainage system.

In the future, it is imperative that we utilize more efficiently the nutrients present in agricultural production wastes. This means that total systems must be designed with initial constraints being put on the type of waste treatment system. Geographical location will be extremely important; some areas in the northern United States require 180-day storage, while other areas will be able to utilize freshwater flush systems for manure transport and as a source of irrigation water. Effective planning will be necessary for efficient utilization of waste products.

ACKNOWLEDGMENTS

This paper is Journal Paper No. J-8826 of the Iowa Agriculture and Home Economics Experiment Station, Ames, Iowa, Project No. 1842. This report is based on research conducted while the author was located at Iowa State University, Ames, Iowa.

REFERENCES

1. Hinesly, T. D., E. L. Ziegler and R. L. Jones. "Effects on Corn by Application of Heated Anaerobically Digested Sludge," *Compost Sci.* 13:26-30 (1972).
2. Dean, R. E. "Disposal and Reuse; What Are the Options?" *Compost Sci.* 14:26 (1973).
3. Chaney, R. L. "Crop and Food Chain Effects of Toxic Elements in Sludges and Effluents," Proc. Joint Conf. on Recycling Municipal Sludges and Effluents on Land, University of Illinois (1973), pp. 129-142.
4. Spotswood, A. and M. Raymer. "Some Aspects of Sludge Disposal on Agricultural Land," *Water Poll. Control* 72:71-77 (1973).
5. Patterson, J. B. E. "Metal Toxicities Arising from Industry. Trace Elements in Soils and Crops," *Minn. Agric. Fish. Food Tech. Bull.* 21:193-207 (1971).
6. Hunter, J. G. and O. Vergnano. "Trace Element Toxicities in Oat Plants," *Annals Applied Biol.* 40:761-777 (1953).

7. Willrich, T. L. "Primary Treatment of Swine Wastes by Lagooning. Management of Farm Animal Wastes," Am. Soc. Agric. Eng. Publ. Sp. 0366, St. Joseph, Michigan (1966).
8. Smith, R. J. "Manure Transport in a Piggery Using the Aerobically Stabilized Dilute Manure," M.S. Thesis, Iowa State University, Ames, Iowa (1967).
9. Smith, R. J. "A Prototype System to Renovate and Recycle Swine Wastes Hydraulically," Ph.D. Thesis, Iowa State University, Ames, Iowa (1971).
10. Koelliker, J. K., J. R. Miner, T. E. Hazen, H. L. Person and R. J. Smith. "Automated Hydraulic Waste-Handling System for a 700-head Swine Facility Using Recirculated Water," Cornell Agricultural Waste Management Conference Proceedings (1972), pp. 249-61.
11. Bower, C. A. and L. V. Wilcox. "Soluble Salts. Part 2," *Methods of Soil Analysis*, C. A. Black, Ed. (Madison, Wisconsin: American Society of Agronomy, Inc., 1965), pp. 933-51.
12. Chapman, H. D. and P. F. Pratt. "Methods of Analysis for Soils, Plants and Waters," University of California, Division of Agricultural Sciences, Riverside, California (1961).
13. American Public Health Association. *Standard Methods for the Examination of Water and Wastewater*, 13th ed. (New York: American Public Health Association, 1971).
14. Booram, C. V., T. E. Loynachan and J. K. Koelliker. "Effects of Sprinkler Application of Lagoon Effluent on Corn and Grain Sorghum," in *Processing and Management of Agricultural Waste,* Proc. of the 1974 Cornell Agricultural Waste Management Conf. (1974), pp. 493-502.
15. Swanson, N. P., C. L. Linderman and J. R. Ellis. "Irrigation of Perennial Forage Crops with Feedlot Runoff," *Trans. Am. Soc. Agric. Eng.* 17(1):144-47 (1973).

ADDITIONAL REFERENCES

Bear, F. E. *Chemistry of the Soil* (New York: Reinhold Publishing Corporation, 1955).

Pratt, P. F. "Potassium, Part 2," in *Methods of Soil Analysis*, C. A. Black, Ed. (Madison, Wisconsin: American Society of Agronomy, Inc., 1965), pp. 1022-30.

A FERMENTATION PROCESS FOR THE UTILIZATION OF SWINE WASTE

B. A. Weiner
Northern Regional Research Center
Agricultural Research Service
U.S. Department of Agriculture
Peoria, Illinois

INTRODUCTION

Problems with disposal of animal waste in confined livestock production facilities[1,2] and the availability of useful nutrients in this material[3-7] have led to research with cattle waste to produce feed and industrial products.[8] Cattle waste nutrients also have been combined with feed grains in simple fermentations to make an animal feed.[9]

The present work describes a feed made by an aerobic culture process of solid substrate fermentation using swine waste combined with cracked corn. Sequential patterns of growth for microbial groups and development of selected metabolites together with animal feeding trials of fermentation product are described.

MATERIALS AND METHODS

Swine waste in wash water was obtained from a concrete storage container under a slotted floor or from a feedlot with a concrete surface, both located in central Illinois. Coarsely cracked corn was obtained from a local commercial elevator.

Laboratory fermentations were done in Erlenmeyer flasks at 28°C. The flasks (0.3 and 2.0 liter) were arranged in a horizontal position by snap-ring holders fastened to a circular wooden board held at an angle of 10° from the horizontal. The board was rotated by a 0.02 hp motor and reduction gears provided 0.6 rpm. This posture gave a slow tumbling action to the corn-waste mixture, and cotton plugs or gauze were generally not in contact with the material. Small flasks were charged with 25 ml waste diluted with water and combined with 50 g cracked corn to give preferred moisture levels of 40%; useful levels ranged from 34 to 43%. Large flasks were similarly prepared with 200 to 300 g diluted waste and 400 to 500 g corn. Fermentation equipment, though clean, was not sterilized and aseptic techniques were used only with procedures for microbiological analyses.

Pilot-plant fermentations were carried out in a small cement mixer, the inside coated with paint having an epoxy resin base to prevent rusting. The mixer had a volume of 130 liters and was charged with varying amounts of diluted waste, 13.6 to 17.3 kg, and 22.8 kg cracked corn. Ambient temperatures ranged from 18 to 33°C and heat from fermentaion raised temperatures above ambient by 2 to 6°C. Tumbling of fermentation substrate was done with a 1/4-hp motor with belt drive, and rotation of the cement mixer was decreased to 0.5 rpm with reduction gears. Fermentation was stopped at 36 hr by blowing hot air (50 to 55°C) into the container for 16 to 18 hr; moisture levels of 11 to 13% resulted. Fermentation product was bagged and saved for mixing in a twin-shell blender prior to feeding tests.

Microbiology

Certain groups of viable microorganisms were determined by blending 5.0-g samples with portions of water in a micro Waring Blender for two 15-sec intervals, decanting liquid of each blend through glass wool and adjusting to a final volume of 50.0 ml. The initial dilution of 1:10 (v/v) was serially diluted. Four previously prepared media were used in triplicate Petri plates for each of four to six dilutions. Volumes of 0.3 ml added to each plate were spread by sterile bent glass rods. The media used included Eugonagar for total counts, LBS agar for lactobacilli, eosin methylene blue (EMB) for coliforms, Mycophil agar with dihydrostreptomycin sulfate (0.2 mg/ml) and penicillin G (330 units/ml) for yeasts. All these media were manufactured by BBL (BBL, Division of Bioquest, Cockeysville, Maryland). Lactobacilli MRS broth, manufactured by Difco, was also used (Difco Laboratories, Detroit, Michigan). Plates and tubes were incubated at 28°C except for those with EMB, which were held at 37°C. The latter were counted after incubating one day and the others after three days.

Chemical Analyses

Samples collected from fermentations were separated into 5.0-g aliquots for all analyses except for organic acids of fermentation. Analysis for these acids used aliquots of 10 g from flasks and a more convenient amount of 100 g from the cement mixer. Distilled water was added to a 5.0-g sample, triturated, and allowed to stand 5 min before pH was determined by electrode. A constant dry weight was obtained on the same sample by drying overnight at 103°C; this information allowed all data to be calculated on a dry-weight basis. Dry samples from the moisture determinations were ground in a Wiley mill to pass through a 40-mesh screen. MicroKjeldahl[10] and amino acid analyses were done on the powder. Amino acids were determined in an automatic analyzer manufactured by Beckman (Model 120 B). Total acids produced by fermentation were determined on aqueous extracts of 5.0-g samples and were obtained by three 10-sec blends in a micro Waring Blender with filtration of liquid portions through glass wool. Filtrates were made up to 25.0 ml and titrated to pH 7.0 with 0.05 *N* NaOH. A similar aqueous extract was centrifuged (5000 rpm, 15 min) and the supernatant was made up to 24.75 ml. It was added to a beaker together with 0.25 ml 10 *N* NaOH, magnetically stirred and analyzed for ammonia by an Orion ammonia electrode, a modification of the method described by Bremner and Tabatabai.[11]

Odor was evaluated organoleptically by collective impressions of laboratory personnel. Fatty acids, which contribute to odor, were determined on samples stored at -15°C. Thawed samples were extracted with distilled water for 15 sec in a Waring Blender, filtered through glass wool, made up to 40 ml and pH adjusted to 8 with sodium bicarbonate. The aqueous sample was extracted overnight with diethyl ether, followed by acidification of the aqueous portion and further extraction with ether. The ether extract was concentrated on a steam bath to 0.2 ml, dried over sodium sulfate and analyzed for fatty acids by gas-liquid chromatography (GLC) in an F&M model 700 chromatograph with dual columns (6 ft x 1/8 in., stainless steel) and packed with 10% SP-1200/1% H_3PO_4 on 80- to 100-mesh Chromosorb WAW (Supelco, Inc., Bellefonte, Pennsylvania). The instrument was equipped with flame ionization detector and a disc integrator. The temperature program was 5°/min from 65 to 190°C. Standards were prepared in ether using 20% lactic acid and 2.0% of the following acids: acetic, propionic, isobutyric, butyric, isovaleric, valeric, isocaproic and caproic. These fatty acids gave retention times for comparison to fermentation acids and single symmetrical peaks except for lactic acid, which provided two components.

Feeding Tests

Four 6- to 7-week-old pigs, weighing an average of 11.8 kg, were fed for 13 days on a control diet of 77.5% corn and 19.0% soybean meal (48.5% crude protein), and a salt-vitamin mixture at remaining percentages composed of: dicalcium phosphate, 2.7; calcium carbonate, 0.1; iodized salt, 0.5;

vitamin mix, 0.1; and a trace mineral mix, 0.1. An experimental diet which consisted of 80.4% fermentation product, 16.1% soybean meal, and the same percentage of vitamin-mineral mixture as the control diet was fed to four pigs for 13 days. The feed consumed and the weights of the animals were recorded.

Twenty hens, in their second year of lay without being culled, were fed a control diet of ingredients in the indicated percentages for a period of 21 days: corn, 63.2; alfalfa meal, 5.0; soybean meal, 19.0 (47% crude protein); meat and bone meal, 2.0 (49% crude protein); riboflavin, 500 μg/g; limestone, 7.0; dicalcium phosphate, 2.5; salt, plain, 0.5; DL methionine, 0.1; vitamin and trace mineral premix 1552, 0.25. Nineteen hens were fed the same diet containing fermentation product in place of corn. Feed consumption was weighed and eggs were counted.

A control diet of 3 kg chopped hay and 8.8 to 10.8 kg corn in dry state was fed to four sheep for 10 days. The same diet with corn replaced by 10.8 kg fermentation product was fed to four additional sheep, and remaining feed was weighed for both groups to determine acceptance and palatability.

RESULTS AND DISCUSSION

The source of the inoculum which results in lactic fermentation was determined by combining swine waste with cracked corn and incubating aerobically in Erlenmeyer flasks (Table I). Lactic and fatty acids produced by organisms from waste and corn dropped pH to 4.3, and a fetid odor was replaced by one resembling silage. Sterilization of the corn component gave the same results, while cultures with a sterilized waste component provided

Table I. Origin of microorganisms in fermentations of swine waste-corn.

Flask Inoculum	pH				Odor			
	1 hr	21 hr	52 hr	75 hr	1 hr	21 hr	52 hr	75 hr
Organisms from waste and corn	4.7	4.4	4.3	4.3	Fetid	Silage	Silage	Silage
Organisms from waste	4.7	4.5	4.3	4.3	Fetid	Silage	Silage	Silage
Organisms from corn	4.8	4.9	5.0	5.0	Wet corn	Wet corn	Wet corn	Malic acid

microbial metabolism with little change in pH and an odor like wet corn. Thus swine waste, not corn, supplied the microbial population that provided rapid acid fermentation with silage-like odor.

Table II depicts production of fermentation acids and change in odor as a function of waste fraction used with corn. The pH ranged from 4.3 at 8.2% waste to 4.5 at 18.9% waste on a dry-weight basis (DWB). Odor is fetid at 45 hr of fermentation with more than 15.6% waste in the corn-waste mixture. For this reason, waste fractions were chosen between 11 and 12% DWB, and this level was associated with one part waste and two parts corn.

Table II. Influence of waste fraction upon pH and odor of swine waste-corn fermentation.

% Waste	pH				Odor			
DWB[a]	4 hr	24 hr	45 hr	172 hr	4 hr	24 hr	45 hr	172 hr
8.2	4.7	4.3	4.3	4.3	Fetid	Silage	Silage	Silage
12.5	4.7	4.4	4.4	4.4	Fetid	Fetid	Silage	Silage
15.6	4.7	4.5	4.5	4.5	Fetid	Fetid	Silage	Silage
18.9	4.7	4.6	4.5	4.5	Fetid	Fetid	Fetid	Fetid
21.5	4.7	4.6	4.5	4.6	Fetid	Fetid	Fetid	Fetid

[a]Dry-weight basis.

Differences in biochemical and microbial patterns of fermentation were observed with stored vs fresh waste, but silage-like fermentation products with lowered pH resulted from both types. A typical waste cracked corn fermentation is shown in Figure 1. After 36 hr with either fresh or stored waste components, the pH dropped two units in response to a tenfold increase in titratable acid with little further change at 144 hr. Original fetid odor with either waste changed to one resembling a silage fermentation by 36 hr, and the silage-like odor continued to 144 hr. Ammonium ion levels were fourfold higher with stored waste components than with fresh waste and may reflect the higher deaminase levels associated with both microbial activity in stored materials of protein nature as well as their nonprotein N compounds.[12] The microbial pattern in flask fermentations of stored waste combined with corn is represented in Figure 2. Total numbers of microorganisms remained fairly steady at 10^8/dry g and increased slightly by 144 hr. Lactics present initially at 10^7 organisms/dry g increased tenfold at 36 hr but returned to initial numbers at 144 hr, while yeast cells remained steady at 10^5/dry g except for a drop in numbers at 36 hr. Coliforms at 10^5 cells/dry g decreased rapidly beyond 12 hr and were not detected after 36 hr.

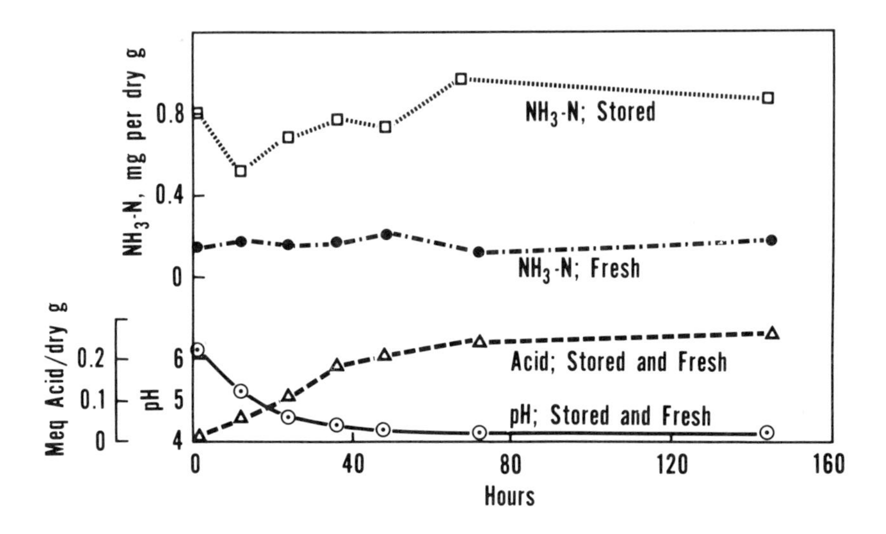

Figure 1. Biochemical pattern of flask fermentations with fresh or stored swine waste combined with corn.

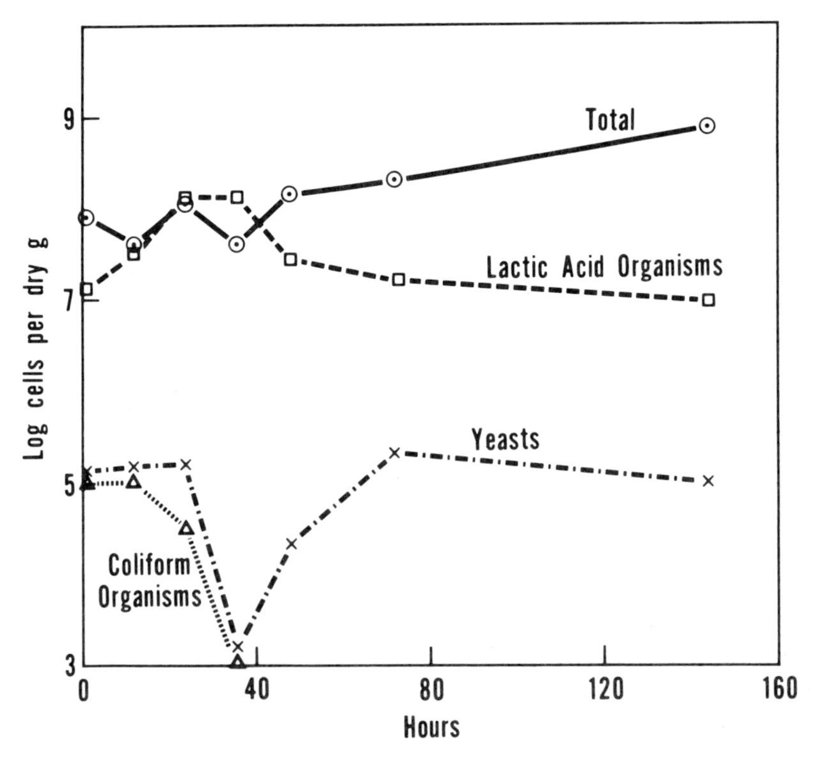

Figure 2. Microbial pattern in flask fermentations of stored swine waste-corn.

The fermentation acids responsible for reduced pH levels in cultures of stored waste combined with corn resulted from carbohydrate metabolism by the dominant lactic group. The initial major acid component of this fermentation was lactic acid (83.0%) with traces of butyric and valeric acids (Table III). By 36 hr 94.6% of the ether-soluble extract was lactic acid with a trace of acetic acid. Homofermentative lactics produce such an acid picture.[13]

Table III. Percent composition of fatty and lactic acids in flask fermentations of stored swine waste-corn.

	Hours of Fermentation				
Acid	1	12	24	36	48
Acetic	0	0.2	0	0.8	1.6
Butyric	0.5	0.4	0	0	0
Valeric	0.03	0	0	0	0
Lactic	83.0	86.4	89.6	94.6	98.9
Unknown	15.4	13.4	10.4	4.5	0

A contrasting growth pattern of microbial groups and fermentation acids was shown by cultures on fresh waste combined with corn (Figure 3). Total organisms were tenfold higher than with cultures using stored waste components (10^9 organisms/dry g), and lactics, present initially at 10^7 organisms/dry g, were the dominant group of microorganisms by 12 hr and increased 100-fold by 36 hr. Coliforms were present at levels of 10^6 to 10^7 organisms/dry g and dimished twentyfold in the first 12 hr, followed by an increase to initial levels. Yeast cells were present at 10^5 and 10^6 organisms/dry g and diminished in number as the fermentation proceeded.

Fermentations of fresh waste-corn produced all homologous acids and certain isomers in the series of acetic through valeric, including an average of 14.2% lactic acid (Table IV). As fermentation proceeded through 36 hr, acids at intermediate times showed decreases in percentage composition for homologous fatty acids from propionic through valeric and the isomers, isobutyric and isovaleric, while acetic and lactic acids increased to 24.9 and 60.3%, respectively (averages). This fermentation acid pattern is associated with heterofermentative lactics.[13]

Further information of the lactic acid bacteria group in animal waste was provided by comparable fermentations using cattle waste fractions combined with corn,[14] which showed that the *Lactobacillus buchneri*-like betabacterium was present in greatest numbers for the first 24 hr and is similar to our fermentation using fresh swine waste. The *L. plantarum*-like

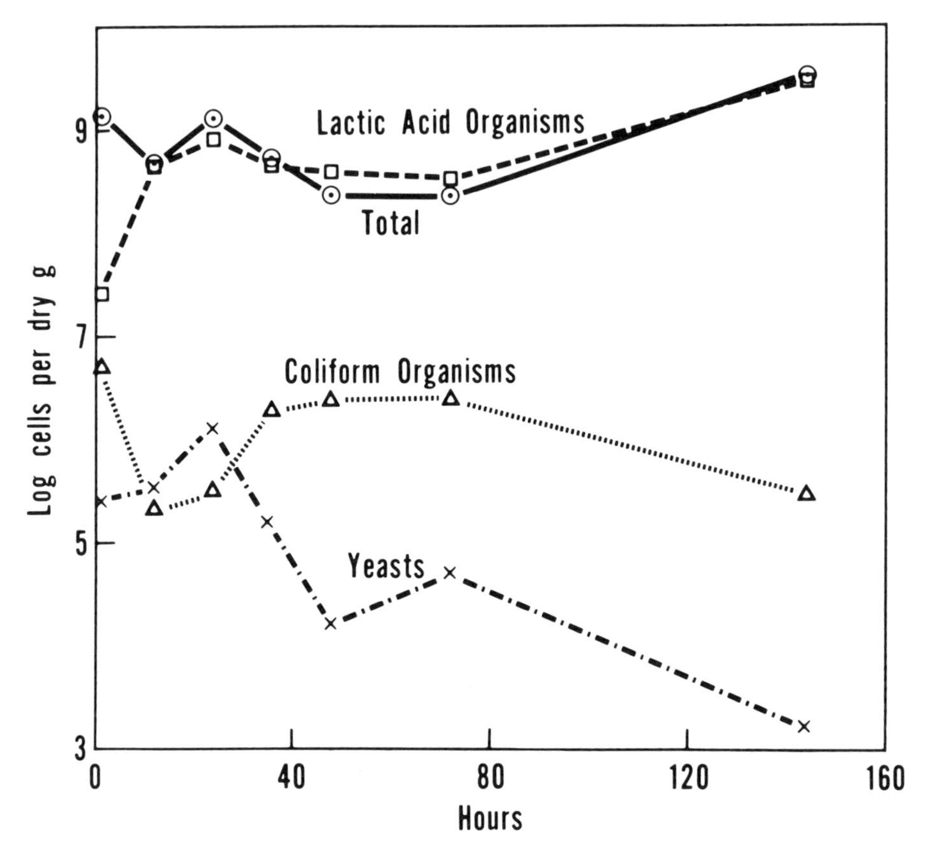

Figure 3. Microbial pattern in flask fermentation of fresh swine waste-corn.

Table IV. Percent composition of fatty and lactic acids in flask fermentations of fresh swine waste-corn.

	Hours of Fermentation				
Acid	1	12	24	36	48
Acetic	13.2	10.0	29.7	24.9	12.4
Propionic	19.0	3.3	7.8	3.0	0.8
Isobutyric	1.9	0	0.2	0	0
Butyric	22.4	4.3	6.1	4.0	0
Isovaleric	7.7	0.7	0.8	0.3	0
Valeric	4.3	0.5	0.4	0.1	0
Lactic	14.2	70.2	54.1	60.3	87.4
Unknown	18.2	9.8	0.9	7.6	0

streptobacterium became the dominant lactic organism from 48 hr on and is comparable to our homofermentative lactic bacteria found with stored swine waste-corn cultures.

Coliform bacteria, present at 0.5% or less of total flora with both types of waste, persisted with fresh waste at levels of 10^6/dry g. Preliminary work with these aerobic swine waste-corn cultures showed growth of coliform bacteria was correlated to concentration of fermentation acids (Figure 4). Batch type fermentations, serving as controls, were compared to others in

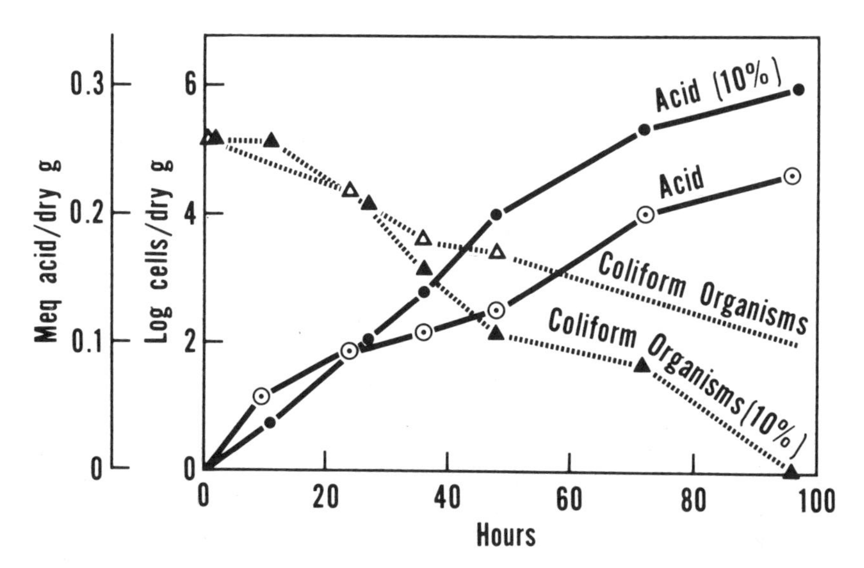

Figure 4. Effect of lactic inoculum in fresh swine waste-corn fermentation on growth of coliforms.

which one-tenth of the zero-time flask contents were replaced by 48-hr swine waste-corn cultures. Total acid levels in flasks with 10% replacement of substrate were over 27% higher than in controls. Coliform bacteria numbers in control flasks steadily decreased, but coliform organisms in displacement flasks diminished more sharply and were no longer detected at 96 hr. Portional substitution in batch fermentations, with 48-hr cultures, augments the inoculum from fresh swine waste with lactic organisms. The effect of fermentation acids can be significant, as shown by McKasky and Anthony,[15] who tested 27 *Salmonella* cultures inoculated into cattle manure combined with Bermuda grass and ensiled for three days at pH levels of 4.0 to 4.5. At this pH level, no *Salmonella* were recovered, while 25 of the 27 cultures were recovered at a pH of 6.0 to 6.5.

The fermentation was scaled up with a cement mixer to produce fermentation product for animal feeding tests. A series of 36-hr fermentations were conducted with fresh waste whose solid content ranged from 18.7 to 29.1% with an average Kjeldahl nitrogen level of 4.1% (DWB). A material balance of cracked corn combined with fresh swine waste in the cement mixer showed recovery of 94.2%. Biochemical characteristics of a representative fermentation with 11.3% waste (DWB), mixed with corn, was comparable to flask fermentations with fresh waste components. However, acid production was half that observed in 2-liter flasks and was reflected by a drop in pH to 5.1; pH values as low as 4.2 were obtained in other fermentations in the pilot plant. The fetid odor changed to one like silage within 36 hr in all fermentations; this transformation frequently occurred in 4 to 6 hr. The ammonia nitrogen pattern was comparable to that found in flask cultures with stored waste components.

The growth pattern of microorganisms in the cement mixer differed from that found in 2-liter flasks. Lactic organisms did not proliferate initially and yeast cells decreased for the same period, whereas coliforms remained steady in numbers (Figure 5). Similarities to flask culture with fresh waste were also observed with fermentation acids for initial samples (Table V). However, some differences were noted in ensuing samples, and by 36 hr over half the percentage composition was acetic acid compared to one-fourth in flask cultures with fresh waste components. A higher proportion of propionic and the presence of isobutyric and isovaleric acids were found in large-scale fermentation. The proportion of lactic acid produced by fermentation in the cement mixer was one-fourth that found in fresh waste-corn flasks at 36 hr.

Amino acid analysis of the fermented product (Table VI) showed that lysine and methionine were limiting for health and growth with starting chicks and young pigs. Other amino acids present in lower than required amounts for chicks were arginine, threonine, glycine and phenylalanine. The product supplied levels of lysine and methionine at 52 and 64% fulfillment of requirements for chicks[16] and corresponding values of 69 and 76% for young pigs.[17] While fermentation product contained 21 to 39% more methionine than corn, it was still less than these animals need. However, amino acid profiles of feed are not as nutritionally important to sheep because microbial syntheses of amino acids will take place in the rumen. For example, lambs on a diet composed largely of carbohydrates with urea gained weight,[18] an average of 0.23 lb per day compared to 0.3 lb for controls on a ration containing casein. Crude protein (DWB) in fermentation product increased 17.2% compared to controls of cracked corn (average Kjeldahl N values: corn, 1.51%; fermentation product, 1.77%). A 14.8% increase in protein content was observed based on detected amino acid levels in fermentation product vs corn. Fermentation product was fed to young pigs

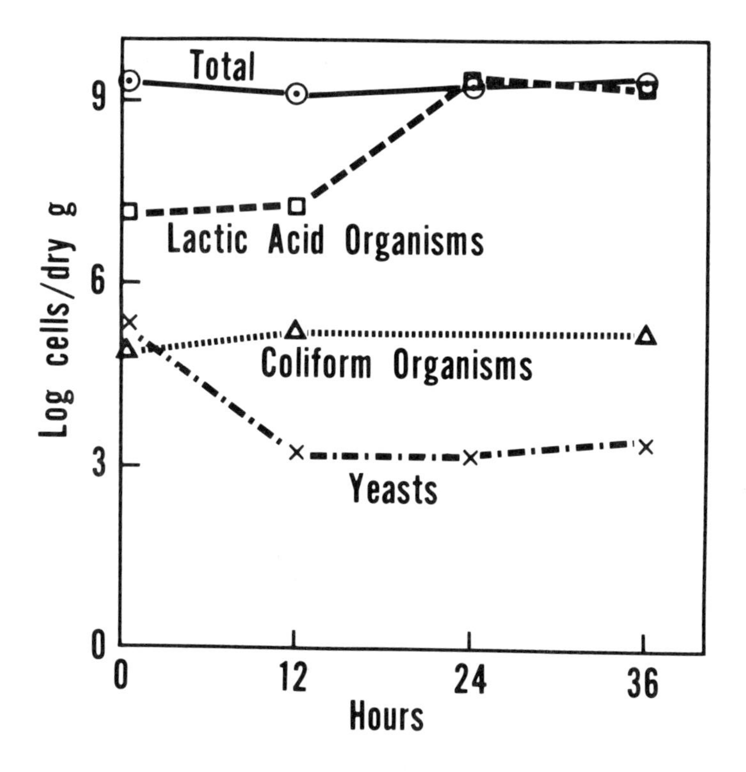

Figure 5. Microbial pattern of pilot-plant fermentation with fresh swine waste-corn.

Table V. Percent composition of fatty and lactic acids of pilot-plant fermentations with fresh swine waste-corn.

	Hours of Fermentation			
Acids	1	12	24	36
Acetic	12.1	25.1	24.2	57.3
Propionic	19.7	0	13.3	12.8
Isobutyric	4.5	17.5	3.2	4.2
Butyric	19.0	2.6	4.7	7.0
Isovaleric	9.0	9.0	3.2	4.4
Valeric	5.3	3.7	3.0	1.2
Isocaproic	2.8	0	0	0
Caproic	2.3	0	0.7	0
Lactic	16.9	14.4	19.6	14.1
Unknown	8.4	27.7	28.1	0

Table VI. Amino-acid composition of fermentation product and percent fulfillment of amino-acid requirement for poultry and swine.

	Product[a] (g/16 g N)	Percent of Requirement	
		Poultry Starting Chicks	Swine 11 to 34 kg
Lysine	2.88	52	69
Histidine	2.69	Exceeded	Exceeded
Ammonia	3.62	NE[b]	NE
Arginine	4.87	82	Exceeded
Aspartic	6.48	NE	NE
Threonine	3.67	85	Exceeded
Serine	4.65	NE	NE
Glutamic	16.85	NE	NE
Proline	7.83	NE	NE
Glycine	4.11	82	NE
Alanine	7.82	NE	NE
0.5 Cystine	1.24	NE	NE
Valine	4.70	Exceeded	Exceeded
Methionine	2.36	64	76
Isleucine	3.69	Exceeded	Exceeded
Leucine	10.82	Exceeded	Exceeded
Tyrosine	4.07	NE	NE
Phenylalanine	4.65	72	Exceeded

[a]N: 1.77%, 17.2% more than corn control; amino acids: 8.36%, 14.8% more than corn control.

[b]Nonessential metabolite.

as a major portion of their diet. A control group of pigs gained a total of 34.9 kg and experienced an average daily gain of 0.671 kg with a feed-to-gain ratio of 1.72. The same number of pigs given fermentation material had corresponding values of 23.0, 0.442 and 2.61, representing a one-third decrease in total and average daily gain. Fermentation product was augmented with 16.1% soybean meal compared to 19.0% in the control diet. This experimental design required fermentation protein to be equal in biological value to soybean meal to achieve equal gain and gain-to-feed ratios.

While equal amounts of feed were eaten by both groups of young pigs and acceptability was not a problem, decreased gains in weight suggest that the process made part of the test protein unavailable to this young an animal. In a previous study[19] which indicates that age of pigs is a factor, a conventional finishing diet had been replaced by 15% swine waste, not processed by fermentation but simply dried. This formulation gave comparable weight gain and feed efficiency data as controls, but 30% replacement gave lower values for these parameters.

In nutritional tests with poultry, layer hens had average feed per bird per day values for 20 controls and 19 experimental feeders of 116.4 and 106.9 g, respectively. Corresponding egg production (%) was 31.2 and 34.6. Test hens ate less experimental feed (8.2%) than controls, but the short feeding trial diminished the significance of this observation as well as of the slight increase in egg production (3.4%). Also, lower egg production by both groups of birds reflected both the absence of culling and the decreased productivity of birds in their second year of lay.

The last nutritional study in which acceptance of fermentation product was not a problem was with sheep. In a control group of four sheep, each was provided with 3,000 g of dry chopped hay plus the following amounts of corn: 8,800; 10,800; 10,400 and 10,600 g. Respective weights of feed left after 10 days were 1,405; 0; 807 and 694 g. The test group of four sheep, each given 3,000 g hay and 10,800 g fermentation product, left nothing to be weighed back for two animals and traces for the last two animals (5 and 61 g), which shows that sheep did not discriminate against this fermented product.

The product appeared to be well accepted by monogastrics, ruminants and poultry. Refeeding swine waste in the form of fermentation product without the drying required for storage suggests a feasible way of decreasing the waste problem of swine feedlots by one-third* and could provide augmented animal diets with attendant economic advantages.

SUMMARY

Aerobic fermentation of swine waste combined with corn produced differences in microbial and biochemical patterns dependent on use of fresh or stored excrement. Lactic acid fermentation and odor control resulted with either waste. Homofermentative lactics were present initially at 10^7 organisms/dry g with aged waste-corn cultures, and total microflora amounted to 10^8 organisms/dry g. Fresh waste-corn fermentations yielded heterofermentative lactics at 10^7 organisms/dry g, and total viable population was 10^9 organisms/dry g. These respective lactic groups dominated from 12 through 144 hr in cultures with either waste, and acid production (0.2 meq/dry g) decreased pH by 2 units to 4.5. The major acid component with stored waste-corn was lactic, whereas fresh waste-corn fermentation

*Mixture ratios of 1:2 for aerobic fermentations of swine waste (18 to 29% dry weight) combined with cracked corn (88% dry weight) will produce a product. When 2.26 kg feed containing 87.6% dry fermentation product with 8.9% soybean meal, 3.5% minerals and vitamins are eaten with excretion of 2.72 kg waste per hog per day, the problem of waste disposal could be reduced by 35.6 to 37.6%.

produced both lactic and homologous fatty acids from acetic through valeric. Coliforms present initially at 10^5 organisms/dry g in stored waste-corn cultures were not detected after 36 hr; coliforms in fresh waste-corn fermentations persisted at 10^6 organisms/dry g. However, acid production in fresh waste-corn flasks was increased over 27% by 0.1 replacement of starting fermentations with waste-corn cultures, 48 hr old, equivalent to an inoculum of lactic acid organisms. Fermentation product from fresh waste-corn cultures was fed as the major dietary component to young pigs, hens and sheep. Pigs showed gain and gain/feed diminished by one-third in 13-day trials. Laying hens performed comparably to controls in a 21-day test, and sheep did not discriminate against fermentation product.

ACKNOWLEDGMENTS

Feeding tests were carried out at the Environmental Quality and Nutrition Institute of the Agricultural Research Service in Beltsville, Maryland. Appreciation is expressed to Lowell T. Frobish, Chief, Nonruminant Animal Nutrition Laboratory (swine) and Research Animal Scientists at the Biological Waste Management Laboratory, C. C. Calvert (hens) and L. W. Smith (sheep). Amino acid determinations were done with the assistance of G. L. Donaldson.

The mention of firm names or trade products does not imply that they are endorsed or recommended by the U.S. Department of Agriculture over other firms or similar products not mentioned. Some of the data describing the fermentation process have been submitted to the European Journal of Applied Microbiology.

REFERENCES

1. Loehr, R. C. "Pollution Implication of Animal Wastes–A Forward Oriented Review," Federal Water Control Administration, U.S. Department of the Interior, Robert S. Kerr Water Research Center, Ada, Oklahoma (1968).
2. Walker, W. R. "Legal Restraints on Agricultural Pollution," in *Relationship of Agriculture to Soil and Water Pollution,* Cornell University Conference on Agricultural Waste Management, Ithaca, New York (1970), pp. 233-241.
3. Taiganides, E. P. and T. E. Hazen. "Properties of Farm Animal Excreta," *Trans. Am. Soc. Agric. Eng.* 9:374-376 (1966).
4. Anthony, W. B. "Feeding Value of Cattle Manure for Cattle," *J. Animal Sci.* 30:274-277 (1970).
5. Harmon, B. G., D. L. Day, A. H. Jensen and D. H. Baker. "Nutritive Value of Aerobically Sustained Swine Excrements," *J. Animal Sci.* 34:403-407 (1972).
6. Smith, L. W. "Recycling Animal Wastes as a Protein Source," in *Symposium on Alternate Sources of Protein for Animal Production,* Am. Soc.

Animal Sci. and Committee on Animal Nutrition, National Research Council, National Academy of Sciences, Washington, D.C. (1973), pp. 147-173.
7. Powers, W. L., G. W. Wallingford and L. S. Murphy. "Research Status on Effects of Land Application of Animal Wastes," National Environmental Research Center, Office of Research and Development, U.S. Environmental Protection Agency, Corvallis, Oregon (1975).
8. Sloneker, J. H., R. W. Jones, H. L. Griffin, K. Eskins, B. L. Buchner and G. E. Inglett. "Processing Animal Wastes for Feed and Industrial Products," in *Symposium: Processing Agricultural and Municipal Wastes*, G. E. Inglett, Ed. (Westport, Connecticut: Avi Publishing Co., 1973), pp. 13-28.
9. Rhodes, R. A. and W. L. Orton. "Solid Substrate Fermentation of Feedlot Waste Combined with Feedgrains," *Trans. Am. Soc. Agric. Eng.* 18:728-733 (1975).
10. Association of Official Analytical Chemists. *Official Methods of Analysis,* 11th ed. (Washington, D. C.: Association of Official Analytical Chemists, 1970).
11. Bremner, J. A. and M. A. Tabatabai. "Use of an Ammonia Electrode for Determination of Ammonium in Kjeldahl Analysis in Soils," *Commun. Soil. Sci. Plant Anal.* 3:159-165 (1972).
12. Looper, C. G. and O. T. Stallcup. "Release of Ammonia Nitrogen from Uric Acid, Urea, and Certain Amino Acids in the Presence of Rumen Organisms," *J. Dairy Sci.* 41:Abstr. 729 (1968).
13. Rogosa, M. and M. E. Sharpe. "An Approach to the Classification of the Lactobacilli," *J. Appl. Bacteriol.* 22:329-340 (1959).
14. Hrubant, G. R. "Changes in Microbial Population During Fermentation of Feedlot Waste with Corn," *Appl. Microbiol.* 30:113-119 (1975).
15. McCaskey, T. A. and W. B. Anthony. "Health Aspects of Feeding Animal Waste Conserved in Silage," in Proc., 3rd International Symposium on Livestock Wastes, American Society of Agricultural Engineers, St. Joseph, Michigan (1975).
16. National Academy of Sciences, National Research Council. "Nutrient Requirements of Poultry," Pub. 1345, National Academy of Sciences, Washington, D. C. (1966).
17. National Academy of Sciences, National Research Council. "Nutrient Requirements of Swine," Pub. 1192, National Academy of Sciences, Washington, D. C. (1964).
18. Loosli, J. K., H. H. Williams, W. E. Thomas, F. H. Ferris and L. A. Maynard. "Synthesis of Amino Acids in the Rumen," *Science* 110: 144-145 (1949).
19. Diggs, B. G., B. Baker, Jr. and F. G. James. "Value of Pig Feces in Swine Finishing Rations," *J. Animal Sci.* 24:291 (1965).

42

THERMOPHILIC AEROBIC DIGESTION OF DAIRY WASTE

R. J. Cummings and W. J. Jewell
Department of Agricultural Engineering
College of Agriculture and Life Sciences
Cornell University
Ithaca, New York

INTRODUCTION

Animal and human wastes contain too much water to be considered solid and too little to be easily handled as a liquid. Water addition or removal is often used to modify the condition of the "as produced" material. In either form, cost-effective management must be able to include evaluation of the handling problems and potential pollutant transfer. The primary concern of a waste management program may be odor control, nutrient conservation, volume reduction, pathogen destruction or recycling. A large number of handling and treatment alternatives need to be available for any specific situation in order to be able to provide the most cost-effective solution for management of difficult-to-handle materials such as wet animal wastes and sewage sludges. This chapter provides a summary of a comprehensive study[1] of a new and simple process that shows promise of significantly increasing the alternatives available for treating organic slurries. The autoheated thermophilic aerobic digestion process described here requires no external energy to raise the temperature of a liquid slurry to between 45° and 75°C, and it can be applied to dilute slurries, about 2 to 10% of the total weight as dry solids.

High-temperature aerobic treatment of organic slurries, *i.e.*, thermophilic aerobic digestion, offers a number of potential advantages over conventional ambient air temperature aerobic processes such as oxidation ditches or aerobic lagoons. Among the advantages would be:

- higher rate of oxidation or stabilization which would translate into small reactor volumes;
- cost-effective odor control;
- increased efficiency of oxidation;
- pathogen destruction;
- weed seed destruction;
- stabilized slurry which is easier to dewater; and
- an opportunity to conserve nitrogen.

Any single process which combined the above advantages would have wide application in management of organic slurries. Obviously, high-temperature treatment cannot be considered a viable alternative if large expenditures of external energy are required to achieve the high temperatures. As will be shown, this is not necessary in this autoheated process, and all the above advantages can be achieved under conditions which would appear to be highly cost-effective compared to other aerobic treatment technology.

OBJECTIVES

The goals of this study were to define the requirements of the process of autoheating during aerobic digestion of a complex organic slurry and to demonstrate the feasibility of the process with a full-scale commercially available unit. The study was designed to define the major variables of operation in such a manner that the data could be extrapolated to other organic slurries. Dilute dairy cow manure was used as the feed throughout this study.

Specific objectives were to:

1. define the effect of volumetric and organic loading rates on the resulting reactor temperature and on treatment efficiency with a unit operated in a cold climate;
2. measure oxygen transfer efficiencies and transfer rates in tap water and in concentrated organic slurries;
3. define a heat balance by measuring the heat inputs and losses; and
4. develop a mathematical model to correlate the biodegradability of an oxygen slurry with the capability to achieve autoheating to thermophilic temperatures.

BACKGROUND

Aerobic oxidation of organics by microorganisms results in the formation of carbon dioxide, water and new bacterial cells. The energy that is contained in the portion of organic matter that is converted to carbon dioxide and water contributes heat energy to the environment. In conventional aerobic treatment processes this heat energy is usually not conserved nor used for autoheating of the liquid environment. Rather it is stripped from the reactor and lost largely through evaporation.[2,3] Autoheating is most common during the composting of concentrated organic wastes. Thus, the main limiting factors involved in autoheating are the amount of energy available for release to the environment (which is a function of the fraction of organics that are biodegradable) and the feasibility of keeping that energy in the system as heat energy.

The amount of heat released during aerobic microbial oxidation has been reported to average about 3.5 Kcal/g of COD oxidized to carbon dioxide and water. Thus, the total potential temperature increase in a 1-liter reactor which oxidizes 1 g of COD would be 3.5°C (when the specific heat capacity of the waste is 1 cal/°C-g and the density is 1 g/ml) provided all heat energy was conserved. In other words, autoheating to 45°C (from an ambient temperature of 20°C) would require the oxidation of 7.1 g/l of COD, providing that 100% of the energy is retained as heat energy. This, of course, is not possible, but it does indicate the efficiency of heat conservation that is necessary.

The rate of oxidation of the organics, oxygen transfer efficiency, total volume of aerating gas and many other factors must be understood before the feasibility of autoheating can be demonstrated. However, a main controlling parameter is the oxygen transfer efficiency, since this will indirectly affect the amount of evaporation from the aerating slurry, and this is one of the main heat losses. Most conventional aeration systems will transfer between 0.6 and 3.0 kg O_2/kWh (1 to 5 lb O_2/hp-hr) at an efficiency of less than 5% (oxygen used divided by the total supplied). Andrews and Kambhu[2] showed in a mathematical model that the transfer efficiency required to conserve enough heat to autoheat slurries to thermophilic temperatures was about 15%, and Wright[4] confirmed this in a laboratory study. Thus most conventional aerators would not be expected to be able to be used for this process. For this reason a number of investigators have suggested that only pure oxygen aeration would be capable of achieving autoheating.[2,3,5,6] As will be shown, a unique air aerator used in this study did sufficiently achieve the high oxygen transfer efficiencies required to conserve the heat energy required for autoheating.

The ability to predict the autoheating capability with any organic slurry will depend on the ability to define the substrate removal rate kinetics, their relationship to temperature, and the thermal characteristics of the system. By combining a mass balance with an energy balance it is possible to develop a temperature prediction relationship for any system. Additional details are in the complete project report.[1]

The rate of heat production in the reactor can be expressed in terms of the effluent concentration of biodegradable substrate as follows:

$$B_T = RK_TS_e \tag{1}$$

where B_T is the rate of heat produced in the reactor (Kcal/l-day), R is quantity of heat released per unit of organics oxidized (Kcal/g COD destroyed), and S_e is the biodegradable organic matter concentration surrounding the organisms in a completely mixed reactor at any given time (g/l biodegradable COD at time, t).

The rate of heat lost as sensible heat in the effluent liquid (H_E, Kcal/l) can be equated to the heat required to warm the influent liquid to the reactor temperature.

$$H_E = D_LC_{PL}\,(T_R - T_O)/\theta \tag{2}$$

where D_L is the slurry density, C_{PL} is the specific heat of the slurry, T_R is the reactor temperature, (°C), T_O is the influent liquid temperature (°C), and θ is the hydraulic retention time.

Thus the reactor temperature can be estimated by combining the heat balance and the substrate kinetics as in the following equation:

$$T_R = \frac{\theta K_TS_eR}{D_LC_{PL}} - \frac{H_L\theta}{D_LC_{PL}} + T_O \tag{3}$$

MATERIAL AND METHODS

Investigation of the aerobic thermophilic digestion process at Cornell University actually began in 1973 with the studies of Koenig[7] and Wright.[4] Their experiments were followed by construction of a full-scale system at Cornell's Animal Science Teaching and Research Center (ASTARC) in Harford, New York. This facility is fully equipped for thorough monitoring of the full-scale process and includes an on-site wet test laboratory. The full-scale studies were conducted with a commercially available two-stage, completely mixed system. It is marketed by the De Laval Separator Company as the "LICOM II" system. The duration of this study was 29 months.

Process Design

A diagram of the digestion system and some of its characteristics are shown in Figures 1 and 2. The cow manure for the study was obtained from a free-stall dairy cow area. This manure contained no bedding and was continuously removed to a sump located just outside the barn. The capacity of the sump (50 m^3) retained three to four days of manure plus an equal

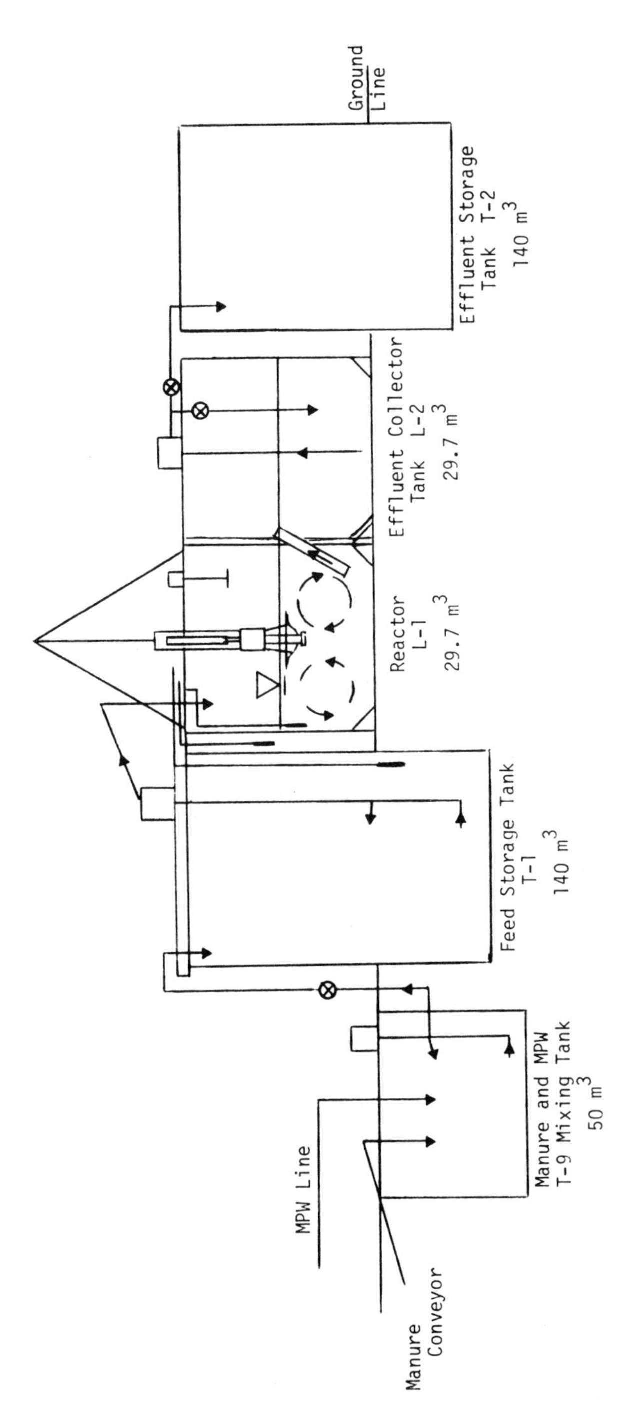

Figure 1. A schematic of the full-scale aerobic thermophilic digestion process at Cornell University.

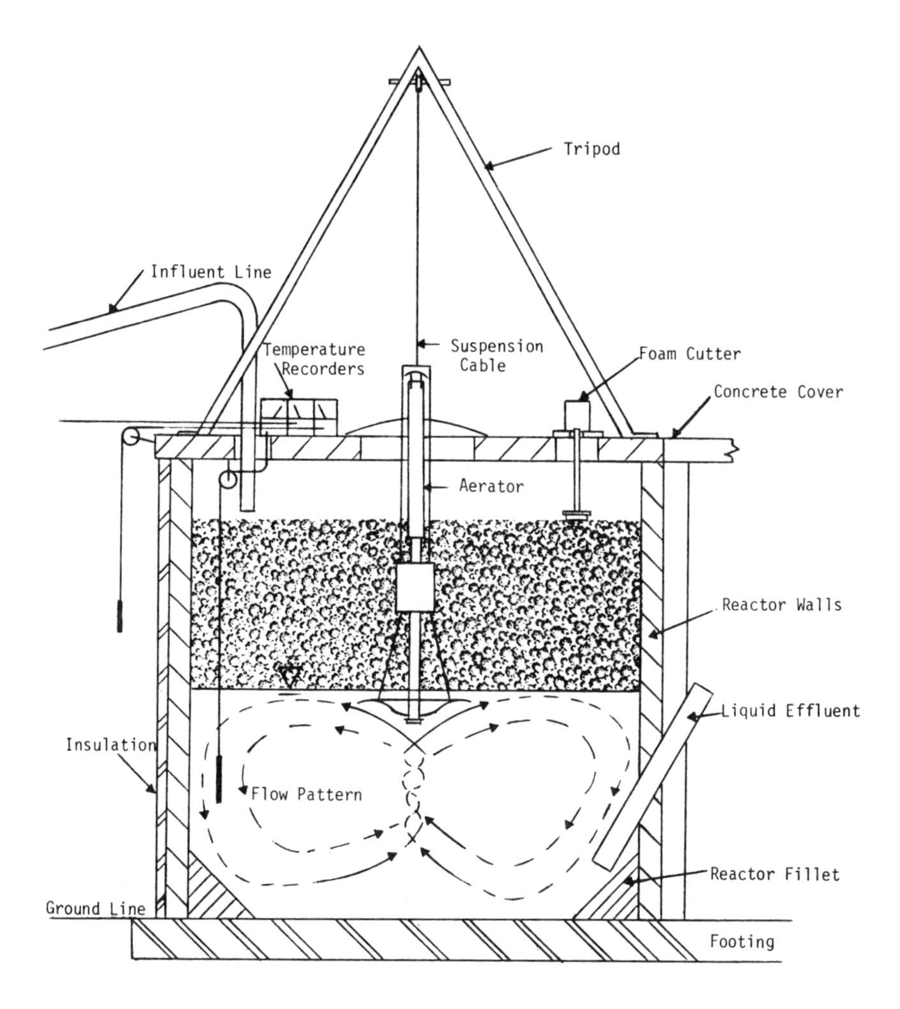

Figure 2. A schematic of the full-scale thermophilic aerobic digestion system.

volume of milking parlor washwaters (MPW). The resulting slurry of approximately 5% total solids was thoroughly agitated and transferred to the digester feed tank about twice per week, depending on the feed rate. The feed composition was measured after each major addition to the feed tank.

The feed tank was a cylindrical, reinforced concrete structure with an effective volume of 140 m^3. The substrate was pumped to the reactor on a semicontinuous basis which varied from once per hour to six times per day. The substrate in the feed tank was thoroughly agitated daily by manually placing the feed pump in the agitation mode. Sedimentation of the solids in the feed tank was not observed during the 24-hr quiescent period.

The cylindrical reactor was connected in series to a duplicate tank. The interior dimensions of each reinforced concrete tank measured 4.3 m in diameter and 3.7 m in height. Both tanks were positioned beside the feed tank totally above ground and constructed with reinforced concrete. The outside walls of both tanks were sprayed with 5 cm of polyurethane insulation.

To provide the proper geometry for mixing, according to earlier studies,[8,9] the liquid depth in the reactor was adjusted to equal one-half the reactor diameter, and the intersection between the floor and walls was filled with concrete as illustrated in Figure 2. The effective liquid depth in each reactor was 2.15 m, allowing 1.55 m of sideboard for foam dissipation. The resulting liquid volume was 29.7 m^3. A concrete cover was provided for both tanks.

Aeration and mixing in the reactor was accomplished with an aerator-agitator (De Laval Centri-Rator 480V, 3ϕ, 3.7 kWh, 1750 rpm) (see Figure 3). This unit is referred to by several manufacturers as a self-aspirating pump, since the vacuum created at the center of the aerator draws air down a hollow shaft and liquid up from the center of the tank, thus providing aeration without the need of a compressor or blower. It was designed to operate suspended from a tripod anchored to the reactor cover. The authors were not aware of information describing the fundamental aeration capabilities of the unit in tap water and organic slurries and attempted to evaluate these capabilities at various temperatures and immersion depths.

Mode of Operation

Full-scale research raises problems in obtaining a continuous and accurate description of materials flowing through the system. In order to obtain a mass balance, it was necessary to utilize a combination of extra tanks for definition of the experiments. Initial substrate was stored where it was easily mixed and sampled. This substrate was transferred to the single-stage, completely mixed reactor without recycle, on a semicontinuous basis. When the reactor was charged, liquid was forced by gravity from the bottom of the reactor to the second tank which was utilized as an effluent collection tank. This effluent liquid was accumulated for one hydraulic retention time, before it was mixed, sampled and taken to the field for land application.

Preliminary Investigation

Preliminary studies were conducted on the reactor using heated and unheated tap water to determine the following specific characteristics of the system:

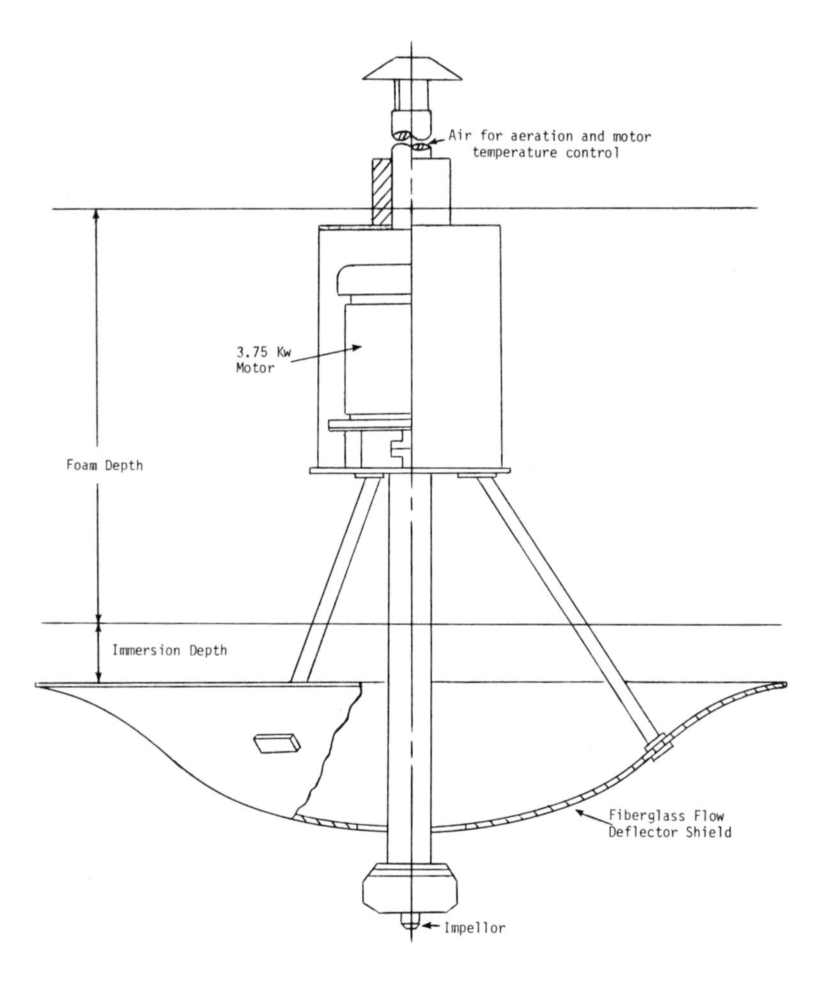

Figure 3. The self-aspirating aerator used throughout the study.

1. heat loss through the reactor surface area (H_A);
2. fundamental aerator characteristics at 20°C;
3. aerator characteristics at thermophilic temperatures; and
4. heat loss coefficient (H_L).

Heat loss to the surroundings (H_A) was defined as that loss through the floor, walls and cover of the reactor while the aerator was in the off position. This heat loss was determined by filling the reactor with tap water heated to 55°C. The ambient air and reactor temperatures were monitored hourly, as the reactor cooled, with a temperature recorder (Bristol Instrument Company).

After the reactor had cooled to 20°C, the fundamental characteristics of the reactor were investigated. This phase of the study was concerned with the determination of the airflow rate power consumption and oxygen transfer of the aerator at various immersion depths.

The nonsteady-state procedure was used to determine the oxygen transfer rate. The rate of increase in the dissolved oxygen concentration (DO) was measured with two DO meters (Yellow Springs Instrument Company, Model 54), calibrated with the azide modification for DO.[10] Simultaneously, power consumption and airflow rates (Datametrics Model 800-2) were monitored. Immersion depths ranging from the surface to 45.7-cm submergence were investigated at a water temperature of 20°C.

The temperature of the tap water was increased to 65°C and the characteristics of the aerator were determined at thermophilic temperatures. The procedure was the same as that at 20°C. The heat loss coefficient (H_L) was defined as the heat loss from the reactor while the aerator was operating in a batch mode.

Biodegradability Investigations

The maximum biodegradable fraction of the dairy wastes was determined with three long-term batch studies, one using a thermophilic bench-scale reactor (4 liters) and two using the full-scale reactor in a batch mode.

Sampling Methods and Procedures

A summary of all test conditions and some associated general characteristics are shown in Table I. Analytical data for the full-scale study were collected through a period of 23 months from March 1, 1975 to February 11, 1977. Throughout this period, temperatures, hydraulic flow rates and power consumption were monitored daily. Two independent methods were used to calculate the average airflow rate to the reactor, one based on air velocity readings (Flow Corporation, Model 800-2) and the other on manometer readings (Merian Instrument Company, Model 34FB2). The airflow rates determined with each method were within 5% of each other. The average of these two values was used in the study to calculate the average oxygen flow rate into the system. The atmosphere was assumed to be 20.95% oxygen by volume and the air density was taken at the ambient air temperature.[11]

The oxygen transfer efficiency was related to the oxygen used in this system measured as COD removed per day (CRR). Oxygen flow rates, transfer efficiencies and oxygen transfer rates were calculated using Equations 4, 5 and 6, respectively, during period of steady-state operation:

Table I. Summary of experimental test conditions for the duration of the study.

Influent characteristics of diluted dairy cow waste and milk waste in g/l			
Total solids	46.7	-	70.0
Total volatile solids	39.9	-	62.8
Total Kjeldahl nitrogen	3.4	-	5.5
Ammonia-nitrogen	1.6	-	2.1
pH	7.3	-	8.6
Chemical-oxygen demand (COD)	42.9	-	57.7
Potential total solids biodegradability		56%	
Potential total COD biodegradability		54.2%	
Biodegradable COD	23.2	-	30.7
Hydraulic Retention Times	1.3 to 11.2 days		
Organic Loading Rates			
2.4 to 42.6 (gTS/l reactor-day) [150 to 2660 lb/1000 ft^3-day]			
Operating Temperatures			
Reactor	30° - 68°C		
Ambient	-15° - 28°C		
Feed	1.0° - 23°C		

$$Q_{go} = (Q_{gA} \text{ x } .2095 \text{ x } \Upsilon_A)/1000 \quad (4)$$

$$OTE = (CRR \text{ x } V)/(10\ Q_{go}) \quad (5)$$

$$QTR = (CRR \text{ x } V)/(24 \text{ x } 1000 \text{ x } P) \quad (6)$$

where,

Q_{go} = average oxygen flow rate into the reactor in Kg/d;
Q_{gA} = average airflow rate into the reactor in Kg/d;
Υ_A = average air density at ambient air temperature T_A in g/c;
OTE = oxygen transfer efficiency in %;
CRR = COD removal rate in g/l-day;
OTR = oxygen transfer rate in kg/kWh; and
P = power consumption in kWh.

Steady-state conditions were examined after continuous operation for five consecutive HRTs ranging from 1.3 to 11.2 days. The unit was operated for a period equal to at least six hydraulic retention periods before steady-state sampling initiated. Extensive steady-state sampling was repeated over a duration of time equal to 3 to 7 consecutive HRTs. A steady-state condition was defined as the condition where the influent and effluent composition from consecutive HRT examinations changed no more than 10%. An average influent and effluent composition was numerically determined from the consecutive HRT values to define a loading condition. In this study, 21 steady-state HRTs were achieved to define 5 different loading rates.

Chemical analysis of liquid samples was conducted according to *Standard Methods*[10] unless otherwise specified. The following analyses were performed on most liquid samples: total solids, total volatile solids, total Kjeldahl nitrogen, total chemical oxygen demand[12] and pH. Nitrogen analysis was not performed on the bench-scale unit.

RESULTS

Aerator Performance in Tap Water

The conventional procedure for evaluating aerators is to measure their capabilities in tap water, and these characteristics are then adjusted for the particular waste for specific applications. Thus, the following data on tap water were obtained to be able to correlate the aerator characteristics in tap water with those in organic slurries. The aerator characteristics examined in tap water were immersion depth and effects of temperature on aerator performance. The aerator characteristics and ranges of immersion depths and temperatures examined are summarized in Figure 4. The average airflow rate was maximum at aerator immersion depths between 5 and 10 cm. However, at greater immersion depths the rate decreased and the oxygen transfer rate remained relatively constant. Therefore, to maximize airflow and oxygen transfer an immersion depth of 8 cm was used throughout the remainder of the study. These data indicate that both the oxygen transfer rates and efficiencies at 20°C were about equal to other aerators. Oxygen transfer efficiency decreased to less than 4% at all immersion depths at 60°C.

Tap water was also used to define the thermal characteristics of the reactor in terms of H_A and H_L. The reactor (T_R) and ambient air (T_A) temperatures were monitored while the reactor cooled, as illustrated in Figure 5.

Substrate Biodegradability

The biodegradability of the substrate (dairy manure and MPW at approximately 5% TS) was determined in terms of COD (g/l). The average of all

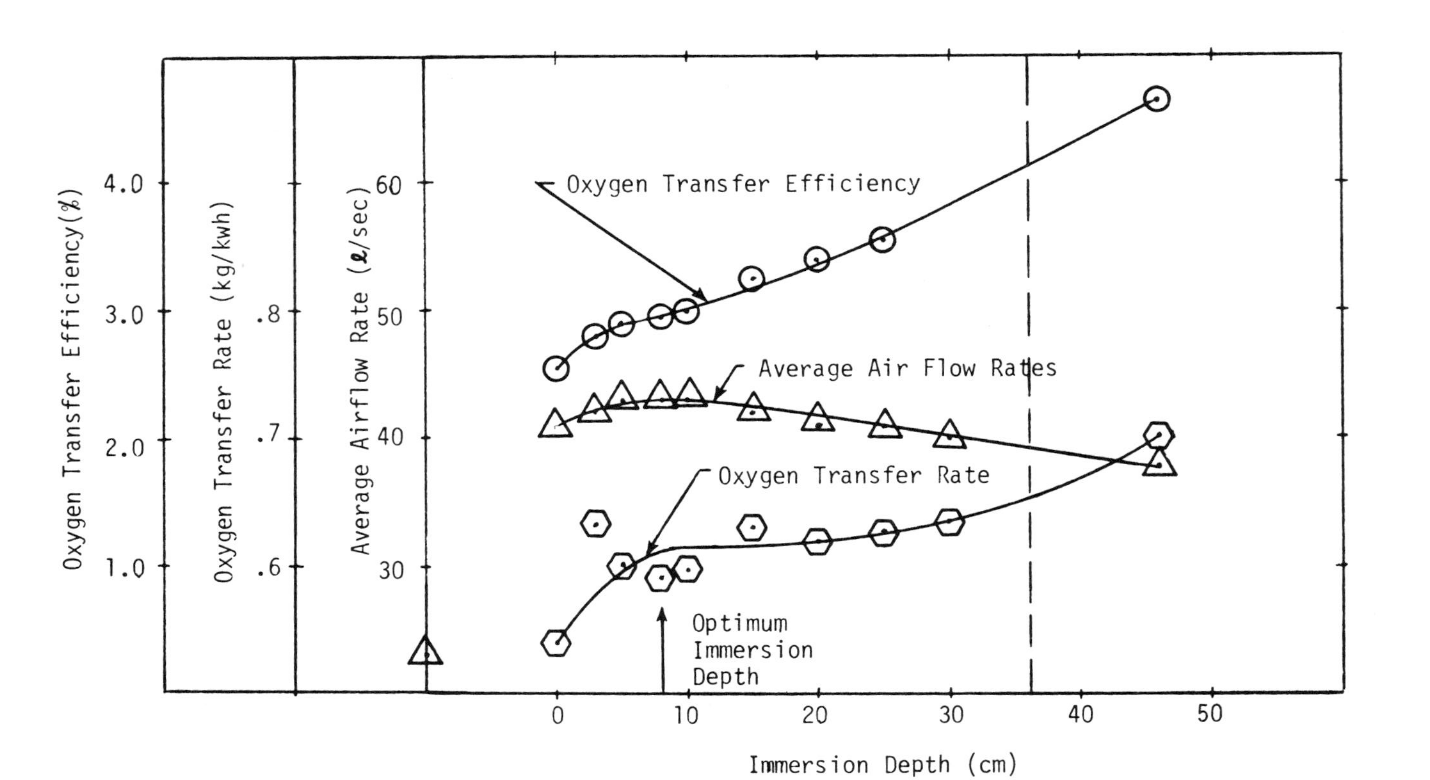

Figure 4. The optimum immersion depth was selected from tap water testing in the reactor at 20°C.

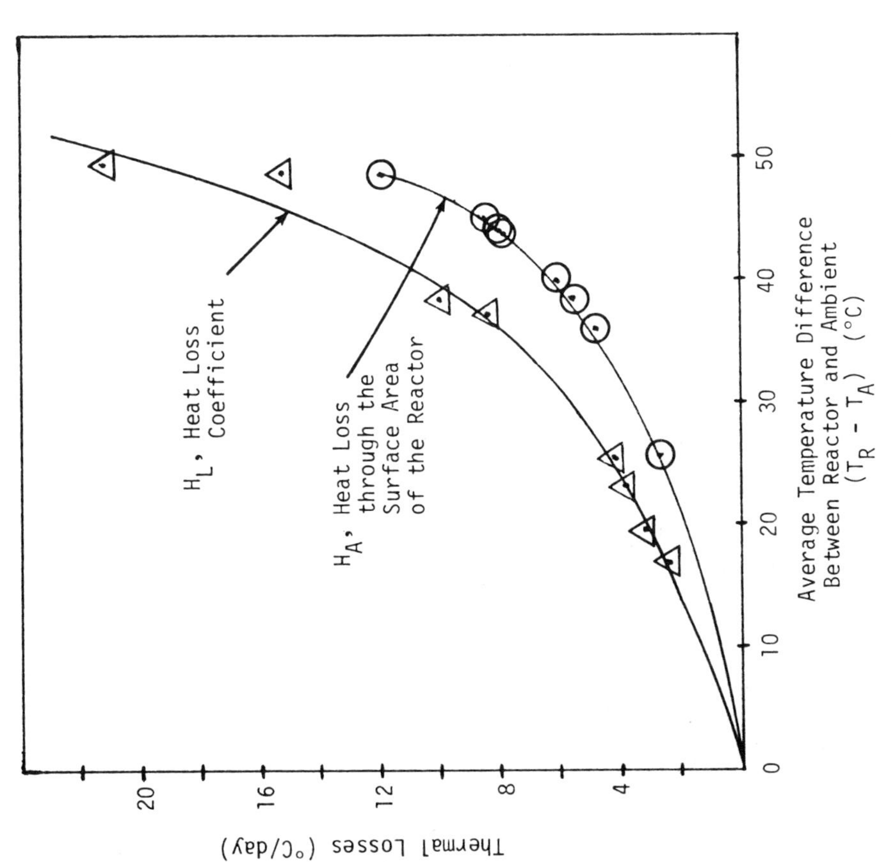

Figure 5. Definition of the thermal characteristics of the aerobic digester tank with tap water.

these tests indicated that 54.2% of the total COD was removed after 35 days of operation under batch conditions. This was assumed to be the maximum biodegradable fraction of the substrate.

Steady-State Performance

From the first start-up of this unit with the dairy waste it was clear that autoheating was easily achieved. Only under severe stress conditions was it possible to cause the temperature of the reactor to fall below the thermophilic range (less than 45°C). The feed substrate concentrations were maintained relatively constant through the steady-state study; however, loading rates over a wide range were examined by varying the flow rate to the system. The single-stage reactor was fed semicontinuously 24 times per day at the highest loading rate (HRT of 1.3 days) and 6 times per day at the lowest (HRT of 11.2 days). It was operated completely mixed without recycle; therefore, the hydraulic retention time (HRT) and solids retention time (SRT) were assumed equal. The data which defined each of the loading conditions were calculated by averaging the results from 3 to 7 HRTs. A loading condition was altered by adjusting the hydraulic flow rate (O_L, liter/day) to the reactor and was determined in terms of the hydraulic loading rate (SRT, days) and the organic loading rate (OLR, g/l-day).The steady-state performance at each loading condition was defined in terms of reaction rate (K_T, day^{-1}), removal rate (RR, g/l-day), removal efficiency (RE, %), temperature (T, °C) and aerator performance.

A summary of the relationship of substrate removal kinetics and its relation to organic loading rates is shown in Figure 6. As the organic loading rate was increased (or the SRT decreased), the reaction rate coefficient also increased due to the increased availability of biodegradable substrate [measured as $(BCOD)_e$]. K_T and the organic removal rate increased to a maximum value of 0.58 (day^{-1}) and 5.7 (g/l-day), respectively, at a loading rate of 9 g of biodegradable COD per liter per day (g BOD/l-day). At greater loading rates, K_T decreased rapidly, resulting in a decreased efficiency of removal.

The reactor operated at thermophilic temperatures throughout most of the study. However, at the highest loading rate (lowest SRT), the steady-state reactor temperature decreased from 49°C at 9 (g/l-day) to 31°C at 20 (g/l-day). K_T decreased with temperature from 0.58 day^{-1} to 0.29 day^{-1}. This decrease in reactor temperature resulted from the large amount of heat carried out of the system at the high flow rates.

A summary of all heat losses is shown in Figure 7. The heat in the effluent waste (H_E) was always the greatest individual loss from the system, while the heat loss due to aeration was the smallest. However, the combined effects of

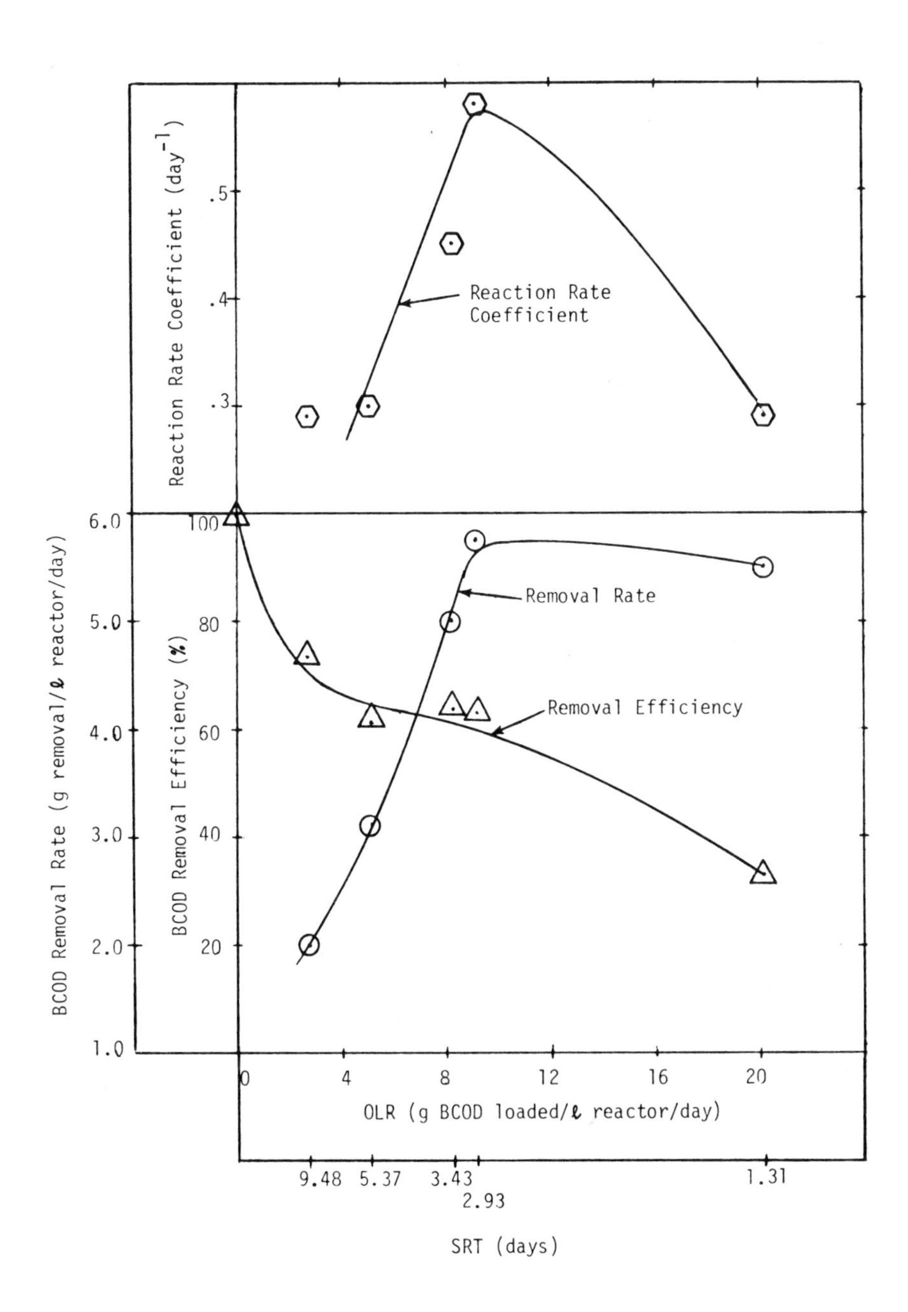

Figure 6. Relationship of loading rate, SRT, substrate removal rate and substrate removal efficiencies.

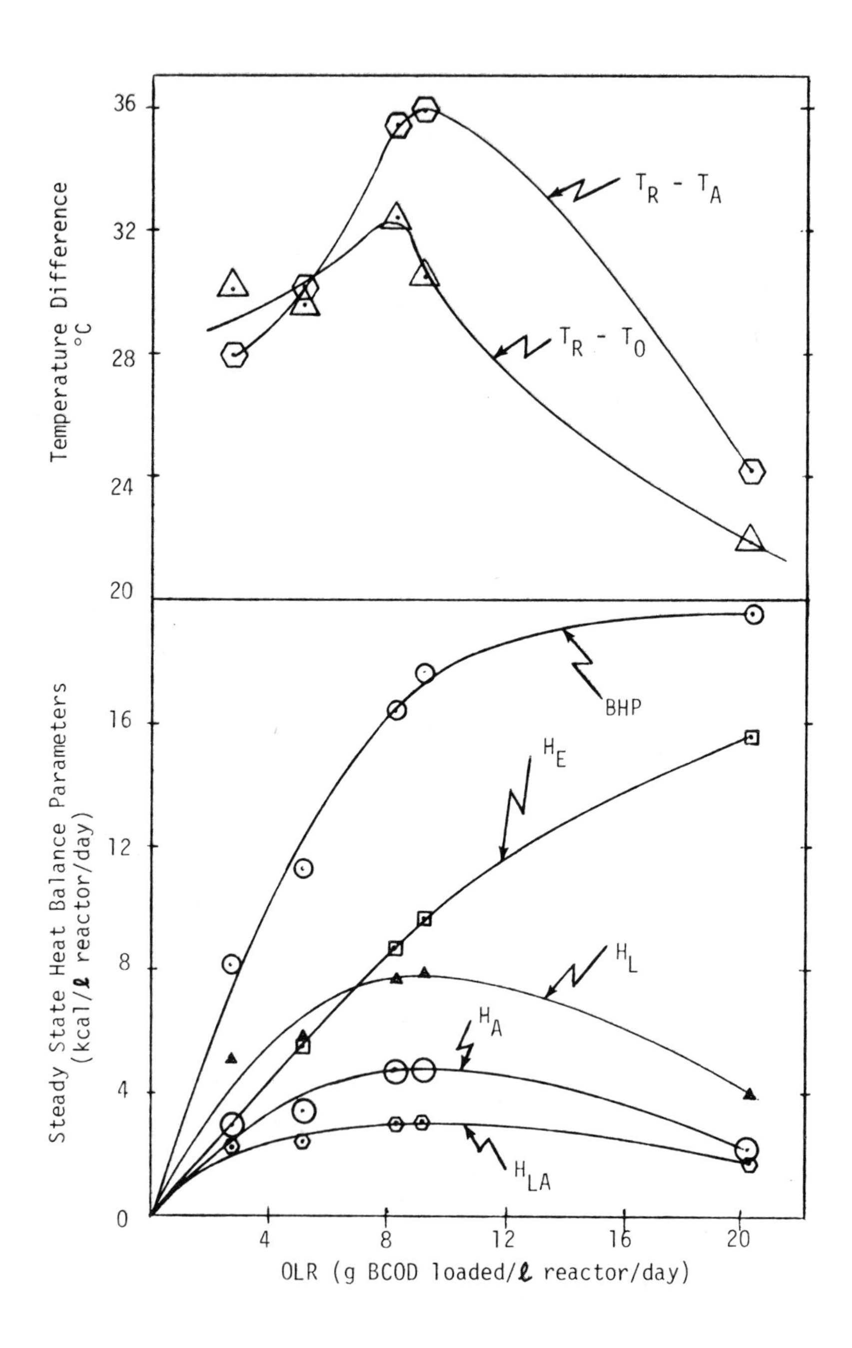

Figure 7. Illustration of the various components in the heat balances in relation to the organic loading rate with the dairy manure substrate.

of H_A and H_{LA} on the heat loss coefficient (H_L) was slightly larger than H_E at loading rates less than 7 (g BCOD/l-day). The highest temperature differences occurred at a loading rate of 9 g BCOD/l-day .

The heat energy released during respiration (R) is related to the substrate removal rate. The quantity of heat released decreased as the loading rate increased, as shown in Figure 8. At a loading rate of 9 (BCOD/l-day), R was 3.03 (Kcal/g). At greater loading rates, the reactor temperature and therefore the reaction rate decreased, which resulted in an increase in the amount of heat released to the surroundings. Accurate definition of the value of heat released per unit of organics oxidized is a critical parameter in determining autoheating potential.

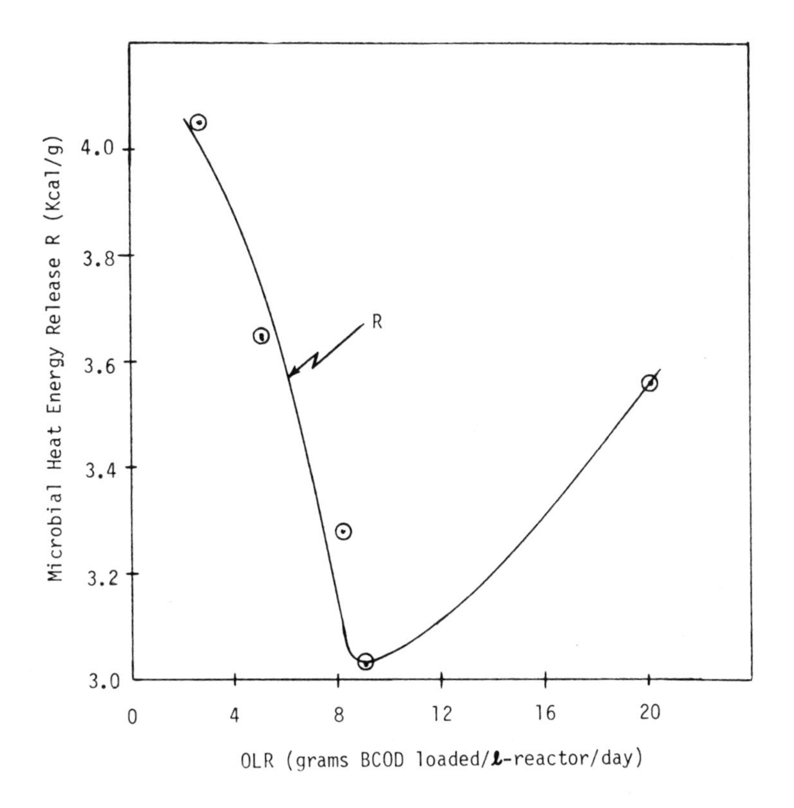

Figure 8. The quantity of heat released during respiration decreased as the loading rate was increased to 9 (g/l-day). The energy transfer efficiency decreased at greater loading rates and more heat energy was released to the surroundings.

The aerator characteristics were determined in terms of efficiency and rate of utilization of oxygen by the microorganisms. Oxygen transfer efficiency (OTE) was expressed as kilograms of oxygen used (measured as COD removed) per kilowatt-hour (kg/kWh). These data are summarized in Figure 9. The average airflow rate during operation with the substrate was approximately half the measured values in tap water. At a reactor temperature of 49°C and a loading rate of about 9 (g/l-day) the OTE and OTR were 36% and 1.6 kg/kWh (1.19 lb O_2/hp-hr), respectively. These values were 11 and 4 times greater than the tap water values at the same temperature and immersion depth.

The nitrogen losses at all loading conditions were low. The maximum removal of TKN was 13% at a 9.5-day SRT.

DISCUSSION

The data presented here illustrate the empirical data needed to describe the autoheated thermophilic digestion process for liquid slurries. The reactor temperature was found to be dependent upon the parameters of the steady-state heat balance equation and the substrate removal kinetics. The rate of heat loss in the effluent waste was the most significant loss at all loading conditions in the system. Minimizing this loss by concentrating the influent substrate or by recovery of the effluent heat using a heat exchanger would result in even higher reactor temperatures. A higher reactor temperature could also be achieved with additional insulation of the reactor. The rate of heat loss due to aeration (H_{LA}) was the least significant loss at the loadings conditions tested in the study.

The rate of biological heat production was directly related to the heat released by the microorganisms (R), the reaction rate coefficient (K_T) and the availability of biodegradable substrate [measured as $(BCOD)_e$]. When substrate availability was nonlimiting (*i.e.*, high loading rates), R and K_T were sensitive to the resulting reactor temperatures.

The magnitude of R decreased from 3.56 to 3.03 (Kcal/g) when the temperature was increased from 31 to 49°C. The change in magnitude of R is a result of the microbial yield that occurs during oxidation of the organics. Slightly more heat was released in the mesophilic regions than in the thermophilic, probably due to the decrease in substrate solubility as pointed out by Andrews.[2] However, this slight decrease was more than compensated for by the doubling of K_T from 0.29 to 0.58 (day^{-1}) at the same temperatures. This doubling of K_T represents one of the most significant advantages of the autoheated aerobic digestion process, since higher reactor rates result in the ability to achieve many other benefits from thermophilic digestion.

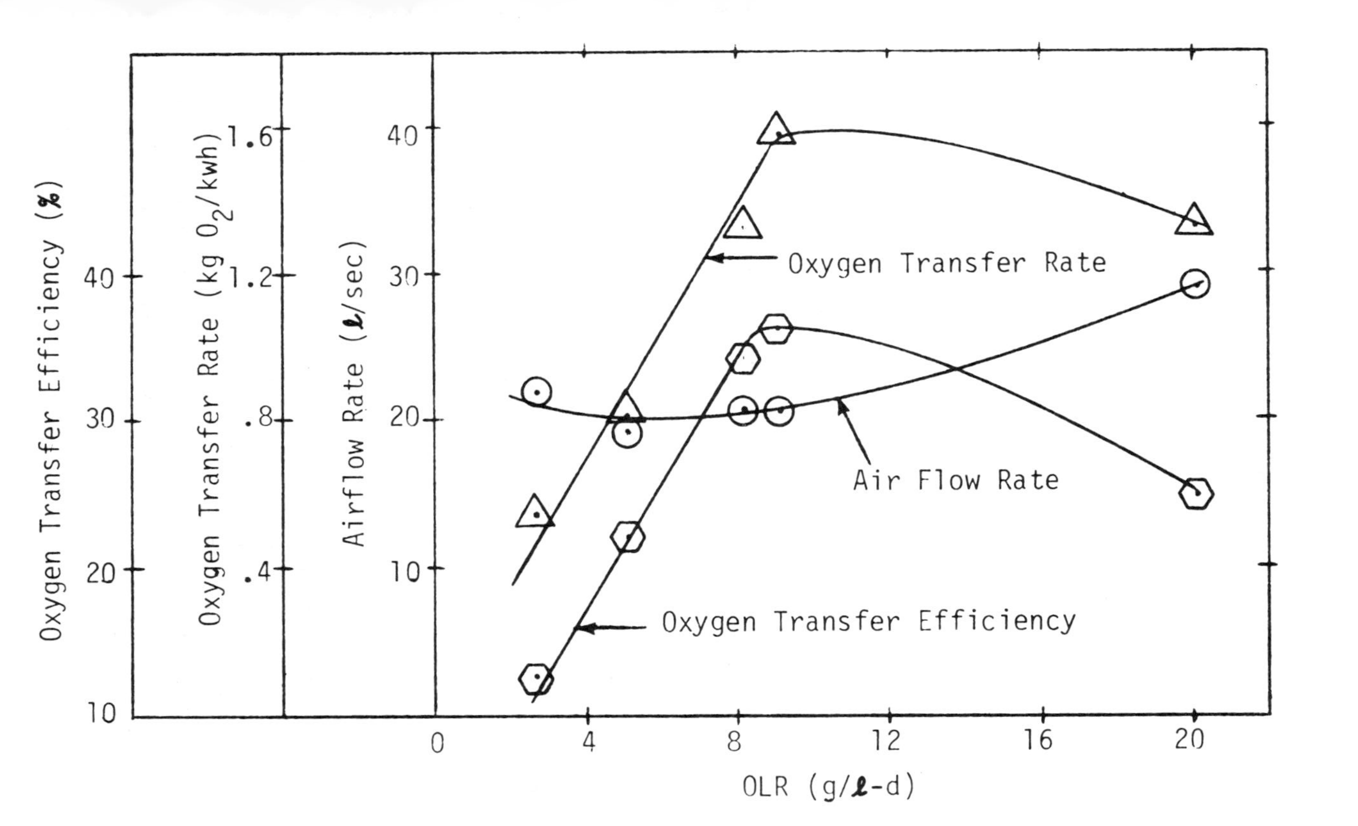

Figure 9. The aerator performance during steady-state operation was significantly improved over operation in tap water. OTE was 11 times greater, OTR was 4 times greater and the inflow rate was half that in tap water.

The tap water investigation of the aerator characteristics indicated that increased temperature would decrease aerator performance at an 8-cm immersion depth. However, as a result of the increased K_T value and efficient oxygen utilization by the thermophilic microorganisms, the oxygen transfer efficiency (OTE) of 36% was 11 times greater than that in tap water. The improved OTE resulted in a simultaneous improvement to the rate of oxygen transfer, which was 1.6 kg/kWh and four times greater than that in tap water. These data indicate that the conventional method of specifying aeration requirements at high temperatures with the equipment is difficult to apply.

CONCLUSIONS

- Concentrated organic slurries could be rapidly stabilized at thermophilic temperatures. The reaction rate coefficient (K_T) doubled when the temperature of the reactor contents was increased from 31 to 49°C.
- Thermophilic temperatures were achieved with autoheating with a simple air aeration system.
- Insulation of the reactor and oxygen transfer efficiencies of 12 to 36% enabled autoheating to be effective.
- Heat losses through the surface area of the reactor and due to aeration were quite small at all loadings due to the insulation and efficient oxygen transfer.
- Heat loss in the effluent liquid (H_E) was the most significant loss from the system. At high loadings of 42.6 g COD/liter reactor-day, H_E was four times the combined effects of all other heat losses.
- High loading rates of up to 9 g BOD/l-day resulted in 60% removal of the biodegradable substrate in 3 days HRT. The reactor temperature was 36°C above the ambient air temperature.

ACKNOWLEDGMENTS

The cooperation of the personnel at the Animal Science Teaching and Research Center was invaluable to the success of this study. The authors wish to express their gratitude to Dr. Samuel Slack, Dr. James Harner, Albert Oltz, A. T. Sobel, Bill A. Harrower, John Riley and the faculty and staff of the Agricultural Engineering Department of Cornell University.

The authors also wish to acknowledge the DeLaval Separator Company for their advice in the construction and operation of their "LICOM II" system.

REFERENCES

1. Jewell, W. J. and R. J. Cummings. "Autoheated Thermophilic Aerobic Digestion with Air Aeration: Full Scale Application to Dairy Wastes," Report of the Department of Agricultural Engineering, Cornell University, Ithaca, New York (in preparation, 1977).
2. Andrews, J. and K. Kambhu. "Thermophilic Aerobic Digestion of Organic Solid Waste," Clemson University, Clemson, South Carolina (May 1971).
3. Matsch, L. C. and R. F. Drnevish. "Autothermal Aerobic Digestion," paper presented at the 48th Conference Water Pollution Control Federation, Miami, Florida, October 5-10, 1975.
4. Wright, P. J. "Thermophilic Aerobic Digestion of Concentrated Organic Slurry," M.S. Thesis, Cornell University, Ithaca, New York (1975).
5. Smith, J. E. Jr., K. W. Young and R. B. Dean. "Biological Oxidation and Disinfection of Sludge," unpublished EPA paper.
6. Grant, F. "Liquid Aerobic Composting of Cattle Wastes and Evaluation of By-Products," Project No. 5801647, U.S. Environmental Protection Agency.
7. Koenig, A. "Heat Generation during Aerobic Oxidation of Concentrated Organic Wastes," M.S. Thesis, Cornell University, Ithaca, New York (1974).
8. Machine Testing Report of the German Agricultural Society, Department of Farm Machinery Tests, Frankfurt am Main, Germany, No. 1971, Group 4d/18.
9. Popel, F. and Ch. Uhnmacht. "Thermophilic Bacterial Oxidation of Highly Concentrated Organic Substrates," *Water Res.* 6:807 (1972).
10. American Public Health Association. *Standard Methods for the Examination of Water and Wastewater,* 14th ed. (New York: American Public Health Association, 1975).
11. The Chemical Rubber Company. *Handbook of Chemistry and Physics,* 50th ed. (1969-1970), p. F151.
12. Jeris, J. S. "A Rapid COD Test," *Water Waste Water Eng.* 4 (1967).

43

PERFORMANCE OF AN ANAEROBIC WASTE TREATMENT LAGOON SYSTEM

R. K. White and R. L. Curtner
Agricultural Engineering Department
R. H. Miller
Agronomy Department
The Ohio Agricultural Research and Development Center and
The Ohio State University
Wooster and Columbus, Ohio

INTRODUCTION

Construction of a new Animal Science Livestock Center at the Ohio State University was completed in 1973. The two-stage lagoon system for waste treatment was monitored from 1974 through 1976. Manure flushed from the swine facility, runoff from beef lots and sanitary sewage from workers' quarters and public restrooms were pumped to the lagoon system for treatment. Beef manure and horse manure were field spread directly or stored until conditions permitted spreading. Figure 1 shows the location of the various livestock units and the two-stage lagoon system. The location of four test wells adjacent to the lagoons and a control well for monitoring ground water are also shown.

In the swine facility, manure is flushed from beneath slotted floors to receiving pits—one pit for the finishing section and one for the farrowing and nursery section. Automatic siphons and tip tanks are used. Fresh water is used and flushing is scheduled twice a day. The wastewater in the receiving pits is pumped to the dewatering building every two or three days. During 1974 and 1975, all livestock wastewaters were pumped over a run-down screen. The solid particles were stockpiled in the manure storage building and the filtrate was pumped to the lagoon system. Beef lot runoff was collected

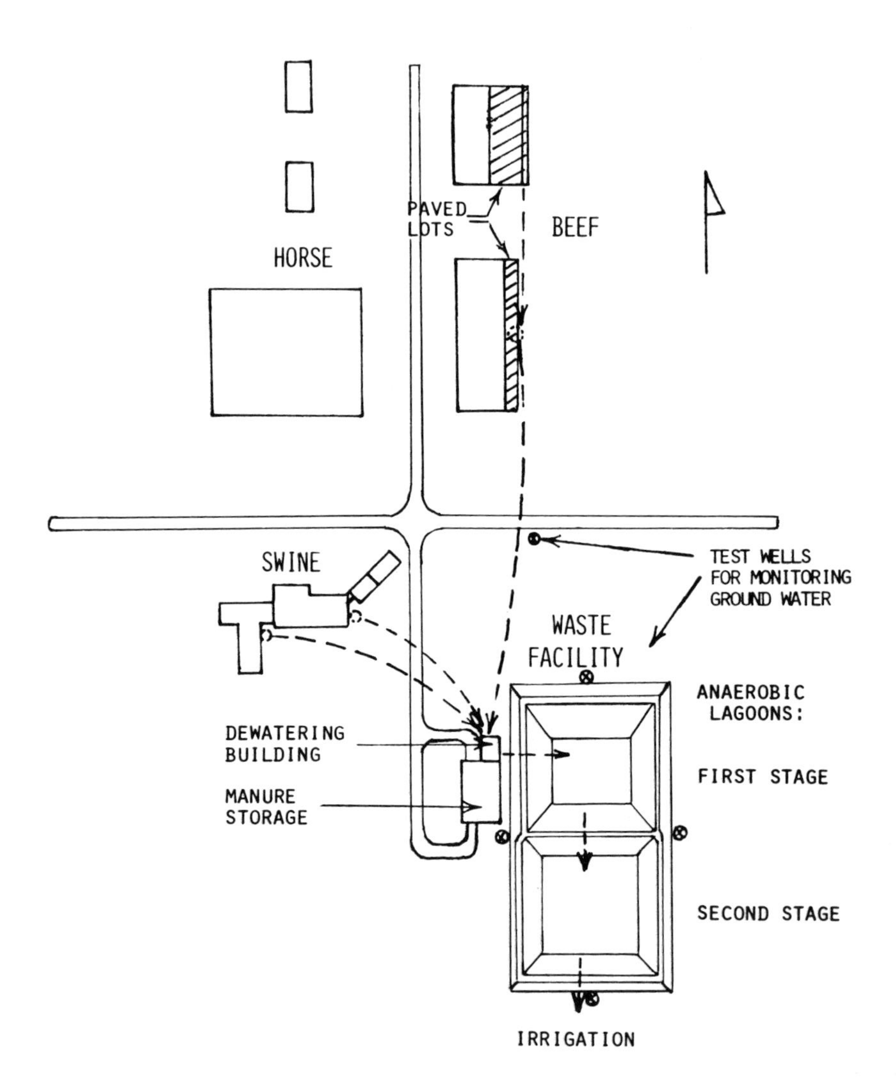

Figure 1. Location of two-stage lagoon system at the Ohio State University Animal Science Livestock Center.

in an underground tank and pumped periodically to the dewatering building and then to the lagoon system. Sanitary sewage had a separate collection system and was pumped directly to the lagoons. Originally, an aeration package plant was designed for treating the sanitary sewage with the effluent to be discharged to a drainage ditch. An engineer with the Ohio Environmental Protection Agency recommended that the sanitary sewage be treated in the lagoon system, which has a planned wastewater irrigation discharge. Table I gives a summary of the livestock population and the wastewater sources, volume and characteristics.

Table I. Size of facility and estimated wastewater production.

Facility	Number	Wastewater Volume (ft^3/day)	Wastewater Volatile Solids (lb VS/day)
Swine (plus flushwater)[a]			
Farrowing	24 crates		
Nursery	168 pigs	275	279
Growing/Finishing	520 pigs		
Beef Runoff			
120 Cow Barnlot	5760 ft^2		
120 Heifer/Feeder Lot	9600 ft^2	110	17 (307)[b]
Sanitary Sewage			
Student Quarter	12 persons	215	4
Public Restrooms (Assembly Area 200 Seats)			
Totals		600	300 (590)

[a]80% animal occupancy.
[b]lb VS in runoff water from a 1.5 in. rainfall event estimated at 0.2% w.b.

The first lagoon has a volume of 9300 m^3 (328,000 ft^3) with a surface area of 0.37 ha (0.92 ac) and a depth of 3.7 m (12 ft). The designed loading rate was 23 g VS/m^3 (1.8 lb VS/1000 ft^3). The second lagoon was originally designed to be an aerated lagoon. However, because of energy cost and adequate odor control due to a low loading rate, it is functioning as a facultative lagoon to date. The second lagoon has a maximum operating volume of 8200 m^3 (291,000 ft^3) with a surface area of 0.39 ha (0.98 ac) at a depth of 2.74 m (9 ft). The volume between the 6- and 9-ft depths was designed as storage for the irrigation system. No wastewater effluent has been irrigated to date, but irrigation will be started either in the fall of 1977 or the spring of 1978.

About 1 m (3 ft) of rainwater was collected in the lagoons before the addition of any wastewater. Because it was planned to use fresh water for flushing, no attempt was made to fill the lagoons to their design depth.

The objectives of this research project were to (1) monitor the performance of the lagoons with regard to the stabilization of the wastes, (2) test ground water to determine if any seepage from the lagoons and contamination occurred and (3) evaluate the performance of the lagoon system with respect to odor nuisance.

EXPERIMENTAL PROCEDURE

The parameters that were analyzed and used in evaluating the lagoon performance were total solids (TS) wet basis, volatile solids (VS) dry basis, total suspended solids (TSS) wet basis, biochemical oxygen demand (BOD), chemical oxygen demand (COD), total Kjeldahl nitrogen (TKN), ammonia nitrogen (NH_3), conductivity, pH, temperature, total coliform (TC), fecal coliform (FC), fecal streptococcus (FS), and heterotrophic bacteria by plate count on nutrient agar and glucose yeast extract agar.

Samples of swine wastewater and beef lot runoff influent were collected periodically in 1975 to establish a range for the loading parameters. No samples of sanitary sewage were collected and tested. The characteristics of the influent samples are presented in Table II.

Table II. Range and mean value of livestock wastewater influent parameters. Samples collected in 1975.

Parameter	Units	Range		Mean	Number of Samples
		Low	High		
TS_{wb}	%	0.33	1.86	1.0	29
VS_{db}	%	56	82	69	29
TSS_{wb}	%	0.21	1.09	0.65	27
BOD	mg/l	1340	3600	2300	29
COD	mg/l	4800	19400	14500	29
Conductivity	μmho/cm	3900	12000	7700	38

Samples of the first-stage mixed liquor were collected at the point of effluent discharge. Samples of the second-stage mixed liquor were collected at a point about 1.5 m from shore and at 0.1 m depth.

All tests were conducted according to the procedure in *Standard Methods for the Analysis of Water and Wastewaters*[1] except as follows: TSS were conducted by centrifuging a 40-ml sample at 4000 rpm for 15 min, decanting the clear supernatant and drying the settled solids. The BOD values were obtained using an oxygen meter, calibrated before each use.

The four test wells and the control well, indicated in Figure 1, were made by using a 3-in. power auger which drilled holes to the shale layer at about 10 ft. Eight to ten inches of gravel were placed at the bottom of the hole, a 3/4-in. pipe was inserted, another 8 to 10 in. of pea gravel and sand were inserted and the hole was then filled with bentonite clay. All wells were pumped out several times before well water samples were collected. A 1/4-in. stainless steel tube, sterilized with methanol, and a hand vacuum pump were used to collect well water sampled in a sterilized flask. Well water samples were analyzed for conductivity and for the presence of fecal indicator bacteria (TC, FC, FS).

RESULTS AND DISCUSSION

The water quality parameters will be considered first. Figures 2, 3 and 4 present the solids data (TS, VS and TSS) of both lagoons for 1975 and 1976. There is no evidence of a significant increase in any of the solids parameters. The use of fresh water for flushing the swine wastes and the dilution effect of the sanitary sewage combine to keep these solids parameters from increasing. There is no indication that seasonal differences affect these parameters.

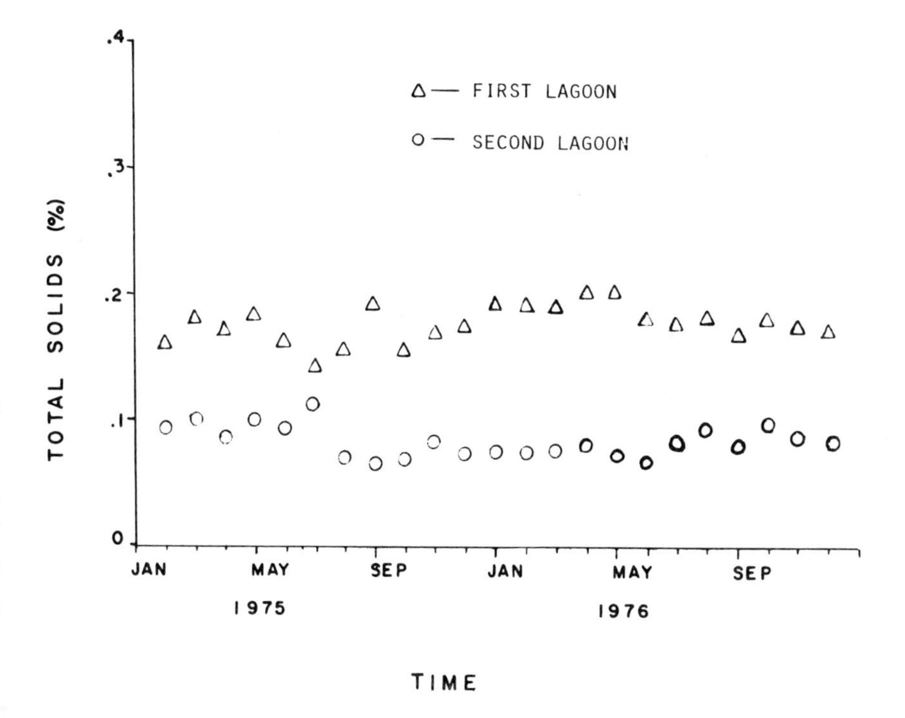

Figure 2. Total solid content of mixed liquor in two-stage lagoon system.

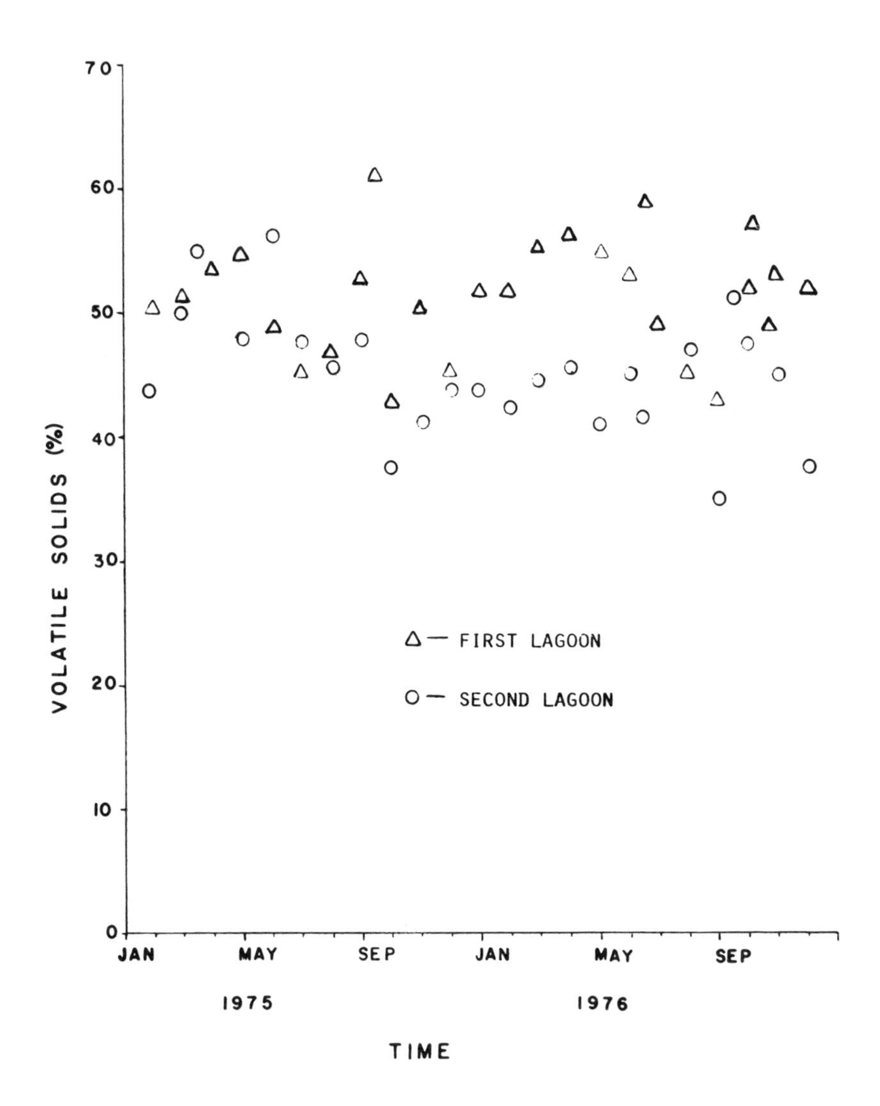

Figure 3. Volatile solids (dry basis) of mixed liquor in two-stage lagoon systems.

There is an indication that the BOD values for both lagoons have a seasonal change. Figure 5 shows that the BOD value in both lagoons increased in the winter and then dropped in the warmer months. The rise of BOD in the second lagoon in September was probably due to algae growth in the mixed liquor. A similar response can be seen in Figure 6 for the COD values of the mixed liquor in both lagoons.

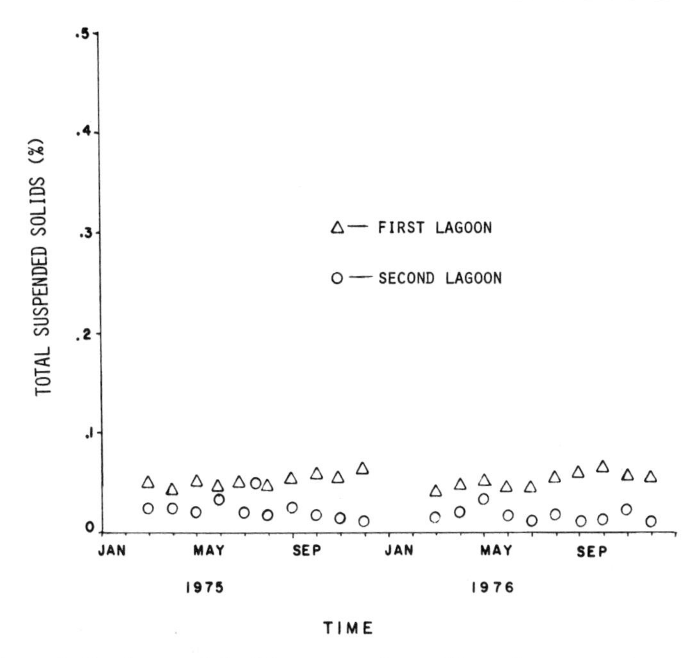

Figure 4. Total suspended solids content of mixed liquor in two-stage lagoon system.

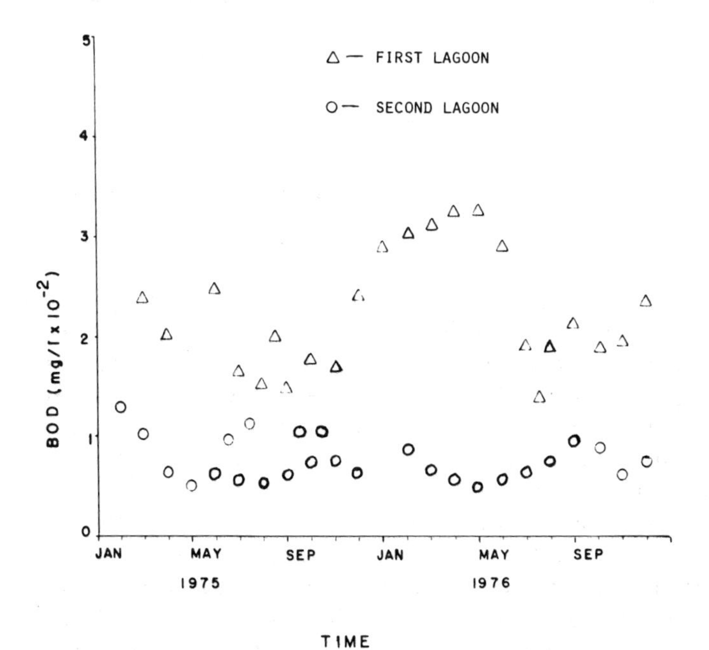

Figure 5. Biochemical oxygen demand (5-day) of mixed liquor in two-stage lagoon system.

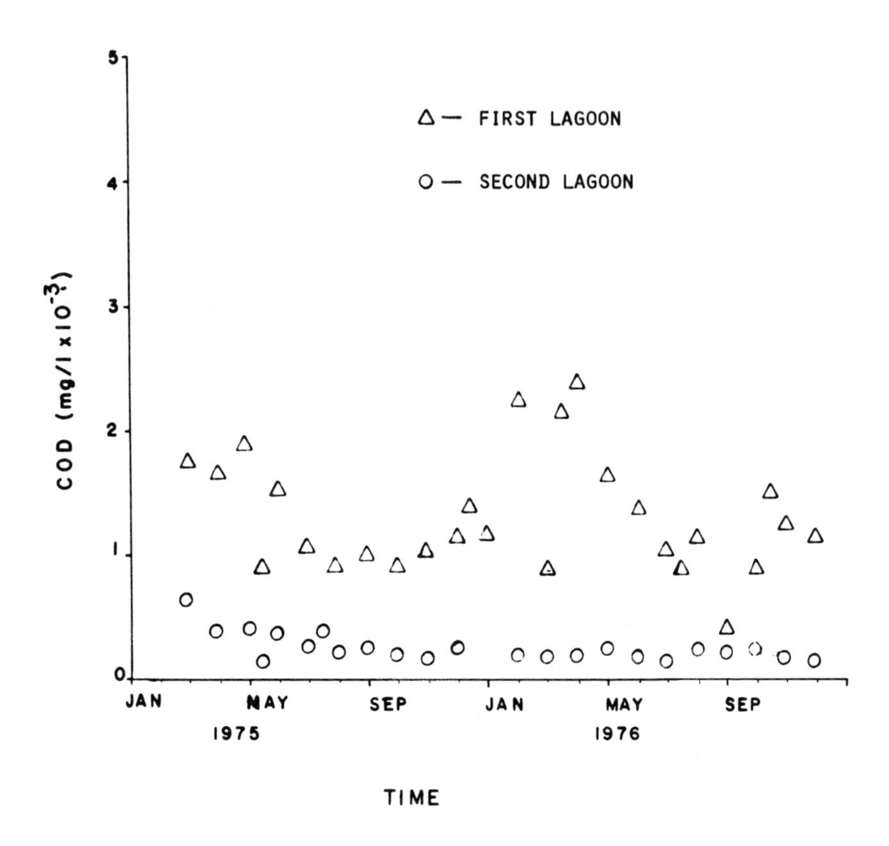

Figure 6. Chemical oxygen demand of mixed liquor in two-stage lagoon system.

The conductivity of the mixed liquor had a mean value of 3700 μmho/cm in the first lagoon. No increase in the overall conductivity of the water occurred in either lagoon (Figure 7). The rise in the conductivity in the late winter is probably due to increased discharge of effluent from the first to the second lagoon at this period of the year. During the summer and fall months, little or no effluent was discharged to the second lagoon due to higher evaporation and/or seepage.

A comparison of the mean values of the influent parameters with those of the mixed liquor in the first lagoon indicate a well-functioning lagoon. Table III gives a summary of the lagoon performance and the percentage reduction of TS, TSS, BOD and COD. The 92% reduction of TSS and BOD is quite good. However, the effluent (ML) of the first lagoon would not meet water quality standards for discharge into a stream. The reduction of VS from 60 to 50% indicates a high degree of stabilization. The reduction

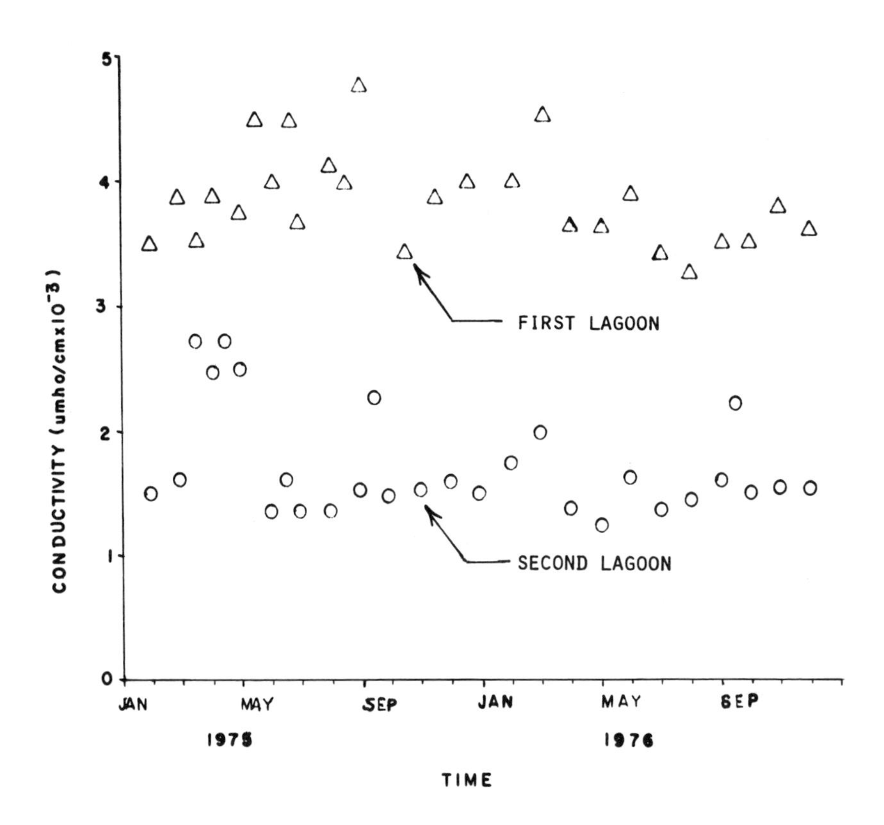

Figure 7. Conductivity of mixed liquor in two-stage lagoon system.

Table III. Summary of first-stage lagoon performance.

Parameter	Units	Influent ($\bar{x}$)	First Lagoon ML ($\bar{x}$)	Reduction (%)
TS_{wb}	%	1.0	0.18	82
VS_{db}	%	69	50	–
TSS_{wd}	%	0.65	0.05	92
BOD	mg/l	2300	180	92
COD	mg/l	14500	2200	85
Conductivity	μmho/cm	7700	3700	–

in conductivity can probably be attributed to dilution and the tying up of dissolved salts by the heterotrophic bacteria of the lagoon.

Figure 8 gives the concentration of total ammonia nitrogen in the mixed liquor of both lagoons. In the first lagoon, between 80 and 90% of the nitrogen is in the ammonia form. In the second lagoon, ammonia constitutes from 60 to 80% of the nitrogen depending on the time of year.

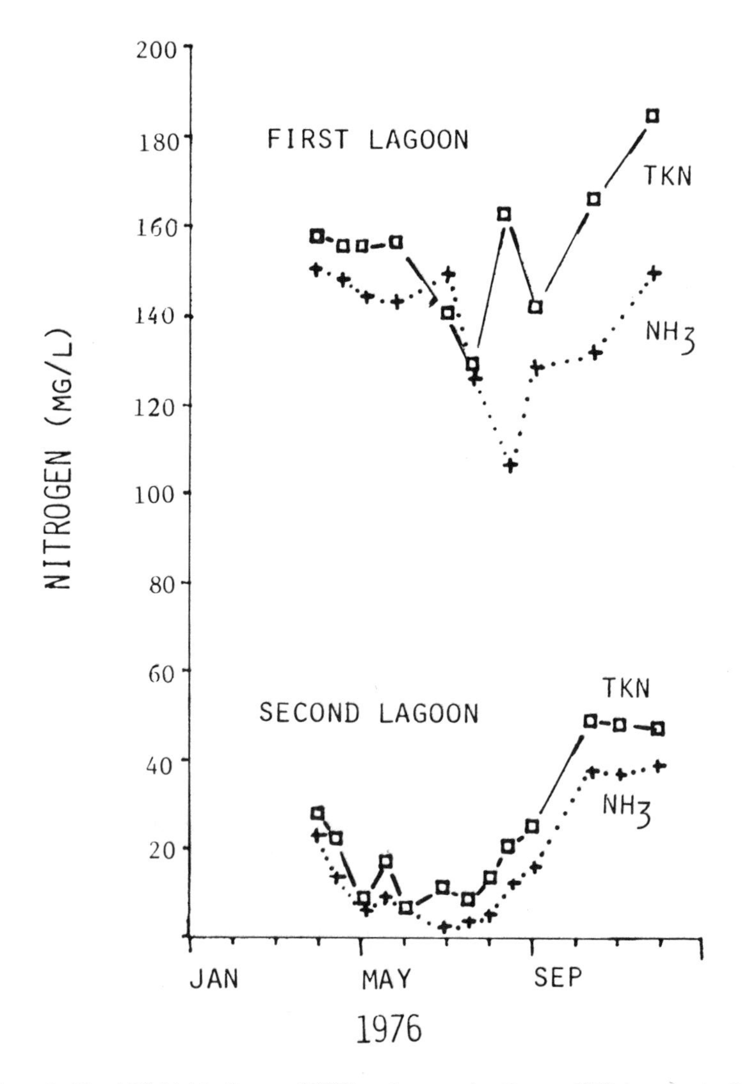

Figure 8. Total Kjeldahl nitrogen (TKN) and ammonia nitrogen (NH_3) concentration in the lagoon facility.

Microbiological analyses were conducted on the mixed liquor of both lagoons. These data are presented in Figures 9 through 13. Samples were collected monthly during the first year of microbiological testing and approximately quarterly thereafter.

The total and fecal coliform data show a trend of higher populations and greater variability during the first year of operation of the lagoons followed by lower populations and decreased variability thereafter (Figures 9 and 10). At the end of the experimental period the population of total coliforms in the first lagoon and second lagoon were 10 to 20 x 10^3 and 10^1 per 50 ml, respectively. Fecal coliform populations were near zero in both lagoons.

The fecal streptococcus population was variable and periodically very high during the first year's operation of the lagoon system (Figure 11). After this time the population stabilized at a value of about 10^3/50 ml in the first lagoon and about 10^2/50 ml in the second lagoon.

The population of aerobic or facultatively anaerobic bacteria was determined in samples from both lagoons using two media, nutrient agar and glucose-yeast extract agar (Figures 12 and 13). The determination of only aerobic or facultative bacteria in an anaerobic lagoon does not provide a measure of total microbial activity. It does, however, provide a simple method of determining the degree of organic loading and operation of the treatment lagoon and the degree of decomposition (polishing) of the wastewater effluent flowing into the second lagoon. Larger quantities of waste organics at either site will be reflected in a high population of heterotrophic bacteria. The two media were selected for different organisms; nutrient agar supports the growth of those bacteria decomposing proteins, peptides and amino acids which are the dominant available organic compounds of animal wastes, while glucose-yeast extract agar supports the growth primarily of bacteria using simple sugars or carbohydrates.

The numbers of bacteria capable of growing on nutrient agar from both lagoons was normally between 1 to 25 x 10^6/ml with slightly less in the second lagoon. Population peaks in the first lagoon were found in August 1974 and in the late summer and autumn of 1975. The late summer and autumn peak could be associated with a period of increased waste loading or alternatively with increased availability of soluble organic compounds because of extensive growth of algae. The population peak in the storage lagoon in March 1975 may reflect some carry-over or organic matter in the effluent of the first lagoon because of washout with spring rainfalls.

The population of heterotrophic bacteria capable of growing on glucose-yeast extract agar was about 10^5/ml in the first lagoon and about 10^4/ml in the second lagoon. The peaks of population increases coincided closely with those described previously for the nutrient agar studies, and the same hypothesis for these increases must be considered.

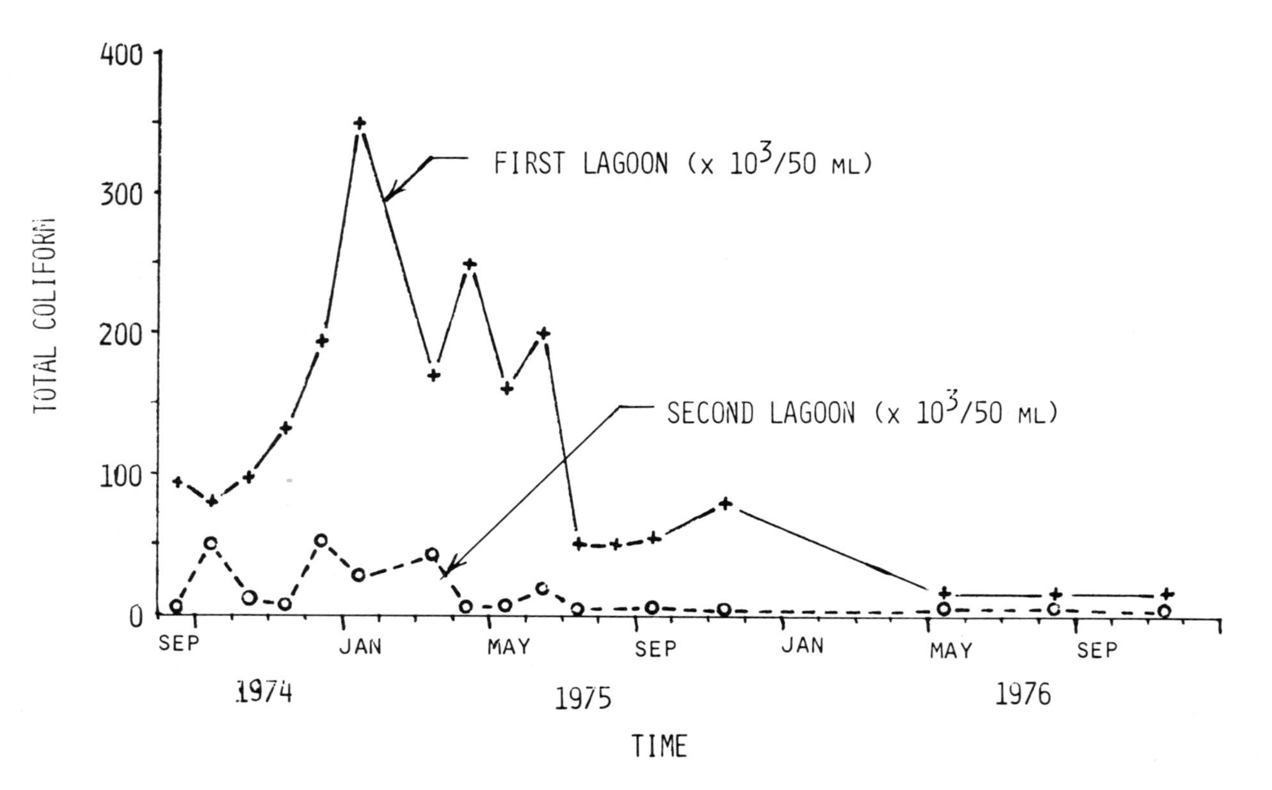

Figure 9. Total coliform population in both lagoons.

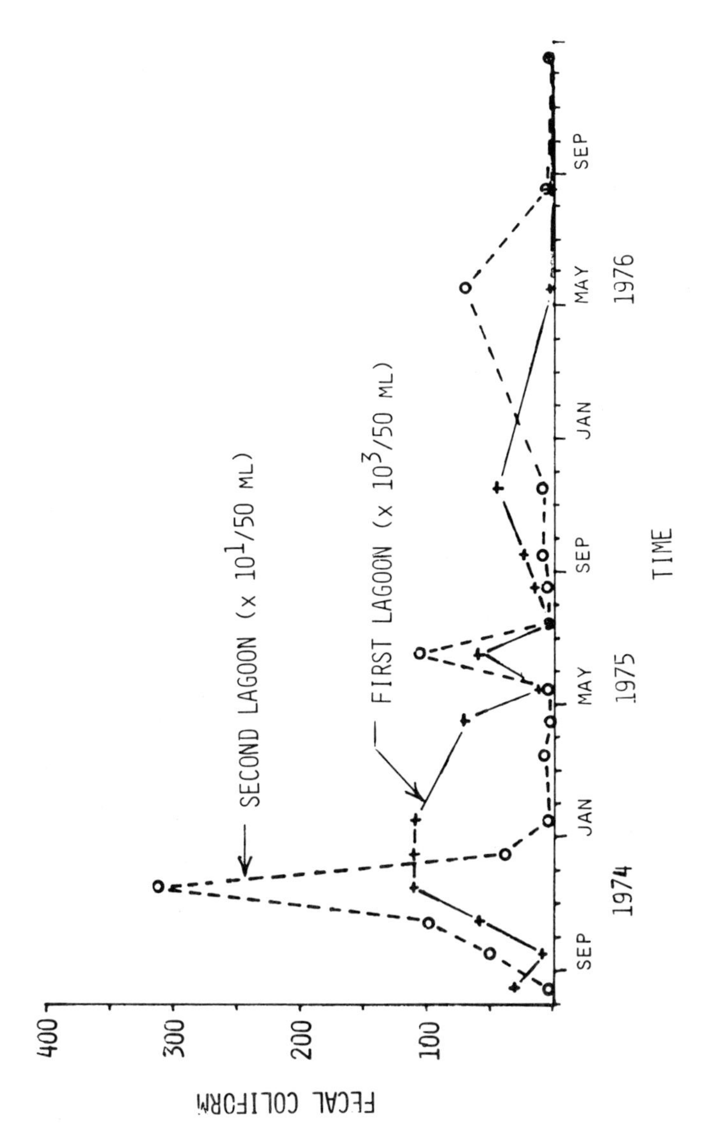

Figure 10. Fecal coliform population in both lagoons.

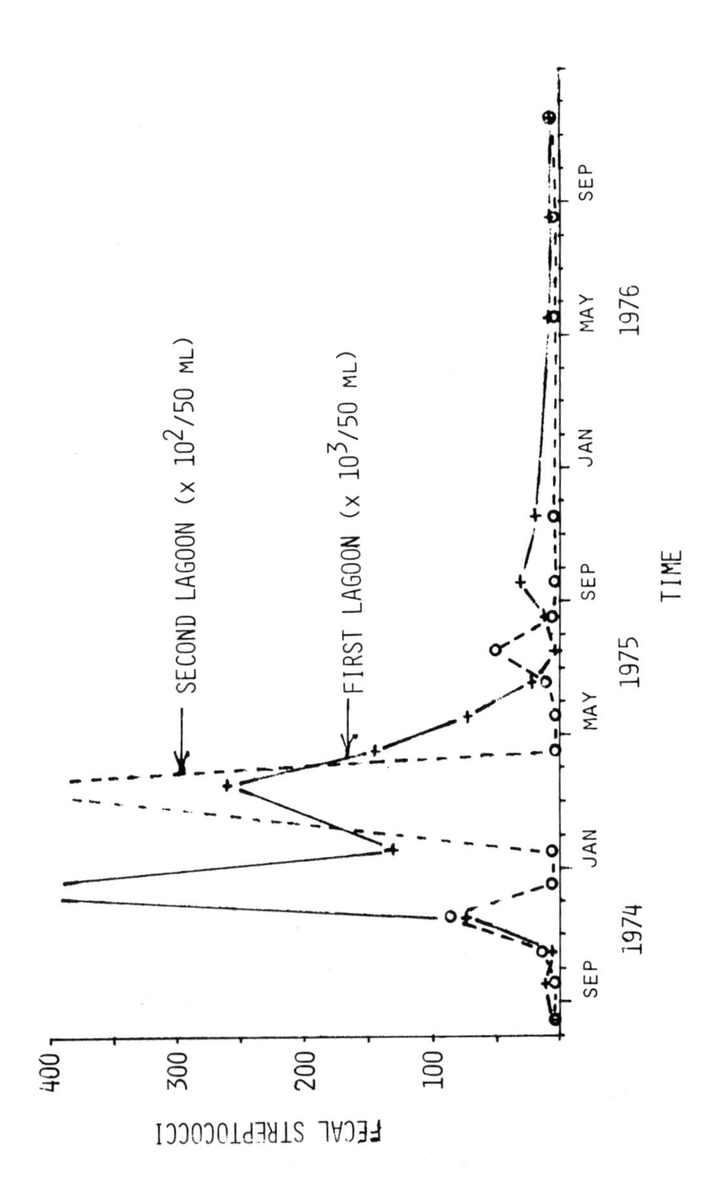

Figure 11. Fecal streptococcus population in both lagoons.

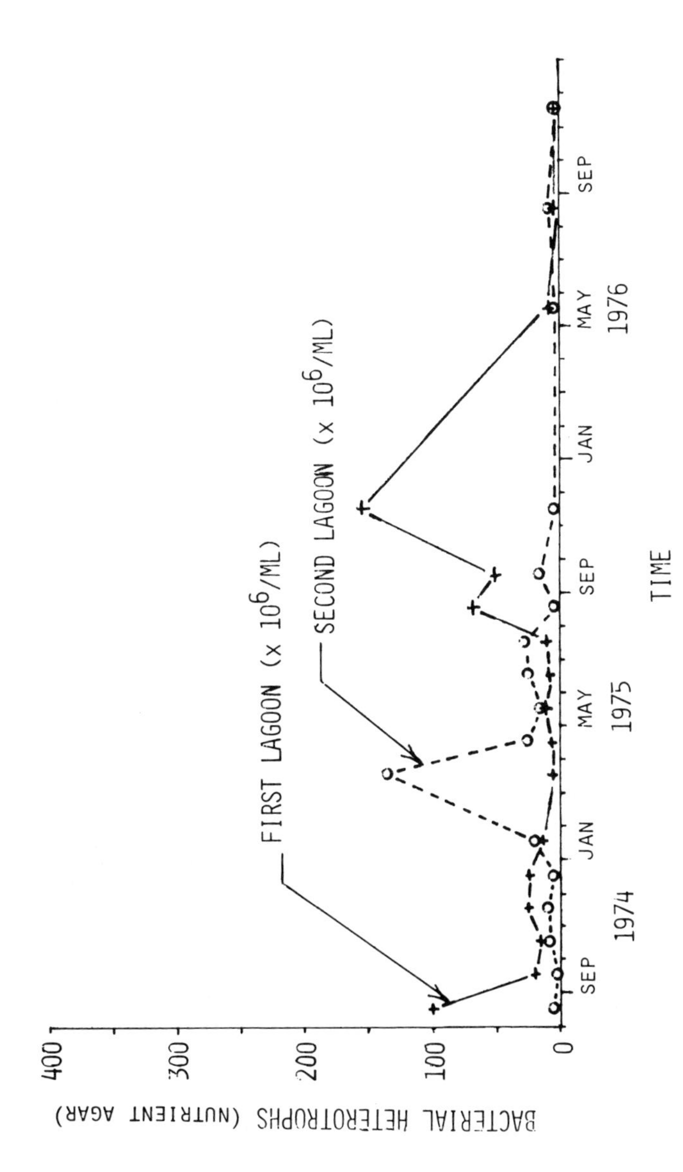

Figure 12. Population of heterotrophic bacteria (nutrient agar) in both lagoons.

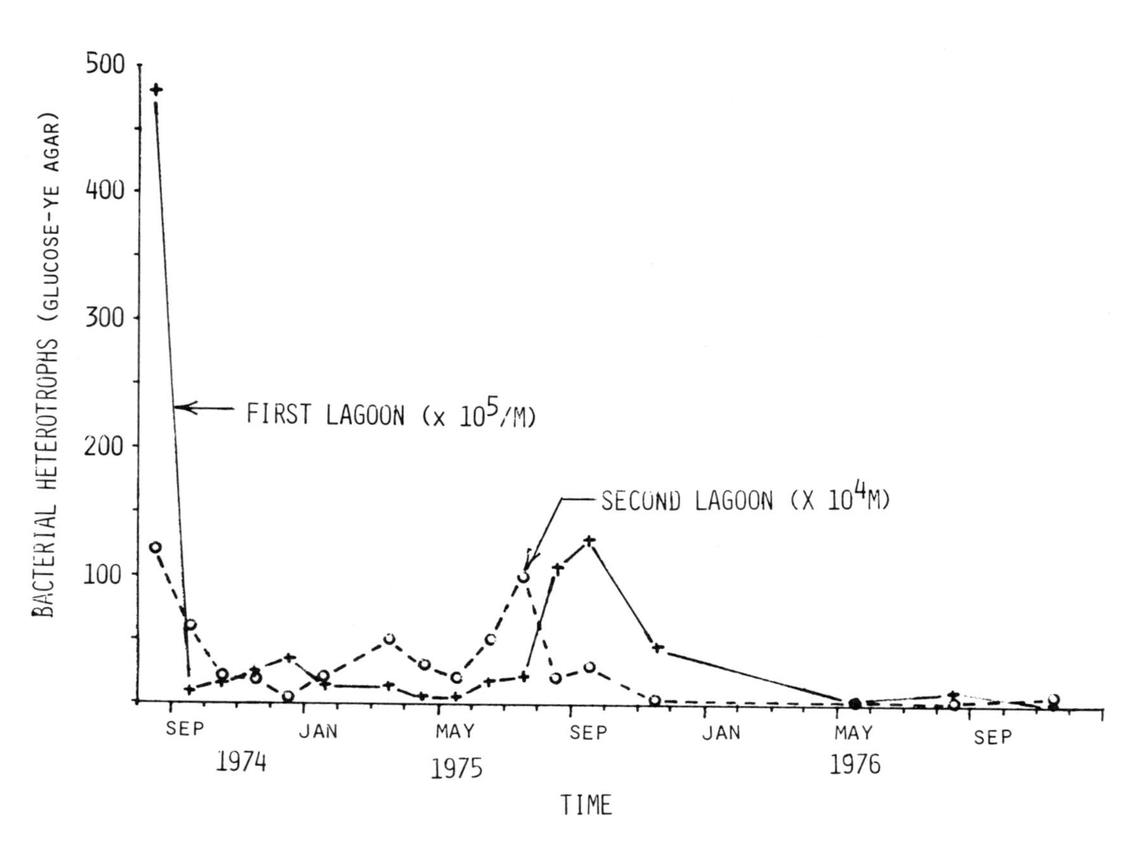

Figure 13. Population of heterotrophic bacteria (glucose-yeast extract agar) in both lagoons.

Sampling was reduced in 1976, but the consistently low populations at all sampling times should be noted. There are two possible explanations: (1) maturation of the treatment lagoon which maximizes the decomposition of waste organics and minimizes peaks of organic accumulation, thus increasing heterotrophic bacterial numbers, and/or (2) the development of a population of floc-forming bacteria which would carry organic matter and residual cells to the bottom of the lagoon.

Seepage of wastewater is a potential problem for all lagoons. Four test wells, located on each side of the lagoon system, and a control well were monitored to determine if ground water might become contaminated. Table IV gives the results of the microbiological analyses on the well waters tested from August 1974 to November 1976.

Table IV. Microbiological analyses of ground water from test wells.

Well Location	Total Coliform	Fecal Coliform	Fecal Streptococcus
	(Number of bacteria per 50 ml sample)		
West	1-25	1-15	1-7
North	0	0	~0[a]
East	1-37	15	1-9
South	~0	~0	~0
Control	0	0	~0

[a]The "~" means that less than 5 bacteria/50 ml were found in a few samples.

There is indication that the west and east test wells had been contaminated with a low level of coliforms throughout the entire period of operation but particularly during the period September 1974 to July 1975. All other test wells were not contaminated.

Fecal streptococci were generally absent or present in very low numbers (maximum number 9.0 per 50 ml) in the water of all test wells during the duration of the project. Consistently low but continuous detection of fecal streptococci at west and east wells between December 1974 and August 1975 suggests some subsurface seepage from the lagoons during this period. This was the same as for the coliform tests.

The conductivity of the test well waters analyzed for the same samples as the microbiological tests is shown in Figure 14. There is variability in the conductivity data but they indicate that the West and East wells were probably contaminated from January 1975 to September 1975, the same period when high bacteria counts were obtained. From January 1976 on, the conductivity of samples from the West and East wells decreased.

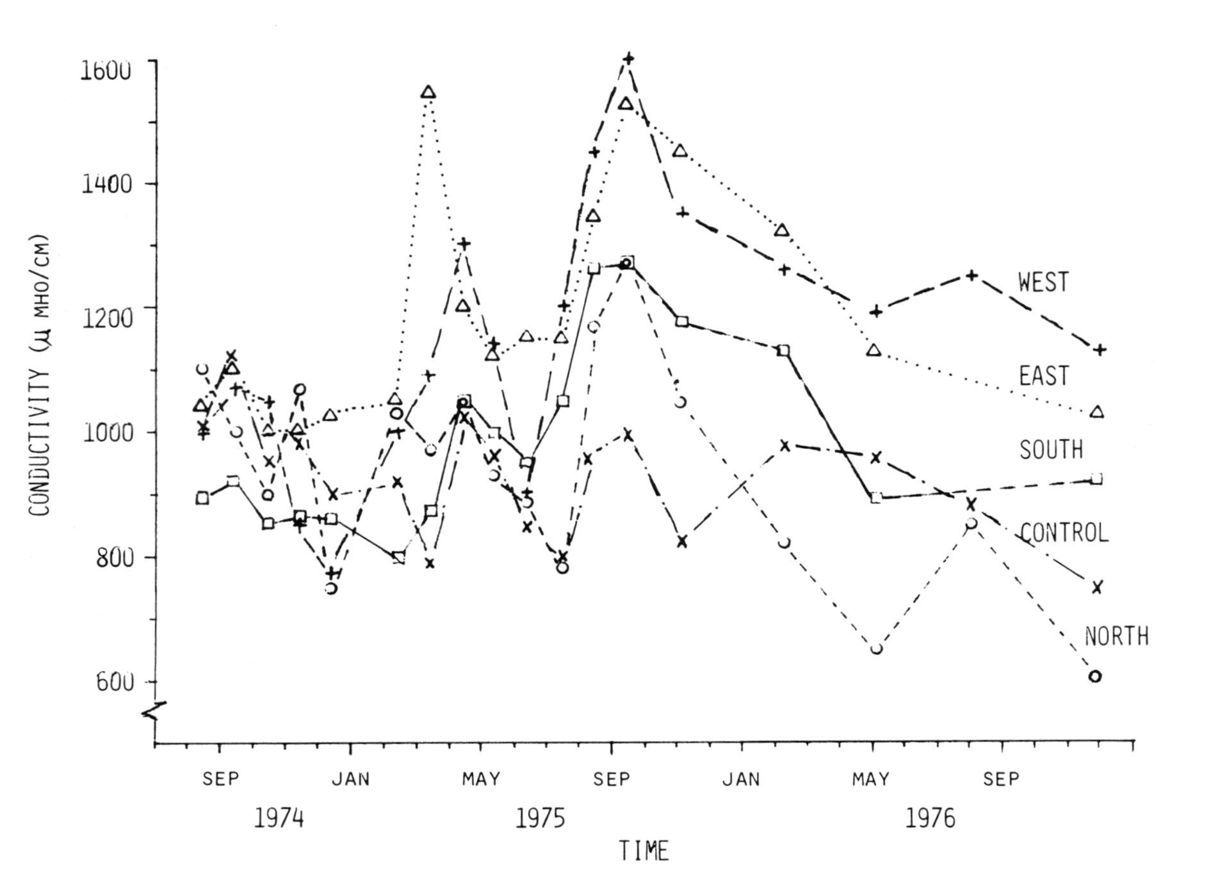

Figure 14. Conductivity of water in four test wells and control (see Figure 1 for location).

Both the bacteria count data and the conductivity data indicate that seepage from the bottom of the lagoons stopped or markedly decreased after 1975. This would give a period of almost three years for sealing of the lagoon bottoms to be effected.

Temperatures were monitored in the first lagoon at 0.1-, 0.5-, 1.0- and 2.0-m depth. Figure 15 shows the temperatures for 0.1- and 2.0-m depth.

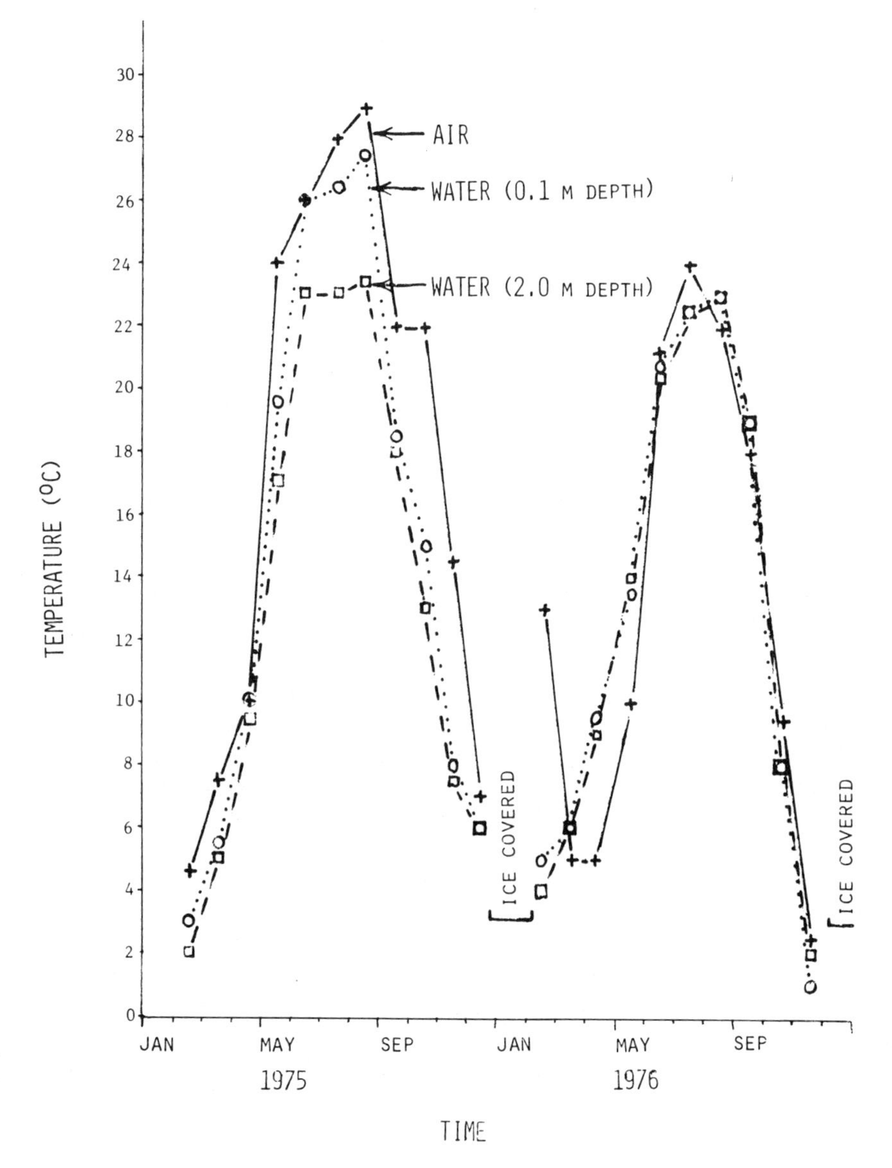

Figure 15. Temperatures at various depths in the first lagoon.

During 1975 there was a larger temperature difference between the top and the 2.0-m depth. In 1976 there was essentially no difference. It is postulated that more mixing of the lagoon wastes occurred because of the increased solids loading (livestock wastewaters were not screened in 1976) and the resulting bacteria action in the sludge layer.

pH values remained fairly constant for both lagoons. The pH of the mixed liquor in the first lagoon varied from 7.0 to 8.3 but had a mean value of 7.6. The pH of the second lagoon varied from 7.3 to 9.2 with a mean value of 8.2.

Sludge buildup in the first lagoon was monitored in the fall of 1975 and 1976. In the first sampling in 1975 a distinct layer of 0.1 to 0.2 m was detected. In 1976 no distinct layer was found.

Odor levels at the lagoon system were noted. In the late spring and summer of 1973, when wastewater was first pumped to the lagoon, an offensive odor was noted. At no time, either spring or fall, was there a highly offensive odor associated with the lagoon system after 1973.

SUMMARY AND CONCLUSIONS

The two-stage anaerobic lagoon system for treating livestock wastewaters and sanitary sewage from the Ohio State University Animal Science Livestock Center was monitored over a three-year period. Swine wastes and beef runoff were pumped to the lagoon system. Sanitary sewage from student workers' living quarters and public restrooms were pumped to the lagoon separately.

Reductions in the concentration of various water quality parameters indicated a well functioning lagoon system. Reductions in the first stage were as follows: 82% for TS, 92% for TSS, 92% for BOD and 85% for COD. The level of these parameters, even in the second lagoon, was still too large to allow direct discharge to a stream. It is planned to begin irrigating wastewater either in the fall of 1977 or the spring of 1978. The total nitrogen in the mixed liquor of the first and second lagoon varied from 140 to 180 mg/l and 10 to 50 mg/l, respectively. From a nutrient recovery viewpoint it would be preferable to irrigate from the first lagoon.

Microbiological analyses of the lagoon mixed liquor over a two-year period indicated a maturation of the system during the last year of testing. The total coliform count dropped to between 10,000 and 20,000/50 ml. The fecal coliform count dropped to near zero in both lagoons. Fecal streptococcus population stabilized at a value of about 1000/50 ml and 100/50 ml in the first and second lagoon, respectively.

Lagoon temperatures at various depths indicated increased mixing during the third year. This fact would also indicate maturation of the lagoon due to increased gas production.

Microbiological analyses and electrical conductivity measurements of water from wells located at each side of the lagoon facility indicated seepage from the lagoon initially. At the end of the second year the number of coliform and streptococcus bacteria in the well water dropped to nearly zero in all wells. The conductivity of the contaminated wells also dropped in the last year. These data indicate that seepage into the ground water was stopped or was at least greatly reduced for the third year.

Odor nuisances did not occur except for the initial start-up of the lagoon.

ACKNOWLEDGMENTS

This paper was approved for publication as Journal Article No. 80-77 of the Ohio Agricultural Research and Development Center (OARDC). The project was supported by State Grant (S-448).

REFERENCES

1. American Public Health Association. *Standard Methods for the Examination of Water and Wastewater*, 13th ed. (New York: American Public Health Association, 1971).

44

NUTRIENT BUDGET IN A DAIRY ANAEROBIC LAGOON– EVALUATION FOR LAND APPLICATION

D. F. Bezdicek and J. M. Sims
Department of Agronomy and Soils

M. H. Ehlers and J. Cronrath
Department of Animal Sciences

R. E. Hermanson
Department of Agricultural Engineering
Washington State University
Pullman, Washington.

INTRODUCTION

Anaerobic lagoons have been used successfully by dairy farm operators to handle milking parlor wastes, excreta and precipitation runoff. Anaerobic lagoons stabilize organic matter and can serve as storage facilities, thus permitting seasonal land distribution of manure slurry. The typical anaerobic lagoon is characterized by high organic loading, a relatively small surface area and low labor and capital costs. One concern with lagoons has been nutrient conservation, especially with regard to nitrogen. A nutrient "budget" of the waste slurry provides a means to assess the effectiveness of the lagoon in reducing waste solids and conserving nutrients and to arrive at a soil loading rate that will support optimum crop production without detrimental environmental effects.

Some studies have reported sizable reductions of total and volatile solids from animal wastes in anaerobic lagoons. Bhagat and Proctor[1] found that over 85% of the total solids (TS) and volatile solids (VS) from dairy wastes were removed through a series of three anaerobic lagoons. Other investigators report VS reductions of 75%[2] and 65%[3] for cattle wastes. However, if all the sludge is not collected from the lagoons, as was apparent in some of the studies, the solids reduction would be overestimated. In a laboratory study with complete sludge recovery of dairy wastes, Barth and Polkowski[4] found an average reduction of only 19% VS in surface-aerated reactors.

Another aspect of anaerobic lagoons is nutrient conservation. Nitrogen is probably the most important nutrient in animal wastes, because it often limits crop yield and can adversely affect water quality. Previous reports indicate poor N conservation in anaerobic lagoons. Jones *et al.*[5] reported a 62% N loss from dairy wastes in a combination aerobic-anaerobic lagoon (lagoon with surface aerator). Poelma[6] reported N losses of over 50% from an underfloor waste storage facility with a floating aerator. Koelliker and Miner[7] found that in a swine anaerobic lagoon, 65% of the N was lost annually by N transformations and desorption. According to Koelliker and Miner,[7] an additional 15 to 30% N loss can occur between the lagoon and land when the lagoon slurry is sprinkle irrigated. Vanderholm,[8] in a summarization of research, estimated that 78% of the N in livestock wastes can be lost during treatment with an anaerobic lagoon and land application by irrigation or liquid spreading.

Other studies have shown different results in nutrient recovery. In a laboratory study, Barth and Polkowski,[4] using surface-aerated dairy waste reactor units, found that nearly 90% of the N was retained. Similarly, Zeisig[9] reported an N loss of only 10% with intermittent aeration of liquid manure. It is apparent from these studies that N losses may not be as high as from other reports. The low N recoveries reported may have been due to inadequate nutrient "budgeting" through incomplete sludge recovery.

Information on the conservation of other nutrients is limited. In swine wastes, Booram *et al.*[10] reported that soluble P was lost by formation of P-containing precipitates (*e.g.*, magnesium ammonium phosphate) in anaerobic treatment or storage facilities. Norstedt *et al.*[3] reported that orthophosphate was reduced 33% by precipitation in an anaerobic lagoon. All the P should have been retrieved with complete sludge recovery. The laboratory study by Barth and Polkowski[4] showed nearly 100% recovery of P and 92% recovery of K.

This study was conducted to "budget" input and output nutrients in a surface-aerated anaerobic lagoon typical of some dairy farm operations in the state of Washington.

SLURRY-HANDLING OPERATION

The system used in this study was a surface-aerated anaerobic lagoon in conjunction with a 65-Holstein-cow dairy lot at the Knott Dairy Center, Washington State University. A flow diagram of the operation is shown in Figure 1. A 63 m^3 concrete collection tank received lot drainage, excreta

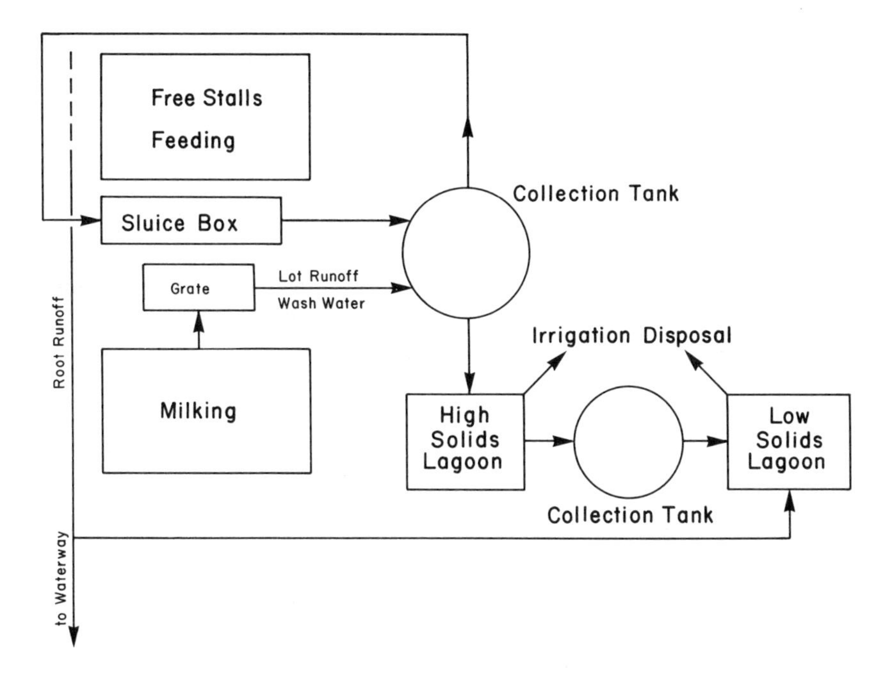

Figure 1. Scheme of slurry-handling facilities.

and other wastes from the milking area. Excreta, bedding and waste feed from feeding and rest areas were scraped daily with a tractor into a sluice box where slurry recirculated from the collection tank was used to suspend and flush the solids through a sewer to the collection tank. Contents were agitated with a 30-hp Mitchell-Lewis-Staver Pump* and then pumped to the high-solids lagoon. The manure represented all that produced by lactating Holsteins averaging 3 yr of age and 620 kg body weight, with "dry period" cows excluded.

The high-solids lagoon was a 430-m^3 earthen storage facility. A floating 2- and a 3-hp Aqua Jet Aerator* minimized odors and kept a surface crust from forming. A center-mounted pump (30-hp Mitchell-Lewis-Staver*) resuspended the solids prior to pumpout of the slurry to pasture through a "manure gun" sprinkler. In absence of rain, the high-solids lagoon held the wastes from the 65 cows for about 5 weeks. However, during rainy periods, effluent from the high-solids lagoon flowed through a collection tank into a 14,000 m^3 low-solids lagoon. Water was supplied from the low-solids lagoon to the high-solids lagoon to aid in sludge dilution during pumpout and to

*Use of brand name does not imply endorsement by Washington State University.

flush the pipeline after pumpout. The low-solids lagoon was emptied each summer onto land with conventional irrigation sprinklers.

METHODS

The study consisted of three experimental periods: May 4 to June 23, 1976 (spring), August 31 to October 4, 1976 (summer), and January 19 to February 22, 1977 (winter). The spring experimental period was longer because of pump failures. The lagoon was empty at the beginning of each experimental period. Because the pumpout procedure was conducted at regular intervals during the past five years, sludge buildup appeared to have reached a plateau. Slurry in the holding tank at sampling represented the total waste during the previous 24-hr period. Three 1-liter grab samples of the waste were taken weekly after agitation when the waste was pumped from the holding tank to the high-solids lagoon. Samples were held at 4°C and analyzed within 48 hr or frozen for later analysis. At sampling, slurry was added at a loading rate of 0.51 kg/m^3/day of VS to six 200-liter drums which were used to simulate conditions in the high-solids lagoon. Three of the six simulators were aerated by continuous pumping of air through a perforated ring placed 0.2 m below the surface of the waste. The simulators were kept at constant volume, and effluent samples of the overflow were taken 6 hr after waste addition and settling. Simulator studies were conducted from March 9 to April 26, 1976 (early spring), May 4 to June 23, 1976 (late spring), and August 31 to October 4, 1976 (summer).

High-solids lagoon overflows were sampled during the experimental periods when runoff was high. At the completion of each experimental period, the high-solids lagoon was emptied to the original level. During pumpout of the high-solids lagoon, samples from approximately 53 m^3 of waste were taken at 1-hr intervals of pumping and agitation. Removal of solids was not complete for the winter period because the slurry was too thick to pump. During the summer period, the entire budget was conducted on the high-solids lagoon during pumpout since there was no overflow to the low-solids lagoon.

In the spring, the simulators were emptied at the completion of the studies by siphoning and sampling the supernatant flow and taking grab samples of the residual sludge. At the end of the summer period, the simulators were emptied by siphoning the sludge and supernatant together after agitation since the sludge came to the surface as a crust.

Samples from the collection tank, simulators and lagoon were homogenized in a blender and analyzed for chemical oxygen demand (COD), NH_4^+-N, total Kjeldahl N (TKN), TS, VS, P, Ca, Mg and K. A 30-min modification[11] of the standard COD test[12] was used. During the spring period,

NH_4^+-N and TKN were determined by steam distillation,[13] whereas during the summer and winter periods, total N was determined with a Technicon Auto Analyzer. Solids were determined according to *Standard Methods.*[12] Aliquots of samples were digested with HNO_3:$HClO_4$ for P, Ca, Mg and K analyses. Phosphorus was determined by vanadomolybdophosphoric acid colorimetry,[12] Ca and Mg by atomic absorption and K by flame photometry. The budget for the waste parameters was determined by computing the weekly 24-hr values over the course of study. Total cow number and average weight were used to compute cow days, with low cow average body weights reflecting the young age composition of the lot. All cows were lactating, with cows entering and leaving the lot at initiation and completion of lactation.

RESULTS AND DISCUSSION

The characteristics of slurries for the experimental periods are shown in Table I. In general, the waste concentrations were similar for the three periods. Hydraulic loadings and input volumes as shown in Table II varied seasonally, depending upon temperature and precipitation. Lower input volumes in summer reflected the low rainfall and the high evaporation during this time. Spring values were considered typical, while winter loading and input volumes were probably low, since winter rainfall was at a record low. Waste volume recovered during the spring pumpout was the highest at 92.3%. Decreased recovery of input volume (78.7%) in the summer period was due to higher temperatures and lower rainfall during this time. The 75.5% recovery of waste volume in the winter period was caused by sludge near the bottom of the pond being too thick to pump. This occurred partially because winter temperatures depressed biological activity, which resulted in greater sludge accumulation. In contrast, solids during the summer periods were more uniformly distributed vertically in the lagoon, which was probably due to gas action from biological activity. During the spring period, over 50% of the wastes overflowed to the low-solids lagoon, whereas in the warmer and drier summer period, there was no overflow.

The VS loading rates were similar for all three periods, eliminating VS loading as a variable in the lagoon treatment process. The VS loadings were fairly high (1.1 to 1.3 kg VS/m^3/day), although Loehr[14] reported that anaerobic lagoons can successfully handle 2.1 to 5.1 kg VS/m^3/day.

Measured and predicted manure characteristics are shown in Table III. Nutrient values were similar, but COD and solids measured were higher than predicted values for 620-kg cows. Considering bedding waste values of 1.3, 1.3 and 1.5 kg/cow/day, the corrected TS were 8.3, 8.5 and 7.3 kg/cow/day,

Table I. Influent characteristics of slurries used during three experimental periods.

	Experimental Period		
	Spring	Summer	Winter
pH	7.5	7.3	7.3
Total Solids (%)	2.4	4.0	3.0
Volatile Solids (%)	1.9	3.3	2.3
COD (mg/l)	23,624	30,376	30,790
NH_4^+-N (mg/l)	269	227	374
Total Kjeldahl N (mg/l)	761	973	1,017
P (mg/l)	119	143	158
Ca (mg/l)	336	344	323
Mg (mg/l)	141	154	137
K (mg/l)	774	959	946
Electrical Conductivity (μmhos)	5,765	5,449	6,396

Table II. Input and output loadings during the three experimental periods.

Experimental Period	Hydraulic Loading (m^3/day)	Volatile Solids Loading[a] (kg/m^3/day)	Volume Input (m^3)	Recovery Volume (%)	Overflow Volume[b] (m^3)	Overflow (%)
Spring	24.8	1.10	1091	92.3	572	56.8
Summer	16.8	1.29	587	78.7	0	0
Winter	19.4	1.08	699	75.5	117	22.2

[a] Based on full lagoon volume.
[b] Overflow volume to low-solids lagoon.

respectively, for spring, summer and winter. While the values were higher than predicted, these reflect the relatively high milk production (26 kg/cow/day) and the incorporation of feed waste not included in the predicted values. If a waste treatment system were designed for the WSU dairy on the basis of the predicted values, the system would be underdesigned with respect to TS.

Table IV shows the complete "budgeting" of solids, COD and nutrients for the three periods. The greater reduction in TS and VS in the summer than in the spring or winter was probably due to higher biological activity. It should be noted that the lagoon was easier to empty during the summer when gas production from microbial activity maintained solids in suspension with less agitation. The winter period was least desirable for ease of solids removal. Thus, the values for solids reduction in the winter period were probably overestimated, since all the solids were not recovered from the lagoon.

Table III. Output of parameters per 620-kg cow for three experimental periods.

Experimental Period	COD	Total Solids	Volatile Solids	N	P	Mg	Ca	K
	(kg/cow/day)							
Spring	9.3	9.6	7.3	0.298	0.046	0.055	0.135	0.289
Summer	7.4	9.8	8.0	0.238	0.035	0.037	0.083	0.232
Winter	9.1	8.8	6.9	0.294	0.045	0.039	0.091	0.277
Mean	8.6	9.4	7.4	0.277	0.042	0.044	0.103	0.266
Predicted[a]	5.6	6.4	5.3	0.253	0.049	-	-	-

[a]Based on ASAE SE-412 Committee Report AW-D-1.[15]

Table IV. Reduction in COD and solids and recovery of nutrients during three experimental periods.

Experimental Period	COD	Total Solids	Volatile Solids	NH_4^+-N	Total Kjeldahl N[a]	P	Ca	Mg	K
	(% Reduction)			(% Recovery)					
Spring	48.6	46.1	53.1	79.8	72.1	70.6	53.6	79.2	90.7
Summer	48.3	54.5	61.5	86.2	68.4	65.7	58.6	74.1	88.0
Winter	41.0	38.5	41.4	66.3	71.1	72.3	73.4	74.9	72.8
Mean	46.0	46.4	52.0	77.4	70.5	69.5	61.9	76.1	83.8

[a]Proportion of TKN as NH_4^+-N was 38.7, 29.9 and 35.7%, respectively, for spring, summer and winter.

Nitrogen recovery was higher than that expected based on previous field studies, particularly since this lagoon was surface-aerated. The TKN recoveries of from 68 to 72% for the three periods were higher than for most reports of full-scale lagoons, probably because of more complete sludge recovery at the end of each period. Barth and Polkowski[4] and Zeisig[9] showed even higher N recoveries with complete sludge recovery. These results are in contrast to reports of N losses exceeding 50%.[5,6,8] The relatively short storage and digestion time may have contributed to differences in N recovery obtained.

Recoveries of P, Ca and Mg ranged from 60 to 80% depending upon nutrient and season. These nutrients may have formed insoluble compounds that precipitated at the bottom of the lagoon. The greater recovery of K was probably due to the greater solubility of salts associated with this ion. Even with nearly complete sludge recovery, nutrients that would be expected to be conserved were dependent on their characteristic separation between the sludge and supernatant fractions. Simulator data from the spring period (Table V) show that nutrients were not equally distributed between the waste fractions. During the complete experimental period, the effluent represented 83.9% of the total waste output and the sludge only 16.1%. However, 44.9%, 50.4%, 10.5%, 89.5% and 41.6% of the N, P, K, Ca and Mg, respectively, were recovered in the sludge. These values provide some estimates of nutrient recovery if only effluent is considered. However, these simulator recovery data were based upon total nutrient recoveries exceeding 95%. Recovery of nutrients in the effluent from actual lagoons may be less. The incomplete recovery for nutrients in the high-solids lagoon indicates that either not all the sludge was recovered or that the nutrients that concentrate in sludge accumulated in the residual sludge always left in the bottom of the lagoon.

Table V. Distribution of nutrients recovered in effluent and sludge in aerated simulators.[a]

	Fraction of Nutrients Recovered in		
Nutrient	**Effluent (%)**	**Sludge (%)**	**Final Sludge:Effluent Ratio[b]**
N	55.1	44.9	4:1
P	49.6	50.4	9:1
K	89.5	10.5	1:1
Ca	10.5	89.5	35:1
Mg	58.4	41.6	4:1

[a]Spring period; total recoveries of nutrients were greater than 95%; values are an average of two simulators.

[b]Remaining in simulator at end of experiment.

The concentration of nutrients in the sludge at the end of the study as compared with the final effluent concentrations is also shown in Table V. These ratios, which were higher than the distribution reported for the complete study, reflect the continued accumulation of nutrients in the sludge throughout the experimental period. Calcium was greatly concentrated in the sludge, while P, N and Mg were also found in greater amounts in the sludge than in the supernatant. Potassium was equally distributed between sludge and supernatant.

These data indicate the need to recover all solids to maximize recovery of nutrients from lagoon waste slurries. Although the simulators did not "simulate" completely the conditions of the high-solids lagoon with respect to solids reduction and nutrient loss, the data do show the distribution of nutrients in the sludge and effluent that can be expected.

EVALUATION FOR LAND APPLICATION

Estimates were made as to quantities of N, P and K that would be added to cropland if the slurry from the lagoon were applied on the basis of providing a specific quantity of available N. The data were calculated from mean output values per 620-kg cows in Table III and from nutrient recoveries in Table IV. These data are summarized in Table VI. To provide 100 kg/ha available N, 333 kg/ha of total N must be applied from the lagoon, assuming 30% of the N would be available the first year.[16,17] This amount of slurry would add 50 and 383 kg/ha of P and K, respectively. While added P would meet plant requirements for some soils during the first year of application, soil P may build up to excessive levels during ensuing years of application. The amount of K would appear to exceed crop requirements on most soils during the first year of application. To optimize the nutrient value in dairy manure, lower rates of lagoon effluent with supplemental N may be necessary to prevent excessive application of P and K. These rates would depend on individual soils and crop nutrient requirements.

To supply 100 kg/ha of available plant N, the total cumulative manure from 4.7 cows/yr is required. On this basis, 13.8 ha of land will be required to accommodate the annual manure production from this 65-cow dairy, assuming constant manure production throughout the season. Nitrogen loss estimates were not made for ammonia volatilization when the slurry is spread on the soil surface. These losses will vary with soil temperature, moisture and the duration on the soil surface before incorporation. Estimates as to surface ammonia losses usually do not exceed 30% for most soil conditions.[8] Estimates for N losses due to denitrification are difficult to obtain but would be minimized at low application rates.

Table VI. Total and recovered N, P and K produced per 620-kg cow year and nutrients required to supply 100 kg available N/ha.

Parameter	N	P	K	Cow Years[c]
	(kg/cow yr)			
Total Nutrients	101	15.3	97.1	-
Recovered Nutrients[a]	71.2	10.6	81.4	-
Nutrients to Supply 100 kg Available N[b]	333	50	383	4.7

[a]Based on average recoveries for the three trial periods.

[b]Based on recovered values and an estimated N mineralization of 30% during the first year of land application.

[c]For 620-kg cows.

This study shows that total land requirements for optimum nutrient utilization from dairy manure are about three times greater than estimates based on published guidelines.[16-18] The greatest differences in land requirements between this study and the published guidelines were due to the differences in lagoon N recovery. Estimates of N recovery in anaerobic lagoons from published guidelines varied from 20 to 22% as compared with 70.5% for our study. Our data suggest that estimates of N loss included in these guidelines may be an overestimate. Consequently, total land area required for optimum nutrient utilization may be underestimated.

SUMMARY AND CONCLUSIONS

An account of nutrients and reduction of waste parameters in an anaerobic lagoon representing the waste output from 65 dairy cows varied little from season to season. Reduction in TS and VS averaged 46 and 52%, respectively, for the three study periods. Slightly higher reductions in TS and VS were obtained during the summer period. While the anaerobic lagoon used at this facility was intended largely as a storage unit, the reduction in solids was significant. This reduces considerably the amount of solids that must be applied to the field. Average recovery of TKN for the three periods was 71%, with little variation between seasons. In spite of continuous aeration, these values were considerably higher than other published field reports. Recovery of P averaged 70% for the three seasons, whereas Ca and Mg recoveries were 62 and 76%, respectively. Average recovery of K was higher at 84%, which reflected the greater solubility of this ion. High recoveries of plant nutrients were attributed to the near complete removal of solids from the lagoon.

From spring simulator data, effluent represented 83.9% of the total waste volume but contained only about 50% of the N, P and Mg. Only 10% of the Ca was found in the effluent fraction. These data show the importance of recovering the sludge remaining in anaerobic lagoons for nutrient removal. Potassium was an exception, where the majority of this nutrient was found in the effluent.

While the simulator data provide some suggestions as to the distribution of nutrients in the effluent and sludge fractions, lagoon performance varied with season. Sludge was removed most easily in the summer where gas production by microbial activity appeared to suspend the solids with a minimum of agitation. Good nutrient recoveries were obtained in the spring from the addition of water to resuspend the solids when the slurry was too thick to pump. Slightly lower volume and nutrient recoveries were obtained during the winter because of an unpumpable slurry remaining at the end of the pumpout. Addition of water and recirculation would have increased nutrient recovery for this period.

ACKNOWLEDGMENTS

This paper is Scientific Paper No. 4795, College of Agriculture Research Center, Washington State University, Pullman, Washington.

REFERENCES

1. Bhagat, S. K. and D. E. Proctor. "Treatment of Dairy Manure by Lagooning," *J. Water Poll. Control Fed.* Reprint 249 (1969).
2. Witzel, S. A., E. McCoy and R. Lehner. "Chemical and Biological Reactions from Lagoons Used for Cattle," *Trans. Am. Soc. Agric. Eng.* 449:451 (1965).
3. Nordstedt, R. A., L. B. Baldwin and C. C. Hortenstine. "Multistage Lagoon Systems for Treatment of Dairy Farm Waste," in *Livestock Waste Management and Pollution Abatement,* Proc., Int. Symposium on Livestock Wastes, Ohio State University, ASAE Publication PROC-271, (1971), pp. 77-80.
4. Barth, C. L. and L. B. Polkowski. "Low Volume, Surface-Layer Aeration Conditioned Manure Storage," in *Livestock Waste Management and Pollution Abatement,* Proc., Int. Symposium on Livestock Wastes, Ohio State University, ASAE Publication PROC-271 (1971), pp. 279-282.
5. Jones, R. E., J. C. Nye and A. C. Dale. "Forms of Nitrogen in Animal Waste," ASAE Paper No. 73-439, Annual Meeting, American Society of Agricultural Engineers, St. Joseph, Michigan (1973).
6. Poelma, H. R. "Aeration of Liquid Manure and Forced Recirculation of Aerated Manure," Paper 74-11-112, VIIIth Int. Congress of Agric. Eng., Elberg, The Netherlands, September 22-28, 1974.
7. Koelliker, J. K. and J. R. Miner. "Desorption of Ammonia from Anaerobic Lagoons," *Trans. Am. Soc. Agric. Eng.* 16:148-151 (1973).

8. Vanderholm, D. H. "Nutrient Losses from Livestock Waste During Storage, Treatment, and Handling," in *Managing Livestock Wastes,* Proc., 3rd Int. Symposium on Livestock Wastes, University of Illinois ASAE Publication PROC-275 (1975), pp. 282-285.
9. Zeisig, H. D. "Surface Ventilation of Liquid Manure for the Purpose of Limiting Emissions of Smell," Paper 74-IT-119, VIIIth Int. Congress of Agric. Eng., Alberg, The Netherlands, September 22-24, 1974.
10. Booram, C. V., R. J. Smith and T. E. Hazen. "Crystalline Phosphate Precipitation from Anaerobic Animal Waste Treatment Lagoon Liquors," *Trans. Am. Soc. Agric. Eng.* 18:340-343 (1975).
11. Humenik, F. J. and M. R. Overcash. "Analyzing Physical and Chemical Properties of Liquid Wastes," in *Standardizing Properties and Analytical Methods Related to Animal Waste Research.* ASAE Special Publication SP-0275, American Society of Agricultural Engineers, St. Joseph, Michigan (1975).
12. American Public Health Association. *Standard Methods for the Examination of Water and Wastewater,* 13th ed. (Washington, D.C.: American Public Health Association, 1971).
13. Bremner, J. M. "Inorganic Forms of Nitrogen," in *Methods of Soil Analysis. Part 2.,* C. A. Black, *et al.*, Ed. *Agronomy* 9:1191-1206 (1965).
14. Loehr. R. C. *Agricultural Waste Management* (New York: Academic Press, 1974).
15. "Manure Production and Characteristics per 1000 Pounds Live Weight," ASAE SE-412 Committee Report AW-D-1, unpublished (1973).
16. Turner, D. O. "Guidelines for Manure Application in the Pacific Northwest," Washington State University EM 4009 (1976).
17. Willrich, T. L., D. O. Turner and V. V. Volk. "Manure Application Guidelines for the Pacific Northwest," ASAE Paper No. 74-4061, Annual Meeting, American Society of Agricultural Engineers, Stillwater, Oklahoma, 1974.
18. Ohio State University 1974-75 Agronomy Guide. Cooperative Extension Service Bulletin 472 (1974).

45

EFFECT OF MANURE ON PLANT GROWTH AND NITRATE N IN SOIL WATER

L. F. Marriott
Department of Agronomy

H. D. Bartlett and M. J. Green
Department of Agricultural Engineering
The Pennsylvania State University
University Park, Pennsylvania

BACKGROUND

The use of manure for its crop-producing potential has had its ups and downs, from the time when the size of a man's manure pile was a measure of his wealth to the present time, when the same pile might be the basis for a lawsuit while also serving as a hedge against rising fertilizer prices. However, good management was and still is the key to the conservation of the plant food in manure and to minimizing air and water pollution.

Part of the current problem in manure utilization lies in determining the maximum "safe" rate for a crop. The peak crop requirements for nitrogen must be balanced with the potential movement of nitrate-nitrogen from the soil profile when the crop has little or no need for N. Thus, a grass crop provides a longer period of active uptake than does corn, but the potential for loss of N is still present even with the grass.

In this study, the source of the manure was a dairy barn at The Pennsylvania State University. Sawdust bedding and other materials from barn cleaning became part of the manure which was dumped into a storage pit at the end of the barn. Water was added as needed to create a slurry which could be readily pumped. The slurry was injected under an established orchardgrass sod at rates to supply 340, 450 and 560 kg N/ha in November 1973 and April 1974. The treatments were repeated on the same plots in the following November and April. The slurry was applied with commercially manufactured equipment for liquid manure tank spreaders. The equipment was modified to locate the injectors behind the wheels with lateral adjustment as required. Rates were determined by the spacing between injection centers of 68, 50 and 40 cm, respectively, for the three rates used. Urea was topdressed at 112 kg N/ha in April 1974 and 1975 as a control. No treatments were applied for the 1976 season. There were four replicates of each treatment.

Soil water for nitrate-nitrogen determinations was collected from suction lysimeters installed at depths of 30, 60, 90 and 120 cm. The samplers were placed near the center of each plot midway between injection centers after the initial manure application was made. The samplers were pulled prior to the second year's application and replaced shortly after. Since it was impossible to follow the initial injection centers, the replacement of the samplers between the fresh manure centers often altered their position relative to the initial centers. To obtain an estimate of the effect of location of the sampler on nitrate-N concentrations, additional samplers were installed in 1975, some directly under the injection center and some between centers. Water samples were taken generally at least once a month, usually following a rain. Nitrate-N concentrations were determined with an Orion Specific Ionmeter Model 401.

RESULTS

The variability in nitrate-N concentrations in soil water emphasized the problem of sampling under injected manure. Even a few inches difference in distance from the injection center resulted in some large concentration differences, particularly at the 30-cm depth. For this reason, the 1974 data in Figure 1 probably present an overly optimistic view of the effect of manure on soil water at the 30-cm depth (see Figures 2 and 3 for comparison). Data from water collected directly beneath and between injection centers in 1975 indicated decreasing differences in concentration with increasing depth. This provides some assurance that the data from 90 and 120 cm in 1974 may be fairly representative of the manure influence at those depths. However, this is an assumption, since not enough data were collected to prove it.

Taking into account the dual effects of nearness to injection center and total N applied, Figure 1 was drawn to show the highest nitrate-N concentration regardless of the rate of manure applied. However, where the values resulting from 340 kg N/ha were somewhat lower, a separate line was drawn to account for the lower application rate. With the altered placement of samplers in 1975 and 1976, the values obtained are considered valid for those two years.

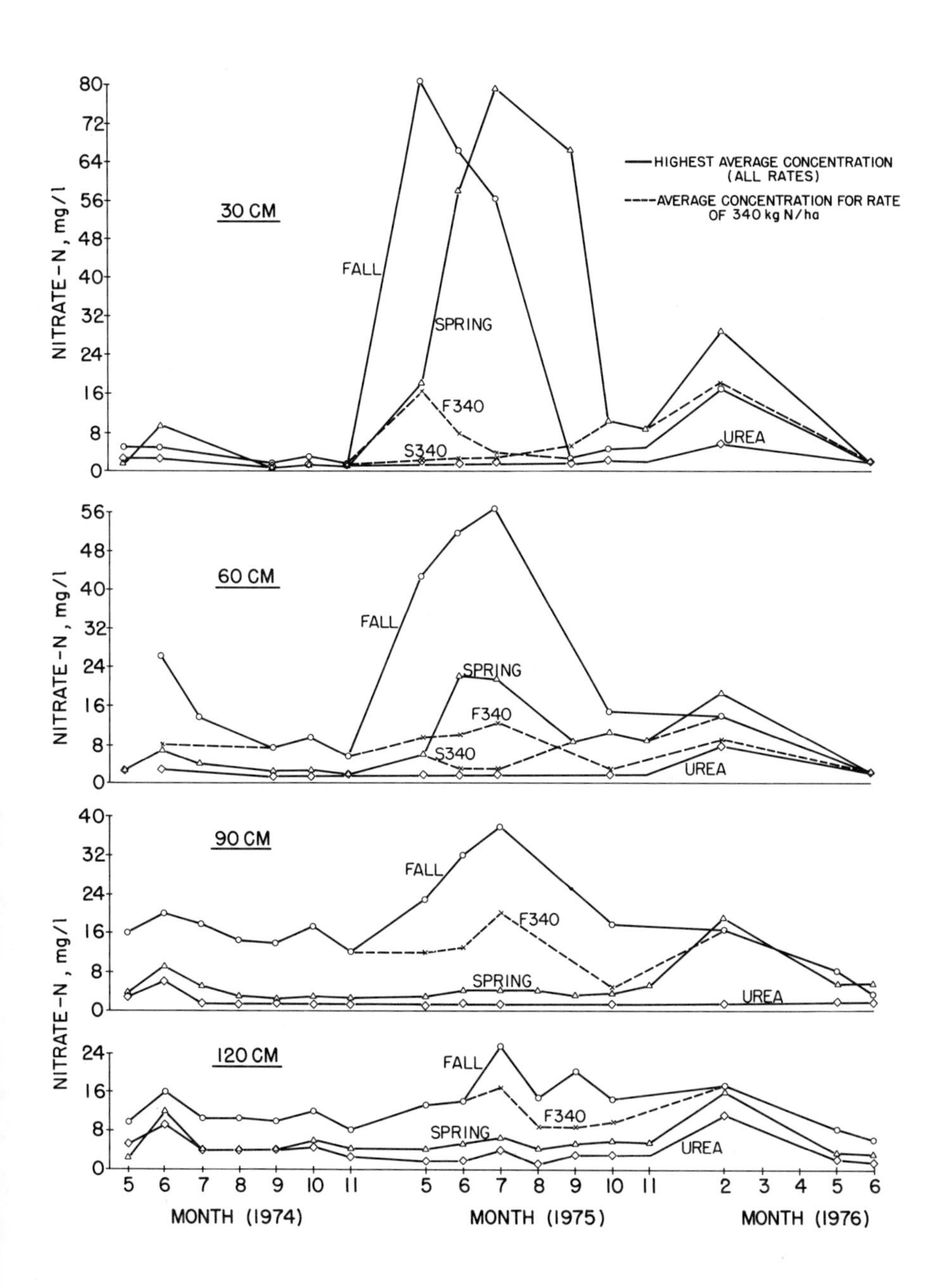

Figure 1. Concentration of nitrate-nitrogen in soil water. Dairy manure slurry was applied in fall or spring at rates to supply 340, 450 or 560 kg N/ha for 1974 and 1975, none for 1976. Urea was applied in spring 1974 and 1975 at 112 kg N/ha.

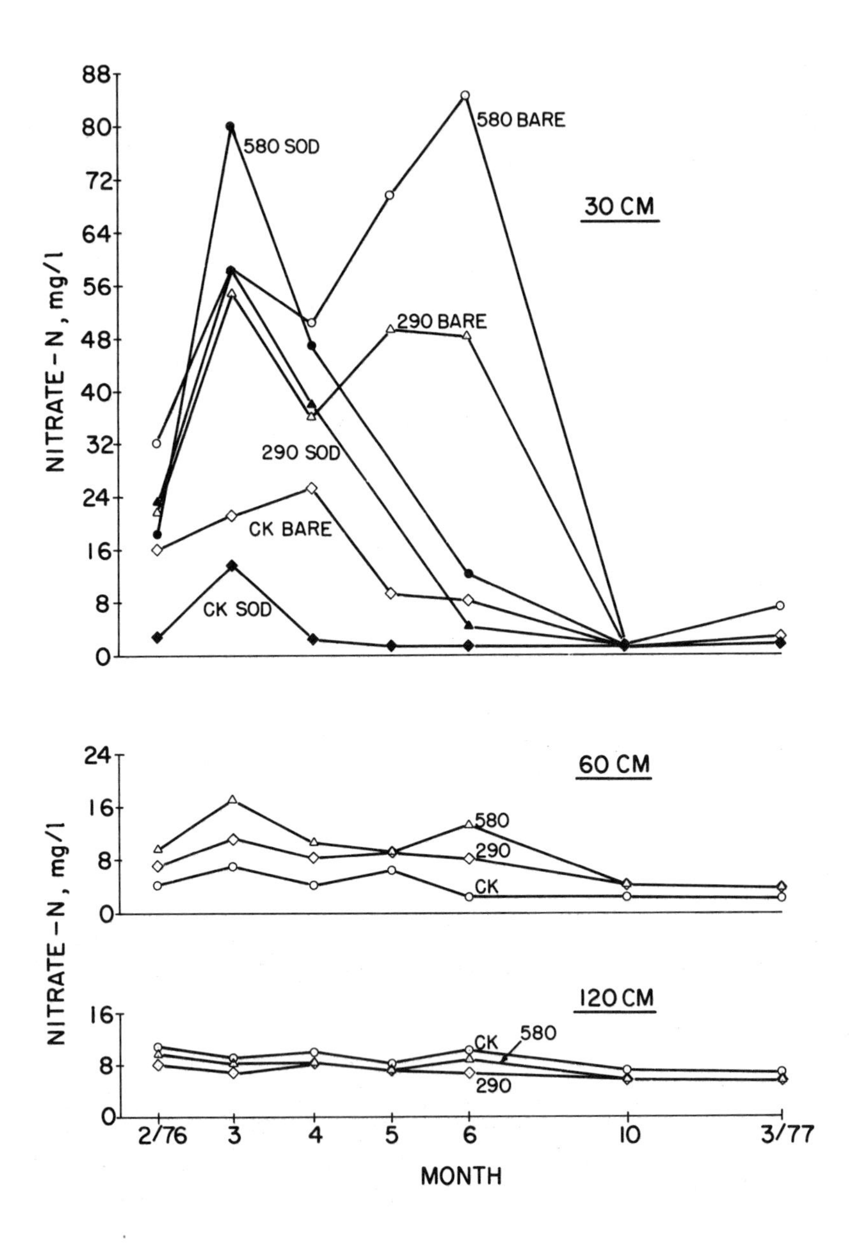

Figure 2. Average concentration of nitrate-nitrogen in soil water. Dairy manure slurry was applied in late November 1975 at rates to supply 290 or 580 kg N/ha on bluegrass sod or where the sod had been removed.

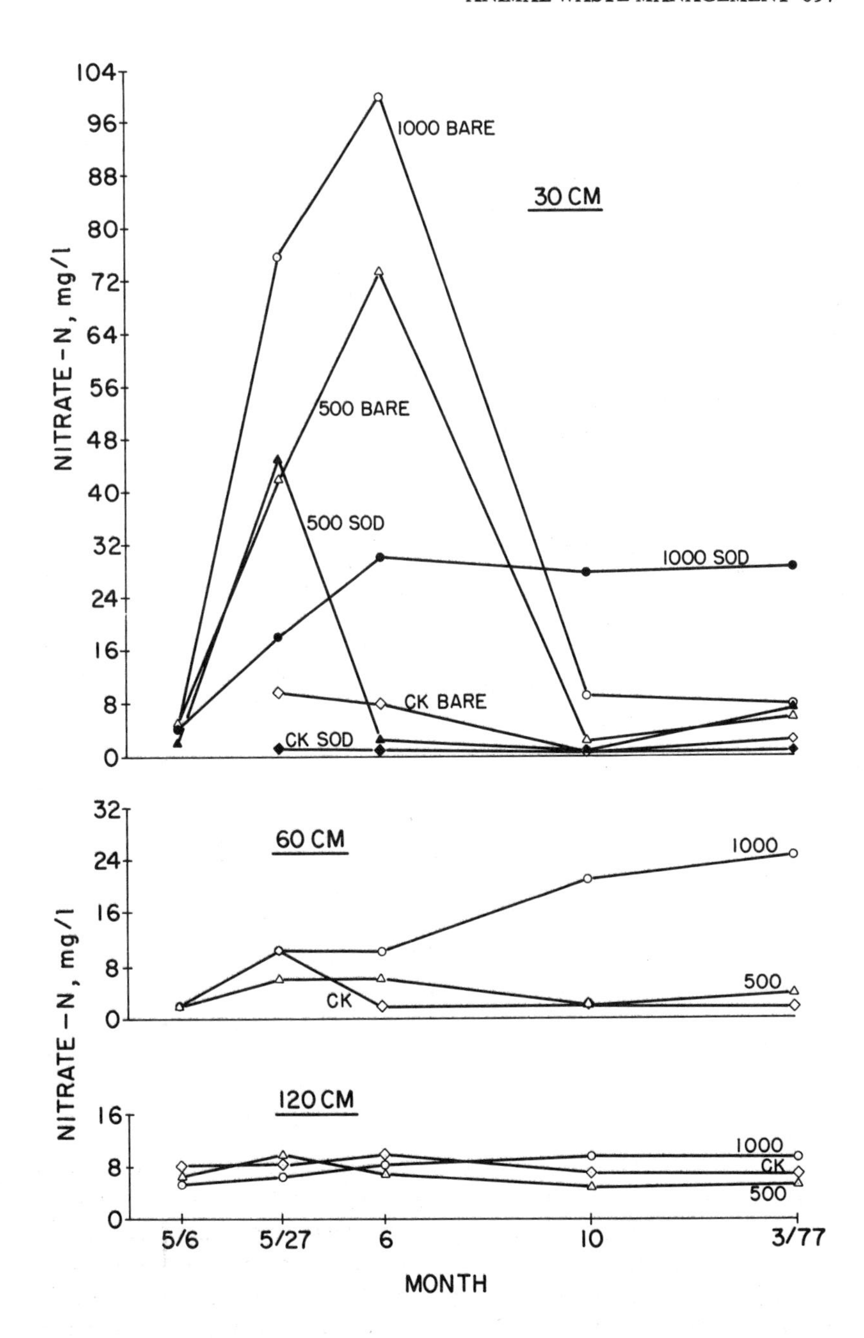

Figure 3. Average concentration of nitrate-nitrogen in soil water. Dairy manure slurry was applied in early April 1976 at rates to supply 500 or 1000 kg N/ha on bluegrass sod or where the sod had been removed.

Fall-applied manure resulted in a higher nitrate level than spring-applied manure at the three lower soil depths throughout most of the period. The greater time span between application and the high N requirement by the grass, along with the greater amount of water movement through the soil in late winter and early spring, could account for this. The lower recovery of the fall-applied N in the forage (Table I) would indicate agreement in that there was less N available for crop use. The lowest rate of manure contributed less nitrate to soil water during the second season of cropping and would appear to be a relatively "safe" rate. Comparatively, the spring application is even "safer," since the nitrate concentration for this rate of manure closely parallels that resulting from the application of urea. Also, the peak of available N from the spring application generally coincides more closely with the time of high crop usage.

Table I. Dry matter yield and nitrogen removal in orchardgrass forage as affected by subsurface injection of dairy manure slurry in fall (F) or spring (S), or by spring-applied urea (U) supplying the indicated rates of N per year for 1974 and 1975, with 1976 residual.

Nitrogen (kg/ha)	Dry Matter (m.ton/ha)			N in Forage 3-yr Total (kg/ha)	Recovery of Applied N (%)
	1974	1975	1976		
340 (F)	6.05	7.19	4.50	396	40
450	6.65	7.46	5.38	460	37
560	6.83	7.24	5.82	475	31
340 (S)	5.20	7.80	6.07	436	46
450	6.14	7.55	6.72	493	41
560	5.62	7.88	6.68	510	34
112 (Urea)	5.13	5.33	2.40	273	65[a]
LSD (0.05)	1.21	NS	1.30		
CV (%)	13.7	16.6	16.2		

[a]Average for 1974 and 1975.

Over the three-year period, the spring manure application resulted in slightly higher yields and 7 to 10% more N removed in the forage (Table I). There was a trend toward higher yields with increasing manure rates in the first and residual years, but the differences were not sufficient to justify a rate higher than the lowest rate used. The high yields and nitrate levels in soil water indicate that 340 kg N/ha appears to be excessive for an annual application rate for this crop.

MANURE ON BLUEGRASS

The variability in nitrate-N concentrations under injected manure prompted an experiment in which the manure was evenly spread over small plot areas.[1] The site of the work was an excellent stand of Kentucky bluegrass adjacent to the orchardgrass study. Individual plots were 1.52 m^2 with a treated area 1.22 m^2. Fall treatments included two manure rates (290 and 580 kg N/ha) applied on the surface of the sod or on bare soil where the sod had been removed and, in addition, approximately 12 cm below the surface of the sod or bare soil. For the latter treatments, the soil was lifted, manure was spread and soil was replaced. There were four replicates. The spring treatments were similar, except the manure rates supplied 500 and 1000 kg N/ha instead of the planned rates, and only two replicates were used. Water samplers (suction lysimeters) were installed near the center of each plot at depths of 30, 60 and 120 cm. The bluegrass was harvested twice in the summer and again in November. Water samples for nitrate determinations were taken periodically in 1976 and in March 1977.

RESULTS

Weather conditions have a great effect on the results of manuring. Fall applications are often subjected to leaching, with consequent movement of nitrate-N out of the root zone. In this experiment, the average daily temperature fell to below freezing about two weeks after the fall manure application and remained low until mid-February. This effectively delayed nitrification and prevented any leaching. The combination of warming and early March rains (3.8 cm in 8 days) resulted in a rapid rise in nitrate-N at 30 cm, with some movement to the 60-cm depth (Figure 2). April was relatively dry and mild, followed by adequate moisture in the summer months. Thus, grass growth was encouraged, and the available N was utilized with no appreciable movement of nitrate-N through the profile.

In contrast, the dry weather following the April manure application delayed the availability of the N, resulting in a reduced first harvest. Following this, grass growth effectively utilized the available N from the 500 kg N/ha application and prevented movement of N below the 30-cm depth (Figure 3). The 1000 kg N/ha application on sod nearly smothered the grass, which did not fully recover until after the rains in the last half of May. Nitrate-N under this rate remained high at 30 and 60 cm but did not move to 120 cm, although there was sufficient rainfall that should have moved the N downward.

Yields and nitrogen recovery generally indicated an advantage for the fall-applied manure (Table II). Another combination of winter and early

Table II. Dry matter yield and nitrogen removal in bluegrass forage as affected by dairy manure slurry applied in fall or spring on or below the sod surface and supplying the indicated N.

Nitrogen (kg/ha)	Dry Matter (m.ton/ha)	N Removed in Forage (kg/ha)	Recovery of Applied N (%)
0	1.21	24	-
Fall			
290 Surface	5.94	127	36
290 Subsurface	5.00	116	32
580 Surface	7.97	197	30
580 Subsurface	7.26	193	29
Spring			
500 Surface	5.87	139	23
500 Subsurface	5.94	168	29
1000 Surface	4.82	119	10
1000 Subsurface	6.65	196	17

spring conditions could have resulted in the opposite effect, as occurred with orchardgrass.

CONCLUSIONS

A valid comparison of movement of nitrate-N from injected and broadcast manure is not possible because of the use of different grasses and application in different years. However, the data from the lower rate on orchardgrass and from rates on bluegrass, excluding the 1000 kg N/ha rate, have much in common at the 60- and 120-cm depths. This suggests that the method of sampling used under the injected manure was providing results as reliable as those under broadcast manure.

The yield, nitrogen removal and soil water data from orchardgrass would indicate that the annual rate of 340 kg N/ha may be higher than necessary. An annual rate of 250 to 300 kg N/ha should maintain yields with minimum pollution of soil water.

REFERENCES

1. Green, M. J. "Effects of Methods and Rates of Dairy Manure Application on Soil Water Nitrate Levels," M.S. Thesis, The Pennsylvania State University, University Park, Pennsylvania (1976).

46

NUTRITIONAL VALUE OF ENSILED CROP RESIDUE-CATTLE WASTE MIXTURES

W. L. Braman and R. K. Abe
Animal Science Department
Fort Valley State College
Fort Valley, Georgia

INTRODUCTION

During the last decade, the intensification of animal production has increased dramatically throughout the entire country. This has resulted in the need for development of new concepts for the effective and efficient handling and disposal of livestock wastes. The amount of waste originating from the livestock industry in the United States far exceeds the combined organic waste output of the human population of the United States. Traditionally, these livestock wastes have been utilized as fertilizer; however, this organic material has high potential value beyond its use as a source of plant nutrients. The reutilization of these wastes by feeding offers an attractive alternative, both economically and ecologically, to its use in crop production.

Several excellent reviews in the area of recycling animal wastes have been published concerning the utilization of processed livestock and poultry waste as a ruminant feed.[1-5] It has been suggested that the value of refeeding animal waste lies in the reutilization of its nitrogen content.[5] There has been a great deal of interest in the use of crop residues (corn stover, milo stover, cereal straws, etc.) in the feeding of ruminants due to the relatively high costs of concentrates and high-quality hays. The low digestible protein content (0 to 5%) is one inherent disadvantage to the feeding of crop residues. Thus, protein supplementation is required when feeding crop residues. Animal waste could be a source of supplementary protein in diets containing crop residues.

Animal wastes are generally too high in water content to process successfully and store without some drying. Combining animal wastes (high in protein, low in dry matter) and crop residues (low in protein, high in dry matter) could result in a satisfactory feed following processing for use in growing or maintenance ruminant diets. As such, a series of investigations were initiated to study the use of ensiled crop residue and cattle-waste (feces and urine) mixtures as cattle feed.

NUTRITIONAL EVALUATION

Previous laboratory research by the author[7] indicated that the predicted nutritional value of cattle feedlot waste was dependent upon the nature of the diet fed. Although the processing of either corn or coastal Bermudagrass hay had little effect upon the nutrient composition of the waste, the concentrate-to-roughage ratio had a great effect upon the waste's nutritional value (Table I). Waste from cattle fed a high-concentrate diet had a much higher predicted nutritional value than waste from cattle fed an all-roughage diet.

Table I. Nutritional composition of cattle feedlot waste from hay or corn donor diet.

	Diet	
Item (%)	Hay[a]	Corn
Dry matter	24.2	32.1[b]
Crude protein	13.6	15.8[b]
Crude fiber	28.9	17.9[c]
Ether extract	1.6	2.9[c]
Ash	13.0	8.5[c]
N-free extract	43.0	54.9[c]
IVOMD[d]	19.2	43.6[b]
Neutral detergent fiber	71.6	40.6[c]
Acid detergent fiber	42.9	23.0[c]

[a]Excludes data from cattle fed long hay plus pasture.
[b] ($P < 0.05$).
[c] ($P < 0.01$).
[d]*In vitro* organic matter digestibility.

A cattle metabolism trial was conducted to evaluate waste from cattle fed either an all-roughage diet (pelleted coastal Bermudagrass hay) or a low-roughage, high-concentrate diet (80% concentrate). Cattle waste (urine and feces) was collected from sheltered pens with solid concrete floors. No bedding was used, and the waste was collected daily. The cattle wastes were sun-dried for 48 hr and then oven-dried at 105°C for 24 hr. The dried wastes were then ground with a hammermill equipped with a 1.27-cm screen. The ground roughage waste (13.6% crude protein) or the concentrate waste (17.1% crude protein) was substituted on a nitrogen basis for cottonseed meal in a high-concentrate diet. A diet containing no supplementary protein was also included. Ingredient composition and chemical analysis of the diets are presented in Table II. Four 270-kg steers were utilized in a 4 x 4 Latin-Square design to evaluate the four diets. Apparent digestibility and nitrogen

Table II. Ingredient and nutrient composition of diets used in digestion and N metabolism trial.

Item (%)[a]	Corn	Corn–CSM	Corn–R Waste	Corn–C Waste
Corn, cracked	88	83	54	61
Cottonseed meal (41% C.P.)	-	5	-	-
Roughage waste	-	-	34	-
Concentrate waste	-	-	-	27
Wheat straw, ground	10	10	10	10
Vitamin and mineral mix	2	2	2	2
Crude protein	8.9	10.4	10.2	10.3
Crude fiber	6.6	6.8	15.8	10.4
Ether extract	5.1	4.9	3.7	4.3
Ash	3.8	4.1	7.1	5.3
N-free extract	75.6	73.8	63.2	69.7
Neutral detergent fiber	19.8	21.6	41.3	26.6
Acid detergent fiber	8.1	9.2	22.0	13.1

[a]Expressed on a dry-matter basis.

metabolism data are presented in Table III. Dry matter, organic matter and protein-apparent digestibility values were similar among the unsupplemented, cottonseed-meal-supplemented and the concentrate-waste-supplemented diets. Lowest values were observed with the roughage-waste-supplemented diet. Nitrogen metabolism data indicated the effectiveness of the concentrate waste in supplementing a high-concentrate-corn based diet; however, it was not equivalent to cottonseed meal.

Calculated values of organic matter and protein digestibilities for the roughage and concentrate wastes are presented in Table IV. It is apparent that the concentrate waste has a much greater potential as a ruminant feed than the roughage waste.

Table III. Apparent digestion coefficients and nitrogen metabolism for steers fed waste-containing diets.

Item	Corn	Corn–CSM	Corn–R Waste	Corn–C Waste
Dry matter (%)	77.5[ab]	82.3[a]	65.2[b]	76.9[ab]
Organic matter (%)	79.2[ab]	82.9[a]	65.5[b]	80.2[ab]
Crude protein (%)	63.3[ab]	72.8[a]	40.6[b]	52.9[ab]
N intake (g/day)	58.5[b]	76.3[a]	61.1[b]	64.9[b]
Fecal N (g/day)	21.5[b]	20.7[b]	36.3[a]	30.6[ab]
% of intake	36.7[ab]	27.2[b]	59.4[a]	47.1[ab]
Urine N (g/day)	22.9	25.2	16.4	18.0
% of intake	39.1[a]	33.0[ab]	26.8[b]	27.7[b]
N retained (g/day)	14.1[b]	30.4[a]	8.5[b]	16.4[b]
% of intake	24.2[b]	39.8[a]	13.8[c]	25.2[b]
Absorbed N retained (%)	38.2[b]	53.7[a]	34.0[b]	47.7[a]

[ab]Means on the same line with different superscripts are significantly different ($P < 0.05$).

Table IV. Calculated apparent organic matter and protein digestibilities of cattle wastes (% digestibility).

Item	80% Concentrate	100% Roughage
Organic matter[a]	77.4	36.4
Protein[a]	36.8	14.3

[a]Values are calculated by difference.

ENSILING EVALUATION

Previous research[1] has indicated that mixtures of cattle waste, corn and/or hay can be successfully ensiled and fed to cattle. With the ever increasing cost of energy, it is generally uneconomical to dehydrate cattle waste artificially for refeeding. Another factor favoring the ensiling of wastes as a means of processing is the availability of automated handling, storage and feeding equipment on many cattle feeders' farms.

Little information is available concerning the ensiling of cattle waste with crop residue. As such, two laboratory ensiling trials were initiated to determine the composition and fermentation characteristics of ensiled cattle waste and oat-straw mixtures. In both trials approximately 500 g of various cattle waste-ground oat-straw mixtures were packed into laboratory silos (3 laboratory silos per treatment) and stored in a constant temperature chamber at

30°C. After 56 days of storage, fresh samples of silage were removed from each laboratory silo and prepared for pH, VFA, lactic acid and Kjeldahl-N analysis. The remaining silage was dried and analyzed for proximate constituents, *in vitro* organic matter digestibility, neutral detergent fiber (cell wall constituents) and acid detergent fiber.

In the first laboratory ensiling experiment, cattle waste originating from cattle fed either an 80%-concentrate diet or an all-roughage diet was ensiled with ground oat straw (50% cattle waste:50% oat straw on a dry-matter basis). Water or a dilute NaOH solution was added to the mixtures to lower the dry matter content to 40%. Sodium hydroxide was added at the rate of 2% (w/w) of the dry matter. Results of this experiment are presented in Table V. Sodium hydroxide treatment was effective in decreasing neutral detergent fiber and increasing *in vitro* organic matter digestibility. However, the fermentation characteristics of NaOH-treated silages were undesirable, as evidenced by the higher pH and butyric acid concentration and the lower concentration of lactic acid. As would be expected, the nutritional qualities of the silage mixtures containing concentrate waste were superior to those of the roughage waste. The fermentation parameters of the mixture containing concentrate waste with no NaOH were characteristic of a desirable silage fermentation (*e.g.*, low pH, low butyric acid concentration and high lactic acid contents). The NaOH treatment demonstrated potential in improving the nutritional value of waste-crop residue mixtures. Overcoming the poor fermentation characteristics may be accomplished by neutralization of the mixtures prior to ensiling. This requires further study.

The second laboratory ensiling experiment was conducted to determine the nutrient composition and fermentation characteristics of ensiled cattle waste (71.2% moisture) and ground oat straw (8.0% moisture) as affected by the ratio of cattle waste to oat straw. Waste was collected from cattle fed an 80%-concentrate diet. Ratios of waste to oat straw ranged from 100:0 to 0:100 on an "as-is" basis. Water was added to all mixtures except 100:0 to bring the dry matter content to 40%. Results of this experiment have been previously reported.[8] In general, putrefaction was found to occur in silages containing 80 or 100% cattle waste as evidenced by the high butyric acid value and the discolored slimy appearance. These mixtures lacked the physical characteristics required for proper handling to utilize mechanical equipment for silage handling, storing and feeding. Mixtures containing 40 to 60% cattle waste gave a more desirable fermentation as indicated by higher lactic acid and lower butyric acid concentrations. Mixtures containing 0 to 20% cattle waste had the highest pH values but the greatest VFA production. However, more butyric acid and less lactic acid were present compared to the intermediate mixtures. As would be expected, the nutritional value of

Table V. Nutrient value and fermentation characteristics of ensiled cattle waste–oat straw mixtures.[a]

	Donor Diet			
	80% Concentrate		All Roughage	
Item	-NaOH (%)	+NaOH[b] (%)	-NaOH (%)	+NaOH[b] (%)
Crude protein	14.3	15.9	10.5	10.6
Crude fiber	34.8	38.3	41.4	47.4
Cell wall constituents	73.7	57.6	83.2	77.9
Acid detergent fiber	45.6	46.3	52.1	53.8
In vitro organic matter digestibility	28.6	51.9	12.0	22.2
Acetate[c]	5.4	7.9	3.7	5.2
Propionate	0.2	0.5	0.3	0.5
Butyrate	1.1	6.7	1.5	3.3
Lactate	4.8	1.2	3.6	0.3
pH	4.3	9.7	5.5	6.4

[a]50% cattle waste + 50% oat straw on a dry-matter basis.
[b]2% NaOH (w/w).
[c]VFA and lactic acid means are expressed on a dry-matter basis.

the mixtures decreased as the amount of straw increased. When considering fermentation, nutritional and handling characteristics, it was concluded that mixtures containing 40 to 60% wet waste would be used in the feeding trials. This is the range of waste utilized in much of the work of Anthony.[9] In addition, all subsequent research would utilize only waste originating from cattle fed high-concentrate (finishing) diets.

FEEDING TRIALS

Forty-eight Hereford heifers averaging 208 kg were randomized to one of two dietary treatments with two pens per treatment. The two dietary treatments compared were waste silage (Table VI) and medium quality coastal Bermudagrass hay (11.4% protein). The waste silage was ensiled in an oxygen-limiting silo for approximately 30 days before feeding. In addition to the silage or hay, all heifers received a limited amount (2.7 kg) of fortified grain mix daily. Heifers were adapted to the experimental diets for 2 weeks prior to the initiation of the trial, and the trial was terminated after 180 days. Performance of the heifers is presented in Table VII. Although daily gains decreased (8%) and feed efficiency decreased (12%) significantly ($P < 0.05$), feed cost per kg of gain was decreased (16%) by using the waste silage. The

Table VI. Composition of waste silage used in the feeding trials.

Item	(%)
Wet cattle waste[a]	60
Peanut hulls[a]	35
Ground Corn[a]	5
Protein[b]	9.5-10
Dry matter[b]	45-50
pH	4.5-5.0
Acetic acid[b]	2.5
Lactic acid[b]	4.1
DM digestibility[b]	35.4

[a]Expressed on an as-is basis.
[b]Expressed on a dry-matter basis.

Table VII. Performance of heifers fed waste silage or Bermudagrass hay with grain.

Item	Hay-Grain	Waste Silage-Grain
Roughage DM (kg/day)	3.9	4.1
Grain DM (kg/day)	2.7	2.7
Total DM intake (kg/day)	6.6	6.8
Initial weight (kg)	210	206
Final weight (kg)	360	344
Average daily gain (kg)	0.83[a]	0.77
Feed/gain (kg)	7.9[a]	8.9
Feed cost/kg of gain (¢)	97	81

[a]$(P < 0.05)$.

heifers readily consumed the waste silage, and there were no health problems which could be attributed to the waste feeding.

A long-term beef cow feeding trial was then initiated (and is currently in progress) to study the effect of feeding cattle waste upon the health and reproductive performance of the beef cow herd. Heifers which were utilized in the previously discussed feeding trial were continued on the waste silage, while the other group of heifers was placed on a conventional feeding program. The heifers were bred to calve as two year olds. Both groups of cattle are presently being maintained continuously in dry lot. Cattle on the conventional diet are being fed hay or crop residue, and those on the waste silage are fed only the waste silage. The waste silage is composed of the same mixture that was previously discussed. Supplementary vitamins and minerals are offered free choice. Cattle are periodically weighed and additional energy and/or protein are fed as required. The heifers are to be maintained on their respective treatments for three consecutive reproductive cycles. Criteria to be measured will include conception rate, weight loss of the heifer/cow during the lactation period, calf weaning weight, and any other reproductive or health problems that may be associated to the dietary treatments. Data from the first year's calving are presented in Table VIII. These preliminary data suggest that beef cows fed an ensiled cattle waste-crop residue mixture can be expected to perform in a manner similar to cows on a conventional beef cow diet. It would appear that the feeding of cattle waste may be beneficial in reducing cow feed costs. However, there is a considerable need for more information concerning the long-term effects upon cow health and reproduction when refeeding cattle wastes.

Table VIII. Performance of cows fed waste silage or a conventional diet.

Item	Conventional Diet	Waste Silage
Number of Cows	24	24
Weight (12-11-75) (kg)	382	371
Weight (3-1-77) (kg)	440	452
Cows Requiring Assistance at Calving (%)	29	54
Calf Crop (%)	91	87
205-day Adjusted Weaning Weight (kg)	183	219

CONCLUSIONS

A series of investigations were conducted to evaluate the nutritional value of cattle feedlot waste. From this work the following conclusions were made:

- Waste from cattle fed high-concentrate diets (finishing diets) excelled waste from cattle fed high-roughage diets in nutritional potential for beef cattle.
- The most desirable combination of fermentation, nutritional and physical characteristics observed with cattle waste-crop residue silage mixtures contained 40 to 60% wet cattle waste originating from cattle fed high-concentrate diets.
- The feeding of an ensiled cattle waste-peanut hulls mixture to growing heifers resulted in decreased daily gain (8%) and feed efficiency (12%), but it

resulted also in decreased feed cost per kg of gain by 16%, compared to heifers fed coastal Bermudagrass hay.

- First-calf heifer performance was not affected by the feeding of an ensiled cattle waste-peanut hulls mixture compared to a conventional feeding program.
- Additional information is needed to determine the effect of the long-term feeding of cattle waste to the beef cow herd upon herd health and reproductive performance.

REFERENCES

1. Anthony, W. B. "Animal Waste Value–Nutrient Recovery and Utilization," *J. Animal Sci.* 32:799-802 (1971).
2. Bhattacharya, A. N. and J. C. Taylor. "Recycling Animal Waste as a Feedstuff: A Review," *J. Animal Sci.* 41:1438-1457 (1975).
3. Bucholtz, H. F., H. E. Henderson, J. W. Thomas and J. C. Zindel. "Dried Animal Waste as a Protein Supplement for Ruminants," *Proc. Int. Symposium Livestock Wastes*, ASAE Publ. Proc-271 (1971), pp. 308-310.
4. Fontenot, J. P. and K. E. Webb, Jr. "The Value of Animal Wastes as Feeds for Ruminants," *Feedstuffs* 46(14):30-31 (1974).
5. Smith, L. W., C. C. Calvert, L. T. Frobisch, D. A. Dinius and R. W. Miller. "Animal Waste Reuse-Nutritive Value and Potential Problems from Feed Additives. A Review," ARS 44-224, ASRD, ARS, United States Department of Agriculture (1971).
6. Ward, G. M. and D. Seckler. "Protein and Energy Conservation of Poultry and Fractionated Animal Waste," *Proc. Cornell Waste Management Conference* (1975), pp. 467-474.
7. Braman, W. L. "Nutritional Potential of Cattle Feedlot Wastes," *J. Animal Sci.* 41:239 (1975).
8. Braman, W. L. "Ensiling Cattle Waste with Oat Straw," *Proc. American Soc. of Animal Sci. Southern Sec.* (1976), p. 80.
9. Anthony, W. B. "Cattle Manure as Feed for Cattle," *Proc. Int. Symposium Livestock Wastes*. ASAE Publ Proc-271 (1971), p. 293.

47

ECONOMIC IMPLICATIONS OF WASTE EFFLUENT REGULATIONS FOR MINNESOTA DAIRY PROCESSING PLANTS

B. M. Buxton
Economic Research Service
U.S. Department of Agriculture

S. J. Ziegler and J. A. Moore
Agricultural Enginering Department
University of Minnesota
St. Paul, Minnesota

INTRODUCTION

Congress has passed legislation designed to reduce or eliminate water pollution by all industries, municipalities, government agencies and/or individuals.[1] The U.S. dairy processing industry was one of several industries specifically identified in implementing the program to clean up the nation's water. Discharge limitations for 1977 and 1983 were established for each of 12 types of dairy processing plants.[2]

The economic implications of wastewater effluent limitations for the dairy processing industry will depend largely upon two factors. First, to what extent do plants in the industry meet effluent limitations? And, second, if some plants do not comply with the limitations, what would it cost them to come into compliance? This chapter presents an evaluation of the economic impact of wastewater effluent regulations on the Minnesota dairy processing industry. The first section discusses what happens to the wastewater discharged from dairy processing plants and reviews some information concerning whether the plants are in compliance with the typical wastewater effluent limitations. The second section presents the estimated investment and increased annual operating costs for typical plants to construct and operate three alternative private wastewater treatment systems. The last section discusses some broader economic implications of pollution control regulations, such as total industry investment required for compliance, increased production costs and expected increases in consumer prices.

CURRENT STATUS OF INDUSTRY

In 1976, Minnesota accounted for almost 8% of the nation's total milk production, ranking it fourth among all states. For the same year, about 22% of the U.S. butter and 11% of the U.S. cheese production was manufactured in Minnesota. In November 1975, there were 271 dairy plants operating in the state, down from 563 in 1965. Although wastewater effluent limitations were probably a factor, the rapid decline in number of dairy plants has been greatly influenced by economies of size in plant operation, concentration of milk production within the state and improved transportation methods for moving milk over longer distances to central processing plants. These plants discharged waste effluent into municipal sewer systems, their own private treatment or disposal system or directly into receiving waters.

Dairy Plants Using Municipal Wastewater Treatment

In 1975, about 68% of Minnesota's dairy processing plants discharged wastewater into municipal sewer systems. These plants are technically in compliance with wastewater effluent limitations. However, this does not mean that these plants will be unaffected by the more general effort to clean up the nation's water. As municipalities are required to install new or remodel existing wastewater treatment facilities, the increased costs will be passed on to the users, including the dairy processing plants.

Discussions with officials in 123 municipalities, where most of the 186 dairy plants discharging into a municipal system operate, indicated that only 16% of the municipalities had installed or remodeled their wastewater treatment systems since the pollution control legislation was enacted in 1972. About 63% had installed their present systems between 1960 and 1972, while over 20% of the systems predated 1960. Almost half of these municipalities reported intentions to construct or remodel their existing wastewater treatment systems in the near future. These changes will impact the sewer-use charges of dairy processing plants. The discussions with officials also revealed that in 1975 the annual sewer-use charge was $1000 or less for half of the plants using municipal waste treatment systems. Twenty-six percent of the plants paid between $1000 and $4000 annually for using the municipal system, while 22% paid more than $4000 annually. The annual sewer-use charge exceeded $15,000 for 13 plants in Minnesota.

The range in cost per 1000 lb of milk received was from 0.2 to 93¢ and averaged 7.2¢ for all municipalities surveyed. There is clearly a wide disparity on sewer-use charges. Generally, the older the municipal waste treatment system, the less the charge per 1000 lb of milk received.

When the municipalities were contacted in 1975, 16 had actually begun planning to construct new or to remodel existing wastewater treatment systems. Ten of these municipalities had determined or estimated the new sewer-use charges for the dairy plants using their system. The average sewer-use bill for these 10 plants was to increase from $8,820 to $17,400, or almost double, after the new or remodeled facilities were installed. The plant-by-plant impact was quite variable. For one large plant, the annual sewer use charge was to increase from $27,900 to $61,740, or 120%. For another plant, it was to increase from $33,130 to $39,400, or 19%. For one plant that, prior to the new construction, had not been charged by the municipality, the annual sewer charge was to be $9920.

Given the current status of wastewater treatment by Minnesota municipalities, dairy plants that are using these systems can expect substantial increases in sewer use charges due to pollution controls imposed on municipalities.

Dairy Plants with Private or No Treatment

In 1975, about 16% of the dairy plants in Minnesota utilized private waste treatment systems.[3] Septic tanks were the most common, while stabilization ponds, aerated lagoons and ridge and furrow systems were used by about 4% of the plants. About 24 plants (nearly 9% of all plants) discharged processing wastewater directly into receiving waters without treatment. These 24 plants processed about 263 million lb of milk in 1974 (3% of all milk processed in Minnesota). Results showed these plants on the average processed less milk per plant than the average size plant in the state.

A range of wastewater discharge limitations of 5- to 100-mg/l concentrations of both BOD_5 and suspended solids have been written into many of the discharge permits issued by the state of Minnesota under the national pollution discharge elimination system. These permits usually require the processing plant's management to sample and analyze, on a monthly basis, the plant's effluent and to send the results of the analysis to the Minnesota Pollution Control Agency. The "discharge monitoring reports" of 38 reporting plants showed that none of them met a 50-mg/l limitation for both BOD_5 and suspended solids during all months of the year. Effluent from only 6 of the 38 plants met a 100-mg/l limitation for both BOD_5 and suspended solids in all months of the year. This means that enforcement of 25-mg/l BOD_5 and suspended solids limitations usually written into the permits for the 1977 deadline would require much higher levels of treatment than presently exist for all 38 plants.

Cast Study Plants

Effluent from four plants in Minnesota was sampled during the fall, winter and spring seasons and analyzed in a University of Minnesota laboratory.* These plants were selected because of fairly up-to-date private wastewater treatment systems. The effluent from a large cheese and butter plant using a two-cell aerated lagoon system had concentrations as high as 945 mg/l of COD and 2912 mg/l of total solids. The period of highest concentration occurred in May. A small butter plant using a two-cell stabilization pond had concentrations as high as 25 mg/l of COD and 86 mg/l of total solids. The period of highest concentration occurred in January. A receiving station using a primary settling tank, small trickling filter and final settling tank had effluent concentrations with as high as 1040 mg/l COD and 888 mg/l suspended solids. The period of highest concentration occurred in February. A small receiving station using a packaged aeration plant had concentrations as high as 326 mg/l COD and 935 mg/l total solids. The period of highest concentration for COD occurred in March and for total solids in May.

Since the four plants had private waste treatment systems, the effluent analysis results suggest that many plants with private waste treatment systems presently could not meet a 250mg/l BOD_5 and suspended solids effluent limitation.**

PRIVATE WASTE TREATMENT ALTERNATIVES

Dairy processing plants discharging wastewater directly to receiving waters or using private treatment that may not meet water quality regulations must consider alternative ways to come into compliance. Plants presently using municipal waste systems but facing substantial increases in use charges may also consider an alternative private waste treatment system. Three alternative systems for typical Minnesota dairy plants were considered in this study: (1) stabilization pond, (2) aerated lagoon-irrigation and (3) ridge and furrow. All three systems require land at least one-fourth mile from the dairy plant and other residences. Therefore, these systems may not be feasible for some plants located in urban areas.

---

*Sampling and analysis of wastewater effluent were done by the Department of Agricultural Engineering, University of Minnesota.

**COD concentration measures run about 100% higher than BOD_5 concentration measures on the same dairy processing wastewater.

Stabilization Pond System

This system will discharge effluent at least during part of the year. The aerobic treatment is achieved by designing a series of three ponds of sufficient total size to provide 1000 sq ft of surface area for each 0.5 lb BOD_5 received per day. The ponds are designed with a 3-ft freeboard on the dike and the bottom 1 ft set aside for solids storage.[4] The major components of the system, in addition to the ponds, include a sump pump lift station and transport pipe to move the waste from the plant to the pond (Figure 1).

Aerated Lagoon-Irrigation System

Surface aerators were assumed impractical for winter operation in Minnesota because of the cold temperature and freezing water. It was assumed that wastewater would be stored over the winter months and then irrigated onto land during the growing season. Aerators would operate for about six months in the summer, primarily to control odor.[5] A seven-month detention time is based more on the volume of storage required for the winter period than on the level of treatment desired.

Wastes are discharged from the dairy plant into a sump pump lift station that in turn pumps the waste underground to the lagoon (Figure 2). The irrigation system is used to dispose of wastewater on land adjacent to the lagoon and is designed so that the sprinkler pipe is moved only once each week. The disposal area is large enough to meet the Minnesota Pollution Control Agency's recommended maximum application rates of 0.25 in./hr and 2 in./week and capacity to handle 12 months of wastewater over an 18-week irrigation season.

Ridge and Furrow System

This system is a final land application wastewater disposal system. Wastewater is discharged into a sump pump lift station where it is pumped underground to a main distribution ditch. The wastewater then flows into furrows which are nearly level and at a slightly higher elevation than the main ditch (Figure 3). When the liquid rises in the furrows to about 1 ft in depth, an overflow into another cell or area is provided. In Minnesota, about twice as much land area is assumed for this system during the winter months as during the summer months. The system is designed to receive 5000 gal of wastewater/ac/day.[6]

What Do These Systems Cost?

Investment per 1000 lb of milk received was about half as much for the ridge and furrow as for the other two systems considered, regardless of the

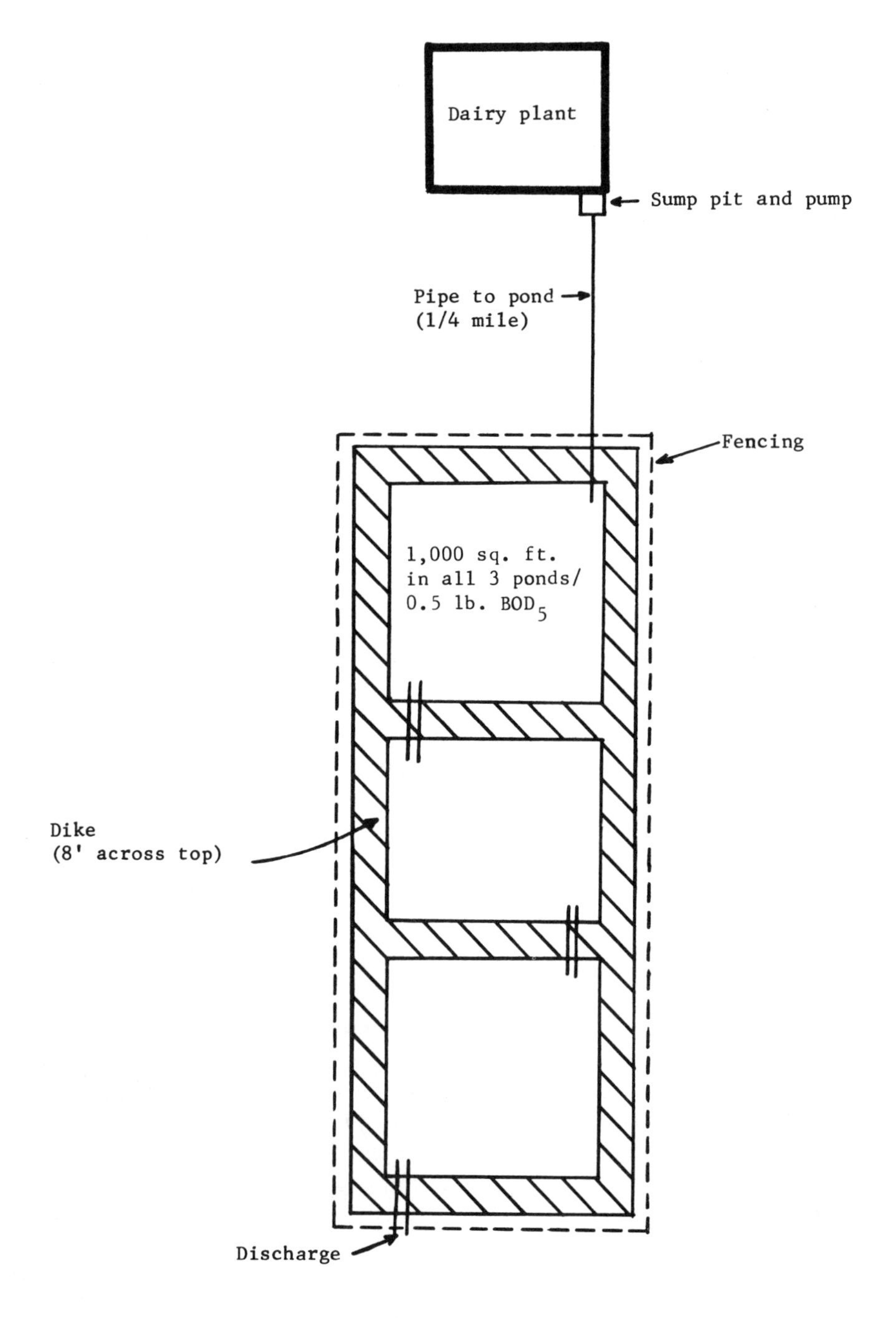

Figure 1. Sketch of stabilization pond wastewater treatment system (not drawn to scale).

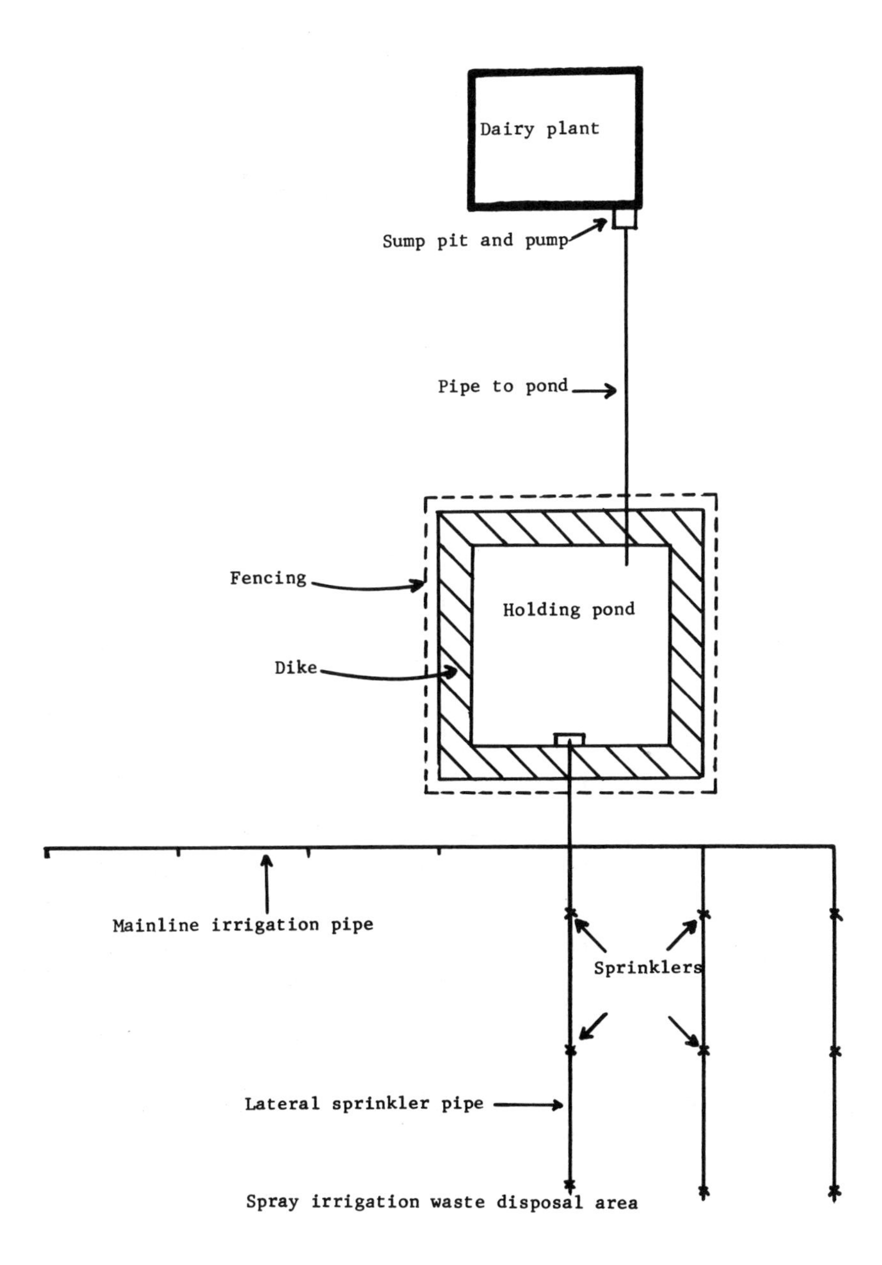

Figure 2. Sketch of a spray irrigation waste disposal system for dairy processing plant (not drawn to scale).

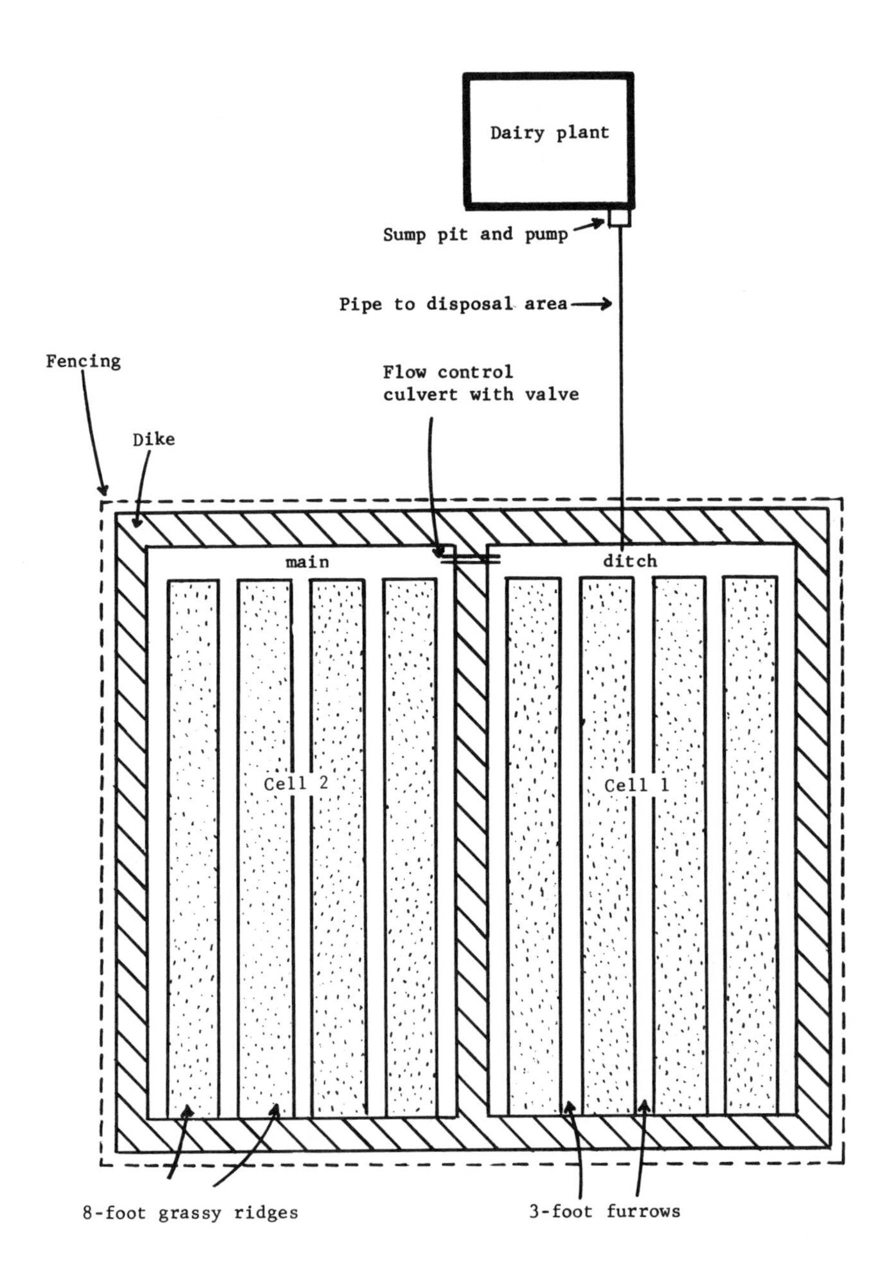

Figure 3. Sketch of a two-cell ridge and furrow waste disposal system (not drawn to scale).

type of dairy plant (Table I). This was also true for the annual costs of operating and maintaining the alternative systems.* Generally, the average municipal sewer charge for given plant types was less than for the three alternative private systems considered. However, as discussed above, the average charges by municipalities could be expected to at least double when municipalities construct or remodel their waste treatment systems.

Table I. Estimated investment and annual cost per 1000 lb of milk received for four typical dairy processing plants and increase in dairy product processing costs, 1977.

		Wastewater Treatment System		
	Average Sewer Charge[a]	Stabilization Pond	Aerated Lagoon-Irrigation	Ridge and Furrow
		¢		
Investment per 1000 lb milk received:				
Butter plant (receiving 40 million lb/yr)	-	118	119	54
Cheese plant (receiving 70.6 million lb/yr)	-	92	72	31
Fluid bottling (receiving 39 million lb/yr)	-	152	103	51
Receiving station (receiving 23 million lb/yr)	-	90	114	68
Annual cost per 1000 lb milk received:[b]				
Butter plant (receiving 40 million lb/yr)	5.0	30.9	36.6	15.3
Cheese plant (receiving 70.6 million lb/yr)	11.1	23.2	22.3	8.9
Fluid bottling (receiving 39 million lb/yr)	11.4	39.0	32.4	14.8
Receiving station (receiving 23 million lb/yr)	7.2	25.7	37.1	20.3
Cost increase per:[c]				
Pound of butter	-	0.34	0.40	0.17
Pound of cheese	-	0.24	0.23	0.09
Half-gallon of milk	-	0.17	0.14	0.06
100 lb milk received	-	2.57	3.71	2.03

[a]Average sewer charge for all plants of given type discharging into a municipal waste treatment system, 1975.
[b]Includes principal and interest on land at $1400/ac.
[c]Assumes 100 lb of milk yields 4.61 lb of butter and 8.96 lb of nonfat dry milk, 9.6 lb of cheese or 23.25 half-gal of fluid milk.

The increased cost of manufacturing a pound of butter or cheese and bottling a half-gallon of fluid milk are estimated to be 0.4¢ or less per unit for

*Costs vary by plant size. This variation is reported in Reference 7.

all three private waste treatment systems. For cheese, these figures compare to a total processing cost of about 8 to 9¢/lb.* Hence, a 0.2¢ increase in cost/lb of cheese would increase total plant costs about 2%, a significant amount. Similar conclusions exist for butter and fluid milk.

AGGREGATE ECONOMIC EFFECTS

Given the number of plants with private or no wastewater treatment systems and the apparent lack of ability of the present systems to meet selected pollution control regulations, as many as 60 dairy plants in Minnesota may be required to construct a waste treatment or disposal system. If this is true, estimated total investment would be from 1.2 to 2.6 million dollars, depending upon the type of waste treatment system used.

About 150 plants are discharging into municipal systems that were constructed prior to 1972. As these systems are reconstructed or remodeled, the dairy plants will face higher sewer-use costs. This could mean that as many as 200 plants in the state using private treatment or municipal systems would face higher wastewater handling costs. Because the annual sewer-use charge by municipalities will be as high or higher than the added cost of the alternative systems discussed in this report, sewer costs for the industry could increase to 0.65 to 1 million dollars annually.

IMPACT ON RETAIL PRICES

Minnesota is only part of the total U.S. dairy industry. If pollution regulations were applied only to Minnesota dairy plants, Minnesota would be placed at a competitive disadvantage relative to dairy plants in other states. Pollution control regulations only in Minnesota would have less impact on U.S. retail prices than if all states were required to comply.

If all dairy plants in the United States were faced with the same increases in cost as those in Minnesota, the effect of these higher costs on retail prices depends on how U.S. consumers of dairy products respond when costs are passed on to them. If consumers respond as expected, about 25 to 45% of the increased cost would be reflected in higher retail prices.[9] Regardless of the actual change in retail prices, society would bear the full amount of the increase in unit costs times the total amount of dairy products produced. This decrease in welfare, borne by all society, should be compared with the benefits associated with cleaner water in order to determine changes in total social welfare.

---

*Based on government allowances used in the dairy price support program and a study by Nicholas B. Lilwall and Jerome W. Hammond.[8]

REFERENCES

1. Federal Water Pollution Control Act Amendments of 1972.
2. *Federal Register* 39(103):18594-18609 (May 18, 1974).
3. Minnesota Pollution Control Agency. "A Survey of Minnesota Dairy Processing Plants" (1975).
4. Minnesota Pollution Control Agency. "Recommended Design Criteria for Sewage Stabilization Ponds," Division of Water Quality (November 1, 1974).
5. Minnesota Pollution Control Agency. "Recommended Design Criteria for Disposal of Municipal Effluents by Land Application," Division of Water Quality (May 1, 1972).
6. Schraufnagel, F. H. "Dairy Waste Disposal by Ridge and Furrow Irrigation," *Proceedings of 12th Industrial Waste Conference*, Purdue University (1957), pp. 28-50.
7. Buxton, Boyd M., S J. Ziegler and James A Moore. "Impact of Water Quality Regulations on Minnesota Dairy Processing Industry," Unpublished Report, Economic Research Service, U.S. Department of Agriculture.
8. Lilwall, Nicholas B. and Jerome W. Hammond. "Cheddar Cheese Manufacturing Costs," *Agric. Experiment Station Bulleton 501*, University of Minnesota (1970) pp. 25-26.
9. George, P.S. and G.A. King. "Consumer Demand for Food Commodities in the United States with Projections for 1980," Giannini Foundation Monograph No. 26 (March 1971)

INDEX